Scientific American

INVENTIONS
AND
DISCOVERIES

Scientific American

INVENTIONS
AND
DISCOVERIES

*All the Milestones in Ingenuity—
from the Discovery of Fire to the
Invention of the Microwave Oven*

RODNEY CARLISLE

WILEY

John Wiley & Sons, Inc.

Library of Congress Cataloging-in-Publication Data:
Carlisle, Rodney P.
   Scientific American inventions and discoveries : all the milestones in ingenuity—from the discovery of fire to the invention of the microwave oven / Rodney Carlisle.
      p.   cm.
   ISBN 0-471-24410-4 (Cloth)
   1. Inventions—History—Encyclopedias.   2. Inventions—United States—Encyclopedias.   3. Technology—History—Encyclopedias.   4. Technological innovations—Encyclopedias.   I. Title.
T15 .C378 2004
609—dc22   2003023258

Printed in the United States of America

10 9 8 7 6 5 4 3 2 1   •

CONTENTS

Acknowledgments *vii*

General Introduction *1*

Part I   The Ancient World through Classical Antiquity, 8000 B.C. to A.D. 330   *9*

Part II   The Middle Ages through 1599   *81*

Part III   The Age of Scientific Revolution, 1600 to 1790   *149*

Part IV   The Industrial Revolution, 1791 to 1890   *223*

Part V   The Electrical Age, 1891 to 1934   *319*

Part VI   The Atomic and Electronic Age, 1935 into the 21st Century   *397*

Index   *481*

ACKNOWLEDGMENTS

Writing the essays for this encyclopedia has provided me with an opportunity to bring together thoughts, information, and ideas that drew from many sources, both literary and personal, to which I have been exposed over many years.

My interest in the history of technology was stimulated by a course taken as a freshman at Harvard that was taught by Professor Leonard K. Nash. As I recall, Natural Sciences 4 or "Nat Sci Four" was suggested by other students and advisers as the appropriate course for a history major to take to meet the college's general education requirements. I did not realize it at the time, but the course had been established by James B. Conant and was later cotaught by Thomas S. Kuhn, who would publish *The Structure of Scientific Revolutions*. Professor Nash and Thomas Kuhn developed many of the ideas together that would later appear in Kuhn's pathbreaking work, including a focus on the scientific revolution initiated by Copernicus and expounded by Galileo.

In later decades, as I was teaching in the History Department at Rutgers University in Camden, our college adopted a similar approach to general education requirements as that established by Conant. To provide a course titled "Science, Technology, and Society," I approached a colleague in the Chemistry Department, Professor Sidney Katz, and together we offered a sweeping history of science and technology, which we often taught in summer sessions, reflecting Thomas Kuhn's focus on the revolutions in scientific thought, as well as investigations into the social impact of innovation.

Of course, a great deal has happened in the disciplines of the history of science and technology over the past decades, and our readings in the subject took us to a finer appreciation of the complex crosscurrents between these two progressing fields. As Derek de Solla Price has remarked, the two fields are sister disciplines, each progressing sometimes independently, sometimes one helping the other. Although it became fashionable among government policymakers after World War II to believe that technology sprang from the advances of science, historical studies had shown a much more complex interweaving of the two fields over the centuries.

My debts of gratitude include not only those to Professor Nash for teaching the course at Harvard, taken nearly half a century ago as an undergraduate, and to Professor Katz at Rutgers for coteaching with me but also to the many students who took our own course in recent years. Although some faculty are loath to admit it, it is often the case that teachers learn more by attempting to answer the questions posed by students than they have gained by preparing their lecture notes. Often what has puzzled students about the subject can lead into the most fruitful courses of scholarly inquiry. More than once the questions they asked led to thought-provoking discussions between Professor Katz and myself over coffee in his laboratory-office. The collaboration of Professor Katz and myself was so interesting and we both learned so much that we looked forward to the courses with pleasurable anticipation. Later, Professor Katz made a number of contributions to my *Encyclopedia of the Atomic Age.* Using ideas we had honed in discussion, I later individually taught a course, "Galileo and Oppenheimer," that again led to new insights from students.

Surprisingly, it was a much older book that I found in a used-book store, Lewis Mumford's 1934 study *Technics and Civilization,* that helped formulate my thinking about the relationship of science and technology. I had the opportunity to work with the ideas stimulated by reading that work when, through a contract at History Associates Incorporated of Rockville, Maryland, I produced a study for the Navy Laboratory/Center Coordinating Group. Due to the wisdom of Howard Law, who served as the executive of that group, I was commissioned to produce a small bibliographic work evaluating more than 150 books and articles in the fields of science and technology. Our intent was to bring many of the insights and perspectives of historians of both fields to the community of naval researchers and science and technology managers.

Working on other studies for the U.S. Navy through History Associates Incorporated contracts helped hone my thinking about the complex interplays among the disciplines of science, engineering, and technology more generally. My studies of the history of naval science and technology shore facilities included the Naval Surface Warfare Centers at Carderock, Maryland, Indian Head, Maryland, and most recently at Dahlgren, Virginia. The work of researchers, past and present, at those facilities required that I think through those changing relationships. Similarly, projects for the Department of Energy, also on contracts with History Associates, shaped my thinking. In particular, a study of nuclear production reactors for the Office of New Production Reactors, and recent work on fuel cells for automotive use for the Office of Advanced Automobile Technology at that department, deepened my appreciation for the practical side of technology policy at work. Ideas formulated decades ago by Mumford and Kuhn helped inform the books I produced for all those clients and others.

Among the many people I met in these tasks who assisted me in my thinking were Dominic Monetta, William Ellsworth, Steve Chalk, Dennis Chappell, Mary Lacey, and Jim Colvard. At History Associates, bouncing ideas off colleagues and gaining their insights were always profitable, and that community of active scholars made working with them a pleasure. They included Phil Cantelon, Richard Hewlett, James Lide, Brian Martin, Jamie Rife, and Joan Zenzen, among many others over a period of more than 20 years. I had the pleasure of working with J. Welles Henderson on a history of marine art and artifacts, based on his magnificent collection of materials, which exposed me in much greater depth to the history of the age of sail and its technologies. Much of the work we produced together reflected a melding of Henderson's intimate knowledge of the materials and my growing interest in technology and its impacts. We tried to illustrate the consequences of 200 years of maritime innovation on the life of the sailor.

More immediately, for this encyclopedia, I was assisted by Bruce Wood, an indefatigable researcher who tracked down literal reams of information about almost all of the 418 inventions and discoveries covered here. Joanne Seitter also helped dig up some interesting material. A reading of part of the encyclopedia by former Rutgers colleague and noted medievalist James Muldoon helped me identify a few Eurocentric ideas that had crept into the manuscript despite my efforts to be more cosmopolitan. I found the picture research a bit daunting, and Loretta Carlisle, an excellent photographer in her own right as well as a lovely wife, provided much-needed assistance in handling electronic picture files and organizing the materials.

I wish to express here my appreciation to all of those other folk who contributed, whether they realized it at the time or not, to the ideas in this work. Whatever errors survive are, I'm afraid, my own.

GENERAL INTRODUCTION

This encyclopedia of invention and discovery is a historical one, dividing the inventions and scientific discoveries of the human race into six periods and reviewing them in the context of their impact on broader society. Organizing significant inventions in such a way, rather than in a single listing, requires some thought as to the periodization, and this introduction provides an explanation and rationale for the organization of the work.

Lewis Mumford, in his classic study *Technics and Civilization* (New York: Harcourt, Brace, 1934), defined ancient inventions such as fire and clothing as *eotechnology*. These ancient arts, he pointed out, were part of the legacy of the human race, much of it developed in prehistoric times. The era of the Industrial Revolution, from the late 18th and through the 19th century, he called the era of *paleotechnology*. He used the term *neotechnology* to define the more modern era in which science and technology advance together, each feeding the other with developments, which he saw beginning in the 19th century and continuing into the 20th century up to the date of the writing and publication of his work in the early 1930s.

We found Mumford's classification thought-provoking, and we have adopted a periodization that builds on his thinking but that uses more familiar terms to designate the eras. In this encyclopedia we have divided the ages of scientific discovery and technological invention into six periods, or eras.

I. The Ancient World through Classical Antiquity, 8000 B.C. to A.D. 330

II. Middle Ages through 1599

III. The Age of Scientific Revolution, 1600 to 1790

IV. The Industrial Revolution, 1791 to 1890

V. The Electrical Age, 1891 to 1934

VI. The Atomic and Electronic Age, 1935 into the 21st Century

This division elaborates on that introduced by Mumford, by subdividing each of his three eras into two (which could be called *early eotechnic* and *later eotechnic* for I and II, respectively, and so forth). It would be roughly accurate to name the six periods using his concepts and terms, except for the fact that his nomenclature is so unfamiliar to the modern reader that it seems more appropriate to adopt more conventional terms for the periods, similar to those often used in historical treatments and textbooks of world history. For example, it is far easier to visualize and more useful to the student of the subject to refer to the era of the "Industrial Revolution" from 1791 to 1890 than it is to think of that century as the *later paleotechnic* period.

However, our division into these six eras allows us to organize the more than 400 inventions and discoveries considered in this encyclopedia in such a way that the interesting intersection of science and technology, a concept explored by Mumford, is revealed rather clearly. In periods I and II, technical progress in tools, materials, appliances, fixtures, methods, and procedures was implemented by farmers, animal herders, cooks, tailors, healers, and builders, with only a very rare contribution by a natural philosopher (or scientist). As the human race acquired an increasing body of ordinary procedures and instruments, from making fire through planting seeds, harvesting crops, cooking food, and living in shelters, specialists emerged, with craftsmen and artists perfecting and passing down through families and apprenticeship systems such special arts as jewelry and instrument making, the fine arts, wood furniture making, plumbing, carpentry, masonry, and metal smithing. These craftsmen and artists flourished in antiquity and organized into craft guilds in many of the societies of the Middle Ages and Renaissance. In these eras, the lasting discoveries of scientists were very few, although natural philosophers speculated (sometimes correctly) about topics later given the names of astronomy, physics, chemistry, biology, and anatomy.

In the third era (the first of what Mumford would call the *paleotechnic* periods), which is more commonly known as the period of the Scientific Revolution, natural philosophers now had new instruments developed by the craftsmen and instrument makers. The scientific questions raised, particularly by the telescope and the microscope, brought a refinement of scientific observation and many new discoveries. Coupled with more accurate timekeeping, thermometers, barometers, and better laboratory equipment such as glass retorts, sealed bottles, beakers, and glass tubes, science made a number of leaps forward in measurement and knowledge of nature, refined into "laws" that often seemed immutable and universally applicable. In the era of the Scientific Revolution, the relationship between science and technology was that technology aided science by providing better tools with which to explore nature.

In the next era, that of the Industrial Revolution, a specialized group of craftsmen emerged: mechanics who developed machines, engines, and different crucial types of electrical gear, mostly with little information derived from science. Known as engineers and inventors, these people changed the nature of production and brought a host of new devices into being, making the part of this book that covers this period the longest of those presented here. In this era, scientists turned their attention to the new machines and sought to develop new and comprehensive laws that would explain their operation. By the end of the 19th century, as the world moved into the neotechnic phase, scientific training now included a body of knowledge about the behavior of machines and electricity. That training began to affect the world of technology. On a regular and organized basis through technical schools and professional training of engineers, science began to feed back its knowledge to technology.

The *neotechnic* era or what we call the Electrical Age that resulted was highly productive, with a burst of inventions that changed human life more drastically than all that had preceded, in the span of fewer than 50 years up to 1934. (That year happens to be the one in which Mumford published his study.) What followed in the next decades, as he predicted, was even more impressive, as the flow of technical innovation was further stimulated by scientific discovery. By the 1940s, governments and laboratories organized regular structures of research and development (R & D), creating horrific instruments of warfare and at the same time introducing a host of technologies in electronics, nuclear, and biological fields that held out the promise of future peaceful progress.

This encyclopedia represents a selection of more than 400 important inventions, discoveries, and systems of inventions that have changed human life, divided roughly equally over the six eras described. The historical approach has the advantage of allowing us to examine the unfolding of progress in the different eras, driven by the different styles of creation and innovation, shaping the world in different ways. The first era, in which the Neolithic Revolution of about 8000 B.C. transformed prehistoric life with agriculture and animal husbandry, led to great early civilizations in the ancient Near East and the Mediterranean world. In the second era, through the Middle Ages and the Renaissance, the complex societies of Europe developed trade, cities, and intensive commerce, and we explore the roots of those developments in the systems of agriculture and the uses of animal, wind, and water power. In the Age of Scientific Revolution, the great discoveries of natural laws were supplemented by exploration and discovery of new lands, with improved ships and navigation equipment. In the Industrial Revolution, the means of production changed from craft and shop work to the large-scale factory, with the beginnings of mechanization of processes and innovations such as the conveyor belt and overhead crane, which would allow for mass production. In the Electrical Age, consumer goods proliferated. And in the Atomic and Electronic Age, the arts of war and the technology of communication transformed the world again.

By examining discovery and invention in this chronological and historical fashion, we look beyond the act of invention itself to social and intellectual impact, providing insights into human progress itself. It is this concern that helps us choose which items to include and which to exclude, or to mention in passing. Some of the developments that shaped human life are difficult to consider as "inventions," yet because they are so much a part of our life and experience, and since their history and impact have been well documented, they are included, such as **agriculture, cities, theater,** and **plumbing.** Other inventions are not a single innovation but represent a combination of dozens of separate technological advances, such as the **steam railroad,** the **automobile,** and **motion pictures,** and they have been included because their consequences were so profound and because they have been subjects of so much study.

In many cases, a system such as **canal locks** or a development such as **steel** was invented in one era and continued to have profound consequences for centuries. We have attempted to spell out such developments that span across the periods of our entries, placing the entry in the period in which a discovery or innovation had its greatest impact. When an entry includes a reference to a discovery or an invention described in a separate entry in the same part of the encyclopedia, the first appearance of the cross-referenced entry title is in boldface type. If the cross-referenced entry is located in another part of the encyclopedia, the part number where it can be found is shown in brackets after the cross-referenced entry's title. To assist the reader in finding the entry for a particular invention, we have provided a detailed listing in the index.

This work also includes some of the most important discoveries in the scientific world, as well as the practical innovations that have changed the life of the human race. Great scientific discoveries are relatively rare. Believing that the universe was ordered by law, reflecting the fact that many early scientists had legal and theological training as well as training in natural philosophy, scientists reduced their findings to laws representing simple descriptions of how the universe operates. In a number of cases the laws were named for the person who discovered them, giving credence to a kind of "great man" view of science. In a few cases several scientists simultaneously came to nearly identical formulations of the laws, often leading to bitter disputes over priority of discovery. Such simultaneity served to demonstrate that scientific discovery was not simply a matter of individual brilliance but also a consequence of a more general process of scientific advance and progress.

Such discoveries of the laws of nature represent a special class of work in which a cluster of natural phenomena, long observed by the human race, is analyzed and reduced to a group of immutable principles that are found to govern the phenomena. In many cases the laws can be expressed in mathematical or algebraic fashion, reducing the complexity of the world around us to a set of numerical constants and immutable relationships. The fields of physics and celestial mechanics include Newton's three laws of motion, the law of gravity, Kepler's laws of planetary orbits, Pascal's principle, Boyle's law, and the four laws of thermodynamics, among others. In each of these cases, one or more natural philosophers contemplated a long-observed phenomenon and deduced a mathematical or mechanical principle at work, reducing it to a statement with universal application. In several cases such laws had to be "amended" when later information, usually developed from improved or new instrumentation, required a change in the law or limitation of the application of the law.

Another class of scientific discoveries derives more strictly from improved instrumentation and observation, many of them resulting from developments in the optics of the telescope and the microscope. Star watchers from antiquity, including astronomers, sailors, and the merely curious, had been able to detect in the night sky the curious phases of the Moon and the seemingly erratic paths of Mercury, Venus,

Mars, Jupiter, and Saturn, as well as of comets. With the development of the telescope, new aspects of those objects were discovered, and eventually, applying the laws worked out by Kepler, Newton, and others, three outer planets were discovered, along with a host of satellites, asteroids, and new comets. Later, more powerful telescopes and other instruments allowed astronomers to make many discoveries about galaxies, the process of star formation, and the universe itself.

The microscope yielded some very basic findings about the cellular structure of living matter and about microscopic organisms. Later technologies of observation took the reach of the human eye further down to the level of atomic and subatomic particles. Some discoveries of natural constants, such as the speed of light, simply represent increasing accuracy of measurement and qualify more as increasing precision of knowledge rather than discovery in the accepted sense of the word. In short, some discoveries resulted from thinking about how the universe worked, while others resulted from measuring and looking more closely at the world. One method was based on thought, the other on observation.

In classic discussions of the nature of scientific learning, these two broad categories of scientific discovery were classified as *deductive* and *inductive* or sometimes as *theoretical* and *empirical*. That is, the great laws, such as those of Boyle, Newton, and Pascal, were generated by theoretical thought deduced from common observations. On the other hand, the outer planets, satellites, comets, and the microbes, cells, and their qualities were observed through empirical observation that relied on advances in the tools of observation, by observers such as Galileo, Huygens, and Herschel. Certain conclusions derived from those observations could be said to be *induced* from the new evidence. In fact, much scientific work represents a combination of theoretical thinking; confirmation through experiment with advanced tools of investigation; and the discovery of guiding principles, relationships, and natural constants. Thus the simple "inductive-deductive" or "empirical-theoretical" distinction is no longer held to adequately explain the various mental processes of scientific investigation and discovery.

In this work we have included about 100 of the great scientific discoveries, including the major laws, the most notable astronomical discoveries, and a number of findings at the microscopic level.

Of course, in common discussion, the term *discovery* is also used to describe the process of geographic exploration and the location of previously uncharted lands. In fact, the discovery of the West Indies and the North American continent by Columbus and the other great explorations by European navigators of the 16th and 17th centuries really represent *cultural contacts* rather than actual discoveries of something previously unknown to the human race; after all, the peoples living in the lands so "discovered" had already explored, settled, and exploited the lands. Hence, for the most part, the new regions were known to some peoples, just not to those resident in Eurasia and Africa, and it is a little Eurocentric to claim that Europeans discovered the Americas. An exception might be made to this statement by including as true discoveries the uninhabited lands found by Europeans, such as Antarctica and Pitcairn's Island, or the Northwest Passage through the islands of extreme northern North America. However, we have not attempted to include geographic discoveries in this encyclopedia but have restricted ourselves to the process of scientific discovery and technological invention.

As a consequence of such geographic exploration, various plants and animals, many of which had been in use by peoples already living in the new lands, became known to European explorers. Many plants "discovered" by Europeans became major commodities, including pepper, cardamom and other spices, quinine, rubber, opium, tobacco, tropical fruits and vegetables (bananas, potatoes, tomatoes), and chocolate. But, of course, local peoples were using all such commodities before the Europeans encountered them. These subjects, while interesting, fall outside the scope of this encyclopedia.

Some very basic observations can be derived from a historical approach to discovery and invention. What a review of these fields reveals when looking at the whole sweep of human history from the Neolithic Revolution to the Atomic Age is that science and technology are two separate human enterprises that resemble each other but that are quite different. They are "sister" endeavors, but neither is the root of the other. For more than 20 centuries before science understood the molecular crystal structure of alloys, practical technical metalworkers made two soft metals, tin and copper, into bronze. People wore eyeglasses before the optics of glass or of the eyeball were even vaguely understood by physical scientists or doctors of physiology. So in many cases, an important invention took place with no fundamental or basic science behind it. Yet the two sister fields went forward hand in hand as technology provided tools to science and as science sometimes provided laws that made it possible to build better machines, make better drugs, and, sadly, to make better weapons for human warfare. This encyclopedia may make those interactions between science and discovery somewhat more explicit.

At the same time, this work helps pin down for more than 400 important inventions and discoveries the basics: when, where, and by whom the innovation took place. It has been a tradition that scholars of one nation or ethnicity seek to give credit to their fellow countrymen in such disputes, but here we try to take a more cosmopolitan view, tracing important innovations to cultures and peoples around the world when the evidence is there.

PART I

THE ANCIENT WORLD THROUGH CLASSICAL ANTIQUITY, 8000 B.C. TO A.D. 330

Before the evolution of *Homo sapiens,* earlier races of hominids dispersed from Africa through Europe and Asia. Knowledge of these prehuman ancestors, including the Neanderthals who roamed Europe, is sparse and still being gathered. Apparently such races may have existed as long ago as 1 million to 1.5 million years B.C., and there have been finds of stone tools and skeletons from the period 700,000 to 40,000 B.C. Although these races chipped stone to make adzes, cutters, knives, and burins (pointed chips apparently for working bones or antlers), it is not at all clear that they had used language or knew how to start fires. These Old Stone Age or Paleolithic peoples definitely belong to the period of prehistory.

Some sources identify the Paleolithic cultures of *Homo sapiens* from 40,000 B.C. to about 14,000 B.C. as Upper Paleolithic. The last Ice Age began to end in about 11,000 B.C. with a warming trend. During a period of 1,000 to 2,000 years, changes began to take place in the Middle East, Asia, and Europe. We start most of our documentation in this volume of human invention and discovery with the Neolithic Revolution, which occurred between about 8000 and 7000 B.C.

During the Ice Ages, humans obtained food by hunting wild animals and gathering wild edible plants. Using flint, bones, antlers, and wood for tools, people learned how to reduce hides and leather to workable materials, painted in caves some excellent depictions of the animals they hunted, and apparently lived in family groups, often clustered together into groups of families. Little is known of exactly where and when most of these developments took place, but they had spread over much of Europe, Asia, and Africa from their starting points, and some had moved with Asian migrants to the Americas long before 10,000 B.C.

With the ending of the last Ice Age, in about 11,000 B.C., the supply of available vegetable food declined with arid seasons, and the number of animals went into decline both with the changing climate and because some were hunted down to extinction by the slowly growing human population. This climatic development set the human race on a path of progress, and most of the human inventions we know today, from the wheel to the computer, have occurred in fewer than 300 generations since that time.

Historians, archaeologists, and classicists have attempted to divide the ancient and classical world into several eras. In this part we explore the inventions that were added to the human culture between the Stone Age and the end of Classical Antiquity.

With the shortage of game animals, humans began to follow migrating herds, and then to teach other animals, such as sheep and goats, to migrate between seasons to obtain pasture. At the same time as this nomadic style of life began in the Middle East, other groups settled down and domesticated some wild plants, beginning agriculture. Agriculture and nomadic herding, both recorded in the Bible, led to a host of accumulated inventions and innovations in what historians have called the New Stone Age or the Neolithic Age. Agriculture and herding were at the heart of the Neolithic Revolution, and much of the human heritage of arts and artifacts can be traced back to this period. As the reader might note in the table below, the Mesolithic and Neolithic periods overlap a good deal, partly because the Neolithic Revolution began at different times in different areas. Some European sites as late as about 3000 B.C. show signs of having the older Mesolithic cultures, while some in the Near East as early as 7000 B.C. had already shown signs of Neolithic agriculture. It took about 4,000 years, from about 7000 B.C. to about 3000 B.C., for agriculture and farming societies to spread across most of Europe. Against a background of often sophisticated hunting and gath-

Eras of the Prehistoric and Classical World	
Upper Paleolithic Age	40,000 B.C. to about 8000 B.C.
Mesolithic Age	8000 B.C. to about 3000 B.C.
New Stone or Neolithic Age	7000 B.C. to 5000 B.C.
Copper and Stone or Chalcolithic Age	5000 B.C. to 3500 B.C.
Bronze Age	3500 B.C. to 1000 B.C.
Iron Age	After 1000 B.C.
Classical Antiquity	800 B.C. to A.D. 330

ering (and fishing) lifestyles of Mesolithic peoples, the introduction of new crops and domesticated animals such as sheep brought the Neolithic changes from area to area in a gradual dispersal.

In the later Neolithic period, improvements to the human tool kit included sewing, fishing, bow-hunting, cooking, advanced shelter-building, and village life. In a period roughly from 5000 to 3500 B.C., with working of copper and the use of improved stone tools, the age has been called the Chalcolithic or Copper and Stone Age. While copper was easy to melt and to pound into decorative items, it was too soft for many useful tools, so when a sharp edge was required, chipped flint or obsidian remained the material of choice.

In the Copper Age, in the ancient Near East, cities, specialized craftsmen, and leisure classes of priests and rulers began to emerge. In this era, systems of writing were developed, and the modern scholar has not only the study of artifacts but also a few inscriptions and later recorded oral traditions as sources. In this period the beginnings of long-distance trade of commodities and metals such as gold, tin, and copper can be found. The social innovations of large cities that came in the Chalcolithic or Copper and Stone Age in the Near East did not at first affect most of Europe, where villages and hamlets continued to represent the mixture of hunting and farming societies, with freestanding wooden houses rather than walled masonry towns. However, as the cities of the ancient Near East grew, they established trade routes, and the cities drew desirable materials, such as metals and precious stones, from hundreds of miles away. Subtle changes spread through these trade routes, as potters in Europe began to imitate forms of metal cups and pitchers made in the urban centers.

Between 3500 B.C. and 1000 B.C., in the Bronze Age, the alloy of copper and tin produced a practical and strong metal useful for weapons, tools, fixtures, and hardware. After 1000 B.C., with increasing use of iron, bronze was still used for specialized purposes and remains into the modern era a useful metal for specific machine parts. In this period, the eastern Mediterranean saw the beginnings of continuous maritime trade, with regular exchanges among Crete, the Near East, Greece, and Egypt.

The division of ancient human progress into ages based on the evolution from stone tools through various metals produces only a very approximate scale. Specific inventions have been traced to ancient Iran or Persia, Mesopotamia (now Iraq), Egypt, and China. Frequently, 19th-century European historians preferred to place credit for an invention or development whose origin was in doubt in Persia or Mesopotamia instead of China or Egypt, suggesting that a kind of racial chauvinism was at work.

There remain many unsolved mysteries about the technologies of the ancient world. There is evidence that in Mesopotamia, craftsmen knew how to use electrical currents for electroplating metals. The ancient Polynesians, without the aid of compasses or charts, navigated the Pacific. The Egyptians not only constructed the pyramids but also were able to lift massive stone obelisks onto their ends by some unknown method. The ancient Egyptians built a canal to link the Red Sea with the Mediterranean, and other technological and mathematical innovations took place in India, China, and central Asia. In pre–Bronze Age Britain and on the continent of Europe, builders somehow moved heavy stones to build monuments with apparent astronomical orientations such as at Stonehenge.

The ancient Greeks used a complicated navigational device that was a sort of early geared analog computer to locate the positions of the stars and planets, known as the Anikythera computer. The workings of that strange machine, found by a sponge fisherman off the Greek island of Syme in 1900, were partially unraveled by 1974 by a historian of technology, Derek de Solla Price. In the Americas, the Mayans, Toltecs, and subjects of the Inca knew about wheeled pull toys, but they never used wheels for vehicles or even wheelbarrows. Yet the Mayans used the concept of the mathematical zero several centuries before the Europeans. It is not known how the stoneworkers of ancient Peru were able to precisely fit together massive stones weighing 5 tons or more. These and other unanswered questions about ancient technology present a fascinating agenda for those who study these peoples.

The fact that widely dispersed nations and races came upon the same idea, in cases of parallel invention, rather than diffused invention, leaves another set of tantalizing mysteries. In Egypt, Mexico, Central America, and the jungles of Cambodia, ancient peoples built pyramids. Did the Mound Builders of Illinois hear of the great pyramids in Mexico and try to emulate them? Did the burial mounds and stone monoliths in Britain represent a diffusion of a Europe-wide idea? With little evidence, a few writers have speculated that the Egyptians influenced the Toltecs and the Mayas. The strange statues of Easter Island bear a haunting resemblance to similar carvings in South America—was there a connection? Out of such guesswork, popular authors such as Thor Heyerdahl have woven fascinating and suggestive theories. More cautious investigators who link their careers to the conservative halls of academia rather than to the marketplace of popular literature have traced a few patterns but generally insist on rigorous evidence of diffusion before asserting a connection of influence and commerce between distant peoples. Some of that evidence is compelling, such as the spread of bronze artifacts from centers in the mountains of Romania to other parts of Europe or the diffusion of drinking beakers across nearly all of ancient Europe in the Bronze Age.

In an age of high technology and laboratory science it is easy to forget that the ability to invent, and the need to inquire into the principles that operate in nature, are ancient qualities of the human race. By the end of the period that historians call Classical Antiquity, about A.D. 330, mankind had assembled a vast storehouse of tools, equipment, processes, appliances, arts, crafts, and methods that together made up ancient technology.

By the era of the classical world, great thinkers had struggled to understand nature in sciences we now call astronomy, biology, chemistry, and physics. While the modern age regards a great deal of ancient science as simply guesswork, or worse, as mistaken, there were several lasting findings from that time that stood up very well under later advances. Even more striking was the permanent addition of technology, leaving us thousands of devices we still use, from needle and thread to the hammer and chisel and the cup and pitcher.

By the time of the Roman Empire, mankind had created such amenities as indoor plumbing, iced desserts, textiles and leather shoes, dyed clothes, jewelry, theater, sports, and the study of the stars. Thinkers had not only mastered some basic laws of machines to build pulleys and even complex theatrical equipment but also had developed geometry and forms of algebra. Engineers led the building of great monuments, bridges, lighthouses, roads, and public buildings. Palaces and homes had glass windows, hinged doors, simple latches, and such everyday items as tables, benches, shelves, shutters, cabinets, bottles, and metal utensils. Concrete, chains, solder, anvils, and hand tools for working wood and stone were part of the craftsman's kit, while horses pulled light carts and chariots, and oxen plowed the fields. For the most part, this *eotechnology,* the term introduced by Lewis Mumford as discussed in the general introduction to this encyclopedia, is very difficult to trace to its precise origins. Even so, the first known appearance of a specific technology in ancient graves, village sites, and ruins of cities often gives hints as to the eras of invention and subsequent dispersal of particular devices, tools, and social ideas.

Many of the arts and crafts such as medicine, cooking, music, tailoring, cabinet making, jewelry, ceramics, and weaving in what Mumford called the *eotechnic* era all grew in subtle step-by-step improvements as the skills were passed from region to region and generation to generation, parent to child, by word of mouth and instruction rather than by handbook or formal education.

From grandparent to grandchild, a period of 3 generations approximately spans a century. Thus 30 generations covers 1,000 years of human history, and only about 300 generations have passed since the earliest signs of the Neolithic Revolution. Over those 300 generations, through migrants, conquerors, families, descendants, and teachers, the skills of farming, cooking, sewing, and many other arts and techniques have been passed down from that remote period to our times. A trick of the trade would be added here and there, and gradually the incremental improvements took technology forward in a kind of accretion of progress and a slow dispersal by colonization, conflict, trade, and travel. Some skills would be forgotten or lost for a period, but the human race has a knack for recognizing the advantages of a better or cheaper or quicker way to do work. So good ideas might be lost for a period but sometimes revived in a later period to take root again.

Most of the tools and machines of the ancient world were made by anonymous craftsmen or craftswomen. Once in a while an architect or engineer would leave his name chiseled in stone on a bridge or building, but for the most part, the achievements left to later ages as a legacy remained as monuments to the ingenuity of the human race, not to the achievement of a particular individual. Usually the best the modern scholar can do to unravel the origin of an ancient innovation is to track it to one region of the world. We can only make informed guesses as to the location of the invention of the wheel, the basket, clay pottery, the smelting of metal, sewing, the needle and thread, the braided cord, and all the rest of the ancient tool kit of our ancestors.

There is no way to memorialize those thousands of creative individuals who first thought of the good ideas and devices. But as we make continued use of the tools and techniques, we can remember that they all trace back to particular unknown men and women who added to our collective store of technology, some 100 or more grandparents back in time.

So this part of the encyclopedia, and to a large extent its second part, cannot focus on the unsolvable mystery of "who" made an invention. Rather, we explore the evidence and the informed guesses about the "when" and the "where" and about the impact of the inventions and discoveries on the history of the human race.

agriculture

A lively debate developed in the field of archaeology in the 19th century over where and when agriculture was first developed. Today scholars tend to agree that there was no single invention of the process of **domestication of animals** and plants and that independent invention probably occurred in the Tigris-Euphrates region, in central Asia, and in the ancient Americas. Sometime in the period 9000 to 7000 B.C., early Natufian residents of Palestine used a form of **sickle** for harvesting crops, but it is not known whether the plants were wild or planted. Remains of einkorn wheat, emmer wheat, and wild barley have been discovered with 7000 B.C. dating on the Iraq/Iran border.

Emmer wheat may have been the first domesticated plant, representing a cross of wild wheat with a natural goat grass in a fertile hybrid. Whether the hybrid was the result of human intervention or occurred naturally has not been determined. Even though emmer wheat is plumper and stronger than the original wild wheat, the seeds are light enough that it can spread naturally. Bread wheat, another hybrid, is even heavier than emmer wheat, and its seeds must be planted by hand. Bread wheat is found associated with 6000 B.C. remains. Because it must be planted it is often regarded as the first truly domesticated crop.

However, other evidence has surfaced far afield of even earlier domestication of some nongrain food plants. There is evidence that gourds, water chestnuts, beans, and peas were grown in Thailand and China as early as 9000 B.C. Pumpkins, gourds, and beans were known in ancient Mexico before 7000 B.C., with peppers, avocados, and the grain amaranth by 5000 B.C.

In the ancient Near East and central Asia, there were several natural existing plants and animals that provided the basis for agriculture and domestication of animals. The ancestors of wheat and barley, together with wild sheep, pigs, and cows, set the stage. With increased aridity in about 8000 B.C., people who had practiced a hunting and gathering lifestyle turned to agriculture in Syria, Mesopotamia, and parts of what are now Iran and Turkey.

The establishment of agriculture in the ancient Near East and in ancient Mexico in the 7th and 6th millennia B.C. led to other developments, such as village life, diversification of crops, domestication of animals, and the development of food-storage systems such as silos and granaries. Irrigation required organization of labor that could be turned to the construction of fortifications and religious monuments and the

Every stage in the domestication of plant and animal life requires inventions, which begin as technical devices and from which flow scientific principles. The basic devices of the nimble-fingered mind lie about, unregarded, in any village anywhere in the world. Their cornucopia of small and subtle artifices is as ingenious, and in a deep sense as important in the ascent of man, as any apparatus of nuclear physics: the needle, the awl, the pot, the brazier, the spade, the nail and the screw, the bellows, the string, the knot, the loom, the harness, the hook, the button, the shoe—one could name a hundred and not stop for breath. The richness comes from the interplay of inventions; a culture is a multiplier of ideas, in which each new device quickens and enlarges the power of the rest.

—Jacob Bronowski,
The Ascent of Man
(Boston: Little, Brown, 1973)

beginnings of city life, with its requirements for some form of centralized authority, finance, and maintenance of law. Thus agriculture was a direct cause of the emergence of civilization in the form of organized society and the first **cities.** The patterns of early civilizations in the Tigris and Euphrates River plains, in the Nile River valley, and in the highlands of Mexico were similar, although differences in crops and domesticated animals certainly support the concept of independent lines of development.

In ancient Egypt, agriculture and domestic animals went together, with the raising of domestic geese, dogs, cattle, sheep, goats, and pigs. Asses and oxen were used as draft animals. Animal breeding for specialized purposes, such as hunting dogs or cattle bred to produce more milk, was known in ancient Egypt. Flax was grown for linen fibers to make **textiles** before 4000 B.C.

A unique feature of the agricultural societies of Mexico was the absence of domesticated animals except for the dog, with at least one breed developed as a source of meat. In Peru, the llama was domesticated as a beast of burden, and the alpaca was raised for its wool. The guinea pig was raised for its meat. Potatoes, peanuts, gourds, chili peppers, pineapples, and cotton were grown in the Inca Empire before the arrival of the Spanish. More than 100 different types of potato, and corn, manioc, kidney beans, and avocados were cultivated, indicating some spread of agricultural crops within the Americas. One mystery is the apparent hybrid blending of cotton in Peru with an Asian type as early as 2500 B.C.

In both the ancient Near East and in the Americas, agriculture led to a chain reaction of innovation, including the invention of the **basket, pottery,** and improved stone tools such as the hoe and the **sickle.** These new stone tools brought about by the Agricultural Revolution created what has been called the New Stone Age or the Neolithic Age.

The Neolithic Revolution had another side to it. Besides settled agriculture, some peoples turned to **nomadic herd tending,** in which groups drove domesticated sheep, goats, donkeys, and later, horses and camels, in regular routes to watering holes in the summers and into the flowering deserts during the rainy season.

alphabet

The alphabet is a system of writing that represents individual sounds of a language with a set of symbols, usually with the most common sounds assigned a single written form. Several alphabets have been independently invented, but the system used in Western languages derives from the "North Semitic" alphabet, which originated in the eastern Mediterranean between 1700 and 1500 B.C. That alphabet represented a simplification of the Egyptian system, which had reduced thousands of hieroglyphs to a syllable-based system with several hundred syllables.

One earlier system of writing evolved from pictographic symbols, leading to cuneiform or wedge-shaped writing, suitable for inscribing on a clay tablet with a stylus. The cuneiform system was a syllabary—that is, each sign represented a syllable rather than a specific sound, as in an alphabet. The earliest cuneiform style appeared in about 2400 B.C., and by the time of the Assyrian Empire, about 650 B.C., had become quite standardized. Apparently as early as 2300 B.C., a system of envelopes was developed for covering clay tablets, set up so the envelope could be sealed against alteration of the tablet. Scribes regularly attended school, with surviving records of writing schools from as early as about 2000 B.C. Many texts in cuneiform survive, even describing school days, student disputes, parental guidance to students, and the routine of tablet instruction in Sumerian. The cuneiform syllabary apparently began with about 1,200 signs, and with constant improvement was down to fewer than 200 symbols by 2000 B.C. Even so, it required extensive training before it could be written or read with ease.

The Semitic alphabet did not have separate symbols for vowels. This system had about 30 symbols, which were later reduced to 22 symbols. The Phoenicians derived a similar system from the Semitic alphabet, and as merchants they spread its use throughout the Mediterranean world. The Greek alphabet, which developed from about 1000 to 900 B.C., represented a modification of the Phoenician, changed the writing so that one read it from left to right, and added separate symbols for vowels.

The early inhabitants of the Italian peninsula, the Etruscans, used the Greek alphabet. From the Etruscans, the Romans learned that alphabet and modified it by dropping certain consonants. The alphabet used in English, French, Spanish, Italian, German, and many other modern

European languages is virtually the same as the Roman alphabet finalized by about 114 B.C. The letters *j* and *w* were added in the Middle Ages.

The modern Hebrew alphabet derives from the "Square Hebrew" alphabet derived from an early Aramaic alphabet, developed in the period from about 580 to 540 B.C., and like the Phoenician, is read from right to left. Most vowels are indicated by diacritical marks rather than by separate symbols. The Arabic alphabet, like the Hebrew, derives from Aramaic. The modern form is flowing in shape, quite suited to handwriting, and like Hebrew is virtually free of vowels, using diacritical marks for most of them. East Indian alphabets are also apparently derived from Aramaic, although like some other independently invented alphabets, they may have been invented by emulating the concept rather than the specific letters.

Similar independent inventions by emulation developed in the 19th century, when the Cherokee leader Sequoyah developed a syllabary in the 1830s, and when the Vai people of West Africa, perhaps influenced by early Baptist ministers, created another system.

The Cyrillic alphabets used to write Russian, Bulgarian, Serbian, Ukrainian, and Belorussian derived from the Greek alphabet.

All of the alphabets greatly stimulated literacy, and by contrast to hieroglyphic or pictographic systems such as the early Egyptian and the Chinese, required less training and hence were open to wider participation.

aqueducts

Although known in several ancient civilizations, including the Assyrian, Persian, Indian, and Egyptian, the aqueduct was perfected by the Romans. The city of Rome was supplied with a water distribution system containing 11 aqueducts, totaling about 260 miles in length, from which water was supplied by lead pipes to city fountains and public baths. The system took about 500 years to construct, from 312 B.C. to after A.D. 220. Much of the system consisted of underground piping systems made of clay, wood, lead, and **bronze** and linked to very modern-seeming **plumbing** systems. The Aqua Appia, the first aqueduct in the system, was completed in 312 B.C., with a total length of about 10 miles. The Aqua Alexandrina was completed in A.D. 226, with a length of 13.7 miles. The longest section, built over the 4-year period 144 to 140 B.C., was the Aqua Marcia, at 56.7 miles. The first aqueducts were made entirely of stone, but in the 3rd century B.C. a form of concrete using volcanic ash with lime, sand, and gravel was used to build sections. The

Aqua Tepula, completed in 125 B.C., used poured concrete. Only a fraction of the total 300 miles, about 30 miles, was on the characteristic raised channel supported by arches. The system worked entirely by gravity, with storage tanks, which, when they overflowed, were used to flush through the sewer system.

Remains of arched Roman aqueducts are found in Greece, France, Spain, and North Africa as well as Italy. Part of Athens is still supplied by an aqueduct built by the Roman emperor Hadrian in about A.D. 115. Along with other structural remains from antiquity, such as **roads** and bridges, aqueducts are regarded by tourists and historians alike as major accomplishments of Roman civil engineering.

In the 16th century, both London and Paris developed systems with **waterwheels** [II] mounted under bridges to pump water to the city system. With the growth of these cities it became necessary to import water from farther away in the 17th century. A system brought water to London from a distance of nearly 40 miles over a series of small bridges, and in France an aqueduct brought water into Paris on a high aqueduct more than 500 feet above the level of the Seine.

aqueduct The Romans developed some of the most extensive aqueducts, and this fanciful 16th century depiction shows the parts of an elaborate distribution system. *Library of Congress*

All of the ancient aqueducts worked on gravity flow and were nonpressurized. Modern systems, using steel and concrete pipes, have included pressurized flow, such as the Catskill system, constructed in the 1920s and supplying New York City. Other modern systems combine gravity flow and pressurized pumping systems.

arch

Although the arch is usually remembered as a Roman invention, examples have been found earlier, particularly in the Egyptian civilization. Perhaps because the Egyptians were concerned with making structures that would be extremely permanent, they tended to use the arch only in auxiliary or utilitarian buildings such as granaries. By its nature, the arch involves structural forces that will cause a collapse if the supporting members are damaged. The Romans, however, adopted it widely, using it for bridges, **aqueducts,** and private and public buildings.

The principle of the Roman arch is a structure composed of wedge-shaped stones to span over a void. The center stone at the apex or top of an arch is known as the keystone, while the supporting wedges to each side are known as voussoirs. During construction, the wedges and keystone were held in place by wooden supports, which would be removed on completion. With finely cut stone, no mortar was needed to hold the arch together. However, the thrust of weight from the arch itself and from any structure above it was transmitted in a diagonal direction away from the supporting pillars. Thus the pillars needed to be buttressed, either by a standing wall or by another archway. A series of arches or a colonnade could be constructed as long as a solid buttressing wall was constructed at each end. Such colonnade forms were desirable for open squares or for interior courtyards or atriums within buildings and were a characteristic of **cathedral** [II] structures built during the high Middle Ages, along the sides of the central nave.

In the Gothic period (1300–1500) a pointed arch was developed. In Gothic structures, some of the most spectacular arches were found, in which the buttress form was external to the building as a flying buttress, or in which arcades of several stories were constructed with wider arches on the lower stories to support those above. Often multiple arches would intersect to create vaulted ceilings. Piercing the stone walls with multiple arches allowed for many windows.

Archimedes' principle

Archimedes, who lived from about 287 to 212 B.C., is regarded as one of most profound of the ancient Greek mathematicians and as a prolific inventor. In a charming legend, which is perhaps the archetype of being struck by a discovered concept, he is said to have discovered the principle of displacement of a fluid. The story goes that he had been called on to determine whether a crown presented to King Hiero of Syracuse was made of pure gold. Lying in his bath and considering the question, Archimedes observed that the bath overflowed by an amount of water equal to the volume of his own body. He had the solution, and shouting "Eureka!" (I have found it!), he rushed into the street without getting dressed.

His procedure was then to immerse the crown in water, collect the displaced water, and then weigh an amount of known pure gold equal in volume to the displaced water. Then he compared that weight to the weight of the crown. The weights differed, and thus the crown was proven not to be pure gold. In effect, he compared what we would now call specific gravities of the two sample materials to determine whether they were identical.

Several other legends surround Archimedes, including one in which he is reputed to have told King Hiero, "give me a fulcrum and I will move the world," and when challenged, he moved a fully loaded ship by means of pulleys. In a work on levers he stated Proposition 6: "Commensurable magnitudes balance at distances reciprocally proportional to their weights," a principle demonstrated when a small child sits at the end of a seesaw, balanced by a heavier one sitting closer to the fulcrum bar.

The principle that bears his name is that of buoyancy, discovered in the solution to the affair of the counterfeit crown—that is, a body immersed in a fluid is buoyed up by a force equal to the weight of the displaced fluid. Archimedes wrote several works on mathematics and mechanics. He designed a number of military machines such as grappling hooks and missile-throwing devices, used in the defense of Syracuse when under attack by Roman troops under Marcus Claudius Marcellus. Archimedes was killed by an enemy soldier during the Roman conquest when, by legend, he was interrupted in contemplation of geometrical figures he had drawn in the dust of the marketplace.

Archimedes' principle explains the flotation of objects in liquids and in gases as well—that is, his principle is demonstrated by the suspension of fish in the sea, by boats, ships, and **submarines** [V], and by lighter-than-air ships such as blimps, **hot-air balloons** [III], and **dirigibles** [V].

Archimedes' screw

The screw pump, supposedly invented by Archimedes (c. 287–212 B.C.) while studying in Alexandria, Egypt, consists of a hollow pipe in the form of a helix, wound around a shaft. With one end immersed in water and the shaft tilted at 45 degrees, water would be lifted as the shaft was turned. In another form, a helix-shaped blade inside a tube can be turned to achieve the same effect. This ancient device, sometimes called the Archimedes snail, continues to serve as a form of water pump in parts of Africa and Asia.

The screw served to inspire 19th-century inventors to attempt to design screw-propelled ships, one of the first of which was named the *Archimedes* in honor of the inventor of the concept. The same principle is used in auger drills and in many other machines and tools relying on a spiral or a helix.

ax

Often noted as the earliest tool, the ax was preceded by other cutting edges and scrapers consisting of simple chipped or sharpened stones. The ax proper combines a cutting edge with the leverage gained from a handle in a striking tool, producing one of the first **woodworking tools,** essential in the first shaping of wood and in gathering firewood from growing trees rather than deadfall limbs and trunks.

Stone axes helped give the name to the Old Stone Age, or Paleolithic period, stretching from about 40,000 B.C. to about 8000 B.C. **Bronze** ax heads were developed after about 3500 B.C. The earliest ax heads were lashed to the handle. In an alternate method, a bundle of sticks (fasces) was bound tightly around the ax head. It was this implement, suggesting strength derived from collecting individually fragile elements together, that provided the symbol and the origin of the term for the Italian Fascist Party in the 1920s. By the Roman period, ax blades were specially created with eye openings into which the haft or handle could be fitted.

In the medieval era the ax became widely used for clearing forests and also became a weapon used in hand-to-hand combat. In the United States, with its vast forests, clearing trees by ax created demands for improved designs. In the 1830s the brothers Samuel and David Collins began to produce a special ax that had a **steel [IV]** bit made from Swedish ore mounted between two iron strips that created an eyehole for insertion of the handle. Stamping machines improved the output of the Collins factory.

American axes became known for their extended poll or flat edge opposite the cutting edge, providing excellent balance for woodsmen. An alternate form, with a double blade, became popular after the Civil War.

basket

Since baskets are made of materials that easily decay, it has been difficult for archaeologists to date the origin of basketmaking with any certainty. However, in desert regions, some baskets have been found in the Western Hemisphere that date to as early as 9000 B.C., while in Iraq and Egypt, basketry can be dated as early as about 5000 B.C. Some remains of baskets and wickerwork apparently used for eel and fish traps have been found from the Mesolithic period, which preceded the Neolithic Revolution. Along with fish-hooks, twine, and some **textiles,** there is evidence that even before the development of **agriculture** and the **domestication of animals,** people were using such improved technologies to harvest fish. In the United States, the ancestors of the Pueblo Native Americans have been designated the basket-maker people because of their fine work in the craft.

Because of their ease of construction and their flexibility, baskets are excellent containers, especially for fragile items, since the structure absorbs and distributes shock. Baskets are used to gather and transport food, and when coated with tar or pitch, can be made waterproof. Because of their ability to protect the contents from damage through jarring, applications of basketwork have been found in such varied items as hats, luggage, baby cribs and carriages, wicker furniture, horse-drawn vehicles, and the passenger containers of **hot-air balloons [III].**

basket   This modern Native American woman keeps alive the traditional method of basket making by interweaving rushes. *U.S. Census Bureau*

The techniques of basketmaking from prehistoric times include two basic methods: coiling and weaving. In a coiled system, a bundle of fibrous plant material is bound together in a long strip then coiled around a center-piece. The coils are held in place by sewing a strip around the coils. The coil method was used extensively by a number of Native American tribes, including the Pomo of California. The weaving method involves intertwining, as in the manufacture of fabric, of fibrous materials at right angles to each other. Some anthropologists have suggested that the woven basket is the forerunner of textile weaving, by setting the principle of alternate interweaving of materials. A very fine weave can be achieved such as that used in Ecuador in the manufacture of hats known incorrectly as "Panama hats." Although textile weaving has been mechanized, the nature of basketmaking has tended to prevent such modernization, and it has remained a handicraft for the most part.

beer

The first alcoholic beverage was either **wine** or beer. Although wine apparently originated in Babylonia, it seems that beer or "wine made from barley" was also known at the same period, as early as 6000 to 5000 B.C. One early method was to place barley in a pottery vessel and then bury it in the ground until it began to germinate. Then it was milled, made into dough, and baked. The cake could then be taken as a lightweight item on travels, and when stopping at an oasis for water, it would be soaked until fermentation began. The very acid-tasting beer was known in the 20th century A.D. as "boozah," apparently the origin of the English word with a similar meaning. Records of beer being served have been found in Babylonia as early as 2225 B.C., and both the Egyptians and the Babylonians used beer as a medicine. The Greeks imported the concept of barley beverages from Egypt.

Since few physical remains or artifacts survive from the beermaking process, very little is known about the diffusion of beermaking. Scattered literary references help document the fact that beer of different varieties was widespread throughout the ancient world, including frontier regions and among peoples generally regarded as beyond the fringes of civilization.

By the end of the classical period, in the fourth century A.D., beer was known throughout northern Europe. Varieties included mead, which involved a fermented mixture of honey and water in Britain. Another type was metheglin, mead with herbs added to it. A dark beer, similar to modern porter, was in use in Britain even before the Roman invasion, and in the 1st century A.D. the Irish had developed a local beverage similar to ale.

board games

The most ancient board games have been traced to Mesopotamia and ancient Egypt in the 1st millennium B.C. One, called the game of 20 squares, involved a race using pieces that were moved according to the roll of dicelike joint bones of oxen or sheep. The game was played not only by the wealthy but also by ordinary people and apparently even by guards during long hours of guard duty at palace gates. One magnificent gate piece with statues of colossal bulls from King Sargon's palace gates at Khorsabad has a small game board inscribed on its base, apparently by the bored guards. A second type of board game, with pegs and holes in a board like modern cribbage, had 58 holes.

Archaeologists have uncovered one of the 20-square games complete with a set of instructions on how it should be played, with descriptions of the pieces and the moves. A game using almost identical rules and known as asha survived in India into the 20th century. Ancient dice were 6-sided and numbered 1 to 6. However, unlike modern dice, the opposite sides did not total 7 but were numbered consecutively, with opposite faces giving totals of 3, 7, and 11. The markers moved on the boards were in the shape of birds, animals, dogs, cones, and pyramids.

brass

Brass, an alloy of copper and zinc, is valued for its light weight, rigid strength, and ability to hold a polish for some time before tarnishing. Although the ratio of the alloy components can vary a great deal, typically brass is about one-third zinc and two-thirds copper. Apparently brass was first used in Palestine in about 1400 B.C., and many of the biblical references to brass in more ancient times are actually incorrect translations for the word for **bronze,** an earlier alloy of copper and tin. Since the distinction between zinc and tin may not have been clear to early metalworkers, some brass may have been produced accidentally with the intention to make bronze.

Depending on the amount of zinc mixed with the copper, the malleability of the metal will vary. If the zinc content exceeds about 45 percent, the resultant metal cannot be worked at all. Such high-zinc-content brass, in powdered form, has been used as a form of **solder.**

The Romans used brass for vases and dishes, jewelry, and for brooches and clasps on clothing. Brass was expensive to produce, and some coins were made of brass during the Roman period. Brass plates were used in cemeteries to commemorate the dead, engraved with details of the person's life. The British often imported stone and brass plate markers or "tomb slabs" from France in the 13th century A.D., in which the brass pieces would be fixed to a marble backing with lead plugs. Brass dishes were common in the 16th century A.D. throughout Europe, although silver plate tended to replace brass as the metal of choice after the opening of silver mines in the New World. Brass continued to be used widely for candlestick holders, sundials, **clocks [II],** and musical instruments.

Various brass alloys in which the copper and zinc are varied in percentage or are supplemented with other metals have differing qualities, including coloration. Adding tin in low percentages helps reduce the tendency of brass to tarnish. Minute amounts of less than 1 percent of arsenic, phosphorus, or antimony, in addition to the tin, further increase the resistance to tarnishing. By increasing the zinc proportion with copper to about 40 percent, brass reaches a color approximating gold, and this alloy mix is known as Muntz metal. The best strength is achieved at 70 percent copper and 30 percent zinc, in an alloy known as cartridge brass. By adding lead to the alloy, the machineability of brass can be improved.

bronze

The earliest known alloy, or blend of two metals, is bronze. Developed about the year 3500 B.C. in the Middle East, the alloy consisted of **copper** and tin, resulting in an alloy that was stronger than copper but still easily worked. Perhaps surprisingly, tin is also a soft metal, but the alloy of the two, because of the mixture of crystalline shapes, is harder than either.

Bronze is still made, with a ratio of one part of tin to three parts of copper. In ancient bronze artifacts, the proportion of copper varied from about two-thirds to more than 90 percent. In bronze **cannons [II]** built in Europe in the 12th century, the proportion was 8 parts copper to 1 part tin. Bronze intended for large bells, or "bell metal," has a proportion of about 20 to 25 percent tin. In the modern era, bronze alloys with small amounts of other metals have included phosphor bronze, used in making valves and other machine parts, as well as nickel and aluminum bronzes, which are more corrosion-resistant and used in ship propellers and pipe fittings. So-called copper coins are usually a bronze alloy of 95 percent copper, 4 percent tin, and 1 percent zinc.

Bronze is harder than copper yet easier to cast, more readily melted, and is harder than pure, unalloyed iron. Bronze resists corrosion better than iron. When iron was substituted for bronze in about 1000 B.C., it was simply because it was more abundant and available than tin, needed to make bronze, and hence was cheaper. But iron had no metallurgical advantages over bronze for use in tools and weapons.

In 1865, British archaeologist John Lubbock (1834–1913) proposed an idea in the textbook *Pre-Historic Times,* based on the prevailing notions of a Darwinian social evolution, that the history of the world could be divided into four ages: the Stone Age, the Bronze Age, the Iron Age, and the Steam Age. Lubbock's periodization has been modified, with a division of the Stone Age into the Old Stone and New Stone or Paleolithic and Neolithic (with a transitional period between, known as Mesolithic) and the addition of a period when copper and stone were both used for tools, known as the "Chalcolithic." His text was republished in several languages and used widely around the world, establishing the periodization he suggested as a standard assumption. The working of bronze was only one of several advances in metallurgy that characterized ancient civilizations from about 3500 to about 1000 B.C.

Bronze, as distinct from copper, as a durable and hard alloy began to have an immediate impact on the environment. Bronze could be used for tools, fittings, and appliances where copper would have been too soft. Furthermore, bronze was hard enough to hold an edge. Thus bronze could be used to replace flint knives and spear points and could be shaped into swords. Fighters equipped with bronze weapons would rapidly defeat equal numbers of an enemy armed with stone and flint weapons. Bronze axes and knives could be used in clearing timber and in woodworking. Copper, because it was so soft, had found most uses in decorative items, along with gold.

canals

Canals were built in ancient times. Among the earliest was one about 50 miles long built in the 7th century B.C. to bring fresh water to Nineveh. An early Suez Canal was constructed in about 510 B.C. to connect

canals Invented in the Ancient Near East, canals moved cargo and passengers far more efficiently than roads. This mural depicted the ceremony at the opening of the Erie Canal in 1825. *C. Y. Turner mural, De Witt Clinton High School, New York*

I commanded this canal to be dug from the Nile River, which flows in Egypt, to the sea, which goes from Persia. This canal was afterward dug as I had commanded, and ships passed from Egypt through this canal to Persia as was my will.
—Inscription, Emperor Darius, c. 500 B.C., at Behistun, Persia (Iran)

the Red Sea and the Nile River. The canal, ordered by Darius, followed the course of the modern canal from the Red Sea to the Great Bitter Lake and Lake Timsah, where it forked west, following a natural watercourse, joining the Nile. Apparently the canal constructed under Darius was a reconstruction of an even earlier canal built as early as 1470 B.C. by Egyptians, preceding the modern Suez by about 3500 years.

These early canals were all constructed through nearly level territory and simply consisted of a ditch sufficiently deep to accommodate shallow-draft boats, connecting two waterways and supplied with water from them. The **canal lock [II]**, developed in Holland, allowed the construction of canals through low passes through hills and resulted in many 18th- and 19th-century canals in Europe and the United States.

cemeteries

The first cemeteries, as distinct from individual burials, have been dated to the Mesolithic period, with many identified sites in Europe from the period about 4500 B.C. Cemeteries seem to represent the growth of complex societies and were found in coastal areas or adjacent to lakes or rivers. Some investigators assumed that the cemeteries were used to mark a fertile area as belonging to a particular group, through linkage to local ancestors.

Some early Mesolithic cemeteries had as few as 20 burials, and others had as many as 170. From the skeletons, modern scientists have been able to determine the causes of death of Mesolithic peoples, with common ailments including arthritis, rickets, and dental decay. Now and then the remains indicate a case of murder or death from hunting accidents, individual fighting, or possibly organized warfare between groups. The graves also included a wide variety of artifacts and showed diversity in burial customs. In some cases, hunters were found buried with their dogs. Some cemetery burials from the Mesolithic era show the emergence of a ranked society, indicated by the burial of children with items of wealth, suggesting that the goods in the grave could not have been acquired by the child in his or her lifetime but rather were an indication of inheritance.

Some cemeteries have revealed other social phenomena, such as both inherited wealth and acquired wealth, and some differentiation in the types of artifacts found with male and female skeletons. Pierced tooth pendants from elk, bear, and beavers appeared to represent status or hunting ability.

Archaeologists have found individual burials, some with ceremonial items in them, dating to the Paleolithic era, but the grouping of burials into cemeteries is an indication of the growth of the more complex social arrangements that characterized the Mesolithic and Neolithic peoples.

central heating

Central heating systems were installed in China during the Han dynasty (202 B.C.–A.D. 220), in which a raised floor for sleeping was built over the location of an oven. Pottery models of homes from that period reflect this method of construction.

Systems of central heating have been found in the remains of ancient Greek structures, but central heating reached a high state of development in Rome in the 1st century B.C. The Romans constructed systems with raised floors supported by **columns.** In a central location, a fire would be built of wood and the heat conducted through the passages under the floors. The hypocaust provided heat for homes, public baths, and for other public buildings. According to one story, the hypocaust was invented by an oyster farmer, Caius Sergius Orata, in about 100 B.C. He kept fish and oysters warm in large tanks by raising the tanks on pillars and bringing in hot air through the space under the tanks. From this, it is assumed, the idea of heating tanks of water for public baths, and then for homes, was a natural evolutionary extension. The public baths of Constantinople were heated with petroleum contained in lamps.

With the decline of the Roman Empire, central heating fell into disuse and was not revived in Europe until the 19th century, when coal-fired boilers and steam systems were introduced. Piped hot water systems came into use in the 1830s. However, in Turkey, the tradition of public baths was kept alive through the centuries, and when built in other countries, they were frequently known as Turkish baths.

chain

The exact date and location of the invention of chain is unknown, but one assumption or speculation is that ancient fishermen may have noted the effect of tangled or interlinked fishhooks. Some of the earliest known chains were found in tombs in Mesopotamia dating from before 2500 B.C. Egyptian tombs show chained prisoners, and references in the Bible to ornamental chains suggest that chains were common in the Middle East by 1000 B.C. Most of the ancient chain appears to have been made from **bronze** rather than from iron. Iron tended to stretch under strain, and it was not widely used in chain manufacture until the development of stronger wrought iron in the 19th century.

During the period of the Roman Empire, chains were used to moor ships, to restrain slaves and animals, to secure chests, and to hang oil lamps and cooking vessels. During the Middle Ages and into the modern era, the making of chains was considered a regular part of the work of blacksmiths. In ancient Rome, a form of armor, mail, was first made by sewing metal rings to a fabric or leather garment, and later medieval **chain mail** [II] armor was made by interlinking the rings without sewing them to any underlying garment. Plate armor was gradually added over the chain mail, eventually fully supplanting it. Ancient and medieval jewelers made chains of gold and silver for ornamental wear.

Chains began to undergo various improvements in the 18th century, and they found new uses in machinery and industrial settings. In 1801 James Finley built a suspension bridge in Uniontown, Pennsylvania, using chains, and a patented flat link chain found uses in bridges and piers in Britain early in the 19th century. Samuel Brown, who patented the British flat link chain, developed a patented anchor chain whose use rapidly spread. Stud-link chain had a short crossbar in each link, adding weight and strength and helping to prevent the chain from tangling. In the United States, industrial chainmaking centered around Pittsburgh and York in Pennsylvania. In 1864, James Slater in Britain patented a **drive chain** [IV], although drawings of nearly identical chain had been made by Leonardo da Vinci (1452–1519). The drive chain, ideal for connecting a power train over a distance too great to be spanned by gears alone, soon found application in the **bicycle** [IV] and even in the early **automobile** [IV].

cities

The ancient development of cities has been regarded as synonymous with the development of civilization. Depending on the definition, some early walled towns, such as Jericho in southern Palestine, built between 10,000 and 7000 B.C., might be regarded as the first evidence of urban life. In what is now Turkey, the city of Çatal Hüyük was built in about

6700 B.C., covering about 30 acres. Consisting of mud-brick buildings built around narrow alleys or streets, the town was about 4 times larger than Jericho. Ancient Mesopotamia had Uruk, with a population approaching 50,000 by the year 3,000 B.C. The early cities of the Tigris and Euphrates region known as Mesopotamia were characterized by a temple and surrounding walls pierced by fortified gates, some of which were cast from bronze or made of heavy wood studded with and held together by bronze bolts. Babylon, Nineveh, Ur, and several other major cities left ruins still only partially excavated and explored in modern Iraq.

It is difficult to estimate the population of ancient cities, but Rome may have had as many as 900,000 inhabitants in the first century B.C., while Alexandria in Egypt may have had about 600,000.

Ancient cities in the Indus Valley in India and Pakistan were laid out in regular rectangular grid patterns. Evidence suggests that nearly all of these ancient cities had many of the features of modern urban life, including crime, a police force, some form of public lighting of streets, sewers, fire brigades, **plumbing,** banks, money, organized religion, a class structure, and division of the cities into specialized neighborhoods devoted to different trades or crafts.

The relative mix of social forces that caused seminomadic people to gather in cities is a matter of scholarly discussion. Clearly, a fortified city provided a degree of security against attacks by bandits, or even whole armies. Furthermore, the practice of religion could require a social structure that supported a priestly class. Certain psychological preconditions to the city had to exist, including traditionalism, a power structure, and recognition of the difference between an in-group and out-groups. Specialization of crafts such as potterymaking, textile weaving, butchering, masonry, and carpentry could flourish only where large numbers of people could gather to exchange the products, supporting the specialization with a market for the products and with the necessary security to allow the transactions. Whether the craft economies stimulated urban life, or urban life, created for military and psychological reasons, stimulated the development of crafts is a subject of interpretive slant.

It was **agriculture** and the **domestication of animals** in the Neolithic Revolution of 8000 to 7000 B.C. that provided the preconditions for the development of cities. However, the fact that they did not come with the Neolithic technologies to western Europe, but were restricted to the ancient Near East, India, China, and to Crete before 1000 B.C., presents a historical mystery. What factors in western Europe prevented the

> At that time I enlarged the site of Nineveh, my royal city, I made its market streets wide enough to run a royal road. In days to come, that there might be no narrowing of the royal road, I had steles made that stand facing each other. Fifty great cubits I measured the width of the royal road.... If ever anyone of the people who dwell in that city tears down his old house and builds a new one, and the foundation of his house encroaches upon the royal road, they shall hang him upon a stake over his house.
>
> —King Sennacherib of Assyria, c. 685 B.C.

emergence of cities along with agriculture? Conversely, what factors in the Near East, Asia, and Mesoamerica stimulated the growth of cities? Even as the technologies of the Bronze Age moved into Europe from 2500 to 1000 B.C., the largest human settlements were little more than villages or hamlets by the standards of Mesopotamia.

Only a few of the ancient Babylonian cities ever prospered for more than a short period. They would grow affluent when their ruler conquered other peoples and brought back booty, with which his generals would be enriched and the city's temples decorated. When they flourished, cities became economic magnets, drawing luxury items such as spices, wines, metals, and ornamentations from a wide region and along distant trade routes. However, when a city was attacked and its army conquered, the city would fall into disrepair. For centuries, descendants might live among the ruins, prey to raids from more nomadic peoples.

The development of utilities such as water supplies, sewer systems, and modern means of transportation greatly stimulated the growth of cities during the Industrial Revolution. It has been estimated that before 1850, fewer than 7 percent of the world's population lived in cities or urban centers made up of more than 5,000 people. A century and a half later, 30 percent live in such centers, and in some industrialized countries, the percentage is more than 60 percent.

column

The use of the column in architecture grew out of the use of wooden post and lintel structures, in which a pillar or a post would support a crossbeam. With stone construction of large temples and pubic structures, a stone pillar constructed of sections could be erected. Columns were used by the Egyptians as early as 2600 B.C., but the ancient Greeks made extensive use of them and perfected the various types. Some Egyptian columns were constructed to imitate bundles of papyrus or to resemble other plants.

The columns at the Parthenon in Athens were constructed of 11 marble sections each weighing up to 9 tons, with a total height of about 33 feet. Greek architects developed three different styles of capital or column top that would support the lintel. The Doric was the simplest, with a flat slab at the top, while the Ionic is surrounded by four scrolls arranged on the sides and the front and back. The Corinthian column top was more ornate, with leafwork, often representing the acanthus plant. Doric columns tended to be shorter, ranging in height from 4 to 5½ diameters as measured at the base. The height of all columns could be emphasized by fluting, or vertical grooves. The Roman architect Vitruvius in the 1st century A.D. laid out a set of rules for columns, dictating that the Doric should be 8 diameters in height, the Ionic 9 diameters, and the Corinthian 10. Composite columns combining Ionic scrolls and the acanthus leaf of the Corinthian should also be 10 diameters in height. Later architects varied the forms with spiraled or sinuous columns characterizing some Baroque period columns of the 17th century.

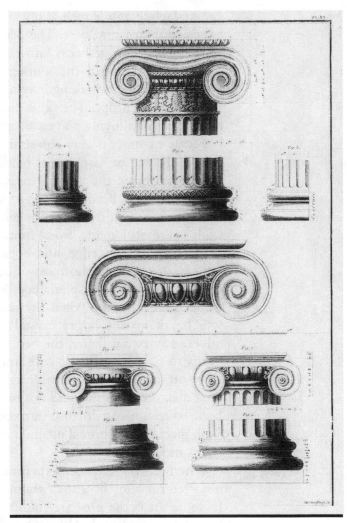

column The classic column became a staple of Roman revival architecture in the United States. *Library of Congress*

copper

Copper was perhaps the first metal worked, in approximately 8000 B.C., about the time that settled communities dependent on **agriculture** developed. However, to regularly work copper, there had to be a regular system of extraction of the metal from the ores, and that began in about 5000 B.C. in Iran and Afghanistan. Copper is found in the green ore malachite and is melted out of the ore at temperatures that can be achieved with wood fires. The period in which copper tools and stone tools were used together is known as the Chalcolithic Era.

Copper was mined and worked in Europe later than in the ancient Near East, and its working may have been independently discovered in southeastern Europe. The skills did not include at first the sophisticated techniques of alloying and casting in use in the Near East.

Because copper is very soft and malleable, it could be molded, hammered, drawn, or cast. It was readily made into tools, cups, and ornaments. However, the very softness that made it easy to handle was its main drawback. Because it was so soft, it did not take a sharp edge, and for this reason, in the period when copper was the only metal, cutting tools were still made from flint and obsidian stone that could be chipped to a sharp edge, and hoes and plows were made from wood, antlers, or stone. The softness of copper derived from its atomic structure, forming parallel plane crystals that slip past one another easily rather than gripping when the metal is pulled. Although hammering will break up the larger crystals and make them jagged enough to adhere somewhat better, the best way to increase the hardness of copper is to blend or alloy it with another metal. Copper and gold, both very soft metals, were desirable as status symbols and often traded over long distances, but they did not bring about a revolution in tools, since both were too soft for use in axes for wood clearing or as weapons. Both metals were more a medium of display than a means to change the environment.

Whether by luck or experimentation, early workers in copper invented the alloy **bronze** in about 3500 B.C. by mixing tin with copper. In some areas the ores are found together, so the mixing could easily have been accidental. Although tin is even softer than copper, the two metals together produced a tougher alloy. The resultant metal was hard enough to hold an edge, and bronze weapons and tools supplanted stone in the following centuries.

crossbow

Although some Western scholars trace the origin of the crossbow to 10th-century European improvements on the Greek siege ballista, crossbows were common in ancient China. Pictures of crossbows in China have been dated to the 4th century B.C., and by the reign of Emperor Shih Huang Ti (221–210 B.C.), both heavy crossbows mounted on carts and individual crossbowmen were found. Among the terra-cotta figures in the emperor's tomb at Mount Li were some 200 crossbowmen. It is estimated that in Shih Huang Ti's reign the imperial army had 50,000 crossbowmen.

The advantages of the crossbow over the longbow were that the crossbow could be carried in a cocked or armed position, it shot with extreme power, and it took little training to use. It was so powerful that the Lateran Council of A.D. 1139 attempted to ban its use by Christians against Christians. With the addition of **steel [IV]** in its construction in the 15th century it became even more powerful, capable of piercing **chain mail [II]** and with a range of up to 300 feet. The steel-bowed crossbow was known as the arbalest.

Both crossbows and arbalests were usually cocked by placing the tip on the ground, where it would be held by the archer's foot in a **stirrup [II]**. Then the bow would be drawn, either with a hook attached to the archer's waist, by a lever-crank known as a goat's foot lever, or with a windlass **crank [II]**. The crossbow remained the most powerful infantry weapon well after the introduction of **gunpowder [II]** in Europe, not replaced until the late 15th century, when it was supplanted by the harquebus.

dams and dikes

Dams and dikes perform the same function, using similar materials. Earth, stone, or concrete structures separate a body of water from lower land. The economic object of the two types of structure, however, differs. Dams hold water to preserve it as a resource, either for agriculture, to maintain navigability of the lower stream, to create a recreational lake, to impound it for hydroelectric generation, or to combine several of those purposes. A dike, on the other hand, can preserve the land as a resource by holding back river water, estuaries, or the sea itself, as well as serving to enhance the navigability of the retained water.

The first dams constructed in antiquity were in Babylonia and Egypt, and ancient earth-fill dams still exist in India and Sri Lanka (formerly Ceylon). One in Sri Lanka, some 11 miles long, was built in the 4th century B.C. A masonry dam, perhaps the oldest in the world, was constructed in about 2700 B.C. across the Nile, while the Orontes Dam, constructed in about 1300 B.C. in Syria, still impounds a large lake, some 3 by 6 miles in extent. The oldest known dam in China was built before 2200 B.C.

In the late A.D. 1400s the Dutch began a land reclamation system, using dikes to hold back seawater, coupled with **windmill [II]**-powered pumping systems to maintain the land for agriculture. The reclaimed areas, known as polders, greatly increased the land available to Holland.

Modern dams, often combining hydroelectric generation with recreational uses and water impounded for agriculture, have become the largest structures in the world. There are more than 10,000 dams higher than 50 feet in the world, with more than 200 added every year. The Hoover Dam, completed in 1936 at the Nevada-Arizona border, was the highest in the world at that time, at 726 feet. Since then, dams that are higher and dams that impound greater volumes of water have been built. Among the largest is the Aswan High Dam, in Egypt, with a reservoir of 137,000 acre-feet, completed in 1970, and the Bratsk Dam in Russia, completed in 1964, with the same reservoir volume.

For most dam construction, the greatest challenge tends to be the temporary diversion of the waterway while the dam is under construction. In some modern dams, the diversion requires the construction of tunnels through surrounding walls. The invention of **dynamite [IV]** for blasting through rock has made possible the building of modern dams in steep rock canyons, such as the Hoover Dam and the Glen Canyon Dam, both on the Colorado River.

days of the week

The seven-day week was invented by the ancient Babylonians, and it was based on their astrological observations. They named each of the days after the Sun and the Moon and the five planets visible to the naked eye: Mars, Mercury, Jupiter, Venus, and Saturn. When the Romans adapted the Babylonian calendar, they changed the names to the Latin equivalents. The names of the days in European languages reflect their origin in these planet names, with the Romance languages following the Latin pattern. When the Anglo-Saxons and the Vikings adapted the names, they chose the names of gods that were roughly equivalent to the Latin gods represented by each of the celestial bodies, and those were adapted into English. An exception in the Latin-derived Romance languages was the conversion of Sunday to the Lord's Day, with names derived from the Latin *Dominus* for Lord. See the table "Language Equivalents for Celestial Body Names."

The order of the days was a more complex pattern, derived from the Babylonian system. The order of the celestial bodies according to the Babylonian astrologers (not at all in accord with the **heliocentric solar**

Language Equivalents for Celestial Body Names

Celestial Body	Languages			
	Latin	**French**	**Saxon**	**English**
Sun	Sol	Dimanche	Sun	Sunday
Moon	Luna	Lundi	Moon	Monday
Mars	Mars	Mardi	Tiw	Tuesday
Mercury	Mercurius	Mercredi	Woden	Wednesday
Jupiter	Jove	Jeudi	Thor	Thursday
Venus	Venus	Vendredi	Frigg	Friday
Saturn	Saturnus	Samedi	Saturn	Saturday

system [II] as understood today) was Saturn, Jupiter, Mars, Sun, Venus, Mercury, and Moon, but the days of the week were not simply put in that order. Having divided the days into two 12-hour segments totaling 24 hours, the Babylonians named each hour of each day after the celestial bodies in that order, rotating through the 7 and starting over. Thus Saturday, the first day, began with Saturn's hour. After going through three cycles of seven, Saturday's night hours would come out as follows:

9:00 P.M. (21st hour)	Moon
10:00 P.M.	Saturn
11:00 P.M.	Jupiter
12:00 A.M.	Mars
1:00 A.M.	Sun

Thus the Sun's hour would start the day following Saturn's day, making it Sun day. A similar calculation would bring the Moon's hour up at midnight on that next night, leading to the next day beginning with the Moon's hour, and hence Moon day. And so the cycles would proceed through the week, coming out with the next Saturday beginning appropriately with Saturn's hour.

Although since the discovery of the heliocentric nature of the solar system, science has described the order of the planets, the Sun, and the Moon in a completely different fashion, and although custom has abandoned the original names for hours in favor of a more simple system of numbered hours, the complex Babylonian astrological system with its order of days has been retained in modern calendars, even though its origin has vanished from public memory.

domestication of animals

The earliest domestication of animals has been very difficult to establish, and it is generally assumed that the first animal successfully born and raised with humans is the dog. Guesswork has suggested that dogs may have domesticated as early as 10,000 B.C., but the earliest remains of domesticated dogs date from about 6500 B.C. in several locations, including Britain, Denmark, and the Middle East. Breeds resembling terriers were found dating from that period in Jordan, while a greyhound-type dog is represented in Egyptian art by about 4500 B.C. It has been assumed that the reason why dogs were the first animals domesticated is because they voluntarily began to associate with humans, although that is simply speculation. In one sense, dogs may have adopted humans to gain access to food and shelter, and in that sense dogs can be said to have domesticated humans.

Other animals caught and raised in captivity include cats, cattle, and fowl. Hens may have first been domesticated in Southeast Asia for gamecock fighting rather than as a source of food. Sheep bones dating from about 9200 B.C. indicate that sheep were raised at about the same time that grain was cultivated in what is now Iraq and Iran, and domestication of those herd animals went hand in hand with the development of **agriculture** as part of the Neolithic Revolution. Apparently various types of wild sheep and goats in the region were captured and then bred in captivity in that period. Sheep, goats, asses, and horses were not naturally migratory, like the reindeer, so early herdsmen, to increase their flocks, moved them from location to location for wild forage, forcing them into a migratory pattern. The **domestication of the horse** was itself a major achievement, changing the nature of transport and warfare.

Cattle may have been domesticated before 4000 B.C., but clear evidence that these animals were herded and not simply hunted is not available before about 3000 B.C. in both Egypt and Iraq. The domestication of cattle, with the use of neutered males as oxen, contributed to the development of the **wheel**.

domestication of animals The domestication of naturally herding animals, including goats, was a defining invention of the late Neolithic period. *U.S. Department of Agriculture*

Prior to herding, humans adopted a nomadic transhumance lifestyle of **migratory herd following** that still characterizes the lifestyle of the Lapps, who move with their reindeer.

A humped breed of cattle was domesticated in what is now Pakistan in about 3000 B.C. At about that time the donkey was also domesticated, from a wild type of onager that lived in the deserts to the south of Egypt. In about 5000 B.C., the horse was domesticated somewhere between the Ukraine and the Caucasus Mountains of central Asia. The later invention of the **stirrup [II]** made the horse-mounted cavalryman a formidable weapon system.

In other regions of the world, various animals were domesticated as beasts of burden, including the two-humped Bactrian camel in central Asia, the one-humped dromedary camel in Arabia and North Africa, the reindeer (or caribou) in northern Europe (only semidomesticated in the migratory herd-following style), the elephant and buffalo in South Asia, and the llama in South America. In general, the native peoples of North and South America had very few domesticated animals, including only a species of duck, turkeys, bees, the llama as a pack animal in a restricted area, and the related alpaca for wool. Native peoples of North America had no large domestic animal but were known to use dogs to pull sledges and sleds.

domestication of the horse

In the **domestication of animals,** the sequence was first the dog, probably before 10,000 B.C., then goats and sheep, the wild ass, and the ox. In about 5000 B.C. the draft horse begins to appear in central Asia and the ancient Near East. Apparently horses were first used to draw carts with **wheels** and were soon employed in drawing chariots to carry fighting men and to make displays in processions.

In about 2000 B.C., humans started to ride horses. Horseback riding represented a major revolution in technology, with dramatic effects, particularly in warfare. Horses had to be bred larger to carry a person, and the practice seemingly grew up first among nomadic tribes that herded horses in central Asia, including what are now Iran and Afghanistan. Some of the great conquering peoples of central Asia, including the Huns and the Mongols, achieved their victories via the horse-mounted cavalry soldier.

The horsemen of central Asia developed organized warfare to plunder, a tradition that pitted the mounted horseman against settled agricultural villagers and against defenders of **cities.** The techniques of horseback warfare evolved, becoming truly formidable after the introduction in the Middle Ages of the **stirrup** [II], which allowed the cavalryman to put the full force of the horse behind the strike of the handheld lance.

The only surviving species of originally wild horses is *Equus caballus przewalskii*, named after a Russian-Polish explorer, Nikolai Mikhailovitch Przewalski (1839–1888), who identified it in 1880 in the wild, mountainous regions between Tibet and Mongolia. Smaller than the domestic horse, the Przewalski horse is assumed to be the ancestor of the modern horse. Horses were not useful for drawing plows or other heavy hauling until the development in the Middle Ages of the **horse collar** [II] and the **horseshoe** [II].

Genghis Khan (A.D. 1162–1227) led horse-mounted cavalry, equipped with stirrups, in the conquest of northern China in 1213, then ranged over what is now India, through to the Crimea on the Black Sea.

Since the Middle Ages, horses have been bred with numerous specializations, including the Thoroughbred for racing, the Arabian for riding, Percherons for heavy hauling, the Lippizanner (from Croatia) and Ten-

domestication of the horse Domesticated and bred for riding, horses often return to the wild. These were captured in Utah.
U.S. Bureau of Land Management

nessee walking horse for performance, and a variety of cowponies for herding and ranch work. When a male ass is bred with a female horse, the result is the usually sterile hybrid mule, a hardy pack animal.

dye

The craft of dyeing and the development of various dyes, like the discovery of pigment for **paint,** were developed in the ancient world, including Egypt, China, and India. Saffron yielded a yellow-orange color, as did henna. Among the earliest dyes were blues derived from indigo and madder plants and Tyrian purple derived from a vein in a mollusk found near Tyre on the shore of the eastern Mediterranean. Tyrian purple, one of the rarest colors, made purple a sign of royalty in the Mediterranean region. It is estimated that it took about 8,500 shellfish to produce a single gram of Tyrian dye.

Such animal and vegetable dyes were best used when the fabric was pretreated with a mordant, a metal solution that would allow the dye to adhere better to the fibers and enrich the color effect. Ancient mordants included aluminum sulfate, and tin in the form of cream of tartar. Tannic acid, from bark, was also used as a mordant, for browns and tans.

In the ancient world, various colors became associated with particular locales or even specific villages because of the interaction of local metal salts with locally available vegetable dyes.

As commerce expanded to include new regions, dyes were exchanged and new colors added. In the 13th century A.D., a new purple dye, achil, derived from a lichen, was discovered in northern Italy. After the discovery of the Americas, cochineal, quercitron, and logwood added new colors. Cochineal was derived from a small insect that Native Americans used to create a rich red powder, harvested by drying the female bugs in sunlight and then grinding the bodies. Cochineal is produced in the modern era in the Canary Islands and is used, among other things, to dye maraschino cherries red. Quercitron was derived from the inner bark of the North American oak and was patented in 1775. Brazilwood was known as early as the 4th century A.D., and the country of Brazil was named for the wood already used for dye, rather than the wood being named for the territory, as one might assume. In the 18th century, cloth was often bleached with kelp, before the discovery of **chlorine bleach [III].**

The first artificial dye was created by the British chemist William Henry Perkin (1838–1907), who experimented with coal tar in hopes of making an artificial source for the medicine quinine. His aniline dye, derived from petroleum naphtha, was later named mauve by a French textile manufacturer, and it soon became popular. By the 1890s, the color was so widely adopted that the period was sometimes known as the mauve decade. Perkin was only 18 years old when he patented the process to make the color in 1856. Perkin's company produced more than a ton a day of alizarin, a yellow-red color derived from coal by 1871, and he was able to sell his company and retire to academic pursuits and research at age 36. In effect, Perkin was the creator of the synthetic dye industry.

In 1905 Adolf von Baeyer (1835–1917) received the Nobel Prize in chemistry for discovering the molecular structure of indigo and developing a process for synthesizing it in 1880.

Eratosthenes' sieve

Eratosthenes (c. 276–c. 194 B.C.), who was also known for establishing a preliminary definition of the **latitudinal zones** and a suprisingly accurate estimate of the **size and shape of the Earth [III]**, invented a method for calculating prime numbers known as Eratosthenes' sieve. It was called a "sieve" because his method consisted of sorting out nonprime numbers from all numbers by eliminating those that were not prime, just as a sieve separates the desired product by allowing the unwanted material to fall through a screen or mesh. He developed the sieve in about 200 B.C.

A prime number is a number that can only be divided evenly by itself and by 1. Thus the first ten prime numbers in order are 2, 3, 5, 7, 11, 13, 17, 19, 23 and 29. The number 2 is the only even prime number, since all other even numbers can be divided by 2. Eratosthenes' method consisted of the following process. After listing all numbers, beginning with 2, up to the desired amount, every second number is crossed out. Then, beginning with 3, every third number is crossed out. Then, beginning with 5, every fifth number is crossed out. Numbers that are already crossed out are included in the count. The process is continued with 7, 11, 13, and so on. The numbers that are left uncrossed out are prime numbers.

With the modern **computer [VI]**, researchers have used similar methods to try to determine the highest known prime number. In 1989 a group of researchers at Amdahl Corporation derived the then highest known prime number, which itself had more than 65,000 digits in it. On November 14, 2001, a young researcher, Michael Cameron, from Canada, proved that $2^{13,466,917} - 1$ is a prime number. That number has more than 4 million digits.

fire

The questions of the first human control of fire and the manufacture of fire have been much debated in the study of prehistoric man. It is assumed that the first control of fire developed when humans took advantage of naturally started fires, sparked by lightning, and transported embers to be able to provide fire for heating and cooking. Prehistoric Peking man clearly controlled fire in this sense as early as 500,000 B.C. There is evidence that prehuman races of hominids may have controlled fire as early as 1.4 million or 1.5 million years ago. Pre-*Homo sapiens* hominids may have used fire to defrost carcasses found under snow, and from this practice may have evolved the first cooked meat.

It is clear from archaeological finds that during the Upper Paleolithic period, from 40,000 to about 8000 B.C., humans used fire, apparently for cooking, heating, lighting in caves and at night, and for drying fish and meat. The clearest early evidence of controlled use of fire is the discovery of hearths in the remains of human camps and shelters. However, the ability to sustain a naturally started fire (from lightning or volcanic sources) by carrying embers is quite a different matter from the ability to actually start a fire.

Although archaeological discoveries constantly change the estimate, it appears that the ability to manufacture or start a fire developed among humans sometime before 7000 B.C. Some of the first firemaking devices included drills, fire saws, and fire-bow drills that would produce heat by friction and spark-igniting flint struck against pyrites.

Various types of fire drills and fire plows remained in use among primitive peoples of Oceanea and Indonesia into modern times. A bow-driven drill for making holes in wood, an early **woodworking tool,** may have accidentally led to its alternate use as a firemaking device. A fire plow consisted of a grooved board with a second piece of wood that the user would repeatedly rub through a groove to increase heat by friction. Ancient Egyptians, Eskimos, and Asian peoples used fire drills. The fire saw consisted of a notched or nicked piece of bamboo through which a thin bamboo sliver would be repeatedly drawn to obtain friction heat. The fire piston, also found in Southeast Asia and Indonesia, was a bamboo tube with a piston that would heat air by compression. The variety of primitive firemaking devices in the region of Indonesia and the Pacific Ocean has raised the possibility that the first innovations in fire manufacture occurred there and dispersed to other parts of the world.

Flint-steel sparking devices remained in common use until the development of **matches [IV]** in the 19th century. Steel-sparking lighters were trade items used by Europeans in the 17th and 18th centuries as they traded with Native American peoples.

geometry

The invention of geometry as a method can also be classified as the discovery of eternal principles. As a branch of mathematics, geometry is concerned with the mathematical properties of space in two and three dimensions. The methods are attributed to Euclid (c. 330–c. 260 B.C.), who wrote 13 books on mathematics. Nine of his volumes dealt with geometry and 4 with the theory of numbers. His writing was so compelling and logical that many copies were translated into Arabic, and they survived through the destruction of other classical works. He systematically arranged prior discoveries, classifying them into axiomatic truths, definitions, and theorems. Prior mathematical findings developed in surveying land as well as methods developed by mathematicians Thales (c. 624–c. 547 B.C.) and Pythagoras (c. 580–c. 500 B.C.) were incorporated by Euclid into his work.

Euclid used a synthetic method, in which the user proceeds from known to unknown mathematical principles through several logical steps, and the analytical method, in which an unknown is posited and then worked toward from the known facts.

Euclid taught in Alexandria, establishing his own school of mathematics in about 300 B.C. The first translation of Euclid's works into Latin was made in about A.D. 1120. A number of other works, on spherical geometry, music, logical fallacies, and conic sections, have been attributed to Euclid but usually are not included in the listing of 13 works on geometry and the theory of numbers.

In 1637 René Descartes (1596–1650) developed analytic geometry, using algebraic notations for the description of geometric objects. Isaac Newton (1642–1727) and Leonhard Euler (1707–1783) extended analytic geometry.

glass

The origins of glassmaking are lost in antiquity, although improvements in the manufacture, design, coloration, and production of glass can be dated to different regions, periods, and even specific artists in more recent times. Glass beads have been found in ancient Egyptian settings from as early as 2500 B.C. Vessels and containers of glass were made by melting block glass in molds as early as 1500 B.C., and bottles could be made by coring—hollowing out a core of limestone or clay around which a coating of glass had been molded. In Alexandria, mosaic glass was produced in which glass canes of different colors were cut to make decorative patterns.

Glassblowing, using a blowpipe, appears to have originated in Syria in the 1st century B.C., whence it spread through the Roman Empire. The Romans developed a technique of blowing glass into a mold and then cutting away a layer to leave a relief, or cameo. In the 13th century, glassmaking reached a high art form on one of the islands of Venice, where artists made richly decorated forms. Later Venetian glassmakers developed a clear crystal or *cristallo,* which became a steady Venetian export. Other glassmaking centers developed in Bohemia and London. In 1675 George Ravenscroft discovered that adding lead to the glass produced a heavier, crystal type. Lead or leaded crystal became prized for tableware.

Glass is notable in the history of invention in that it was made by craftsmen long before science could explain its qualities, including transparency or the effects of different minerals in changing its hardness or color. In the 19th century, Michael Faraday (1791–1867) characterized glass as a solution rather than a compound, and it is now treated as a supercooled liquid mixture.

The centers for glassmaking in Venice and Holland contributed to the development of **eyeglasses [II]**, **telescopes [III]**, and **microscopes [III]** in those locations. The European production of glass retorts, tubing, flasks, and other shapes in the 17th and 18th centuries made possible many of the advances of chemistry and were essential to the discovery of **oxygen [III]** and other elements. Glass art objects are often classified with fine art, with production of decorative pieces by Louis Comfort Tiffany, by the French firm of Daum Frères and René Lalique, and by the New York firm of Steuben Glass, among many others, becoming highly prized by collectors and museums. Glassmaking in the United States often centered in districts with silica-rich sand, such as southern New Jersey. In 1739, Caspar Wistar began the American glass industry near Salem, New Jersey.

harpoon

The basic hooked spear apparently was invented by the Magdalenian peoples of southern Europe in about 13,000 B.C. The first spears were not barbed, but by the time of the flowering of cave art in what is now Spain and France, harpoons were barbed with a double row of hooks.

The barbed harpoon was used to spear animals, not just fish, and was a primary tool of nomadic hunting.

The hunting and gathering lifestyle of prehistoric man required that small groups be constantly on the move, for as population increased, peoples soon exhausted the local stock of edible animals and had to relocate to another area or starve. The spear or harpoon no doubt contributed to the near-extinction of many large mammals hunted for their meat by early humans in the Ice Ages.

ice cream

The ancient Romans flavored ice with fruit and honey, and Marco Polo (1254–1324) brought several recipes for flavored ice from the Far East to Italy, where it caught on in the 15th and 16th centuries. Water ice and sherbet shops sprang up in France in the 17th century. In the late 18th century, cream was added to water ices at shops in Paris. Ice cream was popular in the American colonies and the early republic. The hand-cranked ice-cream maker was invented in 1846 by Nancy Johnson in New Jersey, and the first known commercial ice-cream factory was established in Baltimore in 1851. With the advent of **refrigeration [IV]** it became possible to manufacture, store, and ship ice cream in larger amounts.

The ice cream soda was invented by Robert Green in 1874 at a celebration at the Franklin Institute in Philadelphia. The sundae was apparently invented in Wisconsin in the 1890s, as a means of getting around laws that prohibited the sale of ice cream sodas on Sundays. The soda-fountain operators served the ice cream with the flavored syrup but no soda and called the resultant dish a *sundae* to avoid any religious backlash from the improper use of the word *Sunday.*

The ice cream cone was invented twice—once by an Italian immigrant to the United States, Italo Machiony, who patented a conemaking machine in 1903, and the second time with an impromptu use of a Syrian waffle known as a *zalabia* to contain ice cream at the Louisiana Purchase Exposition in St. Louis in 1904.

ink

Writing ink was separately invented by both the ancient Egyptians and the Chinese, as early as about 2500 B.C. Using carbon suspended in water with some natural gum, the ink would adhere to the paper as it dried. India ink was a thicker form of the same type of carbon-black pigmented ink, stabilized with shellac, soap, glue, gum arabic, and other materials and used primarily in drawing. The Chinese used colored earths and lampblack in inks for printing with blocks as early as A.D. 500.

With the development of the printing press using **movable type [II]** in the 15th century, printing ink consisted of finely ground carbon black cooked in a mixture with linseed oil and a natural resin or varnish to allow the ink to adhere to the typeface while wet and to dry quickly on the paper. Colored inks were patented in 1772 in Britain.

latitudinal zones

The division of Earth's surface into five zones—two arctic, two temperate, and one torrid or tropical—is characteristic of all modern world maps. The concept was developed in about 240 B.C. by Eratosthenes (c. 276–c. 194 B.C.). Although none of Eratosthenes' work survived in complete form, he is regarded as the father of geography.

He suggested that two latitude lines parallel to the equator, and about 24 degrees to the north and south of the equator, measured the end of the tropical zone. The two lines defined the tropics. The northern parallel of the tropics is called the Tropic of Cancer, so named because the Sun makes its turn on June 21 from its yearly northward passage in the sky to then setting more southerly every day, and June 21 is the date when the constellation of the Crab, or Cancer, makes its first appearance in the sky. The southerly tropic latitude line is named the Tropic of Capricorn, for the date when the Sun is at its most southerly point, and when the constellation of Capricorn first appears, December 22. Eratosthenes further correctly noted that the Sun stood directly overhead on the Tropic of Cancer on June 21 and directly overhead on the Tropic of Capricorn on December 22.

Born in what is now Libya, Eratosthenes was educated in Athens and then moved to Egypt, where he was tutor to the son of the pharoah and where he became the librarian of the **library** at Alexandria. Using his observations about the angle of the Sun, Erastothenes was able to calculate the circumference of Earth with rough accuracy, coming within about 15 percent of the correct figure, using a method that was a logical way to determine the **size and shape of Earth [III]**.

leather

The production of leather is one of the most ancient arts, practiced in Paleolithic times. Although the nature of the material is such that archaeology can yield very few clues to the origin of the processes, it has been assumed that the first curing of skins came as primitive people sought to restore flexibility to the hides of animals for their own use by rubbing them with grease and oil. Skins could also be cured by smoking to prevent them from deteriorating. A hard leather could be produced by rubbing with salt, and thus the curing of meat and the preparation of the first leathers may have developed together.

It is not known where or when the first tanning of leather with the tannic acid found in wood, wood bark, seeds, and certain roots was developed. Inscriptions and a few artifacts provide evidence of tanned leather in ancient Egypt, Babylonia, and Assyria. In the Roman Empire, soldiers wore armor, sandals, and items of clothing made from tanned leather. The art of vegetable tanning was reintroduced to Europe, perhaps by several routes, in the Middle Ages. It is known that by the 700s, the Muslim invaders of Spain brought the method with them, and the production of red-dyed vegetable-tanned leather in Cordova and Morocco made cordovan and morocco leather highly prized. By the late Middle Ages, Europeans were widely using leather for aprons, buckets, clothing, sandals and shoes, and straps, among other applications.

Both oak and chestnut wood were widely used for vegetable tanning of leather in Europe, and after settlement of colonies in the Western Hemisphere, other woods included hemlock, quebracho and myrabolam (in South America), and mangrove tree bark. In 1884 Augustus Schultz, an American chemist, developed a process that had earlier been noted by Ludwig Knapp, the use of chromium salts in tanning. Various chromium compounds have been employed since then to produce leathers that can be immersed in hot water without damage. Both vegetable tanning, which produces a heavier and stronger leather, and chrome tanning, which produces softer and more flexible leather, remain in modern use.

library

The first library or place in which records were kept and stored is that of King Assurbanipal, from the 7th century B.C. in the city of Nineveh in ancient Assyria, and located across the Tigris River from the present location of Mosul, Iraq. He built the library at his palace and sent agents out to find tablets from all the fields of education. When the tablets were brought to the library, they were copied exactly, and if a section were missing or obscure, the scribe would note the gap or illegibility in his copy. It is estimated that Assurbanipal's library contained about 1,500 tablets, each with 80 to 200 lines of text. Ninevah, including the library and its contents, was destroyed in about 612 B.C. by the Medes.

The tablets were stored in several ways, either by wrapping each table in a piece of cloth or reed matting or by storing them in a box, jar, or basket. The collections would be stored on wooden shelves or on mud-brick benches along the walls of library rooms. Private libraries of wealthy individuals from the next centuries have been found throughout ancient Mesopotamia, and temples sometimes had separate archive or library rooms with built-in pigeonholes or square shelves for tablets.

Often the library rooms and the individual boxes or containers of tablets would be sealed, apparently to protect the documents from theft. From examining some of these surviving private archives it has been found that they were organized for reference purposes and filed by content, date, and place. Attached to the jars or boxes would be tags indicating the contents. In the ancient city of Ur, tablets have been found with a notation or memo along the edge, much like the title of a modern book on the spine, so that a particular tablet could be identified from a group stacked with their faces together.

The first known public library was established in Athens in 330 B.C. The Alexandrian Library, built in the same year in Alexandria, Egypt, reputedly contained 700,000 scrolls, representing the body of Greek literature and science. After several raids, it was burned in A.D. 640 by a Christian mob. The French Bibliothèque Nationale was established in Paris in 1520, and the New York Public Library in 1895.

> Hunt for the valuable tablets which are in your archives and which do not exist in Assyria and send them to me. I have written to the officials and overseers . . . and no one shall withhold a tablet from you; and when you see any tablet or ritual about which I have not written to you, but which you perceive may be profitable for my palace, seek it out, pick it up, and send it to me.
> —King Assurbanipal, instructions to agents, c. 650 B.C.

lighthouse

In ancient times, lighting bonfires on promontories and harbor islands was a common way of helping returning fishermen and other sailors find their way safely into port after dusk. Although many structures were built to lift the lights to a higher position for greater visibility, the most famous early lighthouse was built in the period 300–280 B.C.

lighthouse Constructed along the lines of ancient light-houses, most lighthouses in the United States have been automated or abandoned. Opened in 1917, this lighthouse on Navassa Island in the Caribbean was shut down in 1996. *U.S. Geological Survey, Department of the Interior*

during the reign of Ptolemy II of Egypt on the island of Pharos. During Roman times, smaller lighthouses were built at almost every port. The Colossus of Rhodes, a **bronze** statue at the harbor of Rhodes, served as a lighthouse. Its bonfire light, reflected with mirrors, was visible up to 35 miles away. It was destroyed in 224 B.C. by an earthquake. During the Middle Ages lighthouses were built at Genoa in 1130 and at Leghorn in 1304.

With the increase in shipping in the early modern era, construction of lighthouses was revived. In Britain, John Smeaton developed a method of hardening concrete underwater in 1759, and lighthouse construction flourished in the second half of the 18th century, using parabolic reflectors to enhance the light. By 1820 there were some 250 major lighthouses around the world. The first lighthouse in the United States was built at Boston, on Little Brewster Island in 1716. The second was built on Nantucket Island in 1746. State lighthouses were all ceded to the federal government in 1789, with more than 10,000 built in the United States in the 19th and 20th centuries.

maritime commerce

The earliest evidence of maritime commerce or trade shows up in the Bronze Age, from 2000 to 1500 B.C., with lively trade on the islands of the Aegean Sea, involving Crete, Greece, and the eastern Mediterranean to the coast of what is now Israel, Lebanon, and Egypt. Bronze Age shipwrecks and their cargoes give a fairly complete view of the sorts of commodities traded along the coastal and island routes. With the Mycenaean culture of Crete, which flourished between about 1400 and 1200 B.C., regular maritime trade became well established. Such long-distance trade, itself representing a combination of technical and social inventions, was crucial in the spread of other Bronze Age technologies and culture from the ancient Near East and the palace culture of Crete to the fringes of western Europe.

Contact between Crete and Egypt appeared to be infrequent before 2000 B.C. Once the Mycenaean Bronze Age trade became more regular, it included valuable items such as silver vessels; large, decorated bowls called kraters, which had pictorial scenes of bulls and chariot groups; ingots of **copper;** and whole tusks of ivory. Some of the **bronze** objects present a mystery for archaeology, since there are no nearby sources of tin, required for the alloy, suggesting that it must have been imported from as far away as Spain or central Europe. The routes for this first tin trade remain to be traced in detail.

Some pottery vessels were small containers, presumably carrying valuable liquid contents, such as rare oils or unguents. Amphorae, which were big ceramic containers 3 or more feet tall and containing resin, have been found intact in some of the wrecks. Additionally, cargoes included ingots of glass, ingots and objects of tin, wooden tablets for writing, and tools such as axes. Other finds have included short swords and bronze spearheads.

The ships themselves were propelled by a combination of sails and oars and were not capable of long-distance cross-ocean trips, like the improved **sailing vessels [II]** of later centuries.

Although this Bronze Age trade helped spread technologies, inventions, and a wide variety of products over routes connecting the eastern Mediterranean, there was little evidence that the first traders worked into overland routes. This early maritime trade influenced a few coastal areas in what is now southern Italy and the island of Sardinia, but for the most part, western Europe developed very slowly in the Bronze Age, apparently through the gradual spread of technologies by imitation and less organized overland and riverine trade. In the last years of Mycenaean maritime trade, routes through the Adriatic began to make contact with overland trade through the northern Italian Po River valley.

> Once seafarers, whether Aegean islanders or easterners, had learnt the sea routes, winds, and currents in the eastern Mediterranean, there was nothing to prevent the expansion of trade as far as economic and political circumstances allowed. Most of this trade remains invisible to us and it is only with the discovery of Bronze Age shipwrecks off Cape Gelydonia and Ulu Burun near Kas that we have any picture of how trade was conducted, what the ships were like, and the kind of cargoes they carried. Invisible influences such as the standards of weights and measures and the use of writing and systems of administration must have been equally important.
> —K. A. Wardle, "The Palace Civilizations of Minoan Crete and Mycenean Greece, 2000–1200 B.C.," in *The Oxford Illustrated Prehistory of Europe,* edited by Barry Cunliffe (Oxford, Eng.: Oxford University Press, 1994)

A later maritime trade expanded with the rise of Greek city-states after 800 B.C., apparently building on some of the earlier outposts and trade centers established in the prior millennium.

metalworking

Evidence of working in softer metals, particularly gold and copper, has been found in ancient Egypt and Mesopotamia. **Bronze** (an alloy of tin and copper) was introduced as early as 3000 B.C., leading archaeologists to suggest a division of the ages of man since the Stone Age into the Copper Age, the Bronze Age, and the Iron Age. Iron was introduced in the Middle East in about 1200 B.C. Since iron required higher temperatures than bronze for smelting, and since useful iron had to have impurities removed through reheating and hammering, a variety of new tools and processes had to be introduced.

In the early Bronze Age, from 2800 to 1200 B.C., metalworking quickly spread in the Carpathian Basin, in the region from Poland to Romania. During the 3rd millennium B.C., metal use spread via migratory nomadic peoples as far as the Urals, and local peoples through what is now Germany and other areas of central Europe began to learn such techniques as casting and hardening of metals and the making of alloys. Long-distance trade, mostly overland or by canoe on rivers, led to the spread of copper, gold, and bronze artifacts manufactured in central Europe as far afield as Denmark, Italy, and Greece. In some regions the design of locally made flint weapons and pottery vessels began to imitate the patterns of such items made from metal. Apparently the concept of casting a hilt to a lengthened dagger to produce a short sword developed first in what is now Germany.

The early metalworking skills of central Europe led to a largely north-south overland trade involving bronze tools, ornaments, dishes, and weapons, as well as amber, spearheads, and jewelry. Burials and gravesites across Europe from about 2000 B.C. reveal the spread of awls, knives, rings, and bracelets, as well as more sophisticated **pottery**, jewelry, woolen cloth, and wooden furnishings.

Early ironworkers discovered that iron worked with charcoal could produce **steel [IV]**, which could be hardened by repeated quenching (rapid reduction of temperature by immersing in water or another liquid) and tempering (slow raising of temperature), to reduce the brittleness of the metal. The Greeks developed hardened steel as early as 800 B.C., and the Romans had tempered steel weapons.

Tools for working metal, like **woodworking tools,** evolved to follow the basic functions of cutting and holding in place. However, metal has other characteristics that make it highly desirable for working into a variety of shapes. When heated and softened it can be bent, hammered, or drawn. Softer metals such as copper and gold can be worked in these ways at very low temperatures. In addition, since metal could be melted to a liquid and then take on the shape of its containing vessel, casting became a way of shaping metal objects.

Goldsmiths practiced the so-called lost-wax method, in which a wax shape would be made then covered with clay. When baked, the clay would harden, and the wax was "lost" or drained out through a minute hole. Then the clay mold, holding the impression of the original, would be filled with molten gold, which would harden to the original shape. This art was practiced in both ancient Egypt and in the Aztec world. A simpler method was sand casting, in which a shape would be impressed into a bed of sand that contained binding materials, then lifted out and the other side impressed into a second bed. After the two impressions hardened, they would be joined together and filled with the molten metal to take the shape of the original. Using such techniques, repeated castings were ideal for decorative yet utilitarian items such as tableware and other household utensils, appliances, and fixtures.

In the Middle Ages, metalworkers developed and perfected many further skills and tools, including shaping metal on hand-turned lathes, stamping, wire-drawing, shearing or snipping softer metals such as tin, grinding, and rolling. Techniques of joining metal pieces together, not only with fixtures such as bolts but also with welding, soldering, and riveting developed long before the modern age. An alloy of tin and lead made ideal **solder,** since it has a wide melting range, between 361 and 460 degrees Fahrenheit. Before the age of the Scientific and Industrial Revolutions, metalworkers had many of the processes and tools that would continue into the later periods, developed over centuries of craftwork. The invention of **movable type [II]** and the printing press allowed for the wide dissemination of craft skills through illustrated manuals. Long before the introduction of the **steam engine [III],** metalworkers used power from **windmills [II]** and **waterwheels [II]** to provide forced-air bellows for heating and power for stamping, hammering, or turning metal shapes. Metal shapes made by such methods allowed for the earliest cylinders, pistons, gears, and other movable parts employed in the first steam engines.

migratory herd following

The coming of the Ice Ages, with the last Ice Age beginning in about 15,000 B.C., forced hunting and gathering peoples to rely more on animals for food than on plants. When peoples in the prehistoric period hunted with stone and carved bone tools such as the **harpoon,** they would move from place to place as animal stocks were depleted. However, in what is known as the transhumance way of life, or migratory herd following, some peoples adapted by moving along with migrating herds of wild animals. This development was a practical invention, for it meant that, like hunting, it involved pursuit, but it did not lead to local extinctions of the food animal. Rather, the wild animals would set the pace of travel, and the lifestyle had some of the characteristics of herding. Although the animals were not truly domesticated, the animal would be somewhat tended, and the wild herd, like the domestic herd, would become a mobile food reservoir for humans.

Migratory herd following continues in the modern world only among the Lapp peoples of northern Scandinavia. The Lapps apparently migrated northward as the glaciers retreated from about 11,000 to 10,000 B.C., perhaps from as far south as southern France, and today they continue to follow the natural migrations of the reindeer. Some archaeological finds and cave drawings of reindeer suggest such a pattern.

Although sometimes called herdsmen, the Lapps are actually not herding the reindeer but simply following the herds where they move. The Lapps, however, did adopt a few methods that characterize **domestication of animals,** such as castrating some of the reindeer bulls to make them more manageable as draft animals. The Lapps rely on the reindeer for many products, including meat and milk. They use the sinews, fur, and hides for clothing and shelters, and they make tools and buttons from the antlers.

The transhumance lifestyle can be seen as a transition between the hunting and gathering of pre–Ice Age and Ice Age peoples and the Neolithic Revolution, which came out of **agriculture** and domestication of animals.

mining

Although humans used **stone tools** made of flint before they began **metalworking,** there is evidence that the earliest mines were dug to mine **copper** and that flint mines were only dug later. The earliest copper mines, dug in about 4500 B.C., have been found in Serbia, while a series of flint mines dug in about 4000 B.C. have been found more widely in western Europe. Flint mines were dug using picks made of deer antlers, while the shoulder blades of cattle were used for shovels to remove debris.

Mines to obtain turquoise and copper have been located in the Negev desert in what is now Israel and on the Sinai peninsula, both taken over by the Egyptians from prior inhabitants before 2000 B.C. King Solomon's mines were in fact copper mines. The Egyptians mined for gold in the eastern desert and south into Nubia, often worked by prisoners in chain gangs.

The Chinese are known to have mined for coal as early as the Han dynasty (202 B.C.–A.D. 220). One ancient mine in Hebi has a vertical shaft 150 feet deep. The Romans mined for coal and tin in Britain and for copper and silver in Spain. Several of the copper and silver mines in Spain were more than 600 feet deep.

Mining was such a source of wealth that it drove other technologies. At Rio Tinto, in Spain, archaeologists have discovered a Roman system from the 1st century B.C. of treadmill-operated **waterwheels [II]** that were used to pump water to the surface from nearly 100 feet down. Using eight wheels on **bronze** axles, stepped one above another in separate chambers, water would be raised from chamber to chamber to the surface. Similar waterwheel systems were employed by the Spanish in the construction of silver and gold mines in Central and South America after A.D. 1500. Mining also drove technologies of lighting, ventilation, hauling, and surveying. The first vehicles to move on rails were carts in mines, and improvements in iron tools and pumping systems for removing water eventually led to the first **steam engines [III]**.

mirrors

Polished bronze and copper mirrors have been dated to about 2900 B.C. in ancient Egypt and to the civilization of the Indus Valley, in what is now India and Pakistan, from 2800 to 2500 B.C. Chinese mirrors have been found from about 1500 to 1000 B.C.

In the Han dynasty, from about 202 B.C. to A.D. 220, highly polished mirrors with inlays of gold and silver have been found. During this period the Chinese experimented with the principle behind periscopes, of using two reflecting surfaces to see over objects.

The ancient Greeks developed small boxes with mirrors, very much like the modern compact, while the Romans had both hand mirrors and wall mirrors. According to the Roman encyclopedist Pliny, glass mirrors were first developed in the Phoenician city of Sidon. However, during the Middle Ages, the mirrors found in Europe were made of polished metal rather than glass. Glass mirrors were reinvented in the 13th century A.D.

nomadic herd tending

With the **domestication of animals**, particularly sheep and goats, herders took up the practice of leading the animals from summer to winter grounds. Unlike the earlier **migratory herd following**, the nomadic herd tenders had to force the animals into migration to keep them supplied with food. This method evolved in the first millennium of domestication, about 8000 B.C.

The nomadic herd style persisted in some areas, notably in North Africa and Arabia, where alternation between water holes and surrounding fertile lands in the summer and wider pastures in the desert that flowered during the rainy season allowed for the growth of larger flocks. The nomadic lifestyle led to a number of adaptations and inventions, but only those that could be readily carried on the long marches. Bread would be made by baking flour on hot stones over a fire, producing the flat bread and the nan bread of the Near East and South Asia. Yogurt would be made by churning milk in a goatskin bag. Thread would be spun on hand spindles. Being mobile, nomadic peoples did not engage in metalworking but traded their animals and animal products such as wool for metal utensils made by village and urban peoples. In general, all nomadic herders were engaged in the same work, with no specialization or craftsmen, since tending the animals and making shelter and preparing food engaged all members of the nomadic tribe. As a consequence, the total society would be devoted only to the one task, unlike those who lived in villages and in market towns, in which specialists could emerge.

paint

Paint consists of a coloring matter combined with a liquid carrier or vehicle that, when dried, leaves the color in place where the paint was applied. The concept dates to prehistoric times and is found throughout the human-inhabited world. Coloring with pigments from natural ores and vegetable dyes was developed before the first civilizations. Cave paintings in Lascaux, France, and Altamira in Spain date back to as early as 15,000 B.C., and ancient peoples in Africa and the Americas also used colors to decorate religious sites and homes. By 1000 B.C. the

Egyptians were using sap from the acacia tree, known to the Europeans as gum arabic, to fix colors, and other liquids to fix pigments were derived from beeswax, animal fats, and egg whites. Some primitive peoples simply used mud as a carrier for pigment, as among the Sepik River people in New Guinea.

Pigments were largely derived from either of two sources: minerals and natural organic substances. Minerals often produced "earth colors" such as browns, reds, and ochers. Ochers came from iron oxides and could be changed in tint from red to reddish-orange or sienna depending on the amount of heating or calcination. Umbers from iron oxide would have higher manganese content and have a brownish color. Roasted white lead produced a bright red known as minim. Blue was derived from ground lapis lazuli, found in Armenia. In Europe the color came from beyond the sea and hence was called "ultramarine." Copper could be used to prepare a range of green colors. A metallic pigment manufactured in Germany in the 18th century from a blue iron ore became known as Prussian blue. Organic dyes derived from insects; shellfish; various charcoals; and the bark, stems, and leaves of plants. Carmine and madder were two colors derived from plants.

Paints are classified according to the vehicle or binder. Thus beeswax paint is known as encaustic and was practiced by the ancient Greeks and Romans. Egg yolk–vehicle paint is known as tempera. Some vehicles, such as oils, needed a volatile thinner. Turpentine was used by the ancient Egyptians, and oil paint thinned with turpentine was characteristic of 15th-century Flemish **oil painting** [II]. Linseed oil and other oils would assist in drying. Dutch painters used a thicker style of oil paint in the form of a paste, or impasto, in the 17th century. Pigments bound with gum arabic and thinned with water are simply watercolors or aquarelles. When a white color is added to the wash, it is known as gouache. Widely used to illustrate manuscripts and books, gouache can be readily reworked, and great precision of work can be achieved.

In ancient China, lacquered work was found in which the liquid coating with resin would evaporate, leaving a solidified and hardened surface.

plumbing

Although often associated with ancient Rome, the development of plumbing has been found in more ancient civilizations, including the city of Kish in ancient Iraq as well as in the Minoan civilization on Crete, which reached a peak before 1000 B.C. The ancient civilization of Mohenjodaro in the Indus Valley in about 2500 B.C. had drains installed in private homes. The Babylonian and Cretan civilizations developed cemented drain tiles, swimming pools, bathrooms, and even pottery piping tapered to be fitted together.

The Romans introduced lead piping, and the word "plumbing" derives from the Latin word *plumbum,* for lead. The Romans understood the problems of moving water, and undertook civil engineering projects such as **aqueducts** to supply fresh water, large public baths, and extensive sewer systems to carry off wastewater. Cesspools, lead pipes with workable faucets or valves, and even a type of **flush toilet** [III] have been found in ancient Roman ruins. The ruins of the town of Pompeii, covered in volcanic ash in A.D. 79, have yielded particularly rich evidence of sophisticated plumbing. The Baths at Caracalla, in about A.D. 210–220, had both hot and cold water, and elaborate facilities such as separate rooms for shampooing and manicures, a library, lecture halls, and an art gallery.

In Rome, the main sewer, known as the *cloaca maxima,* was partially vaulted over by the 6th century B.C., and by the 3rd century A.D. it was entirely vaulted. The sewer carried storm water, effluent from the public baths, and some household drainage. In general, the development of cesspools to carry off household sewage and human waste fell into disuse in the Middle Ages. Although privies—simple earthen pits that would be periodically cleaned—were in common use, where possible a garderobe or toilet room would be structured over running water to carry waste downstream.

More common in cities until the 18th and 19th centuries was the simple practice of dumping human waste from chamber pots in the streets, whence it would be carried off with storm water to local streams and rivers. As cities such as London and Rome grew in size, the pollution of the river reached truly epidemic proportions, as the discharge would create the threat of cholera and other wasteborne diseases.

pottery

Pottery is usually regarded as a product of the Neolithic Revolution, occurring after 5000 B.C. However, a few finds associated with Mesolithic cultures suggest that there was some overlap in the development between the eras. Some pottery remains, for example, have been found associated with Mesolithic settlements in Scandinavia, dated to about 3600 B.C.

Ancient Egyptians regularly used the potter's wheel and fired clay pots in kilns or ovens as early as the Third or Fourth Dynasties, from

2900 to 2750 B.C. Drawings from tombs show preparation of the clay and workers placing clay on a mobile disk, shaping it with one hand while turning the wheel with the other. Most authorities agree that the concept of turning the wheel with a foot treadle, which would allow the potter to use both hands to shape the pot, did not develop until the Middle Ages. Some ancient illustrations show a helper turning the wheel while the potter worked the object with both hands.

It is difficult to date the earliest use of firing to harden pottery, but remains of kilns from ancient Mesopotamia survive from about 4000 B.C. and in Egypt from about 3000 B.C.

Even earlier fired clay objects have been found, such as shaped animal forms or cult figures, and clay pots made by spiraling cords of clay have been dated as early as 9000 B.C. in Japan.

Copper ore glazes to give the pottery a blue tint, and tin and other metallic glazes to impart white or other colors were used in ancient Mesopotamia. Burnishing the pot before it was fired and dipping the pot in a slip of fresh clay after it had been fired were also ancient techniques found in Mesopotamian ruins from the 17th century B.C. The Chinese perfected a technique using feldspar and kaolin fused at high temperatures to produce a fine grade of pottery. Known as chinaware or china, the technique was not successfully copied in Europe until the 18th century A.D.

pulley

Ancient pulleys have been dated as early as 1000 B.C., and there are depictions of pulleys that go much further back in Babylonian times, to as early as several thousand years B.C. The Greeks and Romans used pulleys in shipbuilding and in theatrical performances. In its simplest form, a pulley is simply a wheel supporting a rope or belt that transfers the direction of motion. Thus a pulley is ideal for raising a water bucket from a well or in hoisting construction material to a higher level while standing on the ground.

However, when several pulleys are combined by attaching them in a block, a block-and-tackle system can be used to achieve a mechanical advantage. Archimedes (c. 287–212 B.C.) detailed the principles that allowed compounded pulleys to provide magnification of effort. In general, the ratio for a mechanical advantage is governed by the number of pulleys involved and how they are arranged, less a minute amount lost to friction. When pulleys are free to move, with one end of the line fixed to an anchor or beam, each pulley can provide a mechanical advantage of 2 to 1. A compounded system with 5 free pulleys reaches a mechanical advantage of 32 to 1. Archimedes is reputed to have used a compound system to pull a ship onto dry land.

pyramids

The finest pyramids of Egypt were built in the period of the Old Kingdom, from about 2686 to 2181 B.C., in the Third through the Sixth Dynasties. Altogether some 90 pyramids in Egypt survive into modern times, and the most famous are at Giza, near the Nile, and south of Alexandria. Almost all of the pyramids from the great age of construction were built on the fringe of the desert to the west of the Nile.

As tombs, the pyramids apparently had the purpose of preserving the body of a dead ruler, protecting his body from disturbance or destruction, and providing a collection of equipment to sustain the body in its afterlife. This so-called cult of the dead appears to account for the origin of Egyptian pyramids.

In predynastic Egypt, rulers were buried in shallow pits surrounded by personal objects and ornaments. The ornaments, of course, made such graves great targets for grave robbers, so more and more elaborate structures were conceived to prevent digging into the grave. Some of these early graves were roofed over to create a tomb, and in later periods the walls were plastered or lined with Sun-baked bricks. When a complete structure of heavy bricks was built, it was known as a mastaba, and some of the mastabas of the First and Second Dynasties have been excavated in modern times, revealing various details of construction, decoration, and practices. It seems likely that as mastabas grew more elaborate and were covered with mounds of brick or stone, the concept of constructing a pyramid to cover the tomb developed.

Burying bodies deep underground to protect them both from robbers and from the weather had the effect of insulating them from the heat that otherwise would have dried and preserved the bodies. As a consequence, the Egyptians developed techniques of embalming the bodies, including wrapping the limbs in linen pads soaked with resin. The first intact mummies to survive into modern times came from the Fifth Dynasty. Often the bodies would be buried with models of individual objects, as well as composite scenes representing the life of the deceased and showing the production of food. From these depictions, archaeologists have learned a great deal about such aspects of that culture as hunting, agriculture, brewing, and baking.

The oldest true pyramid is the Step Pyramid, built of stone in the Third Dynasty, on high ground at Saqqara, overlooking Memphis. The pyramid rose in six stages to a height of just over 200 feet, and the base was about 411 feet by 358 feet. At the core of the pyramid is a stone square mastaba, about 200 feet on a side.

The group of pyramids at Giza, the most famous of the Egyptian pyramids, is assumed to have been started by Khufu (Cheops), followed by two other kings of the same Fourth Dynasty, Chephren and Mycerinus. The pyramid of Cheops is known as the Great Pyramid and represents the culmination of pyramid building in Egypt. It is very precise in its measurements and alignment, within a few inches of 755 feet on each side at the base and oriented almost exactly in the directions of north and south, east and west. It was originally 481 feet high, but now the last 31 feet are gone. Interior galleries, chambers, and channels are still being explored, although detailed descriptions of some of the interior have been available since the 19th century.

Around the world, other peoples also constructed pyramid structures. A complex of pyramids at a ceremonial site near Mexico City includes the so-called Pyramid of the Sun, the largest in the world, at 990 feet on a side at the base. The Mayan ceremonial center at Palenque included pyramids used as tombs.

Pythagorean theorem

The Greek mathematician Pythagoras lived from about 580 to 500 B.C. The "theorem" named after him is more in the nature of a universal truth he discovered. In mathematics, a theorem is a universally true law that can be expressed as a formula but one that is sometimes difficult to prove on mathematical grounds alone. The Pythagorean theorem states that in a right triangle, the square of the hypotenuse is equal to the sum of the squares of the other two sides. With this principle in hand, it is possible to calculate the length of any side of a triangle that has a right angle in it if the lengths of the other two sides are known. For example, if the lengths of the two sides that intersect at right angles are known, those lengths, when squared and the total squares are added together, yield a sum whose square root is the length of the hypotenuse. To express the theorem algebraically, if the two sides of the triangle that meet at a right angle are A and B, and the diagonal hypotenuse is side C, then

$$A^2 + B^2 = C^2.$$

roads

The first roads were probably animal pathways that humans chose to follow, as they offered a way through obstructing vegetation and around other obstacles. The first roads built by human work appeared to follow on the development of the **wheel,** probably in ancient Mesopotamia in about 3000 B.C. By 2000 B.C. roads of varying quality connected major points in Europe, from Italy northward to what is now Denmark.

Long trade routes from Europe eastward that consisted of a combination of roadways and marked trails through open country and desert regions were developed between 1900 and 300 B.C. Known as the amber routes, they ran across central and eastern Europe and were used to transport amber and other ancient trade commodities. Darius I built the 1,500-mile-long Persian Royal Road in about 500 B.C. in ancient Iran. The so-called Silk Road or silk route, a series of trails and routes extending some 4,000 miles, connected ancient Rome and China after about 100 B.C.

The Romans built a durable network of roads, first in Italy, and then more widely into distant regions of the empire. The road-building period began in 312 B.C. with the 350-mile Appian Way, connecting Rome to Brindisi, a port in southeastern Italy. This route was the way to Greece. Another road connected Rome and Naples and was the first leg of the trip to North Africa.

Roman roads, built over a period of several hundred years, had several characteristics. Using local materials, engineers would lay a rock foundation or bed, slightly raised in the middle to provide drainage. Often paved over with carefully fitted stones, sections of Roman roads survive into modern times. Bridges were constructed using cement to hold stones together and **arch** construction, with some of the bridges still in existence in the 21st century.

After the fall of the Roman Empire, road building and maintenance fell off. In the middle of the 17th century, the French government instituted a system of forced labor for road building and care, while the British began allowing the construction of toll roads. Tolls would be paid to enter a road, and a long staff or pike would be turned out of the way to allow passage—hence the term *turnpike* to designate such a toll road. By the 1830s there were more than 20,000 miles of turnpikes in Britain. However, the competition from the **steam railroad [IV]** led to the decline of this system in Britain.

With the development of **asphalt paving [IV]** for surfacing roads, and the coming of the **bicycle [IV]** and the **automobile [IV]**, the ancient art of road building was revived and much improved.

rope

Rope or cordage of various kinds was made in ancient times and apparently invented by various independent primitive peoples. Evidence of spinning vines, hair, strips of leather, grasses, or other fibers into cords has been found in ruins from before the invention of **textile** weaving, and the concept of spunwork probably had to precede weaving or making **baskets.** Long and strong rope was clearly used in the construction of Babylonian and Egyptian stone structures for hoists, ladders, and scaffolding. Lines were used aboard ancient sailing vessels.

Making long ropes required that the fibers be spun together over a long distance. Pictures from Egyptian tombs dated to 1500 B.C. show men engaged in walking while making ropes, in an early ropewalk. Ropewalks were usually level spaces, with posts at intervals to support the work. Roofed-over ropewalks in the Middle Ages and early modern age were often constructed up to several hundred yards in length and frequently were found at seaports. Hemp provided excellent fiber for heavy-duty lines and cords, and hemp came in great demand with the expansion of long-distance sailing from 1400 to 1899. In colonial North America, ropewalks represented some of the earliest industrial establishments.

A ropemaker would walk backward, paying out fiber from a bundle around his waist and spinning as he backed up. In 1793 Joseph Huddart, a British sea captain, invented a machine for laying or spinning rope. In France, in 1810, Phillipe de Girard adapted cotton-spinning methods to hemp and flax. John Good of New York City patented several machines that would allow for long spun rope, and by 1900 his machines were used throughout the world.

shelter

Evidence from the Upper Paleolithic period when *Homo sapiens* first began to replace the prehuman hominids of Eurasia, about 34,000 B.C., shows that they began to construct shelters in open areas away from caves and pits. In what is now the Czech Republic and Russia, there is evidence of the construction of substantial houses made up of either depressions dug into the subsoil with a central hearth or a circular or oval arrangement of stones or postholes. Some of these foundations involved collections of tusks, jaws, and leg bones of mammoths. These pieces were major components of the structure when the local supply of wood was inadequate.

Such shelters, whether of tusk and bone or of wood, were clustered with several families in a single settlement. The development of such shelters suggests several social patterns. First, the need to build shelters rather than occupy existing caves and hollows in the ground would indicate a growth of population. The clustering of several shelters into hamlet-size communities suggests the beginnings of specialization among craftsmen and the formation of large and stable social units such as extended families or village groups. Coupled with elaborate burials, it is possible to imagine that such peoples had social status, ranking, and demarcated social territories.

Although shelter was invented in the Upper Paleolithic period, it took a much greater variety of forms in the Mesolithic and Neolithic periods, with encampments showing repeated styles of housing. In the Danube River valley, for example, from the Mesolithic period, it was common to find one-family huts consisting of a lean-to of poles, possibly covered with hides, supported by poles inserted in the ground, with the interior paved with flat stones or limestone plaster, and with a hearth. Mesolithic settlement patterns suggest communities ranging from 25 to 100 individuals, representing congregations of separate family groups. The communities appear to have been associated into networks extending over a wide region. Often the Mesolithic groupings of shelters were linked to a nearby **cemetery.**

During the Neolithic period, after about 5000 B.C. in parts of Eurasia, the basic unit of settlement was often the rectangular, post-framed longhouse made from locally cut timber. The longhouse would be 18 to more than 100 feet in length and mostly about 18 to 20 feet wide. Most of the longhouses would be about 45 to 90 feet long. Clearly the longhouses were built by a cooperating group, and they were intended to be permanent. Possibly both humans and livestock lived in the same buildings, and apparently goods and food would be stored in raised attics. Longhouses have been found in what is now Poland, Germany, and Denmark.

At some settlements, as in areas of Germany, longhouses were dispersed, with distances between them of 150 to 300 feet, with other arrangements in small hamlets and villages. Again, such clusters of shelters often had cemeteries with them. By examining the remains of Neolithic housing and graves, palisades, and ditches, modern archaeologists have concluded that population increase led to territorial competition and to the penning and control of larger herds. It also appears that certain longhouses were devoted to ritual purposes.

In the ancient Near East, by contrast, the Neolithic Revolution came earlier, and shelters were made of stone and Sun-dried mud brick. The emergence of large, tightly settled communities and **cities** there apparently preceded the development of Neolithic shelters in Europe by at least 1,000 years.

sickle

The first sickles to cut wheat and other grain crops were made from the 9th to the 7th millennia B.C. Used by the ancient Natufian peoples of Palestine, they were made out of pieces of sharpened flint set in a curved horn, bone, or wooden handle, held in place with pitch or hardened asphalt tar (bitumen). Sickles from the era have been recovered by archaeologists, and the small size of these sickles suggests that harvesting of the grain crops was usually done by grasping one bunch of grain plants in one hand and cutting the tops off with a sawing motion. With the development of emmer wheat after 8000 B.C., and then bread wheat in the next millennium, the sickle evolved slowly. With the development of metals such as **bronze** and later **iron,** larger and more efficient sickles that could cut with a single stroke developed.

silkmaking

Silk fiber is made by the silkworm, which feeds on the white leaves of the mulberry plant and then makes a cocoon of a single long silk thread or filament. The ancient Chinese learned how to harvest this naturally produced fiber as early as 3000 B.C., and the regular harvesting of silk cocoons and thread from specially planted mulberry bushes has been found as early as 1500 B.C. Depictions of the gathering of mulberry leaves for feeding cultivated silkworms show that the industry was well organized by the 5th century B.C. By the time of the Han dynasty (202 B.C.–A.D. 220) silk fabric was in regular production, and remains have been found in tombs from the era.

Silk was traded to the West, but the Chinese maintained the secret of its production until A.D. 552, when the Byzantine emperor Justinian in Constantinople obtained smuggled silkworms and the secret technology. Thereafter, Muslim craftsmen began to develop the technique, and it dispersed slowly through the Near East.

soap

Soap was apparently first developed in Babylonia and was one of several substances used when bathing in the ancient world, including sand, ashes, pumice stone, and olive oil. When the Romans imported it, they first used it as a hair dye, to bleach hair to a reddish or sandy color. Roman author Pliny (A.D. 23–79) wrote in his *Natural History* that *sapo* was made of beechwood ash combined with fat from goats. In the second century A.D. Galen (130–200), a Roman physician, recommended its use for cleaning.

Soap is created by the action of caustic soda found in ash on animal or vegetable fats, producing fatty acids such as palmitic, stearic, and oleic acid. By the chemical action of the soap molecules, one end of the molecule will penetrate into greasy dirt, while the other end remains soluble in water. The droplets become ionized and repel each other, remaining in suspension in the water, and are then washed away with rinsing, carrying off dirt.

During the Middle Ages in Europe, bathing was no longer popular. However, in about A.D. 1000, soap, usually made from olive oil and ash, was reintroduced from Muslim countries. Soaps made from ash and animal or fish oil were offensive in odor and, for use on the body, were often perfumed with flower or fruit essences. By 1300, locally produced and scented soap was being widely manufactured in Europe, with Castile in Spain and Venice known as the centers for fine soap manufacture.

solder

Soldering is a method in which two pieces of metal are joined together by heating, using a separate metal with a lower melting point than the pieces to be joined together. To make the joint effective, a separate substance known as a flux, made of a rosin or alcohol, is introduced. The purpose of the flux is to remove oxides and to promote the spreading and wetting of the solder material. One method of making solder is to alloy the host material with a metal with a lower melting point, such as tin. When the solder is applied, the pieces to be joined have to be held immobile.

Soldering can be done with great accuracy, leaving very little trace of the work, and hence it has been used extensively in making jewelry and in the construction of small statues or other fine metalwork. Evidence of soldering has been found as early as 4000 to 3000 B.C. in ancient Ur and Egypt, in which copper-gold alloys and tin-lead alloys were used for the material. From the ancient period through the Middle Ages solder was used extensively in making jewelry, and with the development of electrical work in the 19th century, solder provided an excellent method for connecting wires to terminals in smaller devices, in which the heat produced by the electrical current would not exceed the melting point of the solder. Metals that can be soldered with relative ease include gold, silver, copper, tin, lead, and brass.

The need for intense heat at a very local point can be applied in many ways, the simplest of which is to heat a pointed iron bar in a flame to above the melting point of the solder but below the temperature at which the host metals would melt or deform. In modern solders the flux is often packed inside a small-diameter tube of solder material, and the soldering iron is heated electrically.

stone tools

The first tools used by the human race may have been made of wood, antler, bone, or other perishable objects. But since stones are more durable, the earliest known tools were those made of flint, quartz, quartzite, obsidian, shale, or other workable stone. In the 1930s, in the Olduvai Gorge in Tanzania, the anthropologist L. S. B. Leakey found evidence of toolmaking that has been dated to 1.2 million to 1.8 million years ago, indicating the use of tools by prehuman hominids. Known as the Olduvai Industry, the stones found were simple choppers.

Overlapping in period was another stone tool culture, the Acheulian, with more sophisticated tools such as hand axes; chisels; awls; hammer stones; scrapers; hafted knives; and others with straight, sharp edges. A birin was a pointed tool used to work bone or antler. To make a true stone knife, with a blade at least twice as long as it was wide, was a relatively late development, and continuous finds constantly readjust the dates for such developments.

By the so-called Late or Upper Paleolithic Age (Late Old Stone Age), from about 34,000 B.C. to about 8,000 B.C., finely worked stone tools were used for hunting, butchering, making clothes, and as weapons. The addition of wood, bone, and antler handles provided leverage. There developed a whole technology of working bone, antler, and ivory, reflecting the mastery of new procedures. Some of these involved the deep grooving of large pieces of antler and bone to produce splinters that could then be worked. The surfaces were often sawed, ground, or polished to achieve regular forms. Continuing archaeological finds reveal that tools in that period may have been far more complex, with overlapping development of several technologies, including making **baskets, textiles** and sewing with needles, and making **ropes.**

The spread of these technologies, known as the Aurignacian phe-

The knife is thought to have had its origins in shaped pieces of flint and obsidian, very hard stone and rock whose fractured edges can be extremely sharp and thus suitable to scrape, pierce, and cut such things as vegetable and animal flesh. How the efficacious properties of flints were first discovered is open to speculation, but it is easy to imagine how naturally fractured specimens may have been noticed by early men and women to be capable of doing things their hands and fingers could not. Such a discovery could have occurred, for example, to someone walking barefoot over a field and cutting a foot on a shard of flint.

—Henry Petroski, *The Evolution of Useful Things* (New York: Vintage, 1994)

nomenon, before the Neolithic Revolution raises a number of mysterious questions. It is not known what spurred the changes or how they spread. The unraveling of the techniques of stoneworking in the Upper Paleolithic period has been controversial. Some authorities have claimed that the ideas and techniques spread by colonization, while others have seen the pattern as part of a gradual evolution on the local scale. Still others see the spread of techniques as a result of trade and contact rather than either colonization or evolution. A good deal of recent evidence has pointed in the direction of population replacement or colonization, suggesting that many of the Upper Paleolithic changes came about with the spread of *Homo sapiens* groups such as the Cro-Magnon into areas previously populated by Neanderthal peoples, perhaps bringing language with them as well as the Aurignacian tool kit.

Modern efforts to emulate the skills required in stone chipping of flint and quartz have shown that the work itself was highly difficult and sophisticated, requiring extensive experience to get the right combination of angle of and pressure on the tool. Evidence at toolmaking sites of many discarded or spoiled pieces of work give testimony to the diligence of these ancient craftpersons.

textiles

The weaving of textiles appears to have derived from the weaving of mats and **baskets,** as early as 6500 B.C. Several natural fibers proved excellent for the manufacture of textiles. The first for which we have evidence in cloth manufacture is linen, made from flax. Oddly, flax is difficult to prepare for weaving, requiring several steps: soaking in water for weeks, drying and beating to loosen the fibers, and then separating into coarse and light fibers. Flax was domesticated in ancient Egypt, as was cotton.

Weaving of textiles, a core skill of the Neolithic Revolution, apparently moved from the ancient Near East into Europe along with **agricul-**

ture, but different regions developed their own traditions and styles. Woven materials were slow to replace hides and furs in the more remote regions of Europe. There, sheep produced so little wool that linen and other vegetable fibers were used in fairly sophisticated upright, weighted looms.

Apparently cotton cultivation was independently discovered in ancient Peru as well as in Africa. Cloth can be made by knitting, braiding, and felting. In felting processes, the fibers are simply pounded together in a matted mass and subjected to moisture, heat, and drying, resulting in a firm material that will hold its shape, suitable for heavy clothing, hats, slippers, tents, and shoes. Felt manufacture probably preceded woven textiles.

Wool fibers from sheep, goats, camels, alpacas, llamas, and vicuñas also made for excellent textiles. At the microscopic level wool is scaly, and as the scales on the fibers interlock, they capture air, making textiles from wool excellent insulators.

Silk, another natural fiber, was first harvested in China in about 3000 B.C. The silk thread is derived from the cocoon of a moth that lives on mulberry trees. If the cocoons are harvested from the trees before the worm emerges, the worms can be killed by immersing them in boiling water. The unraveled cocoon can yield a single thread up to 1,000 yards long.

Weaving of textiles in ancient cultures appeared to develop independently. The oldest looms for interweaving fabric in a frame probably were invented before 5000 B.C. Miniature lap looms, with a width of about 5 inches, were characteristic of African weaving, a system still practiced as a handicraft in Nigeria and the Ivory Coast.

Spinning the fibers into threads suitable for textile manufacture was done by hand and improved by the **spinning wheel [II]**. Later inventions, such as the **flying shuttle [III]** in 1733 and the **spinning jenny [III]** in 1767, contributed to the Industrial Revolution in Europe and America.

theater

The origins of Greek drama have been traced to ceremonies devoted to the worship of the god Dionysius, who was the god of **wine** and drinking. It has been assumed that to enter the proper state of ecstasy or enthusiasm required in Dionysian worship, drinking wine may have played a part. Some of the older Greek dramas reflect just the sort of ribald jokes and broad allusions to sexuality that characterize an unrestrained drinking party. Although these rites may have served as origins of drama, two annual festivals in Athens became associated with some of the early performances. In the 6th century B.C., a festival known as the Lenaea was celebrated at a location assumed to be near the central marketplace, usually in late January.

In the Lenaean ceremony, performers wore masks and depicted common incidents of daily life. In another form of masquerade, known as the *comus,* performers wore costumes representing birds, horses, and other animals and would dance with a large phallic symbol, often carried from house to house with a musical accompaniment. This revel song was the origin of the word *comedy.* In Icaria, a village north of Athens, performers may have originated the practice of holding the dances in a set dancing place or *orchestra,* and then that innovation was taken to Athens and incorporated in the Lenaean ceremony in about 580 B.C.

A more elaborate ceremony, the "City Dionysia" or Great Dionysia, appeared to have originated in the village of Eleutherai and then introduced into Athens. A small temple was built in recognition of the Eleutherius cult in Athens, and next to that location an orchestra was established, built in the reign of Pisastratus, in the period 540–528 B.C. The origin of regular performances in a set location is often traced to this ruler. A contest originating there and held annually in March first awarded the prize for a tragedy to the writer Thespis of Icaria, after whom the term *thespian* originated to describe an actor. The annual contests continued for more than 600 years.

Over the next centuries, numerous playwright-poets became well known for their works, performed at the City Dionysia contests. In most cases only scattered texts remain of the playwrights Aeschylus (who wrote plays in groups of three or four), Aristophanes, Sophocles, and Euripides, among others. Each member of the audience purchased a small lead token that represented a ticket of admission, and the proceeds went to the administrator of the theater for its maintenance. Before the contest, the playwright and actors would appear in costume in a ceremony held in the Odeum, adjoining the theater, to provide audiences with the equivalent of a playbill, so that the details of who acted which part and the titles of the plays to be performed could be announced.

Other performances, in smaller villages, were known as the Rural Dionysiac performances and soon took on the trappings of the plays written for the City Dionysia and for the Lenaean ceremony in Athens.

Troupes of actors would travel from town to town, and they formed a guild known as the Artists of Dionysius.

The plays originating in Athens spread throughout the Greek world. However, in Roman culture, theatrical performances were not associated with the worship of Dionysius, and the Romans developed a wide variety of public religious celebrations, triumphs, public funerals, and festivals. Performances involved a great many other events, more in the nature of gladitorial shows, parades, processions, and chariot races, but also including theatrical performances. The theatrical work often consisted of pantomimes. Greek plays were introduced as early as 240 B.C. Some plays were presented at night, illuminated by numerous torches, giving the performance more of a modern air than the more religious setting of the original Greek theater contests. Much of the actual technology of the theater originated in the Greek setting, and many modern terms derive from the structure of the amphitheaters built there. The *auditorium,* the *scene* shop, and the *proscenium* (or area in front of the background scene where the actors usually performed) were all established in the first City Dionysius plays and were imitated and improved throughout the Greek and Roman worlds. Although no two theaters were identical in all respects, they incorporated similar architectural arrangements, some with ramps. Where construction on a hillside to achieve the banked seats of the amphitheater was not possible, complete structures, roofed over or covered with awnings, and with passageways and staircases to the upper tiers of seats, were built in later Greek and Roman times. The terms survived into modern theater, and much the same arrangement continues into the present.

Arranging for more and more spectacular effects meant that playwrights and performers developed elaborate scenes, often pulled into place by **pulleys,** requiring considerable engineering to achieve. During the Renaissance, details of ancient theater construction were emulated and new plays were written along the basic lines established in Greek times, with tragedy, comedy, and satire all tracing their origins to Athens. The influential Italian architect Andrea Palladio (1518–1580) revived classical architectural forms in many types of structure, including a public theater, the Teatro Olimpico, which he began in the year of his death for the Olympic Academy of Vicenza. Palladio and his successors studiously emulated elements of the classical theater, helping to contribute to the continuation of the form and structure of the theater into modern times.

umbrella

The first known depiction of an umbrella is found on a victory monument dated about 2400 B.C. to Sargon, king of the city of Akkad, in Mesopotamia. The principle of an umbrella or parasol, to shade from the Sun, spread to Egypt, and by 1000 B.C. they were found from India to the prehistoric Greek culture of Cyprus. The word *umbrella* derives from the Latin *umbraculum* (*umbra* meaning shade). Royal umbrellas for protecting the monarch from the Sun's rays were found throughout North and West Africa in the 16th and 17th centuries, apparently diffusing from the Egyptian use.

During the Wei dynasty (A.D. 386–535) the Chinese developed an umbrella that could shield its user from both the Sun and rain. The first waterproofed umbrellas were made of paper made from mulberry leaves then oiled. Silk umbrellas made from heavy silk were introduced in the 14th century A.D. In India, paper umbrellas were used in the same period. In the 18th century, Europeans began to imitate collapsing paper umbrellas that were made in China.

wheel

Often regarded as the symbol of the beginning of inventiveness of the human race, the origin of the wheel is unknown. It has been suggested that it may have derived from the wheel used in **pottery,** itself possibly derived from earlier work in the crafting of the **basket.** The earliest evidence of the use of the wheel in transportation can be traced to about 3500 to 3200 B.C. in Mesopotamia (modern Iraq). It is possible that the wheel may have been introduced as early as 4000 B.C., but no direct evidence has been found to demonstrate that. The physical nature of the wheel and axle combination is such that its rotary motion can be continuous in direction and magnitude and that a pulled or pushed object on a wheeled base greatly reduces friction. With the addition of connecting rods, the wheel found many applications in later generations beyond transportation in the construction of machinery. Wheels and axles were the essential components of the **pulley,** developed before 1500 B.C.

The popular picture of a "caveman" chipping out a stone wheel derives from a misconception that the wheel dates from Paleolithic times. In fact, the wheel is a relatively recent invention, of the late Neolithic period associated with **agriculture** and the development of the first **cities.**

Artifacts and models of two- and four-wheeled carts have been dated to about 2600 B.C. The earliest wheels were constructed from three wooden planks clamped together with copper or wooden fixtures (clamps or battens) and then carved into a rounded shape. A hole in the center for an axle was often surrounded by a raised hub for support. Such a construction of the earliest known wheels can be suggested as an argument that the wheel did not derive from adaptation of roller logs. However, the transition may have come as craftsmen attempted to reduce the need to smooth wide sections of ground by narrowing the center of a log roller, leaving wider diameters at the ends and clearance in the middle. However, a section of a log, when used as a wheel, would be far less durable than the braced plank structure found in the early Sumerian wheels. The first wheels with spokes, used on chariots, appeared in about 2000 B.C. The fact that the advanced civilizations of the New World, including the Toltec, the Maya, and the Inca Empire, did not develop the wheel for transportation but made use of it for pull toys for children has represented a major puzzle in the history of the application of technology. The Mexican toys were made of clay, with miniature clay rollers. The presence of large numbers of slave laborers has been adduced as a reason that no practical development for land transport, even the wheelbarrow, was required, but the ancient civilizations of Mesopotamia, Sumeria, and Egypt, where wheels were widely adopted, were also characterized by slave labor systems.

It is not excessively determinist to suggest that certain prerequisites are needed for the development of wheeled vehicles. The vehicles will be invented only in societies that have a need to move heavy or bulky loads considerable distances over land that is fairly flat and fairly firm. A suitable raw material, such as timber, must be on hand for building the vehicle. And a prime mover stronger than a man must be available to make the wheels turn. In the Old World the power problem had been solved at least 7,000 or 8,000 years ago by the domestication of cattle. Once it was realized that castration produced a docile, heavy draft animal, oxen were available for traction. The strength and patience of oxen compensated for their slowness. Timber was available in quantity in parts of the Near East that were neither desert nor steppe.... The earliest examples of wheeled vehicles have all been found within a region no more than 1,200 miles across, centered between Lake Van in eastern Asia Minor and Lake Urmia in Northern Iran. Presumably the first wheeled vehicle originated somewhere within this region. The oldest evidence dates back to the final centuries of the fourth millennium B.C., indicating that wheeled transport came into existence somewhat more than 5,000 years ago.
—Stuart Piggott, "The Beginnings of Wheeled Transport," *Scientific American* 219, no. 7 (July 1968)

wine

The origins of winemaking (vinification) and grape growing for the production of wine (viticulture) have been subjects of close study by classical scholars. The cultivation of vines for making wine occurred between 6000 and 4000 B.C. on the border of the modern states of Iraq and Iran. It is possible that the first wine was discovered by accident from storing grape juice made from wild grapes, and this discovery may have occurred as early as 10,000 B.C. However, such guesswork, common to the attribution of ancient inventions, cannot be substantiated. Much more is known about the spread and distribution of winemaking and grape growing than about the first invention of the concept.

Classicists derive their knowledge of the spread of viticulture and viniculture from a variety of sources, including inscriptions and stone reliefs, writings such as the books of the Bible and the history of Herodotus (about 440 B.C.), the Code of Hammurabi (about 1790 B.C.), and the *Epic of Gilgamesh* (about 2100 B.C.). By the time of the Sumerian culture in what is now southern Iraq, in about 3000 B.C., there were vineyards in the region, although the most common alcoholic beverages were **beer** made from barley and wine made from dates. Evidence from the 7th century B.C. suggests that Armenia had become a major wine-exporting region by that period.

The ancient Egyptians made wine from dates and other fruits but also apparently had wine from grapes as early as 2600 B.C. Egyptian tomb paintings, statues, and papyrus records provide quite certain evidence of extensive grape growing and wine consumption. Details of the methods of winemaking with winepresses, sealing in jars, transporting in *amphorae* by **maritime commerce,** and some information about consumption can all be derived from such sources. In addition to local vineyards and winemaking, the Egyptians imported wine from Greece and the eastern Mediterranean. Records indicate that by about 1000 B.C., the cities of Phoenicia (in what is now Israel, Palestine, and Lebanon) exported wine throughout the Mediterranean, and by 100 B.C. as far afield as Arabia, East Africa, and India.

wine The cultivation of vines for wine has been traced to the Ancient Middle East. This page from a 1493 text describes methods of cultivating the vines. *Library of Congress*

One indication that mead, a beer or wine made from honey, had penetrated trade routes to western Europe is the appearance, as early as 2800 B.C. in a whole area from Scotland to Sicily, of what is called the Bell-Beaker phenomenon. This term refers to the discovery of large, bell-shaped drinking vessels, often found in graves with adult males. Although there is no proof that the beakers were used exclusively to drink alcoholic beverages, the fact that they were often found associated with weapons in the graves of adult males and associated with certain other artifacts, such as leather jerkins and belts and flint, copper, or bronze daggers, has suggested the rise of a warrior culture with values associated with drinking and combat. Although it is difficult to infer cultural values from graves, circumstantial evidence supports the idea of a wide change in attitudes glorifying the individual, the warrior, and the ostentation of wealth though personal ornamentation.

The first users of Bell-Beakers did not practice metallurgy, and the earliest daggers were made of flint though they soon came to be cast in copper, and then bronze. This martial image was perhaps completed by leather jerkins and later by woven fabrics, held by a belt with an ornamental bone ring to secure it. . . . The imagery of third-millennium Europe was replete with such symbols, and Bell-Beaker graves expressed the warrior values appropriate to a more mobile and opportunistic way of life. This presented indigenous groups both with a challenge to established values and an opportunity to join a new community with access to exotic items of material culture.
—Andrew Sherratt, "The Emergence of Elites: Earlier Bronze Age Europe, 2500–1300 B.C.," in *The Oxford Illustrated Prehistory of Europe,* edited by Barry Cunliffe (Oxford, Eng.: Oxford University Press, 1994)

Apparently wine was extensively used in religious ceremonies, and in some ancient cultures it was officially restricted to such use, while beer remained a drink for the common populace. However, in Egypt, it is known from inscriptions that wine was consumed at feasts and banquets.

Whether winemaking and grape culture were independently invented in Spain or derived there from imports from Phoenicia has not been clarified.

wooden furniture

The earliest examples of wooden furniture are known from actual surviving pieces, from pictorial representations, and from cuneiform and hieroglyphic written records from ancient Assyria and Egypt. Furniture styles and patterns from the 3rd millennium B.C. continued to be used for more than 1,000 years, and there are records of wooden frame furniture upholstered with **leather** and held together with cords, sinews, and glue. Stools were used for menial work in about 3000 B.C. They were usually constructed of reeds on a wooden frame, with some folding stools very much like modern ones.

Chairs from the same period had legs, backs, and arms, with frames of hardwoods. Assyrian records show 17 different types of wood employed in the manufacture of furniture. More elegant and expensive chairs had inlays of **copper, bronze,** ivory, and gold, and all-wood chairs were sometimes covered with **paint.** Padded seats covered with leather were stuffed with palm fiber, river rushes, or felt. There is even evidence that some chairs had linen dust covers in this period. Chairs included sedan chairs for transporting royalty, thrones, and armchairs. Often the arms or feet would be decorated with depictions of griffins, sphinxes, human figures, or winged gods.

Other wooden furniture from the Assyrian Empire includes tables, tray tables, offering tables or stands, and beds. Tables would be decorated with metal or ivory, with legs often decorated and carved like a lion foot or an ox foot. Details in inscriptions include tablecloths and cloth napkins that would be washed after meals. Wooden beds were frames with rope, reeds, or cross-slats supporting a mattress stuffed with wool, goat hair, or palm fiber, while the poorer classes slept on straw mats. Beds became more common by 2000 B.C.

Only gradually, with the spread of Bronze Age culture to Europe, did wooden furnishings of such quality find their way into Europe. Fewer examples survived the damper climates there, but from the spread of artifacts and the development of **shelter** in Europe, it can be assumed that tables, chairs, benches, and beds were found in villages and hamlets in most of Europe in 1500 B.C.

woodworking tools

Woodworking tools evolved in the ancient world, with three "functional lineages" descended from basic handcraft needs: striking, cutting, and holding. Thus the earliest **stone tools** included blunt stones for hammering and simple cutting devices such as blades and scrapers. The workman braced the work with his own hands or feet or was assisted by another.

Many of these basic woodworking tool shapes were inherited from the era of stone tools and carried through from the Upper Paleolithic era to the Mesolithic and the Neolithic. By the time that **copper** and **bronze** were introduced in the Chalcolithic and Bronze Ages, **ax** and knife shapes, for example, had many precedents in stone. Interestingly, after the introduction of the metal forms, workers who used stone tools and weapons often tried to emulate the emerging shapes of the metal devices seen in trade from more advanced metallurgical cultures. Thus, for example, a stone ax head, after the introduction of metal, might be found with a hole for the insertion of the handle rather than simply a notched end for binding with **leather** thongs to the shaft.

Ancient woodworking or carpentry tools represented specialized lines of descent to perform these handcraft functions, with new tools created and passed down according to the nature of the material worked and the product evolved. Striking tools evolved into a variety of hammers, mallets, sledgehammers, and wedges. Cutting tools evolved into knives, planes, saws, two-handled scrapers, drawknives, files, and gouges. Combining the striking blow with a cutting edge yielded chisels, hatchets, **axes,** and adzes. The anvil, at first no doubt simply a heavy stone, provided a working surface. A wide variety of clamps for holding the work developed. The emergence of the screw shape in ancient Babylonia and Egypt yielded screw clamps to hold work in place. The bow drill, which may have first been introduced as a method to start **fire,** evolved into augers, drills, gimlets, and brace-and-bit drills for cutting perfectly circular holes. With the addition of standard measuring rules, the use of right angles, and simple dividers or compasses, very accurate and fine woodwork could be produced. By Roman times a carpenter had tools that would mostly be recognized in the 21st century.

Few advances in woodworking tools took place in the Middle Ages, but carpenters and woodworkers organized themselves in guilds. Some guilds were found in England as early as A.D. 856, and the Fraternity of Carpenters was formed in 1308. In general, apprentices would study under a master workman for a set time of 7 or 8 years, during which he was bound to employment and had dependent status like that of a child or slave, subject to corporal punishment and confinement for disobedience. After his term of service he would be liberated to work as a journeyman, free to engage his own time. A master would be accepted into the guild from the ranks of journeymen, as qualified to instruct apprentices in return for their labor. Such systems allowed the passage of skills in a structured fashion.

Woodworkers who made *carpents,* a type of heavy wooden cart, yielded the term *carpenter* in English. They made not only carts and wooden buildings but also a wide variety of equipment and furnishings, including beds, cabinets, plows, looms, cider presses, chests, tables, and chairs. Specialized classes of woodworking craftsmen included joiners, who did the finer work of doors and windows; cabinetmakers, who built furniture and chests; coopers, who made barrels; and shipwrights, who built wooden boats and sailing vessels. Related but independent crafts included millworking, upholstery, metalsmithing, and painting.

The continuity of woodworking tools from ancient times to the present is due to the basic needs of the craftsman in working with wood, to hold it while cutting and shaping it and while fitting separate pieces together with glue, nails, screws, or other fasteners such as hinges, braces, clamps, or clasps. For this reason, once the basic tool kit and related metal appliances developed, only very slow evolution improved the basic tools until the application of power to assist the human hand. With **waterwheels [II]** and **windmills [II]**, shaping of lumber and the development of mechanical lathes and drills led to an early preindustrial "Industrial Revolution." Some woodworking factory equipment became highly complex, as in large devices to drill and cut **pulleys** for ship block and tackle. The application of **steam engines [III]** in the 19th century allowed for even larger lumber mills, and their design was only modified, not revolutionized, with the application of other prime movers such as electric motors and internal combustion engines.

Early in the 20th century, the application of electricity to hand tools yielded a new generation of woodworking tools. Again, however, it was not so much the tool that changed but the motive power driving the tool. Thus electric handsaws, sanders, routers, and drills were added to the woodworker's kit.

yoke

Developed in the ancient Near East at about the same time as the **wheel,** the yoke consisted of a simple wooden bar or frame used to join together two or more draft animals at the head and neck so they could pull a vehicle or sledge at the same time. Oxen were yoked across the horns or neck. Depending on the design, the yoke had the drawback that it could choke the draft animal. Evidence from drawings and inscriptions show yokes in ancient Mesopotamia and Egypt, whence they spread gradually with the Neolithic Revolution to Europe and to the rest of Asia. It was not until the development of the **horse collar [II]** in the Middle Ages that horses could be harnessed effectively to pull heavy vehicles, as the ox yoke was quite inappropriate to the anatomy of the horse. The yoke was unknown in the New World, where no draft animals larger than dogs were domesticated before the colonization of the American continents by the Europeans in the 16th century A.D. The South American llama had not been harnessed or yoked to wheeled vehicles but was used only as a pack animal.

PART II

THE MIDDLE AGES THROUGH 1599

The period covered in this part of the volume, from about A.D. 300 through A.D. 1599, includes what historians of European history usually call the Middle Ages (or the medieval period) and the Renaissance. The early part of this period is surrounded with numerous popular myths and misconceptions, including the old-fashioned idea that the centuries from about 300 to about 1000 were the "Dark Ages," during which Europe was dominated by ignorance, superstition, and lawlessness. These assumptions are quite untrue, for churchmen and scholars laboriously copied written texts, perpetuating many of the scientific, legal, and literary achievements of the ancient period. Furthermore, some of the inventions that stimulated the greatest social changes and that led to technical progress occurred in this period. The Middle Ages are the subject of difficult research in rare texts, languages no longer spoken or written, supplemented with the study of artifacts, coins, etymology (the origin of words), art, and archaeology. While better written sources exist for the Renaissance, roughly 1400 to 1599, some unsolved historical mysteries remain for this period as well. For example, we know when eyeglasses first were in use, but we have no certain knowledge of exactly who invented them or where. The same is true of several other inventions, such as the hand crank, playing cards, and the windmill.

These centuries have fascinated historians living in modern times for several reasons. During the Middle Ages, the Roman civilization declined, and what had once been the Roman Empire dissolved into many small states, some no larger than towns or counties in the modern United States. North-central Europe, from what is now the northern region of Spain across France and Germany to Poland and Austria on the east, was united under the reign of Charlemagne as Holy Roman emperor from 804 until his death in 814. But over the next centuries, Charlemagne's empire also declined. Local kings, dukes, bishops, and (here and there) elected leaders ruled and warred to protect their small territories. In some locales, a monastery would provide armed monks who helped maintain order.

Despite the apparent political chaos of the Middle Ages and the Renaissance, a new civilization grew up in Europe during the period, a civilization that held the seeds of modern society. By the end of the 16th century (A.D. 1600), Europe seemed on a path that we now see as defined by technical, scientific, and intellectual progress, with the beginnings of the emerging modern nations. Historians have looked into the factors that explained the transformations from the petty and warring regimes, some held together by the institutions of feudal obligation, to the premodern societies of cities, commercial trading systems, and central and bureaucratic monarchies, with flourishing cathedrals and universities. Along with these social changes, deep-seated intellectual developments also came out of the period, with a new emphasis on the individual and a willingness to challenge traditional explanations and accepted authority in favor of ideas derived from experience.

At the heart of these grand transformations lie a group of technical inventions and discoveries. Even during the 9th century, when Charlemagne ruled, some of the most important changes of tool and technique were at work. Improvements in agriculture led to food surpluses, so that farmers in various regions could produce more than their families could eat in a season. That simple change meant that surpluses could be traded, creating market towns and transportation networks (by coastal, river, or canal boat, or by wagon or cart), and soon led to population growth, as more people could survive on the annual crops. Control of trade and protection of cargoes from brigands and bandits required that the towns and the trade routes have improved systems of law and order. The inventions that led to the agricultural surplus, for these reasons, seem to be at the heart of the social changes of the late Middle Ages. Some inventions that we now regard as so ordinary and obvious that it is hard to imagine that they had to be "invented" made the crucial difference in increasing agricultural production.

The three-field system, in which a field was left fallow to grow in weeds for a season, while another field was planted with grain, and a third with peas, beans, and lentils, with rotation among the three fields, was at the heart of increased food production. To plow the fields in parts of Europe, a heavy plow was required. To use horses for deep and heavy plowing or for hauling heavy carts, a new horse collar that did not choke the animal was invented. And to keep the horses' hooves from breaking up under the constant work, horseshoes were invented. With the three-field system, the plow, and the harnessed and shod horse, some areas produced so much barley that they could neither consume nor sell all of it. The Scots solved that particular problem by making barley into whiskey, while French wine growers produced cognac and brandy with a similar distillation process.

The intermittent small wars and jockeying for power led to a whole series of inventions that gradually changed the nature of warfare. One crucial development was the invention of the stirrup, imported from central Asia. The stirrup allowed the mounted horseman to brace himself, and when charging with a lance, to use the full force of the forward motion of the horse's charge to strike his enemy. With the stirrup and lance, heavier protective armor was developed, including chain mail and hinged metal plate armor, stimulating the growth of metal industries. The era of the chivalric knight could never have been possible without the large horse (with horseshoes), a saddle with stirrups, and a heavy set of armor.

Yet the lone horseback knight was no match for other developments, including the longbow, the crossbow, and later firearms using gunpowder, including the matchlock, the flintlock, and the musket. Heavily fortified castles and city walls were attacked first with improved siege machines that hurled heavy stones, then with cannon fire. The longbow, the trebuchet rock-throwing siege machine, and the cannon put distance between attacker and defender, making the face-to-face or helmet-to-helmet warfare of the chivalric knight largely a thing of the past by the 16th century.

To support an army of longbow men and the improved armaments, larger and larger treasuries were required. Market towns and cities sought the protection of well-financed princes and kings, fostering the growth of larger principalities, dukedoms, and kingdoms. Within the protected cities, craftsmen perfected ancient arts inherited from earlier eras, passing on the skills from master to journeyman and apprentice in guilds that tried, often unsuccessfully, to keep certain craft skills secret. Thus, when one area developed an invention, such as an improved method of making wire, for example, the wiremakers sought to keep the knowledge from spreading, to maintain a monopoly. But in case after case, groups of specialized craftsmen would immigrate to another region in Europe, taking their knowledge with them, to set up shop where the price for their product was high. In this way, skills and technical knowledge spread, sometimes slowly and sometimes quite rapidly. While guild oaths tended to restrict the spread of technologies through secrecy, the system of guilds could ensure also the gradual perfection of method and the passage of technique from generation to generation. Only in a later era would state-granted patents begin to replace the guild-oath system, protecting a monopoly but at the same time publicizing the specific patented design, which would encourage progress and change as later inventors sought to improve on the first patent.

Many of the arts and crafts used tools little changed from the classical era, but the luxury of surplus food and surplus population allowed much finer work in such arts as glassware in Venice; steel in Toledo; textiles and embroidery in Flanders and Belgium; and metalworking, cabinetry, jewelry, and pottery in centers scattered all across Europe. Using old tools but new techniques, craftsmen made such new devices as clocks, watches, guitars, and reed instruments. The glassmaker's skill soon led to the production of lenses, used first in eyeglasses; but in the next period, lenses would play a major part in new instruments such as telescopes and microscopes.

Some of the cities hosted the church of a bishop or archbishop and competed with one another in constructing fine cathedrals. Surrounding the cathedral schools, clusters of students provided the base on which universities were built. Cathedrals and universities flourished from the 11th century through the rest of this period, laying the ground for better-organized scholarship and learning. The challenges of building more imposing churches and cathedrals stimulated architects to develop new forms, including high, buttressed walls, and peaked domes that were held together by the force of their own weight.

Agricultural surpluses meant that more and more grain had to be milled, and the simple mills turned by oxen, mules, or horses had their limits. In the search for better sources for power, medieval millers developed two remarkable, industrial-scale machines: the waterwheel-driven mill and the windmill. Harnessing waterpower and windpower soon meant that these new sources of energy became available for other purposes, such as pumping water from low-lying lands, stamping out metal pieces, crushing ore, or doing other work that previously had been done by human or animal musclepower.

Agricultural surpluses stimulated long-distance trade, as well as improvements to the sailing vessel. The lateen rig, which allowed sailing more directly into the wind than the older square sail, soon produced new types of ships. Additionally, long-distance sea travel was made much safer with other developments in navigation and instrumentation, including the compass, astrolabe, cross-staff, and ship's log. By 1510, European sailors regularly reached the Western Hemisphere, and by 1599, the flourishing Atlantic trades brought new products and the beginnings of burgeoning new industries.

Other instruments and small tools from the era also had lasting social impacts as they spread across Europe, including eyeglasses, the crank, the spinning wheel for making thread, the fork, and the surveyor's theodolite. Improvements to the lathe allowed for the manufacture of screws and then nuts and bolts. The cultural impact of the guitar, the musical reed instrument, musical notation, playing cards, engraving for the illustration of books, and the printing press with movable type all meant that the material culture of 1500 began to show the shape of modern times.

The lone scholars and then the university scholars continued to seek knowledge, with some of the pursuits of the Middle Ages resulting only in faulty conclusions. When scientists tried to design a perpetual motion machine—a machine that once started would continue forever—they showed that they did not understand what we now know as the conservation of energy. But that principle was not defined until the 19th century. Even though the pursuit of such an impossible machine was fruitless, it showed that interest in making machines that could do work, such as the windmill, the waterwheel, the pump, and the stamping mill, was a central and important concern of the age. Out of the work on engines would gradually emerge the engineers of a later era, those who would construct the machines that built on the earlier efforts.

Similarly, other scientists pursued an elixir, the "philosopher's stone." The word *elixir* came from the Arabic *al iksir*, a magical stone that would prolong life and transmute various metals into gold. Alchemists also sought the *fifth essence* or *quintessence*. Considering earth, air, fire, and water as the four elements proposed by the ancient Greek philosopher Empedocles (c. 493–433 B.C.), the fifth essence or element was regarded as a heavenly matter, out of which the planets and stars were made and that invisibly permeated the physical world. If it could be extracted, the quintessence, like the elixir, would have extraordinary powers and effects.

Although no elixir existed and although there are about 90 natural elements, not 4 or 5, the search for better medicine and the search for the knowledge of the chemical nature of materials were valid quests. Over the following centuries, pharmacy, biology, medicine, chemistry, and nuclear physics would bring mankind closer to solutions to the problems considered by the medieval alchemists. Their quests, which later science relegated to the realms of magic and mysticism, were in fact useful starting points for investigating the nature of life, health, and materials.

European science was stimulated and given a fresh start by imports of knowledge, technique, and information from the Muslim world, including the Arabic numerical system and the use of the zero in calculations. With the conquest by Christian monarchs of parts of Spain previously dominated by North African Muslims (the "Moors") in the 12th and 13th centuries, the flow of information increased. While Arabic-speaking scholars had translated and preserved many Greek texts lost to northern Europe, Muslim science had flourished on its own and had borrowed and learned from peoples to the east, including the Hindu and Chinese cultures. Along with the texts and mathematics, other imports to Europe from the Muslim world included such tangible items as the magnetic compass; gunpowder; the lateen sail; and a variety of spices, medicines, and edible products such as sugar, tea, and coffee.

A further stimulus to scientific curiosity came with the voyages of exploration and discovery of the 15th and 16th centuries. Explorers brought back to Europe new peoples and a variety of new materials, including pepper, chocolate, ivory, tomatoes, potatoes, corn, and tobacco. To raise, buy, or steal these products in Africa, North America, or South America required that Europeans develop new technologies, many of which began to flourish in the 1500s, including shipbuilding, gunmaking, sugar refining, and rum distilling. Although not regarded as an industrial revolution, because the shops and the mills were small and localized by later standards, these 16th-century industries helped account for the shaping of the whole Atlantic trading world.

The lands of the New World were rapidly exploited, first with the stripping of gold, silver, and gems, and then with the planting of crops that would find a ready market in Europe, such as the high-priced and highly sought sugar, tobacco, and, to a lesser extent, chocolate, coffee, and tea. As Europeans developed the habits of consumption of these items, their price remained high, and the incentive to find new lands and new sources of labor increased. By the mid-16th century, Europeans traded guns, gunpowder, metal products, rum and other liquors, and textiles to Africans to purchase their prisoners of war as slaves in the beginning of a slave trade that lasted about 300 years. In the Caribbean islands, along the coast of North and South America, sugar and tobacco plantations flourished over the next centuries, worked by African slaves. In West Africa, local industries of metalworking and textile production declined in the face of the cheap European imports. The Europeans made products for trade or conquest in their mills and ran the shipping and plantation enterprises. By 1599, the changes in relationships among Europe, Africa, and the Americas had begun to set the world on a new course.

Over the 15th and 16th centuries, navigators and sailors enhanced their skills and techniques, learning to improve their maps with information received from explorers, rather than simply copying ancient and outdated maps based on false assumptions and erroneous information. This triumph of experience over accepted authority, incorporated in the maps developed by the institution of Prince Henry the Navigator of Portugal, and by later mapmakers and atlas publishers, represented the core of the new scientific approach that would produce the Scientific Revolution of the following two centuries.

By the 15th and 16th centuries, some individual thinkers began to challenge many aspects of traditional and accepted doctrine. While mapmakers began to rely on the experience of explorers, some scientists, too, began to challenge ancient traditions of knowledge. Paracelsus, who died in 1541, and Copernicus, who died two years later, both challenged existing viewpoints, the first in medicine and the second in astronomy.

The inventions and discoveries of this period, unlike the inventions of the ancient and classical eras, produced a rapid set of transformations. While many of them reached fruition and had great impact in Europe, it would be Eurocentric to view the changes as due to some unique characteristic of the European peoples. We need to keep in mind that many of the innovations, from fields such as navigation, sugar cultivation, mining, and textile- and leatherworking, came to the Europeans from the Far East and the Muslim Near East. Many of the most crucial inventions in the period developed first in China, India, central Asia, or the Muslim world.

What happened was that because of the driving force of population increase brought on by improved food production, older crafts were perfected, and new ones generated new products. Ideas, techniques, methods, and machines developed in Asia and North Africa caught on, sometimes centuries later, in the competitive multiple cultures of Europe, spreading and being imitated and eventually having great social impact. Improving on the techniques of warfare over the centuries, the

Many of the important inventions in this period were originally developed in China, India, or the Muslim Middle East. The date of European adoption is shown separately for the period prior to 1400. For the period after 1400, the inventions listed were first introduced and adopted in Europe. Many dates in this period are approximate and are indicated with "c." for *circa,* meaning "about."

Chronology: The Middle Ages through 1599

Date Invented	Invention	Origin Location	European Adoption Date
		31–499	
c. 31	crank	China	834
c. 100	waterwheel	Rome	c. 100
c. 400	astrolabe	Persia (Iran)	c. 500
		500–899	
c. 500	plow (heavy)	northern Europe	c. 500
c. 524	stirrup	China	c. 800
c. 550	cathedral	Europe	c. 550
c. 650	fork	Middle East	c. 1100
c. 760	three-field system	northern Europe	c. 760
c. 800	decimal numbers and zero	India and Uzbekistan	c. 1100
c. 800	horse collar	central Asia	c. 800
		900–1099	
c. 900	sugar refining	India	c. 1000
c. 900	horseshoe	unknown	c. 900
c. 1000	gunpowder	China	1242
c. 1000	lenses	Arabia	c. 1000
c. 1030	musical notation	Europe	c. 1030
		1100–1399	
1128	cannon	China	c. 1314
c. 1150	reed instruments	unknown	c. 1150
c. 1150	chain mail	Europe	c. 1150
c. 1170	horizontal axis windmill	Europe	c. 1170
unknown	spinning wheel	China	1298
c. 1286	eyeglasses	Italy	1286
1337	longbow (as military weapon)	Wales (Britain)	1337
c. 1370	playing cards	Middle East	c. 1370
c. 1373	canal lock	Holland	c. 1373
		1400–1450	
1438	unsupported dome	Italy	
c. 1440	moveable type and book	northern Europe	
c. 1450	oil painting	Flanders (Belgium)	
c. 1450	matchlock gun	Europe	

Chronology: The Middle Ages through 1599 *(continued)*			
Date Invented	**Invention**	**Origin Location**	**European Adoption Date**
1451–1550			
c. 1470	guitar	Spain	
c. 1500	watch	Germany	
c. 1510	engraving	Europe	
c. 1528	musket	Spain	
c. 1530	screw lathe	Italy	
c. 1530	ship's log	Italy	
c. 1540	iatrochemistry (specific medicines)	Europe	
1543	heliocentric solar system	Poland	
c. 1550	corset	Spain	
c. 1550	cross-staff	unknown	
c. 1550	flintlock	Europe	
1551–1599			
c. 1565	lead pencil	unknown	
1569	Mercator projection	Germany	
1570	atlas	Holland	
c. 1570	theodolite	unknown	
1577	comet paths (as orbits)	Denmark	
1588	shorthand	Britain	

Christian Europeans were able to militarily dominate first the Celtic and Slavic fringes of Europe. Then, using improved crucial techniques and devices of navigation such as the astrolabe, compass, and lateen sail, learned from other cultures, they moved out into the Atlantic world to make geographic discoveries and to bring in new wealth and new products. With their position at one angle of the triangular trades that emerged over the 16th century, the rulers and elites of Europe built immense wealth, pouring much of it back into the moneymaking enterprises of industry, shipping, and trade. At the heart of their emerging world domination were sets of tools in agriculture, warfare, navigation, and the use of windpower and waterpower. With the exploration of the world and the flood of new knowledge, transmitted by the printing press, Europe stood ready, by 1600, to move into a revolution of scientific thinking.

For all of these reasons, it was the tools, techniques, and technology and the ideas that surrounded them that put in motion the flow of modern history.

algebra

The modern mathematical procedures called algebra evolved over the ancient period, with contributions by the ancient Egyptians, Babylonians, and Greeks, and later by the Hindus and Arabs. However, the use of letters to represent quantities that characterizes modern algebra was developed in the 16th century.

What we know of the earliest systems is quite limited, but it appears that the Egyptians and Babylonians phrased what we would regard as algebra problems with words rather than with symbols. The Greek mathematician Diophantus, who lived in about A.D. 350, introduced what is known as the syncopated style of algebra, with equations, rather than rhetorical or geometrical descriptions of the problems.

The word *algebra* comes from the Arabic word *al-jabr* in the title of a work written in about A.D. 830 by Mohammed ibn-Musa al-Khowarizmi, *Hisab al-jabr w 'al muqabala*. The word *algorithm* derives from a corruption of the author's name. A Latin translation of al-Khowarizmi's work appeared in the 12th century in Europe, stimulating contributions and the advancement of mathematical studies there.

Algebra as a mathematical method continued to evolve, with 16th-century contributions by Italian mathematicians Scipione del Ferro, Niccolò (Fontana) Tartaglia, and Girolamo Cardano. François Viète (1540–1603), a French Huguenot nobleman, is reputed to be the first mathematician to substitute letters for known and unknown quantities, giving algebra its modern look, and he is sometimes regarded as the Father of Algebra. He was the first mathematician to use the cosine law for triangles, and he also published the law of tangents. Viète was interested in cosmology and also served as a cryptanalyst for Henry IV of France, applying his mathematical skills in extremely practical fashion.

Algebra continued to evolve over the next centuries. In 1637 René Descartes (1596–1650) published *La Geométrie*. Book III of that work very much resembles a modern algebra text.

anatomy

In the mid-16th century, several instructors of medicine began the practice of careful dissection of human cadavers, leading to a range of discoveries such as the **Fallopian tubes,** the **Eustachian tube,** and the **veinous valves.** The leader among the dissectors was Andreas Vesalius (1514–1564), a Belgian physician who led the revolution and who not only dissected bodies, but used detailed illustrations to teach anatomy to students.

Vesalius discovered a number of anatomical errors current in thinking derived from ancient texts. For example, doctors had assumed that men had one less rib than women, and also assumed that emotional processes took place in the heart rather than in the brain. For four years he worked on compiling his information, employing the assistance of accomplished artists and craftsmen using copperplate **engraving.** In 1543 he published his masterwork, *De humani corporis fabrica* (On the Structure of the Human Body). The fact that the work was published in the same year as *De Revolutionibus,* by Copernicus, which spelled out the concept of a **heliocentric solar system,** has often led observers to associate the two as leading the Scientific Revolution.

Vesalius taught Fallopius, the discoverer of the Fallopian tubes, and had great influence with his own 1543 text. However, the work caused such a storm of controversy that Vesalius abandoned anatomy and became court physician to Charles V and Philip II of Spain. The approach of relying on experimentation rather than on inherited but often incorrect "knowledge," initiated by the anatomists, characterized the Scientific Revolution of the following century in many disciplines besides anatomy.

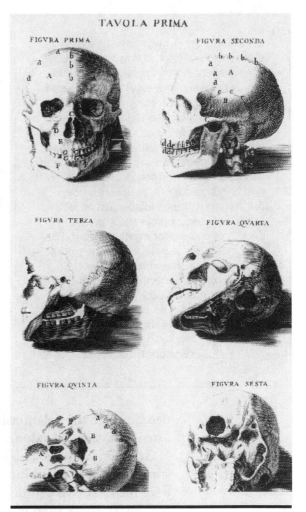

anatomy These drawings of skulls were the work of Bernadino Genga and were originally published in 1691.
National Library of Medicine, U.S. National Institutes of Health

astrolabe

The astrolabe was a navigational device used to determine latitude by mariners and by surveyors to determine the altitude of mountains. Several publications of the 15th century describe in detail how to make an astrolabe, but the instrument itself possibly may date to the 3rd-century B.C. Arabs perfected the astrolabe in the Middle Ages, using a geared mechanism and fine metalwork to improve the accuracy of the instrument.

One of the Greek inventions that Islam elaborated and spread was the astrolabe. As an observational device, it is primitive; it only measures the elevation of the sun or a star, and that crudely. But by coupling that single observation with one or more star maps, the astrolabe also carried out an elaborate scheme of computations that could determine latitude, sunrise and sunset, the time for prayer and the direction of Mecca for the traveler. And over the star map, the astrolabe was embellished with astrological and religious details, of course, for mystic comfort. For a long time the astrolabe was the pocket watch and the slide rule of the world. When the poet Geoffrey Chaucer in 1391 wrote a primer to teach his son how to use the astrolabe, he copied it from an Arab astronomer of the eighth century.
—J. Brownowski, *The Ascent of Man*
(Boston: Little, Brown, 1973)

Introduced from the Muslim world to Europe in the 6th century A.D., most astrolabes were of similar design. They consisted of a set of movable discs, usually made of brass or iron. The base plate, or *mater,* had lines representing heavenly coordinates; an open-pattern disk, the *rete,* contained a map of the major stars that would rotate on the mater on a central pin. A straight rule or *adilade* would be used to sight objects in the sky or the angle of tall objects. From the angle of the Sun (at its high point, or local noon) or a major star above the horizon at a specific hour past noon, the latitude of position of the observer could be determined; this was quite useful for navigators aboard ships or overland travelers in the desert. Conversely, at a known latitude it was possible to use the astrolabe to determine the time at night by sighting to a star.

The mariner's astrolabe tended to be simpler and less ornate than models used to demonstrate the relationship between celestial objects and latitude in astronomical instruction.

Geoffrey Chaucer published a *Tretis of the astrolabie* in 1391, presumably as an explanation of the instrument for his son. Other works over the next two centuries described the instrument. Only one of several instruments used in astronomical instruction, others from the era included the *dioptra,* for measuring the diameter of the Sun or Moon; the *triquetrum,* used to measure meridian transits of the Moon and the stars; and the *meridional armillary,* used to determine the Sun's altitude. Manufacture of these instruments by skilled craftsmen required great precision, a necessary forerunner to technical progress in the mechanical arts of all kinds.

atlas

The first atlas, as a compilation of maps in a single book, was published in 1570, the work of Abraham Ortelius (1527–1598), an associate of Gerardus Mercator (1512–1594), the Flemish-born geographer who developed **Mercator projection.**

Ortelius was born in the Netherlands and at an early age had become a dealer in maps, which he would mount on linen, color, and then sell in markets and at fairs. As the business thrived, he traveled to Britain, Germany, Italy, and France, purchasing maps that were locally produced, and selling his own products. He brought back his purchases to Antwerp. At the request of a fellow merchant, Aegidius Hooftman, Ortelius compiled 30 maps, to be printed in identical sizes, about 28 by 24 inches.

Previous maps had been difficult to consult, with the largest being rolled up, and the smaller ones, although flat, often including such fine print that place names were barely legible. Hooftman complained of the odd collection of miscellaneous sizes, and Ortelius provided the solution in his folio-size volume. The result was the first modern atlas, in an edition of one copy.

Ortelius then seized on the idea of a compilation of maps in books and produced a new volume, using 53 maps, including some provided by Mercator, and began printing a regular edition of an atlas in 1570 on the press of Christophe Plantin. Ortelius's work was reprinted many times, and by 1612 the work had gone through more than 40 editions. The accuracy and quality of the maps were enhanced by the use of copperplate **engraving,** a craft that reached its peak of excellence in Holland. Ortelius was careful not simply to copy maps but also to attribute each to the original mapmaker.

Meanwhile, Mercator worked at producing a definitive 3-volume atlas containing the best and most up-to-date maps of the world and completed 2 volumes before he died in 1594. Mercator's son, Rumold, published in 1595 the 3-volume collection of maps titled "*Atlas: or cosmological meditations upon the creation of the universe, and the universe as created.*" This book was the first to use the word "Atlas" in the title, and the term stuck. Rumold Mercator's work quickly ran through 31 editions.

Later collections were reduced in size, with convenient "pocket-sized" atlases published in the 18th century.

canal lock

Canals [I] were built in ancient times, including one to bring fresh water to Nineveh, and an ancient Suez Canal, built in about 510 B.C. to connect the Red Sea and the Nile River.

The ancient canals were all constructed through nearly level territory. In about 1373 the Dutch developed a system of impounding water, known as a pound lock, or simply a canal lock. A tightly closed chamber is flooded with water, or drained of water so that a vessel in the chamber or pound is raised or lowered with the water level. This system allowed canals to be built through sloping terrain, with the locks used to raise or lower a ship or barge from one level to another. This improvement led to the rapid growth of canal systems in Britain and the mainland of Europe, bringing many lands within convenient reach of market towns and cities, further stimulating the transport of food and commercial products.

The first true canal in the United States was constructed in South Carolina beginning in 1793, connecting the Santee River and the Cooper River and designed by Christian Senf, a Swedish engineer. Another canal started in the same year, connecting the Merrimac River and the Concord River to provide access to Boston Harbor, was completed in 1803. Canal building reached a peak in the United States in the early 19th century. In 1800 there were about 100 miles of canal, and by the end of the century there were more than 4,500 miles.

Among the more important canals in the United States was the Erie Canal (later improved and renamed the New York State Barge Canal). That canal, begun in 1817, eventually connected New York City, by way of the Hudson River and the canal, to Lake Erie. Several marvels of engineering for the age characterized the Erie Canal, including a steep ascent through seven double locks at Lockport, New York. When completed, Lockport became a point to visit for European tourists. The canal was in general 4 feet deep, 42 feet wide at the surface, and 28 feet wide at the bottom. Eighteen **aqueducts** [I] had to be constructed to carry the canal over lower rivers. The Genesee Aqueduct, another "must-see" on the tour, was 800 feet long. The canal opened with elaborate ceremonies on October 25, 1825.

The Erie Canal stimulated trade, allowing New York to become a port for the Midwest. The sleepy town of Buffalo grew 100 times larger within a few years of the canal's opening.

Canals were an important method of transportation because one or two mules could transport an extremely heavy cargo in a canal boat by walking along the towpath that bordered the canal. Instead of carrying a few hundred pounds in a pack load, or 1 ton in a wheeled cart, a single mule, at the same rate of food consumption, could transport 20 tons of cargo by canal boat. Once the cargo boat was in motion and continuing by inertia, the slight friction of the water demanded minimal energy output to maintain the motion.

cannon

In Europe, the importation of **gunpowder** and the knowledge of how to make it in the 13th century soon found application in various forms of artillery. The first cannons were fire tubes, simply forged metal cylinders with a touchhole for ignition of the powder charge at the breech or rear of the weapon, which were introduced in Europe sometime in the period 1314 to 1324 and at least 200 years earlier in China. Until the development of the **flintlock,** the gunner would ignite the charge with a slow-burning match. For many cannons, the match-ignition system remained in place well into the 19th century. Loading was done through the muzzle, with either loose or bagged powder, wadding, and a solid missile slightly smaller in diameter than the bore of the weapon.

Cast **bronze [I]** cannons were introduced by 1450. Improvements in powder came with the discovery in about 1425 that after powder had been wet, it could be forced through a sieve or mesh to form granules that when dried would burn faster, creating a more powerful force. Numerous improvements to the cannon over the next centuries led to lighter and more mobile weapons for land warfare and heavier, longer-range, and more powerful weapons for shipboard use and for defense of fixed fortresses on land. Early mortar weapons were simply heavy metal pots from which a spherical stone could be shot in a very high arc to rise over a defensive wall and inflict damage from the force of its fall from a great height.

In the wars between the British and the Dutch in the 1600s, the Dutch introduced a mobile weapon designed to throw projectiles in a high arc, called the howitzer after the Dutch word *houwitser* for slingshot.

A major improvement to naval guns came with a design introduced by John Dahlgren, an American naval officer, in 1851. Reasoning that the greatest force of the burning charge was at the breech of the weapon, he designed a gun with a thick breech around the bore, but a thinner barrel toward the muzzle. If stood on its breech, the shape resembled a bottle. The smooth shape in the cast iron guns, it was later learned, tended to align the metal crystals, making the weapon far safer than other models in that the metal would be relatively free of weaknesses and irregularities. A process developed by T. J. Rodman at the West Point Foundry in the 1850s, in which the guns were cast on a hollow core, allowed the metal to cool from the inside, greatly strengthening the inner side of the barrel. Rodman-cast Dahlgren guns proved highly effective over the next few years.

Breech-loading weapons, in which the breech could be opened and both charge and shot inserted at the rear of the weapon, were introduced as early as the end of the 14th century. However, a secure, screw-breech system was first patented by B. Chambers in 1849 in the United States. Both breech- and muzzle-loading systems were used during the American Civil War (1861–1865). With the development of **smokeless powder [IV]**, designers sought to improve the breech-loading system to allow for rapid firing and reloading, with several rapid-fire designs coming into use after 1895.

cathedral

Although not strictly speaking an invention, the cathedral, especially the cathedrals of the High Middle Ages (1100–1300) and the Gothic period (1300–1500), have been extensively studied as architectural forms and as ranking among the great technological achievements of the human race. In Christian churches that are structured with an episcopal, or bishop/archbishop hierarchical structure, a cathedral is the church of a resident bishop. For this reason it was generally more imposing than local churches and chapels.

Although there were a variety of cathedral designs, many followed a basilica plan, with a large central hall or nave, flanked on each side by smaller aisles. The origin of the basilica plan has been traced to early Roman forums, or to the early Roman home, with an open atrium surrounded by halls.

Beginning in about the 6th century, cathedrals were laid out on an east–west axis, so that the priest would stand in front of the altar at the east end, facing east, with his back to the congregation, and the entrance would be at the west end. The side aisles were usually separated from the main nave by a colonnade with arches. Above the side aisles, the walls would rise higher and be pierced with windows, to light the nave. Stained glass, embedded in lead frames, often provided richly colored illumination to the cavernous interiors. Over the area in

Perhaps the most striking of all the devices invented to solve the problems created by stone vaulting was the flying buttress introduced in the Île de France in the 12th century. In contrast with English builders, who at first retained the Norman tradition of thick walls, the French reduced their walls to little else than frames for stained-glass windows, and in so doing they had to devise some means of counteracting the thrust of the nave roof. This they did . . . by carrying up a half-arch above the roof of the side aisle to the junction of the roof and the wall of the nave. Later it was realized that the roof thrust extended some way down the wall, and the flying buttress was doubled to meet this thrust at Chartres and Amiens.

—Alistair C. Crombie, *Medieval and Early Modern Science, vol. 1, Science in the Middle Ages, V–XIII Centuries* (Garden City, N.Y.: Doubleday, 1959)

front of the altar, the apse, often a vaulted structure, would allow for a higher ceiling. If the apse extended crosswise beyond the side aisles, that area would be called a transept.

Notable early cathedrals with their years of construction are Pisa (1063–1272); Monreal in Sicily (1174–1189); Milan (11th century); Worms, Cologne, Speyer, and Mainz (11th–12th centuries); and Durham (1133). During the Gothic period, cathedrals reached their climax of architectural splendor and included Chartres and Notre Dame in France, Westminster Abbey in Britain, and Leon in Spain. Often a cathedral took several generations, or more than a century, to complete. In Britain, St. Paul's Cathedral, designed by Christopher Wren (1632–1723), was built from 1668 to 1710. In New York, St. Patrick's Cathedral, designed by James Renwick, was built from 1850 to 1879.

In Washington, D.C., the National Cathedral, operated by the Episcopal Church of the United States, took nearly a century to complete, and faithfully followed many of the design elements of traditional cathedrals of the Gothic period, but incorporated many modern motifs in the stained-glass windows. A complete set of architectural drawings of the National Cathedral was preserved, the only full record of the construction of such a building.

cathedral So-called flying buttresses allowed for much greater height by providing thrust support for outer walls.
Library of Congress

chain mail

The earliest known body armor was discovered in 1960, consisting of a whole suit of plated armor buried with a Mycenean nobleman at Dendra in southern Greece and dated to the 15th century B.C. Made of **bronze [I]**, it had overlapping plates, including a round neckpiece, vertical plates for the shoulders, and three horizontal plates surrounding the body down to the thighs. The Chinese had bronze plate armor mounted on leather clothing dating to about 1400 B.C. At more or less the same time, bronze scale armor was used in Egypt. Ninth-century B.C. Assyrian bowmen wore full-length iron scale armor, with a pointed helmet, and with iron scale armor protecting the neck.

As early as the 3rd century B.C., the Romans created a form of mail armor made from interlocking metal rings. By that time, the Chinese had as many as seven different types of armor.

Other forms of Roman armor included metal plates that would encircle the body and be flexibly connected to allow some freedom of movement, as well as iron helmets. Both the Greeks and the Romans developed leather and metal armor to protect horses against arrows, including headguards and scales made of alloys of copper.

In the Middle Ages, a wide variety of chain mail was developed and worn by European knights, including rings sewn on leather, scales of leather, or rings woven together. Improving on the Roman version of mail in which metal rings were sewn on leather or fabric, medieval armorers developed interlaced rings, as well as solid armor that was welded or riveted. By the 12th century, a complex system of chain mail with fitted sleeves and pants was supplemented by plate armor over the chest. By the 14th century, complete plate armor tended to displace mail.

chimney

The idea of a chimney seems so self-evident that it may seem surprising that it was not invented until the 14th century. Until that time, a fire hearth would be built against the wall, with a hole in the roof directly above the fireplace, but without any flue to confine the smoke and carry it above the roofline. The first chimneys in Britain appeared in scattered locations in the 14th century, in wealthy homes and in monasteries. By 1500, chimney pots made of clay piping extended above brick chimneys across London, but in country villages they remained rare until the mid-16th century.

clock

The keeping of time by sundial was credited to the Greek scientist Anaximander (610–547 B.C.), although there is solid evidence that both the Chinese and the Egyptians used such devices far earlier. The water clock or clepsydra was noted in an Egyptian tomb inscription from 1500 B.C., and remains of one have been dated to the 14th century B.C., in the temple of Amun-Re at Karnak. The Egyptians were the first to divide the day into 12 parts and the night into 12 parts. Near the equator, the slight seasonal variation in the length of the day may have facilitated such a notion of even division.

Water clocks were used in China in the 6th century B.C., and the remains of one have been found in Athens from about 350 B.C. There were different designs of water clocks. Some would have a float, which would indicate the time as water trickled out through a small passage in the bottom of a vessel. Others were more elaborate, in which the water would drive small figurines that would ring bells.

In Europe, mechanical clocks began to appear in the period A.D. 1321–1335, but Chinese models preceded them. Mechanical clocks appeared to descend from earlier water clocks. A cosmic engine built in Khaifeng China in A.D. 1090 consisted of a 30-foot-high building, with a geared mechanism that was driven by a **waterwheel.** On top, a celestial display showed star and planet position, while dozens of miniature figures announced the time of day or night. The cosmic engine ran for more than 130 years.

The father of European clockmaking is Giovanni de Dondi (1318–1389), although his father, Jacopo de Dondi, built a town clock at Padua. Giovanni's major contribution was similar to the Chinese cosmic engine. Known as an *astratrium,* it took 16 years to build, and began operation in 1364. It showed the movements in the sky of the Sun, the Moon, and the five known planets and presented the cycle of religious holidays as well as daily time. To allow for the elliptical orbits of the Moon and Mercury, according to the current Ptolemaic view, he made elliptical gears, and he provided for the irregular orbit of Venus. It was the most complicated machine known to exist since the construction of the Anikythera computer, a geared device built in ancient Greece to represent the motions of the planets. De Dondi's clock operated for some years after his death, but was "retired" as beyond repair in 1530. De Dondi's clock was by no means the first in Europe, where the first to strike an hour is said to have been built in Milan in 1335.

In the 14th century, weights rather than waterpower drove European town clocks in public bell towers. Surviving examples of town clocks in St. Albans Abbey in England, in Prague, Bern, and in Munich all were characterized by astronomical representations and figurines. Most of the large town clocks had iron gears. Springs replaced weights as the driving force in smaller clocks in the late 1430s, apparently derived from the contemporary use of springs in locks and firearms. Peter Henlein, a German locksmith, began making small spring-driven clocks in about 1500. In the 17th century, **pendulum clocks [III]** and **balance**

springs [III] were introduced. The 18th-century development of the **chronometer [III]** was a crucial addition to the tool kit of the ocean navigator.

The psychological and social effects of advances in timekeeping are subjects of interesting speculation. Even the ancient Greeks noted that lawyers and playwrights seemed wedded to the clepsydra. The development of town clocks in the 14th century in Europe may have allowed for more regulation of daily business, and the development of more precision in later timekeeping certainly assisted scientists in recording chemical and physical phenomena. But whether the town clock led to a more regulated daily life, or the regulation of daily life created a market that the town clock filled, is open to discussion.

comet paths as orbits

Although comets had been noted throughout the ancient past, the fact that their pathways through the heavens are in fact orbital ones around the Sun was not discovered until 1577 by the Danish astronomer Tycho Brahe (1546–1601), who observed a comet in that year and five others afterward. In 1680, Isaac Newton (1642–1727) developed a method of calculating the true orbit of a comet from observing its pathway, and he determined that the comet observed in December of that year followed a long, parabolic orbit.

Edmond Halley (1656–1742) discovered that the orbits of several prior comets were almost identical, and he concluded that they represented a single orbit returning every 76 years. He predicted its return in 1758, and since then it has been called Halley's comet. It returned in 1910 with a particularly bright display, but when it appeared again in 1986, its tail was not visible from Earth because at that time it passed on the other side of the Sun.

The physical nature of comets was long a matter of some speculation, resolved in the late 20th century with spectrographic observation of the material in the tail and an encounter in 1985 between the U.S.-launched interplanetary Sun-Earth *Explorer 3* and the P/Giacobini/Zinner comet, and later encounters by separately launched spacecraft with Halley's comet in 1986. From these sources a great deal of material was learned showing comets to be dust-covered loose packs of ice and frozen gases that are heated as they enter the inner solar system, emitting both gas and dust that form the tails. Newly discovered comets enter the solar system regularly, and usually are named after their discoverers or codiscoverers.

compass

Like many inventions used widely in the West, the compass was first developed in China, in the 1st century B.C. A spoon-shaped lodestone could be used to determine north, and south-pointing compasses were made in such a shape from lodestone. The heavier or ladle end of the spoon would swing to the north on a level surface, while the handle or pointer of the spoon would swing to the south. As early as A.D. 1086 the Chinese were manufacturing magnetized steel or iron needles that would be mounted on pivots or floated in water to reduce friction and to indicate north–south direction. A Chinese compass with a pivot was described in the Sung dynasty (1127–1279), in the shape of a turtle. The Chinese first used compasses not as an aid to travel but as a device to determine directions when aligning the construction of buildings and roads.

Yet the Chinese did utilize the compass for navigation well before the Europeans. Records indicate that Chinese pilots used compasses as early as the 12th century to determine direction when clouds obscured the Sun or the stars. The fact that a compass varies from true north because the magnetic North Pole is slightly displaced from the true North Pole (a difference called declination) was known to the Chinese in the 1200s and 1300s, and they produced charts representing the degrees of declination in that period. European navigators began to recognize declination in the 15th century.

The first European writing about the compass was Alexander Neckam (1157–1217). In 1187 he described both floating and pivoted compasses. By the 1300s, European sailors used compasses mounted on a "rose" showing the directions; a glued magnetic needle on the rose would allow the whole card with the directions to turn, giving an easy reading for the mariner.

corset

Although there are records of corsetlike bindings in ancient times, the tightly constricting corset to cinch in the waist became an element of fashion in the mid-1300s, worn outside of other garments. By the mid- and late 1500s, Spanish costumes included a tight bodice that was stiffened with metal ribs, worn by both men and women. The style spread to other countries of Continental Europe as well as to Britain.

Early in the 19th century, style changed to reveal a woman's waistline, and corsets developed as a highly restrictive feminine undergarment. Usually made of heavy canvas and ribbed with whalebone, the mid-19th-century corsets would reach from just under the bosom down to the thighs. Worn daily, such a corset would lead to weakening of muscles, deformity of internal organs, and could contribute to the concept that women were weak or fragile. The demand for whalebone for both corsets and to hold out hooped skirts was so severe that during the American Civil War (1861–1865), blockade runners would risk their lives to ensure a steady supply of the high-priced material through blockaded Southern ports.

With the development of rubberized elastic materials in about 1911, the girdle began to replace the corset. Together with the **brassiere** [V], the girdle revolutionized women's undergarments in the early to mid-20th century.

crank

There are many devices, appliances, and simple machines so common in our everyday life that we rarely think of them as having been invented. The crank and the compound crank are such items. They did not exist in the ancient world but were common by the 1600s, employed in various practical applications such as pulling a bucket from a well on a turned shaft, and a wide variety of situations in which it was necessary to transform continuous rotary motion into reciprocating motion, or to transform reciprocating motion into rotary motion.

The first established use of a crank was in a water-powered bellows dated to A.D. 31 in China. A number of artists' renderings of **Archimedes' screw** [I] show it hand-cranked, but contemporary depictions and surviving remains and texts show it as powered by treading. The first established evidence in Europe of a crank handle to convert a push-and-pull motion into a rotary motion is found in an illustrated work published in Utrecht between A.D. 816 and 834. Coupled with what is probably the first grindstone in Europe, the find raises the curious question of why neither the crank nor the grindstone was found in ancient Greece or Rome, or in the early Middle Ages. It has been suggested that there is something inherently counterintuitive in a device that awkwardly allows a forward-and-back or up-and-down motion to be converted into rotation, and that the human hand, arm, and psychology are more comfortable with reciprocating motion than with circular motion.

Depictions of cranks in Europe remain rare until after 1400, when several illustrations show a hand-cranked **crossbow** [I], which had first been developed in China and some fanciful illustrations of the goddess of fortune cranking the wheel of destiny. In the 1400s, cranks seemed to spread quite suddenly for a variety of applications, apparently originating with a carpenter's brace for a drill, invented in about 1420 in Flanders. From the carpenter's brace, the concept appeared to spread to piston pumps powered by **waterwheels.** Soon applications included the addition of flywheels to get the motion past the dead spot and efforts to construct **paddlewheel boats** powered by hand cranks in the 1430s. Many early cranks were gracefully curved, as if the curve would help impart rotary motion, but in 1567 Giuseppe Ceredi published a scientific analysis of the crank that pointed out that such curves were mechanically useless.

> Next to the wheel, the crank is the most important single mechanical device, since it is the chief means of transforming continuous rotary motion into reciprocating motion, and the reverse. The crank is profoundly puzzling not only historically but psychologically; the human mind seems to shy away from it.
> —Lynn White Jr., *Medieval Technology and Social Change* (Oxford, Eng.: Oxford University Press, 1962)

cross-staff

The **astrolabe,** made of brass or other metal and finely crafted, provided mariners with an instrument to measure the angle of elevation of the Sun or a star above the horizon. Sometime in the 16th century, European navigators began to use a cheaper device for measuring the angle of celestial bodies. The time, place, and inventor of the cross-staff are unknown, but the instrument was in common use by 1600.

The cross-staff was a T-shaped wooden instrument. Navigators would hold the cross-staff to their eye and sight along the long member to the horizon. They would then move the crosspiece along the shaft until the Sun or a star was at its tip. A scale on the long member of the crossbar would indicate the angle. Staring along the cross-staff required looking directly into the Sun, and in the early 1600s, navigators shifted to a back-staff, a predecessor of the **sextant [III];** the back-staff allowed the sailor to stand with his back to the Sun and to move the shadow of the vertical member to the end of horizontal piece, which he held in front of him and pointed at the horizon. The cross-staff may have blinded so many sailors in one eye that it can account for the presence in folklore and art of so many pirates and sailors wearing a single eyepatch.

decimal numbers

Although primitive peoples developed various means of counting, the emergence of a decimal number system is usually traced to the Hindu and Arabic cultures from the 8th to the 11th centuries A.D.

The idea of counting appeared with numbers recorded as marks in clay, in knotted strings, and in beads on wires in Neolithic times. The ancient Sumerians and Chaldeans, as early as 3000 B.C., developed a system of counting with 60 as a base, representing the numbers with cuneiform marks in clay tablets. The Egyptians developed a system using 10 as a base as early as 2000 B.C. The Rhind papyrus, dated to about 1150 B.C., described fairly complicated mathematical problems and how to solve them.

Roman numerals, with the system in which position indicates subtraction or addition (in which LX would indicate 60, but XC would indicate 90) was cumbersome, but convenient in addition and subtraction.

In A.D. 1202, Leonardo Febonacci (also known as Leonardo of Pisa) published a work that captured the decimal notation system developed in the Muslim world and explained it for European readers. A crucial element was the invention of **zero**, attributed to Muhammad ibn Musa al-Khwarizmi in about A.D. 800.

The complete decimal system is characterized by a positional base notation, the use of the zero to indicate nothing in the position or at the end of the number to indicate tens or multiples of ten, and the use of the decimal point to allow the representation of fractions in decimal fashion. In the United States the decimal point is indicated with a period; and in most European settings it is represented by a comma. Before the development of the complete system, simple calculations such as multiplication and division could be handled only by expert mathematicians. As the system evolved, it is not entirely clear exactly when the zero to mark the absence of a quantity in a position was first used, as in the number 503, which represents five hundreds in the third position, no tens in the second position, and three units in the first position, counting the positions from right to left.

During the late Middle Ages, as the decimal system spread, two methods of doing mathematics competed. Those practitioners who used the abacus, known as abacists, and those using the base-10 notation, known as algorists, disputed the speed with which problems could be solved, with the algorists gaining ground by the 1100s.

distilled spirits

The invention of the process of distilling alcohol has not been precisely dated. Using literary references, historians have found it hard to distinguish between simply fermented beverages such as **beer [I]** and **wine [I]** and the more potent beverages produced by boiling off the alcohol from a fermented beverage and recapturing it in a still. References to a rice beer distilled into a rice brandy appear in China as early as 800 B.C. The Greeks, Romans, and Arabs all apparently produced distilled beverages, although the first clear references do not show up until after A.D. 100. In the 12th century, alchemists referred to aqua vitae (Latin) eau-de-vie (French) and *uisge beatha* (Gaelic) as terms for distilled alcohol. Aqua vitae usually referred to a distillation of wine, or brandy, while the Gaelic *uisge* (pronounced wees-geh) is the origin of the word *whisky,* a beverage distilled from fermented barley.

The Scots began distilling in the 15th century and produced three types of whisky. When it was made solely from barley, then malted, dried over peat fires, fermented, and then double distilled in alembics or pot stills, then aged in casks that had previously held sherry, the resulting beverage was known as a single malt or a singleton. When two whiskies, each from a single distillery, were brought together in a large vat, they were known as vatted whisky. Malt and grain whiskies combined were known as blended whisky. When more barley was produced in a district than could be locally consumed, the surplus was turned into whisky for export. One region in the Scottish Highlands that often produced such liquid surpluses was the Speyside, the home of Glenfiddich and Glenlivet, among other whiskies.

In France, the brandy made from Charente wines of the Cognac district was distilled to reduce the bulky wine for transport aboard ship, with the notion that the brandy would be diluted back to a lower alcohol content with water at the receiving end of the shipment. However, the brandy produced was far more palatable than the fairly sour or acidic Charente wines from which it was made, and drinking it neat or undiluted made it a far more potent beverage. The term Cognac is widely used, but properly refers only to distillation from a fairly small district around the towns Cognac and Jarnac on the Charente River.

Italian *acquavite,* Spanish brandy from the Jerez district, and other brandies from Europe can trace their origins to the 17th century and earlier. Early in the 19th century, alcohol production became industrialized, improving on the traditional pot still, which consisted of a closed boiler over heat with a condensing tube to draw off the alcohol vapor, to continuous still systems with the stills arranged in a column. In 1831, Aeneas Coffey in Ireland built one of the first multiple still systems, linking two columnar stills.

Whisky, made from barley in Scotland, was modified in the United States, where other grains were used, notably corn and rye, and where the resultant beverage was spelled *whiskey*. Bourbon derived its name from the corn-based liquor made in Bourbon County, Kentucky. The decision to place a tax on corn whiskey in the 1790s produced a popular uprising in the United States known as the Whiskey Rebellion. In Ireland, a mix of barley with wheat, oats, and rye produced a distinctive Irish whisky. In Canada, where whisky is spelled without the "e," as in Scotland, a wide variety of grains are used, with the goal of producing a smooth beverage similar to the Scotch whisky. In other regions, such as Scandinavia, aquavit is made from either grain or potatoes, and flavored with caraway seeds. Rum is produced from sugarcane, vodka from potatoes, gin from grain, and tequila from the Mexican century plant.

Experience in the production of distilled alcoholic beverages contributed to the advancement of chemistry, and the techniques carried over to the **catalytic cracking of petroleum** [V] to produce lubricants and fuels.

dome

Although there is some record of the dome being used in structures in ancient Mesopotamia and in early Rome, the development of large domes in architecture rightly belongs to the Middle Age and Renaissance periods. In particular, the construction of the unsupported dome, held up by its internal stresses, was an achievement of the 15th century A.D.

The Pantheon in Rome, built in A.D. 112, inspired many of the later structures. That dome has a diameter of 142 feet, with an open space at the top some 27 feet in diameter. The upper parts are made of light pumice and tufa stone, and the massive support is made of concrete. A later dome, built in the period 526–547 at Ravenna, used a system of interlocking chambers topped with a timber roof.

One of the most impressive domes of the Middle Ages was that of Hagia Sophia, built in A.D. 532–537 in Constantinople and still surviving in the 21st century. The dome is supported by a complex system of barrel vaults and partial domes that keep up the main structure. The circular dome sits on a square structure, supported by spherical triangular sections at the corners. All of these domes were known as vaulted domes, and they were limited in shape to a near hemisphere and had to be supported by outside structures. The supported hemispheric dome became quite characteristic of many mosques in the Muslim world.

A breakthrough came with an innovative design developed by Filippo Brunelleschi (1377–1446) from 1420 to 1438, with the construction of the Church of Santa Maria dei Fieori at Florence. The dome covers an octagon and is 138 feet in diameter. Each side of the octagon has a window. The dome itself is steeply pointed, by contrast to the earlier vaulted domes, and is supported by 24 ribs. A hoop of tie bars on the outside prevents the dome from spreading outward at the base. Michelangelo Buonarroti (1475–1564) used the same design for the dome of St. Peter's basilica in Rome beginning in 1547. The use of the hoop of tie bars allowed the dome to rise with a clean and unencumbered silhouette. Christopher Wren (1632–1723) also employed the method in the construction of the dome for St. Paul's Cathedral in London, built from 1675 to 1710. The dome on the Capitol building in Washington, D.C., and smaller domes on some state capitol buildings in the United States follow the Brunelleschi and Wren designs.

Brunelleschi's achievement was so unusual that it is often taken as the starting point for Renaissance architecture. He was able to raise the cupola without any scaffolding or abutments. He conceived of the structure as a series of rings, one above the other. Contemporaries were convinced it would collapse during construction, but he achieved a break with the past in both construction method and in final design without any mathematical calculations. Apparently he had an intuitive understanding of the forces at work.

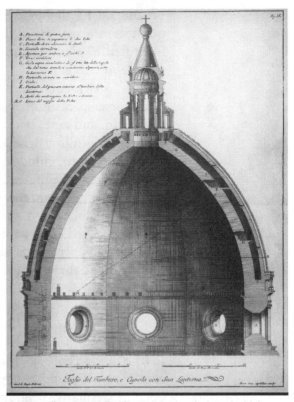

dome Fillippo Brunelleschi's innovative design of a high dome for a church in Florence was widely emulated in the centuries following its creation. *Library of Congress*

elements

The fact that matter is composed of discrete elements had been supposed since antiquity, with accepted science, derived from the ancient philosopher Empedocles (493–433 B.C.), holding that all matter was composed of four essences or elements: earth, air, fire, and water. The search for the magical fifth essence, or *quintessence,* was a quest of the medieval alchemists. They believed that the stars, planets, and other heavenly bodies were composed of the quintessence, and that it was present on Earth but was difficult to extract. In their pursuit of the quintessence and the *al iksr,* or *elixir,* alchemists and metallurgists had identified 13 elements, mostly metals, well before the age of the Scientific Revolution.

The 13 elements known by the end of the Middle Ages were antimony, arsenic, bismuth, carbon, copper, gold, iron, lead, mercury, silver, sulfur, tin, and zinc. It was the hope of the alchemists that the quintessence would allow lead or another inexpensive metal or alloy to be transmuted into gold. A 14th element, platinum, was discovered in 1557 by Julius Scaliger. Gradually over the following three centuries, many more elements were discovered, and in the 1870s, Dmitri Mendeleev discovered the **periodic law [IV]**. Improved tools and methods in the 20th century filled out the table he developed, and in the mid-20th century, physicists achieved the ability to transmute one element into another through atomic fission, fusion, and controlled nuclear reactions.

engraving

In the 16th century, European artists improved on techniques of engraving on metal and printing from metal plates. Engraving developed in Germany, Belgium, the Netherlands, and, later, in Italy. Many of the earliest engraved prints were of religious subjects, but increasingly artists represented scenes of nature, producing detailed illustrations of plants and animals, and of scenes and peoples discovered in overseas exploration. Some of the first accomplished engravers were goldsmiths, and the richest early prints were illustrations for patterns that could be used by artists and craftsmen such as carvers and embroiderers. A few engravers are known only by their initials or by some other name attached by later generations of connoisseurs. The "Master of the Playing Cards" produced many cards with illustrations of plants and animals. Others included "Master E. S.," "Master I. A. M.," and "Master of the Amsterdam Cabinet." The first engraver to etch the design first on copper plates was Lucas van Leyden, a Dutch painter from 1510 to 1520. Copperplate etching and engraving allowed for rich illustration of books, which contributed to the spread of scientific and geographic knowledge through the following decades. By the era of the Scientific Revolution, rich illustration allowed the transmission of the new findings not only in text but also in pictorial representations of equipment, findings, and procedures. Fine copperplate engraving was used by mapmakers of the 16th century to produce high-quality maps reflecting the discoveries of explorers and navigators, compiled in the first **atlas.**

Eustachian tube

The Eustachian tube, connecting the middle ear and the pharynx, was discovered in the 16th century by the Italian doctor Bartolommeo Eustachio (1520–1574). Eustachio was a practicing student of **anatomy** in an era when the dissection of bodies was still controversial and outlawed in some regions. Eustachio's *Tabulae anatomicae* was published in 1722, about a century and a half after his death. A valve in the heart that he discovered is also named after him, the Eustachian valve.

eyeglasses

The invention of eyeglasses is surrounded with mystery. They were known to appear in Europe between 1280 and 1286, and their invention has been variously attributed to specific **glass** [I] workers in Italy and to importation from Asia. Ancient records indicate that some form of magnifying glass was in use in ancient Chaldea, in about 4000 B.C. But it was the development of glassmaking and glass blowing, particularly in northern Italy, that made the invention of eyeglasses possible.

The idea of mounting lenses in frames to be worn on the face may have originated in China, as Marco Polo is reported to have seen spectacles in use there. Although Marco Polo returned to Europe in 1295, his father, Niccolò Polo, and his uncle, Maffeo Polo, had returned from a trading trip to China in 1269, possibly bringing the concept of eyeglasses with them.

Roger Bacon (1214–1294) is said to have provided Pope Clement IV with a type of reading glasses, and Bacon suggested that convex lenses would be appropriate for farsightedness in his *Opus Majus,* published in 1266–1267. The actual invention of mounting lenses in a frame appears to have originated in Italy in about 1285 and to have been popularized or spread by a friar of Pisa, Salvino D'Amato. In Venice, a tombstone reads: "Here lies Salvino D'Amato of the Armati of Florence, inventor of spectacles, God pardon him his sins, A.D. 1317." Other records point to a Florentine, Alessandro di Spina, as having introduced eyeglasses to Europe. The word "lens" may derive from the Italian word for lentils, or glass lentils: *lente de vitro.* In the early 1300s, eyeglasses were sometimes called *roidi da ogla* or "discs for the eyes."

> The outstanding advance made in the West was the invention of spectacles. That weak sight and particularly the difficulty of reading in the evening was felt as a serious affliction is shown by the number of salves and lotions prescribed for this complaint, but although lenses had been known for some centuries in both Christendom and Islam it is only at the end of the 13th century that there is evidence of spectacles with convex lenses being used to compensate for long sight.
> —Alistair C. Crombie, *Medieval and Early Modern Science,* vol. 1, *Science in the Middle Ages, V–XVIII Centuries* (Garden City, N.Y.: Doubleday, 1959), p. 231

The artist Tommasa da Modena painted a portrait of Hugh of Provence in 1352 that showed eyeglasses, perhaps the first representation in art of the invention. Another painting, by Dominico Ghirlandaho in 1480, showed St. Jerome at a desk with eyeglasses in the picture. St. Jerome became the patron saint of the guild of spectaclemakers. Although early lenses were convex in shape to aid farsightedness, by the early 1500s, concave lenses to correct nearsightedness had been introduced. Benjamin Franklin (1706–1790) claimed to have invented the bifocal lens system for both near and distant vision in 1760, although such lenses may have been in use before he publicized the idea.

The fact that eyeglasses, such a major invention, cannot be attributed to a specific inventor is indicative of a basic change in the history of invention. Prior to the Renaissance, few inventors made the effort to claim priority of invention, even of major devices. However, with the rise of individualism in the Renaissance and post-Renaissance world, inventors and discoverers bitterly fought each other for priority of claim to a new idea. Only after the death of D'Amato and di Spina did their advocates come forward to claim that they had invented spectacles. The change in attitude about invention and the individual is thus represented by the anonymity of the eyeglass inventor in about 1285–1290, and the intense rivalry over claims to priority of invention of the same device that surfaced early in the next century. By the time of the invention of **movable type** in the mid-1400s, the issue of priority of invention had become contentious and remains so into modern times.

Fallopian tubes

Fallopian tubes, which connect the ovaries and the uterus in the human female, were discovered by Gabriele Falloppio (1523–1562). Also known by his Latinized name, Gabriel Fallopius, he studied in Padua under Andreas Vesalius (1514–1564), becoming a professor of **anatomy** at Pisa in 1548 and at Padua in 1551. Falloppio studied and described much of the reproductive system, as well as the brain, eyes, and inner ear. He was the first to describe the clitoris, and he named the vagina. Although he described the Fallopian tubes, he did not properly understand their function.

flintlock

With the introduction of **gunpowder** into warfare in Europe in the 13th century, the problem of how to ignite the powder represented what modern engineers would call a "technical barrier." Early methods involved the use of a slow match, a constantly smoldering piece of rope or cord that would be lowered to a touchhole in the breech of the weapon. Although such a system worked fairly well for early forms of cannons, the problems with the system became apparent with the development of the Spanish arquebus, a hand-carried weapon. The soldier would have to remove his eyes from the target to bring the match to the touchhole. To solve this problem, a slow-moving clamp would lower the smoldering match to the firing pan as the soldier pulled the trigger, keeping his eyes aligned on the target and steadying the barrel, in a weapon known as a **matchlock.**

The matchlock system was briefly superseded in the early 16th century with a wheel lock, in which a clockwork wheel would be wound up and, when released by the trigger, would strike a steel wheel against a piece of iron pyrites, spraying sparks into the firing pan. The flintlock was developed about 1550, with a piece of flint that would be cocked back and released or snapped forward with the pull on the trigger. The lid to the firing pan had to be raised separately to allow the spark to ignite the powder in the pan. In about 1625 a Frenchman, Marin le Bourgeoys, made an improvement that made the lid to the firing pan and the striker move together so that with one motion the lid was raised and struck the flint, showering sparks into the firing pan and igniting the powder.

For nearly 200 years one or another form of the flintlock remained the dominant form of firing mechanism for handheld guns. The long "Kentucky rifles" as well as many artillery pieces were equipped with variations of the flintlock.

The flintlock was later outmoded by a percussion lock system developed in 1805 by a Scotsman, Alexander Forsyth (1768–1843), and a percussion cap invented in 1816 by Joshua Shaw (1776–1860), in the United States.

fork

The fork as an eating utensil represented an improvement over spearing a piece of food with one knife and cutting it with another. Although first introduced in the Middle East between A.D. 600 and 700, forks were not introduced into Italy until about 1100. They spread rather slowly, until by the reign of Charles V of France (1364–1380) they began to be listed on inventories of the royal household. The two-tined fork had the particularly useful function of holding a piece of food in place, held by one hand, while the other hand could wield a knife for cutting off bite-size sections.

Forks were regarded in Britain as a Continental affectation, even after they were introduced to the British in 1608 by Thomas Coryate, an Englishman reporting on his travels in Italy, Switzerland, Germany, and France. The first forks had two tines; a third tine was introduced in the late 1600s. Two-tined forks in Britain were still being regularly produced as late as the 1740s. By the 18th century the Germans introduced a four-tined fork that resembles the modern version. With the widespread use of the fork, the table knife evolved, losing its point, which had previously been used to spear pieces of food. By the 19th century, the table knife with a blunt tip and parallel sides appeared as the fork took over the function, not only of spearing and holding food, but also of lifting, scooping, and carrying it to the mouth. The evolution of the fork can be traced through collections of antique tableware and also through literature of table manners, which offered advice about the handling of utensils and food and even the picking of teeth.

guitar

The plucked musical instrument known as the guitar apparently had its origin in Spain in the 15th and early 16th centuries. In that period, the *vihuela* or viola referred to three types of instruments: one played with a bow; one with a plectrum or pick; and the third, the *vihuela de mano,* played with the fingers. Spanish, Portuguese, and Italian publications from as early as 1470 and the 1480s depict or refer to the *vihuela de mano.* Music historians have found many Italian references in the same period but generally conclude that the instrument originated in Spain.

Guitar manufacture spread through Europe, with centers in Paris, Lyons, Naples, and Venice. In Paris, Robert and Claude Denis made several hundred guitars in the late 1500s. Jean and René Voboam were noted makers in the 17th century in France. Like many other crafts of the era, guitarmaking could be supported only because of the general economic prosperity and improved food supply of the European continent in the period.

The guitar originally had four courses of strings, three of which were double. A fifth course of strings was added before 1600, and by the 18th century, six-string guitars appeared. In that period, earlier gut frets were replaced with ivory or metal frets. The later evolution of the guitar can be attributed to specific artists and inventors. Antonio Torres developed the classical guitar between 1850 and 1870, with three gut and three metal spun strings.

In the United States, the C. F. Martin Company of Nazareth, Pennsylvania, manufactured some 8,000 guitars between 1833 and 1890. C. F. Martin (1796–1873) had immigrated to the United States from Austria, and he designed a new shape for the American model. Orville Gibson (1856–1918) began to compete with Martin in the early 20th century, with strong X-bracing and steel strings. In Spain, Salvador Ibañez of Valencia led the way for Valencia exports to South America in the 20th century. In Madrid, José Ramirez III developed a guitar much favored by Andres Segovia, who popularized classical Spanish guitar music from 1960 to 1987.

gunpowder

Gunpowder was first developed by the Chinese as early as the 10th century A.D., and the Chinese were actively using gunpowder in a variety of weapons well before the Europeans. Eurocentric historians tended to place the invention of gunpowder for use in weapons with Europeans, often citing the fact that Roger Bacon (1217–1292) described a Chinese firecracker as proof that the Chinese used the material only for sport in that culture before the Arabs and the Europeans developed it for warfare. However, an 1187 chronicle by Yuan Hao-wen recorded the use of a grenade, and Chinese temple carvings discovered in 1985 but dated to A.D. 1128 from inscriptions depict one demon-warrior with a grenade and another holding a bulbous, handheld **cannon**. Actual guns made in China, as distinct from representations or depictions of guns from there, have been dated to 1290.

The first illustration of a cannon in European art is from a manuscript by Walter de Milamette of 1327, remarkably similar in design to that depicted in the temple carving 200 years before in China. Bacon had recorded a recipe for gunpowder, apparently learned from the Chinese by way of the Arabs, in 1242, calling for 5 parts charcoal, 5 parts sulfur, and 7 parts saltpeter (potassium nitrate). There is no evidence that Bacon ever made any gunpowder himself.

Although commonly thought to explode, in actuality gunpowder combusts rapidly. To make use of it to propel missiles or **rockets,** it must be confined tightly. Europeans developed cannons and handheld tubular weapons such as the **matchlock, flintlock,** and **musket** and using the **rifled gun barrel [III]** from the 1300s through the 1600s, revolutionizing warfare.

Bacon's role in introducing gunpowder has been the subject of a great deal of hyperbole. In fact, Bacon recommended the use of many devices and innovations, and he proposed the experimental method for scientific learning, which seemed far ahead of his time. Yet it appears that he conducted very few actual experiments himself and did not construct any of the devices he described. Among innovations he mentioned but never built were a **hot-air balloon [III]** and **eyeglasses** as well as gunpowder weapons.

heliocentric solar system

The discovery that the solar system consists of planets rotating around the Sun rather than around Earth is a classic case of a theoretical construct or theory being eventually borne out by further explanations, by discovered laws, and eventually by direct evidence. Thus, when Nicolaus Copernicus (1473–1543) propounded his explanation of a Sun-centered or heliocentric solar system, he was putting forth not so much a discovery as an explanation that was later confirmed by experimentally gained knowledge of Galileo and others, and whose mechanics were described by Isaac Newton working out the **laws of motion and gravity [III].** Copernicus was not the first to suggest such an explanation. The ancient Greek astronomer and philosopher Aristarchus of Samos (320–250 B.C.) had proposed the same concept. In 1377, Nicholas Oresme (1320–1382), a French bishop, in an introduction to Aristotle's *Treatise on the Heavens,* had given "several fine arguments" to show that Earth revolves, rather than the sky revolving around Earth. However, it was the work of Copernicus that became most widely known and stimulated new thinking about the question of the structure of the solar system.

Copernicus was born in Torun, Poland, and studied mathematics, astronomy, law, and medicine at several of the leading universities in Europe: Cracow, Bologna, and Padua, receiving a doctorate in canon law at Ferrara in 1503. He served in the clergy at the **cathedral** in

Frauenberg, not far from his place of birth. He also practiced medicine, worked on monetary reform, and continued to study older texts on astronomy.

By 1514 he had completed work on a short study of the planets, *Commentariolus,* which he circulated privately. In this work he challenged the long-standing view of Aristotle and the 2nd-century astronomer Ptolemy that the planets, the Sun, and the Moon rotated around Earth. To account for the motions of the planets, Ptolemy had devised a complex explanation that required the planets to rotate in cycles, and then cycles within cycles (or epicycles). From Earth, the position of a planet such as Mars varies from night to night, appearing to move across the background of the fixed stars. However, its motion forward from one night to the next stops, then goes backward for a period of weeks before progressing forward again. This *retrograde* motion had puzzled astronomers, and Ptolemy's explanation, although complex, fit the phenomenon.

Copernicus based his study on a variety of ancient texts, possibly including the work by Oresme, and combined these with his own observations. He concluded that the retrograde phenomenon could be more simply explained by suggesting that the Sun stood still and that Earth rotated on its axis (accounting for day and night) and that it revolved around the Sun. By taking into account the revolution of Earth around the Sun as well as the revolution of Mars around the Sun, he was able to explain the apparent retrograde motion of Mars against the backdrop of the stars as due to the changed position of the human point of reference on Earth over a period of months.

Copernicus offered his idea as an alternate explanation, and a simpler one, of the motion of the planets. At the urging of friends, he finally had *Commentariolus* published as *De Revolutionibus orbium celestium* (On the Revolutions of the Heavenly Spheres). The work was published in the year of his death, 1543, and he reputedly was presented with a copy on his deathbed. Whether Copernicus had intended to offer the explanation as simply a possible alternative to the accepted Ptolemaic explanation or whether his intent was to assert that his explanation was factually more accurate has been much debated. The published work stated the explanation as an alternative, perhaps at the insistence of friends who sought to protect him from the anger of the religious and astronomical establishment.

In either case, *De Revolutionibus* did indeed put forth a startling new idea, and the word "revolutionary" came into European languages as a consequence, to describe some idea or concept that was a break with the past. The concept that Earth itself moved seemed not only to contradict established Aristotelian and Ptolemaic astronomy but common sense and the words of Scripture as well. From a commonsense point of view, it was difficult to believe that Earth moved, as we sense no motion but see motions in the heavens. References in the Bible to the fact that the Sun moved in the heavens also would indicate that Earth was stationary, with the Sun, the Moon, and the planets in motion around it.

Copernicus asserted that the orbits were perfectly circular (when in fact they were slightly elliptical), and his explanation required some epicycles (that in fact do not occur) to account for observed positions and motions. Thus his heliocentric view, while closer to a modern understanding of the motions of the planets, was only a slight simplification of the intricate Ptolemaic system.

From 1609 to 1632, when Galileo Galilei developed **telescopes [III]** that he used for observation of the heavens, he found various pieces of evidence that tended to support the Copernican view that had suggested that Earth was simply another planet revolving around the Sun rather than sitting at the center of the universe. Galileo's discovery that there were mountains on the Moon suggested similarities between Earth and the Moon that had previously been unknown. Galileo further discovered that Venus went through phases, like the Moon, that Jupiter had a cluster of moons rotating around it, and that mysterious sunspots marred the Sun. These pieces of evidence fit into the Copernican explanation far better than the Ptolemaic one in several respects, especially destroying the distinction between earthly imperfection and heavenly perfection. Later, when Isaac Newton adequately explained the motion of the planets, the suggestions of Copernicus became more logically understood as governed by laws of motion and gravity.

Thus to say that Copernicus *discovered* that Earth is a planet and that the solar system is heliocentric is a bit of a simplification. Nevertheless, his contribution to the advancement of the understanding of the universe was profound, and the core of his explanation stimulated a true revolution in scientific thought. The Roman Catholic Church placed *De Revolutionibus* on its index of prohibited books in 1616 and did not remove it until about 1835.

horse collar

The horse could not readily be hitched to the sort of **yoke [I]** used for oxen because when the horse would begin to pull such a yoke it would constrict the horse's windpipe. The improved horse collar, which consisted of a rigid yet padded U-shaped harness, would rest on the shoulders of the horse. The collar could allow free breathing, and it would be attached to the load by shafts so that the whole weight of the horse could pull against the load in a far more efficient arrangement than the ox yoke.

The introduction of the horse collar helped revolutionize European agriculture. When combined with the **horseshoe,** the **three-field system,** and the heavy **plow,** the inventions vastly increased food production and the efficiency of the peasant. However, it is not certain where the horse collar was invented, nor precisely when.

Some evidence points to the introduction of the invention from central Asia, as English and German words for the horse harness appear to derive from Turkic origins. One of the earliest illustrations showing a horse collar was in a work written in about A.D. 800. Metal parts from horse collars have been found in graves dated to the mid- and late 800s. Records of horses being used in plowing show up in reports of travels published in the same period. Although the horse and the ox could pull about the same weight, the horse could move faster and had more endurance, both of which would help in the agricultural revolution of the late Middle Ages in Europe. Some records indicate that a horse could handle as much plow land as two oxen.

By the end of the 11th century plow horses had become a common sight in Continental Europe, from the Ukraine to Norway, although it may have taken another century for them to be widely used in Britain, where records from the late 1100s show some mixed oxen and horse teams and others with horses only.

horseshoe

The nailed horseshoe first appeared in about A.D. 900, although some have been found in layers of dirt that would suggest an earlier use. However, since a metal horseshoe might have been shed in a ditch, or since it might sink through loose dirt, dating from the depth in ground of such finds is very problematic. There were no references in literature to horseshoes before the reign of Byzantine emperor Leo VI (886–911). By 973, passing references to horseshoes showed up, and by 1100 they were very common. There were references as early as the 2nd century B.C. in Rome to horseshoes and muleshoes, but it is assumed that these were attached as leather boots, perhaps reinforced with an iron plate, rather than true, nailed horseshoes.

> Not only ploughing but the speed and expense of land transport were profoundly modified in the peasant's favour by the new harness and nailed shoes.... Now it was becoming possible for peasants not situated along navigable streams to think less in terms of subsistence and more about a surplus of cash crops.
> —Lynn White Jr., *Medieval Technology and Social Change* (Oxford, Eng.: Oxford University Press, 1962)

The shoeing of horses was less important in the dry climates of the Mediterranean than in northern Europe, where the wet climate would lead to softened hooves that were easily damaged or worn. The combination of horseshoe and **horse collar** made the animal suitable for pulling the **plow** and wheeled carts for transport. Together the advances constituted part of the agricultural revolution of Europe in the late Middle Ages. Together the three inventions led to an increase in population and the growth of cities as agricultural markets and points of commerce. These developments led to a series of other steps in technical progress, such as the search for better sources of mechanical power, leading to greater uses of **waterwheels** and **windmills**. For these reasons, much of Europe's progress can be traced back to whoever in the early 10th century decided to nail a curved metal shoe to a horse's hoof.

lateen sails

The Romans had used lateen sails, shaped in a triangular fashion and mounted with a swinging spar, but their use had died out in Europe after the decline of the Roman Empire. In the 7th and 8th centuries A.D., as they attacked Byzantine ports, European sailors encountered lateen-rigged ships known as dhows, sailed by Arabs in the eastern Mediterranean. European shipbuilders began to emulate the sail, which allowed sailing much more directly into the wind.

One of the greatest problems of sailing vessels was the necessity to tack back and forth when the wind came from the direction toward which the sailor hoped to travel. With a square-rigged sail, such as mounted on the Portuguese *barca,* the navigator would sail at a very wide angle, back and forth, beating into the wind. With the square-rigged barca the ship could not approach closer than 67 degrees to the direction of the wind. With the lateen rig, even on one mast, the sailor could set his course more directly in the direction of the wind, getting within 55 degrees of the wind. While seemingly not a great difference, the closer angle would cut down the number of tack legs in travel, with the lateen-rigged caravel making 3 tacks for every 5 of the square-rigged barca. This simple change was at the heart of 15th-century advances of Portuguese sailors in exploring the coast of Africa. Some historians have credited the adoption of the lateen sail with stimulating or at least facilitating European maritime exploration. The stern rudder, developed in about A.D. 1200, also aided sailing more directly into the wind.

lead pencil

The first record of a lead pencil is found in an encyclopedic work by the German Swiss botanist, physician, and zoologist Konrad van Gesner (1516–1565), published in 1565. Gesner described a writing implement consisting of a piece of lead held inside a wooden casing. Although Gesner described the device, its inventor is unknown. References to such pencils continued to be found sporadically over the next centuries. In the 1760s, the Faber company established a factory in Nuremberg for their production, and in 1795, N. J. Conte developed a pencil made of ground graphite that was shaped into sticks that were kiln-baked.

lenses

Although the magnifying properties of glass balls, some jewels, and water-filled globes were known in antiquity, the first mention of lenses specially prepared for magnification occurs in the 11th-century writings of the Arabian scientist Alhazen. Some other Arabic sources mention the magnifying properties of glass in the 12th and 13th centuries. However, the first widespread practical application of lenses came with the still-mysterious invention of **eyeglasses** late in the 13th century, apparently in or near Venice. The fact that Venice was a center for the manufacture of glassware probably contributed to the first development of lenses in that region.

Later uses in **microscopes [III]**, **telescopes [III]**, and other optical instruments sprang from the uses in eyeglasses. The first telescopes were made by spectaclemakers on a trial-and-error basis. It is a testimony to the fact that technology often precedes science that eyeglasses and telescopes were made and used for decades before the science of optics began to offer explanations for how they achieved their effects. Newton's work on the **spectrum of light [III]** and on other aspects of optics came nearly 400 years after the first eyeglasses were made.

longbow

The longbow was developed in Wales as a hunting weapon and was adopted by King Edward III of Britain as a military weapon in 1337, as he prepared for the invasion of France. Generally about 6 feet long, the longbow was typically made from yew wood that had been carefully tapered. The cord was usually hemp, whipped, or wrapped with linen. An arrow shot by a longbow could pierce **chain mail** at 600 feet and could bring down a horse. In addition, it could be fired much more rapidly than the **crossbow [I]**. The tension on the longbow's line was about 100 pounds, and the use of the bow, a highly accurate weapon, required stamina and extensive training.

An army of longbow-equipped British soldiers defeated Flemish crossbowmen in 1337, and more famously, at the Battle of Crécy on August 25, 1346. In that battle, 11,000 longbowmen, supported by 8,900 other soldiers, defeated a French force of 38,000 men. After at least 16 charges by French knights, about 1,540 of them were killed, together with some 10,000 foot soldiers. The British lost about 100 men.

Later battles at Poitiers (1356) and Agincourt (1415) brought the age of knights and chivalry to an end. The longbow, more than firearms powered with **gunpowder,** was responsible for the change.

matchlock

In the development of handheld firearms, the matchlock, introduced in the 15th century, was one of the first methods for igniting the powder in such weapons. An S-shaped clamp known as a serpentine would be used to hold a burning wick. The user would squeeze a trigger, lowering the serpentine and its burning match into a pan of priming powder attached to the side of the gun barrel. The flash in the pan would penetrate through a minute hole in the breech to ignite the powder inside the barrel. Such an arrangement allowed the user to steady the weapon with both hands rather than using one hand solely to ignite the powder.

Varieties of matchlock weapons included the arquebus, the hachbuss, and the **musket.** Even after the invention of the **flintlock,** the matchlock remained in use, despite disadvantages. A glowing match would give away the position of troops at night, and matches were difficult to keep burning in rainy or damp conditions. However, since the matchlock did not require a snapping motion, it may have been easier to hold the weapon steady while aiming and firing.

medicine

It is difficult to claim that modern medicine was either invented or discovered, but a remarkable advance in the direction of a scientific approach to curing human ills came with the career of Paracelsus, the professional name of Aureolus Phillipus Theophrastus Bombastus von Hohenheim (1493–1541). The technical name for his approach, that of using minerals and chemicals to cure specific ailments, was iatrochemistry. He introduced the use of the painkiller laudanum and rejected accumulated lore in favor of proven cures based on experience. Laudanum was a tincture or solution of opium in alcohol and, in small doses, was extremely effective against a variety of pains. To the extent that medicine and pharmacy have moved in the direction of a scientific approach in which specific cures for separate ailments are sought, the fields owe a debt to Paracelsus.

Born in Switzerland, Paracelsus served as a military surgeon in Venice and the Netherlands and traveled widely though Europe. After treating successfully such famous individuals as Erasmus and the publisher Johann Froben, Paracelsus briefly became a professor at the University of Basel. There he used his teaching post to scandalize the medical community in several ways. Not only did he reject traditional teachings, but he also taught in the Swiss dialect of German rather than in Latin. He was self-taught and did not hold a medical degree. His lack of formal credentials, his irreverent attitude toward the medical establishment, his revolutionary suggestions that folk medicine be closely studied and sometimes emulated, and his use of the vernacular and commonly understood language rather than Latin all combined to stir hostility to his teachings. After the death of his sponsor, Froben, opposition to Paracelsus mounted, and he was dismissed from the university. He went on to Nuremberg, where he criticized the standard treatment of syphilis, which combined doses of mercury with a new drug from the Western Hemisphere, guaiacum.

After dismissal from the university, Paracelsus continued to practice medicine and to criticize the existing theory of disease, which traced ailments to various combinations of the "four humors": sanguine, phlegmatic, choleric, and melancholic. According to the traditional approach, each individual had a temperature and a temperament that resulted from a unique combination of the four humors, and treatments therefore had to be unique to each individual. By contrast, Paracelsus argued that there were common diseases and conditions that could be treated in most individuals with specific chemicals. He believed that the specifics would often derive from a plant or a mineral that had some

> Doctors of medicine should consider better what they plainly see, that for instance an unlettered peasant heals more than all of them with all their books and red gowns. And if those gentlemen in their red caps were to hear what was the cause, they would sit in a sack full of ashes as they did in Nineveh.
>
> Paracelsus, in *The Hermetic and Alchemical Writings of Aureolus Phillipus Theophrastus Bombast of Hohenheim, Called Paracelsus the Great,* translated by A. E. Waite (London: J. Elliott and Co., 1894)

characteristic that resembled the disease and that each disease had a specific cause. He argued that alchemists should turn their attention away from the search for a single philosopher's stone or elixir to identifying minerals and chemicals that would serve to cure specific diseases. He believed that folk remedies should be studied closely, that professionals should study the simple cures that worked for such untrained curers, and that medicines derived from the folk cures should be adopted. Most of his writings were suppressed during his lifetime, but admirers preserved his works and began to publish them in the 1560s, some twenty years after his death.

Mercator projection

The Mercator projection or map, one of the great inventions coming out of the age of exploration, was first published in 1569. Until the world explorations by Europeans in the 15th and 16th centuries, maps of the world circulated in the European world derived from two traditions. One was based on a theological interpretation of the Bible, while the other derived from a compilation by the 2nd century A.D. geographer and astronomer Ptolemy.

The Christian map based on the Bible depicted the world as consisting of three continents, surrounded by the "World Ocean." The three continents were shaped as a round disk, divided by a "T." With Asia and the East situated at the top of the map above the crossbar of the T, Europe was to the left of the center vertical line, and Africa to the right. The vertical division was the Mediterranean Sea, and the crossbar at the top was supposed to be the Danube and the Nile, in a continuous line that separated Asia from Europe and Africa. The names "Asia" and "Africa" were both derived from biblical references, while "Europe" was drawn from classical mythology. With the map showing the East at the top, it was properly "oriented" in the conception of the time, accounting for Western Europe regarding Asia as "the Orient."

Ptolemy had devised a map with reference lines now known as latitude and longitude, and his was an attempt to compile actual discoveries and distances rather than to derive a picture of the world from Scripture. In 1400 a copy of Ptolemy's geography, written in Greek, was brought from Constantinople to Florence, Italy, where it was translated into Latin. The maps associated with Ptolemy's geographic work appeared to be a compilation of some 27 versions authored over the centuries by various scholars attempting to apply his concepts. The Ptolemaic maps included many erroneous ideas, including the concept that Africa was joined to an Antarctic continent, which would have prohibited sailing around Africa to Asia.

However, Arab geographers had been revising Ptolemy, and one Al-Biruni (973–1050) combined precise observation and detailed information from travelers' and explorers' accounts. Al-Biruni correctly believed that the Indian Ocean and the Atlantic Ocean were joined, a fact proven by Portuguese explorer Vasco da Gama in 1497.

Over the next century, mapmakers struggled to improve on the Ptolemaic maps by adding new findings by the Portuguese, Italian, and Spanish explorers who sailed across the Atlantic and around Africa. However, mapmakers faced a problem as they tried to represent the new lands and make practical sailing charts. The round Earth was not easily represented on a flat piece of paper. As sailors tried to align their ships with compass directions, they needed maps that kept north-south longitude lines at right angles to east-west latitude lines. Gerardus Mercator (1512–1594) provided the design solution.

By carefully "stretching" out the representation of the rounded surface of the planet into a rectangle, he arranged his 1569 map of the world so that a navigator could chart a course with a straight path, crossing each longitude at the same angle. Although his map would distort the size of islands and continents at the extreme northern and southern latitudes, it represented every intersection of latitude and longitude as a 90-degree angle, greatly simplifying the problems of navigation.

A refugee from religious persecution, Mercator was born in Flanders and had studied at the University of Louvain. He settled in the Prussian city of Duisburg on the Rhine and worked as the cosmographer to the duke of Cleves. Mercator borrowed a name for the new continents from a previous mapmaker, Martin Waldseemüller, who had decided that "America" would be a suitable term, based on the travels of Amerigo Vespucci. Mercator's 1569 map was the first to use both the terms "North America" and "South America."

Mercator worked with another refugee from religious persecution, Abraham Ortelius (1527–1598), who was the first to publish an **atlas.**

movable type

The invention of the printing press using movable type is attributed to Johannes Gutenberg (1400–1468), in the decade 1440–1450. However, in China, clay movable type had been invented as early as 1045, and lead type was used there prior to Gutenberg. Block printing had been employed widely in Europe since the 1300s, in which a whole picture or page would be carved and then printed. **Playing cards,** for example, were first block-printed in sheets in the 1370s, with the first decks that form the basis of modern tarot decks possibly a subversive means of spreading the doctrine of the "Waldensian heresy," which celebrated a female pope and other characters on the cards. The purchaser would buy the cards on a sheet and then cut them apart, a method still employed for inexpensive playing cards in Mexico and elsewhere.

The actual invention of movable type may have occurred before 1440 in Holland, in the shop of Laurenz Janzoon, (1370–1440), who was a *coster* (or sacristan). On Janzoon's death, an apprentice could have taken a set of type to Mainz, where Gutenberg began to use the system to print Bibles. Evidence that Janzoon had a less-well-developed form of movable type is denied by the advocates of Gutenberg's priority of invention, who claim that Janzoon (if he existed at all) produced only block-printed texts rather than text built letter by letter.

Gutenberg's press used a separate cast type for each letter. Workers inserted them in trays and then used a machine to impress a page of print on **paper.** The press itself derived from already existing machines for pressing grapes or driving the whey from cheese and relied on a screw drive turned by levers that transformed musclepower into great pressure. The tray of type would be moistened with ink from leather balls, and then a sheet of paper was pressed to the type on a metal plate or platen, squeezed tight by the screw-jack mechanism over the paper. Gutenberg's first press could produce several hundred sheets a day, far outstripping the rate of hand copyists, each of whom might complete only a few sheets per day of unembellished text.

The printing press revolutionized literacy in Europe. Fifty years after its introduction, more than 1 million books had been printed, and the press was used for a wide variety of other purposes. Handbills, posters, business forms, and even "indulgences" (documents

> The printed page increased the safety and permanence of the written record by manifolding it, extended the range of communication and economized on time and effort. So print speedily became the new medium of intercourse.... More than any other device, the printed book released people from the domination of the immediate and the local.
>   Lewis Mumford, *Technics and Civilization* (New York: Harcourt, Brace, 1934)

produced by the Roman Catholic Church offering forgiveness of sins) were all widely printed. The printing of indulgences was one of the complaints brought by Martin Luther (1483–1546) against the clergy that led to the Protestant Reformation.

Books themselves created an intellectual revolution. With the printing of books in the vernacular or local languages of Europe rather than in Latin, the literate class rapidly grew from fewer than 5 percent to more than 10 percent of the population, facilitating the spread of new ideas and information in the 16th and 17th centuries. Historians of the period attribute much of the ferment of ideas that constituted the Scientific Revolution to the spread of the book, made possible by movable type.

musical notation

There is some evidence to suggest that the ancient Egyptians practiced some form of musical notation, but modern musical notation can be traced to the 9th to the 11th centuries A.D. in Europe. As music became more complex in Europe with the development of choir singing in the polyphonic style, musicians sought some form of permanently noting their compositions for future use.

The first notation took the form of marks known as neumes. The neumes appear to have represented, like an accent mark, the rising and falling of the hand of a choir conductor. From the 9th to the 11th century, this notation became somewhat standardized. In the mid-11th century, Guido of Arezzo developed a grid or staff on which the neumes could be placed, and the first use of different-size neumes to represent the duration of a note came in at the same time. Thus not only could the pitch of a melody be shown, but its rhythm as well. By the end of the 16th century, musical notation had developed nearly all of the elements used in the following centuries.

musket

The invention of the musket is difficult to date, but early references indicate that it was in use by 1528 and widely introduced in Spain in 1550. The musket is distinguished from other early handheld firearms in that, unlike the arquebus, it could be fired from the shoulder without a supporting stand. With the introduction of the **flintlock,** the musket became widely used in warfare. A single soldier could carry, maintain, load, and fire the weapon.

The musket was a muzzle-loaded weapon that took several steps to prepare for firing. A measured charge of **gunpowder** would be poured through the muzzle, followed by a lead ball and a wad of paper to hold the charge and shot in place and to provide some compression. The weapon had a smooth bore, not a rifled one, meaning that no spin was imparted to the shot, greatly hampering accuracy.

The musket was not only inaccurate in aim but also had very limited range. Thus it could best be employed by a massed group of men marching together and firing in sequence at a fairly close distance from the enemy position. The front rank of a troop would fire, kneel, and reload, allowing the second rank to fire. By alternating two or three disciplined ranks in firing and reloading, a fairly constant and massed fire against an enemy position could be achieved, but only after the group had marched to within 100 yards or less of the enemy position. Thus the muzzle-loading musket was responsible for the development of the close-order drill characteristic of 18th- and 19th-century armies and the nature of land warfare in that era in which masses of disciplined ranks of troops would engage in close and deadly exchanges. The requisite ability to march in unison is still characteristic of demonstrations of military discipline.

Although individually inaccurate, the mass of fire from a rank of musket-equipped troops was devastating at a range of 50 to 100 yards. Thus the spectacle of warfare with musket-armed troops, with opposing forces of well-aligned soldiers marching directly in formation against each other, derived from the nature of the weapon itself. It was only the introduction of **rifled gun barrel [III]** weapons, which allowed for much greater accuracy, that permitted the use of skirmishing tactics rather than frontal assaults. However, the transition was gradual, with many engagements of the American Civil War (1861–1865) representing a combination of the two tactics and resulting in devastating casualties.

oil painting

Oil painting was a development of the late Middle Ages and flourished over later centuries. The first known oil painting has been dated to the 11th century, but oil paintings were relatively rare until the 15th century. In that period, pure egg yolks were used to provide the medium to carry color pigments, in a method known as tempera painting. With the livelier art of the Renaissance, improvements in the quality of linseed oil and the availability of solvents such as turpentine allowed for the use of oil, first as a varnish.

Tempera paintings were fixed with an oil varnish glaze, with the Flemish painter Jan van Eyck using this technique in the mid-15th century. By the mid-16th century, painters in Venice had adopted oil as the main medium, using layers of oil glazes on linen canvas rather than on wood panels.

Oil on canvas became the preferred method in the late Renaissance and Baroque periods. In the 17th century, the "great masters," including the Flemish Peter Paul Rubens (1577–1640), the Spaniard Diego Velazquez (1599–1660), and the Dutch Harmens van Rijn Rembrandt (1606–1669) all worked in oil on canvas. Rembrandt developed methods of showing light and dark shadows through layering and glazing that gave extraordinary excitement and depth to his work. In the 18th and 19th centuries, painters would often coat the canvas with a dark gray or black undercoat of oil paint before starting the image.

Another development in the 19th century was the production of prepigmented oil marketed in small tin tubes. With a supply of colors, a portable easel, and a prepared canvas, the artist could move out of the studio and paint in the outdoors. This technical development contributed to the bright external landscapes of the Impressionist school of the late 1870s and early 1880s, with varied techniques of brushwork and vivid colors reflecting the outdoor settings. The school of Impressionists was so called after the 1872 work by Claude Monet (1840–1926) called *Impressions, Sunrise*.

paddlewheel boat

The successful development of paddlewheels for boats had to await the invention of the **steam engine [III]** and the **steamboat [IV]**. However, the Chinese had developed treadmill-powered paddlewheel boats by the 6th century A.D., and by A.D. 1135, during a protracted civil war in China, the opposing sides developed hundreds of treadmill-powered boats, powered with numerous wheels; each boat could carry up to 300 men. In Barcelona there is a record in 1543 of small tugs powered by large groups of men walking a treadmill.

With the development of the **crank** in about 1420, several inventors and designers realized that a boat could be effectively rowed by paddles attached to wheels and turned by cranks instead of the more exhausting technique of forward and backward pulling of oars or walking in a treadmill. In 1430 a German military engineer described a hand-crank-powered paddlewheel boat. In 1472 Robert Valturio, the private secretary to the Italian Renaissance prince Sigismondo Malatesta of Rimini, published a work on the arts of war, *De re militari,* illustrating such a boat with five pairs of paddlewheels joined with a connecting rod to a single crank system. Valturio's work, first completed in manuscript between 1455 and 1460, was widely popular, and a copy was owned by Leonardo da Vinci (1452–1519), who applied the hand-crank method of propulsion to a projected invention of an armored land vehicle, sometimes seen as a predecessor to the **tank [V]**.

paper

In the ancient world, scribes wrote on cured animal skins, either parchment or vellum. Parchment was made from the skin of a sheep or goat, while vellum was made from younger lambs, kids, or calves, with a finer-grained surface. The Egyptians developed papyrus, a hard, brittle form of paper made from the pith of papyrus reeds. However, none of these types of material was particularly suitable for receiving printed ink.

The development of the printing press with **movable type** would have been of little use without the prior introduction into Europe of linen-based paper, first invented in China in about A.D. 100. The use of cloth-fiber-based paper spread fairly rapidly in China, even to the outmost provinces, such as Turkestan. However, the secret of paper manufacture did not begin to spread beyond the Chinese region until the 7th and 8th centuries A.D. The dispersal of paper has been very precisely traced: Samarkand, 751; Baghdad, 793; Egypt, 900; Morocco, 1100; and Italy, 1276. Paper was used in Cologne in 1320 and in Nuremberg by 1391. Thus when the printing press was perfected in the early and mid-1400s, linen-based paper was in wide use, and the block printing of paper and heavier, cardboardlike paper for **playing cards** was already widespread in Europe. Like several other major inventions of the period, paper was first made in China, then found rapid application in the bustling culture of Europe.

playing cards

The origin of playing cards is surrounded with legends and a few scraps of documentation. It appears that before the use of cards for games, they were introduced for fortune-telling. The 78-card tarot deck was in use in the late 14th century, with the first references prohibiting the use of playing cards showing up from 1377 to 1384. The tarot deck probably originated in northern Italy. The 78-card tarot deck was known as the Venetian or Piedmontese deck, to differentiate it from the 62-card Bolognese deck. The origin of the name *tarroco* or *tarot* is not known, but one theory holds that the deck may have come from the region of the Taro River, a tributary of the Po in northern Italy.

It is possible that the 22 tarot trumps that are used in fortune-telling evolved separately from the 4 suit decks, which originally consisted of 10 cards with markers on them and 4 face cards for each suit (jack, knave, queen, and king), for a total of 56 cards. The original suits were swords (or *spada,* from which the term *spades* derived), coins, clubs (which in the original design looked like heavy wooden clubs), and cups. Between 1380 and 1400, a great variety of games evolved.

The 22 tarot trumps, or "major arcana," have a series of legends surrounding them. One speculation is that they represented a subversive means of distributing information about the Waldensian cult or heresy founded by Peter Waldo, a merchant of Lyons, in 1170. Waldensians believed in a female pope who, according to legend, reigned from 854 to 856, between the reigns of Pope Leo IV and Benedict III. The view that the cards were part of the Waldensian heresy is supported by the existence of a "papesse" card among the trumps of the tarot deck. Furthermore, the Waldensians had their greatest number of adherents in the Piedmont region of Italy, in the very region in which the tarot deck first appeared. However, there are many other unexplained images among the tarot arcana, and the presence of the female pope could simply represent an attempt to represent a persistent but obscure folk legend.

The development of double-headed face cards and the substitution of diamonds and hearts for coins and cups, respectively, took place gradually over the next 300 years. The 52-card deck familiar in Britain, the United States, and northern Europe in the modern period evolved in France in the 17th century.

Playing cards and fortune-telling cards developed together. They created a demand for **paper,** and their printing stimulated the development of block printing, which was a precursor to the development of **movable type** and the printing press of the late 15th century. In this regard, playing cards may represent a more significant contribution to technical progress than one might assume from their apparently incidental use in entertainment and fortune-telling.

plow

The scratch plow was essentially a large digging stick dragged through the earth by a pair of oxen. Developed in the ancient world, it had a conical or triangular shape that did not normally turn the soil but simply created separate furrows. To properly churn up the soil required cross-plowing, leading to fields in the Mediterranean region that were roughly square in shape. However, this form of plow and cross-plowing were not suited to agriculture in northern Europe, where a combination of weather and soil made it inappropriate. With wet and rainy summers and much heavier soil, the scratch plow simply did not work well in lowlands, where the soil offered too much resistance.

Some of the first references to a heavier plow occur in the first century A.D., one pulled by 8 oxen and wheeled. It became common in northern Europe in the Middle Ages, diffusing widely in the period from A.D. 500 through 1000. The heavy plow itself was an innovation that transformed medieval Europe. It consisted of three parts: a coulter or knife-like blade for cutting into the sod; a plowshare or horizontal blade, which would cut the soil at the level of the grass roots; and a moldboard, which would turn the slice of turf and soil to the side. This heavy plow increased the amount of land a peasant could till. It also tended to change the layout of fields, since no cross-plowing was necessary. It was more efficient to proceed with a long furrow to a natural obstacle (such as a ditch or a rock outcropping) and then to turn and return in parallel to the first furrow. The heavy plow also allowed the use of rich bottomlands. Although very hard to track, historians have assumed that the heavy plow came from Germany to Britain in the 5th century, leading to the abandonment there of prior upland cross-plowed "Celtic fields."

In addition to revolutionizing the shape of fields, the heavy plow required some form of organization to share in the use of oxen. Since 8 oxen would be required to pull the plow, various peasant cooperatives and communes developed to share the animals and the equipment. It has been argued that the requirement for shared ownership of draft animals and plow led to the establishment of the manorial system. The correlation between the technology of the plow and the social arrangement of the manor, suggested by historians early in the 20th century, has since been modified as more evidences of complications and variations have been uncovered. It has even been suggested that the causation flowed the other way—that is, that long fields, brought about by division of land through inheritance, led to the introduction of a plow that would be more appropriate to them. The long-strip sections of land still represent the pattern in much of Europe, in French Canada, and elsewhere in the world where the ownership pattern reflected the plowing technique or the plowing technique reflected the ownership.

It has been suggested that the heavy plow represented a change in the relationship of the farmer to his land. Before the heavy plow, the peasant tilled a small patch of land by himself and with his own oxen and tools. Afterward he worked in a cooperative structure, more devoted to exploiting nature than in living by the demands it imposed. This change in relationship foreshadowed and paralleled the exploitative outlook taken in the West toward the environment over the next centuries, but it is difficult to assert that the change in plowing technique was the definitive cause of such an exploitative perspective.

reed instruments

The origin of reed instruments has been traced to Southeast Asia. In its simplest form, a reed effect was achieved by holding a blade of grass or other fiber quite taut and blowing on it. In some primitive Southeast Asian reed instruments, one end of a reed was cut from its frame, leaving the other attached. When not enclosed in a tube, such an instrument would be known as an "aerophone."

Reed instruments enclosed in pipes were introduced to Europe during the late Middle Ages, but the exact route by which they came from the East is not documented. An early form of oboe known as a shawm was mentioned in records from 12th-century Sicily. In the 1600s, experiments with attaching a reed mounted in a mouthpiece attached to a wooden tube led to the clarinet. The invention of the clarinet is attributed to J. C. Denner (1655–1707), a German musician.

Several other reed instruments were made in the period, including the crumhorn, which had two reeds in an interior chamber into which the musician blew. Together with the oboe and the simpler flute, such instruments are known as woodwind instruments. The saxophone was invented by Adolphe Sax in 1846.

The ancient Greek *aulos* consisted of two pipes with a reed strapped in place and was played on both the intake and the outflow of air from the musician's mouth. The bagpipe used a similar principle, with a bag designed to hold a reservoir of air, reducing the strain on the player.

sailing vessels

Using the wind to propel ships was known in ancient times, but most maritime commerce [I] in the ancient world was conducted by ships

The noble spirit of [Prince Henry the Navigator] was ever urging him both to begin to carry out very great deeds. . . . He had also a wish to know the land that lay beyond the isles of Canary and that Cape called Bojador, for up to his time, neither by writings, nor by the memory of man, was known with any certainty the nature of the land beyond that Cape . . . it seemed to him that if he or some other lord did not endeavor to gain that knowledge, no mariners or merchants would ever dare to attempt it, for it is clear that none of them ever trouble themselves to sail to a place where there is not a sure and certain hope of profit.

—Gomes Eanes de Zurara,
*The Chronicals of the Discovery
and Conquest of Guinea*
(London: Hakluyt Society, 1896)

propelled by oars, such as galleys, biremes with two rows of oars on each side, triremes with three rows, and fast Byzantine dromons. All of these ships relied on slaves or, in some cases, voluntary human oarsmen as the main source of power to propel the ships. The Norse used a combination of a large, square sail and manpower for propulsion, sailing onto the open sea and reaching Iceland, Greenland, and North America by A.D. 1000. Strictly speaking, the sailing vessel, unassisted by oarpower, was an invention of the late Middle Ages and Renaissance era.

The age of the sailing ship unassisted with oars began in the 12th century and lasted into the age of steam propulsion in the mid-19th century, a span of about 700 years. Several major developments between about 1100 and 1460 elevated the sailing ship to the forefront of marine transport. In about 1200, possibly in the Netherlands, sailors decided to attach the steering oar to the sternpost of the vessel rather than mounting it loosely to the side. Facing forward, the steersman had previously positioned himself on the right-hand side of the vessel, hence it was known as the "steer board" or starboard side, while the left side, without the steering oar, would be brought to dock, and hence was called the "port side." With the fixed stern rudder, the mariner was able to sail more closely into the direction of an oncoming wind than had previously been possible. Together with the **lateen sail** and the magnetic **compass** adapted from the Arab world (where the compass had come from China), European mariners now could venture farther out into the open ocean. Prior to the development of the compass and reliable maps that reflected reports of travels, rather than mythological or religious conceptions of the planet, sailors rarely ventured far beyond the sight of coastlines and frequently relied on dropping lines over the side to check the depth and to recover samples of the bottom ground to determine their location. With maps, compass, **astrolabe,** or **cross-staff,** navigation over the open ocean became more possible.

Several features of different sailboat designs were blended together by the 13th century, perhaps under the influence of the meeting of different peoples during the Crusades between 1096 and 1291. Shipbuilders began to combine the single, large sail of the Norse, the fixed rudder, and the planked-over hull of the Mediterranean carvel-built ship. Carvel-built ships were constructed by building the frame first and then covering the skeleton with planks laid flush to each other. Caravels constructed by this method carried lateen sails, and they were common in Spain and Portugal in the 13th century. The carrack of the 15th century used the carvel method of construction and had two or three masts, with large, square sails in front and a lateen sail astern.

With the lateen rig, the caravel could travel within 55 degrees of the direction of the wind. The older, square-rigged *barca* used by the Portuguese could only get to 67 degrees of the direction of the wind. The result was that the caravel would only have to tack back and forth 3 times for every 5 of the barca to achieve the same forward motion in the direction of the wind. This saving, amounting to about a third in time and distance, could literally cut off months from a year's voyage.

Carvel construction, fixed rudders, lateen sails, navigation instruments, and the magnetic compass allowed for exploration by Europeans under the coordinating leadership of Prince Henry the Navigator (1394–1460) down the coast of Africa around Cape Bojador and, by the late 15th century, exploratory crossing of the Atlantic to the lands of the New World. The impact of the geographic discoveries and the

sailing vessels The U.S.S. *Constitution* is the world's oldest commissioned warship in service. It sailed under its own power into Massachusetts Bay in 1997. *U.S. Navy*

opening of new lands upon the economies of the European, African, and American continents was profound.

Elevated structures, or "castles," at the bow and stern allowed for better fighting positions in ship-to-ship conflicts and came to be known as forecastles and sterncastles. Spanish shipwrights developed the galleon in the 15th century, an enlarged version of the carrack, and the galleon became the dominant oceangoing ship of the 16th and 17th centuries. Galleons tended to be faster, more heavily armed, and quite seaworthy.

Later refinements of sailing vessels included the 17th- and 18th-century East Indiamen, a combination cargo and fighting ship, and the graceful and fast clipper ships of the 19th century. Through the 18th century, sailing ships designed especially for naval engagements were known as ships of the line, with guns mounted to fire from the side, and classed according to the number of guns they carried. Smaller warships, with fewer than 40 guns and equipped to fight individually, were known as frigates. Sloops or corvettes were two-masted ships with less firepower than the frigates.

screw lathe

The lathe for cutting nuts, bolts, and screws developed in the 16th century A.D. in Europe, and screws were in common use in construction and furnituremaking by the end of the century. Like several other inventions of the period, the lathe for making screws can be traced to ideas developed by Leonardo da Vinci (1452–1519).

Leonardo preserved his ideas in numerous notebooks, some of which were widely consulted after his death. Projects to publish some of the notebooks and their circulation over the two centuries following his death have been closely studied, leading some scholars to conclude that inventions attributed to others derived from ideas that he developed. One that could possibly have derived from his plans is the screw-lathe device for cutting screws out of **bronze** [I] or iron, although no direct link between the circulation of his notes and the development of the screw lathe has yet been firmly established. Nuremberg screwmakers Hans and Leonard Danner, working in the mid-16th century, appear to have used a device that represented an improvement on Leonardo's screw lathe. Leonardo's lathe allowed for the cutting of nuts that could be closely fitted to a screw-threaded bolt.

The screw lathe was greatly improved in the late 18th century by the work of a young machine apprentice, Henry Maudslay (1771–1831).

Working with Joseph Bramah (1748–1814), who had developed an improved **lock [III]**, Maudslay improved a lathe in Bramah's shop in 1794. Maudslay developed a slide rest that allowed the cutting edge to be very closely adjusted and fixed in position. Screws, bolts, and nuts could now be made with much greater precision. By 1797 he made a number of further improvements on the cutting machine. Between 1800 and 1810, in his own shop, Maudslay greatly increased the accuracy and precision of screwmaking. The development of the machine tool industry in the 19th century has been traced to Maudslay's screw lathe and its improvement over the original Leonardo design. The techniques were emulated in the United States at federal armories to move toward the achievement of **interchangeable parts [IV]**.

The use of screws, screwdrivers, nuts, bolts, and screw shafts on presses and other machinery, while possible without screw lathes, would represent "one-off" handcraft construction. With the lathe it was possible to turn out a standard-size screw and, even more importantly, a standard-size nut and bolt. In the latter case, a nut of a standard size could be attached to any of the standard bolts cut to the appropriate dimensions. In this sense, interchangeable parts on the level of interchangeable nuts and bolts were available in the 17th and 18th centuries, well before the introduction of more complicated and smaller interchangeable parts for guns and later machinery in the 19th century.

ship's log

Like the **screw lathe,** the ship's log for measuring speed through the water was notionally developed by Leonardo da Vinci (1452–1519) in his notebooks. The diffusion of the method has not been traced, but by the mid-16th century it was in common usage. It consisted of a small line, knotted at fixed intervals, to which a light piece of wood was attached.

The line was rolled onto a large spindle (resembling a wooden log), and the light wooden float at the end of the line would be cast into the water. By counting the number of knots that passed through the operator's fingers in a fixed period of time, the speed through the water could be calculated. To convert the measurement into nautical miles, the knots would be placed 47 feet, 3 inches apart, and the float would be timed for 28 seconds with a small sandglass. If 6 knots passed, the ship was sailing at a speed of 6 nautical miles per hour, or 6 knots.

The mathematics that led to the result was quite simple. There are 3,600 seconds in an hour (60 × 60). If one divides 3,600 by 28, the result is 128.5714. If one multiplies 128.5714 by 47.25 (the distance in feet between knots) the product is 6075, meaning that if 1 knot passes in 28 seconds, the ship is moving at 6,075 feet per hour, or almost precisely 1 nautical mile (6,080 feet) per hour. If 6 knots pass in that time, one is moving through the water at 6 nautical miles or 6 knots.

The results would be recorded in a logbook. By constantly checking the speed through the water, it would be possible to estimate progress across the trackless ocean. Together with other aides to navigation, including the **compass**, the **sextant [III]**, and finally the **chronometer [III]**, the ship's log made possible the full exploration of the planet by sea.

shorthand

Various systems of shorthand or abbreviated writing have been invented since antiquity. Most of them involved the use of numerous symbols and considerable complexity. So, while using such systems was faster than longhand writing, they were difficult to learn and hard to interpret, and most of them did not survive for very long.

In 1588 a British clergyman, Timothy Bright (1551–1615), developed a system, one of the first in the English language, known as "Characterie." The British lord of the Admiralty and famous diarist Samuel Pepys (1633–1703) used a system of shorthand that was somewhat secret, which may have been due to the fact that his diary included details of his illicit romantic affairs and other episodes that he probably did not care to have revealed publicly. William Tiffin introduced a system in the mid-1700s that used strokes rather than letters and that was faster than earlier systems. Isaac Pitman, who published *Stenographic Shorthand,* improved Tiffin's system in 1837. The Pitman system, based on a phonetic system, emerged as dominant over the next decade.

An American shorthand system, developed by John Gregg (1867–1948) in 1888, became the dominant method in the United States. In 1948 the Gregg system was reduced to 184 symbols. In both the Gregg and Pitman methods, an accomplished shorthand writer can take notes faster than anyone can talk, and both have survived the development of stenographic machines and **tape recordings [V]**.

siege weapons

During the Middle Ages, engineers improved on weapons designed to hurl missiles, usually heavy stones, large arrows, or pitch- or fat-impregnated balls that were ignited before launching, to attack well-fortified positions such as castles or walled cities. The ancient Greeks and Romans had used the ballista, a catapult that used twisted skins or ropes to provide tension. The ballista was found in Greece as early as the 3rd century B.C., and it is assumed that engineers working for Philip II of Macedonia invented that weapon. The ballista consisted of two groups of ropes that were twisted over and over again by wooden arms inserted horizontally into the ropes. When the twisted cords were released, the cords would spring forward, hurling a spear or other missile forward. The Romans used a variation on the ballista, known as a mangonel, that flung a single arm forward until it hit a stop, flinging the missile from a pocket on the end of the arm.

The trebuchet, like the mangonel, had a single long arm but did not rely on tension from twisting, deriving its energy from a seesaw principle. On one end of the throwing arm, a heavy weight held down the arm, while on the other end, a sling held the missile. Several men would pull down the arm at the missile end, against the leverage of the weight at the other end. When released, the weight would pull down the arm, which would fling up the other end with the sling, launching a heavy missile. When properly aligned and aimed, the 60- to 200-pound rock would smash the defenders' masonry walls. Still another type of siege machine was the siege tower, basically a movable tower with several floors connected by ladders. The front of the tower was armored against arrows, and it would be wheeled up to a wall. Shooting arrows or **crossbow** bolts from slots in the tower against defenders on the city or fortress wall, the attackers would move the tower close to the wall, then lower a section of the tower protective front to provide a bridge over which they would run to attack.

Although the ballista and mangonel were developed in the ancient world, the development of several types of trebuchet and siege towers characterized warfare of the later Middle Ages in Europe, and siege weapons are often associated with the period 700 to 1200, before the introduction of **gunpowder** and **cannons** in Europe.

spinning wheel

To make smooth and even textiles, thread must be spun from fiber. Although thread can be spun by hand, it is a slow and laborious process, made much easier and productive by the invention of the spinning wheel, which allows the worker to wind the completed thread regularly on a spindle or quill. The spindle is necessary to mount in a shuttle to be run back and forth through the mounted alternating threads on a loom. While records of the spinning wheel appear in Europe in A.D. 1298 and later, little is known of its origin. It is presumed to have been developed in China and imported to Europe, perhaps during Roman times from the Middle East. A very complete drawing of the spinning wheel appears in a text written in 1480.

stirrup

Although fighting from horseback was common in the ancient world, the device that allowed mounted lancemen to serve as shock troops was the stirrup, not introduced into Europe until about A.D. 800. A simple device, the stirrup was best made of **bronze [I]**, iron, or **steel [IV]** to allow the foot to be firmly held and braced in place on horseback. At the same time, a metal stirrup allowed the foot to fall free in case the horse should stumble or be wounded in warfare. Without stirrups, a mounted soldier could be defeated by foot soldiers; with stirrups, the mounted warrior could use the power of the horse itself, transmitted through the lance, to run down the enemy, either mounted or afoot. The stirrup replaced human musclepower with horsepower behind the weapon, guided by human hand. The introduction of the stirrup has been credited with changing the nature of warfare and European society itself.

The origins of the stirrup have been a subject of intense historical research. Written records show the stirrup in China as early as A.D. 477, while the earliest trustworthy depictions of stirrups occur in Chinese art from A.D. 523 to 554. There is evidence of the rapid spread of the stirrup to Korea, Japan, and India in the 5th and 6th centuries. It was clearly in use in Persia (Iran) by A.D. 700. From there it appears to have spread through the Muslim world. Historians have disputed whether the Magyars (Hungarians) or Bulgars first introduced the stirrup into Europe, dating its

stirrup The simple device of the stirrup revolutionized warfare by allowing the rider to brace himself on the horse and to transfer the forward thrust of both horse and rider to a handheld lance or other weapon. Modern saddles like this one incorporate stirrups little changed since the late Middle Ages. *US Geological Survey*

first arrival in the late 7th or early 8th centuries. One clearly dated depiction of horsemen with stirrups in Europe has been found in panels made in about 840 for a church in Milan, Italy. The manner of the depiction suggests that the stirrup had become commonplace by that year.

The stirrup, apparently a simple device, had a vast impact on warfare and life in Europe. The new mode of warfare meant that heavily armored horse riders were essential to a winning side; the expense of armor and outfitting a horse stimulated a new form of society dominated by an aristocracy of warriors who could fight in the new fashion. In effect, the stirrup stimulated, if not created, the feudal aristocracy of the late Middle Ages in Europe. The Norman Conquest of Britain in 1066 was one step in the spread of a revolution in military and social organization that had been centuries in the making, promoted and made possible by the introduction of the stirrup. The stirrup in turn led to a number of developments in weapons technology, including **chain mail** and the **longbow,** each important as defensive or offensive weapons against stirrup-equipped mounted fighters.

The stirrup remains a feature of horse equipment to this day, little altered from its first introduction more than 1,500 years ago.

> The improvements introduced in the warrior's equipment from the Frankish period onwards made it more costly (and also more difficult to handle), with the result that it became less and less possible for anyone who was not a rich man—or the vassal of a rich man—to take part in this form of warfare. As the logical consequence of the adoption, about the tenth century, of the stirrup, the short spear of former days, brandished at arm's length like a javelin, was abandoned and replaced by the long and heavy lance which the warrior, in close combat, held under his armpit, and when at rest supported on the stirrup itself.
> —Marc Bloch, *Feudal Society, vol. 2, Social Classes and Political Organization.* (Chicago: University of Chicago Press, 1961)

sugar

Sugarcane cultivation and sugar refining were practiced in India, Arabia, and the Mediterranean in the Middle Ages, apparently spread from a point of origin in northeastern India through the Muslim world. By the time of the discovery of the Americas in the late 15th century, sugarcane cultivation had already spread to the Canary Islands. Columbus and other early explorers of the Caribbean recognized that the climate there was ideal for sugarcane cultivation, and Columbus brought the first sugarcane from the Canary Islands to Santo Domingo in 1493. The first sugar mill in the Western Hemisphere was built there in 1509, and by 1511, sugarcane was being harvested in Cuba. It spread rapidly to the other islands and to Mexico and Brazil. The estates granted to

Hernán Cortés in Mexico after his conquest there in 1524 were largely devoted to sugarcane production.

The sugar economy of the Caribbean stimulated the development of the African slave trade, as the indigenous populations of the islands and coastal lowlands had little resistance to diseases from the Eastern Hemisphere and because they had little cultural exposure to organized work systems, being largely hunting-and-gathering groups. The African slave trade in turn stimulated a need for products that the Europeans could barter for slaves in Africa, which included **distilled spirits**. Rum could be made from molasses, one of the by-products of the sugar refining process. Sugar played a part in other basic aspects of trade in the early modern period. As tropical products such as tea, coffee, and chocolate, all mildly habit-forming, found greater acceptance in Europe, the market for sugar to accompany such beverages and to make them more palatable increased.

Thus sugar and rum became essential parts of the triangular trades that brought millions of Africans to the New World over the 300 years from about 1520 to about 1820. Since reducing the sugarcane to raw sugar and molasses required a heating process, the island regions were soon denuded of firewood. Planters would crush the sugarcane to separate the fibers in heavy roller mills usually powered by draft animals or slaves. The crushed cane would be rinsed repeatedly with water to extract the sugar juices. After the exhaustion of firewood, planters used the fibrous remains of the canes, called bagasse, as fuel for the crude boiling pits to crystallize the raw sugar juices.

Although raising and harvesting the sugar and reducing the cane to a brown raw sugar was a relatively simple process that did not require great numbers of skilled laborers, refining to higher-quality sugar required a more industrial process in which economies of scale and improvements of machinery could yield greater profits. Hence, by the 1600s the refining process had been separated from the plantations. Raw sugar was shipped to refineries in Lisbon, Marseilles, London, and Amsterdam, and later to New York and other cities in colonial North America.

Some economic historians have traced the industrialization of Europe and the northern American colonies to their place in the sugar-induced triangular trade and the reduction of the Caribbean region to a plantation economy based on cheap labor to the same factors. When France became isolated from its sugarcane colonies following the French Revolution, the loss of Haiti (the French part of the island of Santo Domingo) to a slave-led revolution led the French to begin sugar beet production. The process of refining sugar from beets had been developed in about 1750 by German chemist Andreas Margraf.

A small sugarcane plantation economy developed in Louisiana Territory between 1750 and 1800 and remained in place after the purchase of that territory by the United States in 1803.

theodolite

A surveying instrument, the theodolite consists of a **telescope [III]** mounted on a frame with a **compass** to determine direction, a spirit level to establish the horizontal, and a marked quadrant to measure angles above and below horizontal. A theodolite mounted on a tripod with a hanging plumb bob beneath it to identify the precise location of the measuring point is known as a transit. The exact inventor of the theodolite is not known, although in 1571 an English mathematician, Leonard Digges, gave the existing instrument that name in a work, *Pantometria,* published by his son Thomas Digges. A similar device had been illustrated by woodcut in 1512, where it was called a *polimetrum.*

Improvements to the theodolite included a precise **vernier [III]** scale invented by Pierre Vernier in 1631; spirit levels, introduced in about 1700; a telescopic sight and level, by Jonathan Sisson in about 1730; and a precise instrument developed by Jesse Ramsden in 1787 and used in the British ordnance survey. Ramsden's theodolites had micrometers for readings down to 5 seconds of arc. In addition, Ramsden's theodolite was highly portable yet stable and greatly advanced the art of surveying. It is difficult to determine whether the need for more accurate land surveying led to the invention of the device or whether the device contributed to a desire for more accuracy. In any case, the theodolite and its improvements went hand in hand with more precise land records and transfers through the following centuries.

three-field system

Some historians have regarded the innovation of the three-field agricultural system in western Europe as the greatest agricultural improvement in the region in all of the Middle Ages. In one sense it can be seen as the beginning of a chain of technical, agricultural, and commercial consequences that converted European culture into a hothouse of institutional and technological change.

The three-field system developed rather suddenly, with the first certain reference to it in A.D. 763, with later references in 783, and many after 800. Some have suggested that it was thought of by Charlemagne himself. In a capitulary or set of regulations known as *De villus,* he attempted to rename the months: June was to be Plowing Month, July was Haying Month, and August was to be Harvest Month. Prior to the three-field system, plowing was done in October and November, and the harvest was collected in June or July.

Under the three-field system, the land was divided into thirds, with one section planted in the fall with winter wheat or rye. A second section would be planted in the spring with barley, peas, lentils, beans, or oats, while the third section of land would be left fallow—that is, to grow in wild weeds. The next year, the previously fallow field was put in the winter crop; the barley, peas, lentils, beans, or oats field was left fallow; and the previous winter-crop field was planted with the barley, peas, lentils, beans, or oats. Coupled with increased plowing of the fallow fields, agricultural production per working peasant increased by nearly 50 percent. The system increased the land that a peasant could handle with his labor, leading to reclamation of swamps, clearing of forests, and the creation of polders or reclaimed land by construction of dikes in locations such as Holland. Inclusion of oats in the crop-rotation system provided food for horses, which were increasingly used in agriculture in Europe, a practice made possible by the introduction of the **horse collar** and the **horseshoe.**

The three-field system apparently spread from an origin in the region between the Seine River and the Rhine River, then in France. It spread to Poland, southern Sweden, and to the Balkan region in the 13th century and to Hungary by the mid-1300s. It reached England in the mid-1100s and was taken from there to Ireland. The ancient Romans and medieval farmers alike recognized that in some fashion, the planting of legumes such as lentils and beans helped the soil. Not analyzed until centuries later, these plants helped fix nitrogen in the soil, making them a form of natural, self-fertilizing system. Their addition to the diet provided proteins, to which has been attributed a great expansion in population. With beans and peas, the diet and energy of western Europe improved, influencing the growth of **cities** [I] and commerce. With the growth of cities and many in the population freed from providing food, the crafts, arts, and sciences could begin to flourish, with specialists making thousands of improvements to tools, equipment, methods, and technique.

veinous valves

The fact that the veins of the body have valves that regulate the flow of blood to the heart was discovered in 1579 by Girolano Fabrizio

(1537–1619), whose Latinized name was Gironomo Fabricius. In the same tradition in the study of **anatomy** as Vesalius, Fabrizio studied under Fallopius, who himself was a student of Vesalius. Fabrizio followed Fallopius in the chair of medicine at the University of Padua and was in turn a teacher of William Harvey (1578–1657), who developed the understanding of the **circulation of blood** [III]. Fabrizio also described the location of the lens of the eye, worked out the anatomy of the larynx, and investigated the mechanics of respiration and the action of the muscles.

Fabrizio demonstrated the working of the veinous valves in 1579, but worked for more than two decades on his studies, not publishing his results, in *De Venarum Ostiolis,* until 1603. He also studied the late fetal stages of various animals, described the placenta, and speculated on the process of human conception.

Like others in the tradition begun by Vesalius, Fabrizio was willing to use dissection and experimentation to challenge traditions, contributing to the growth of the scientific spirit that emerged more fully in the century following his death.

watch

The invention of the watch or clock watch has been attributed to Peter Henlein, a 16th-century clockmaker who lived in Nuremberg. Little is known of his exact contribution, and the attribution is based on a statement quoted in a book published in 1511. The distinction between a clock watch and a clock is that the former would be driven by a spring and could be moved from place to place or even carried on one's person, while the latter was driven by weights and was fixed in location.

The first clock watches were usually spherical and exceeded 3 inches in diameter. Often ornate, they may have been carried more as an item of ostentation than as true timepieces, although modern research indicates that they may have been generally accurate to within 15 minutes of the true time, far more accurate than could be achieved with sundials. The faces of the earliest clock watches were covered with a metal lid, as glass crystals did not become common until after 1610.

A problem with the spring-driven watch was that the power of the spring would lessen as the watch wound down, requiring a system for regulating the power of the spring. Between 1525 and 1540 a Swiss watch mechanic, Jacob Zech, resident in Prague, developed a controlling device known as a fusee (although apparently Leonardo da Vinci had anticipated the invention of the fusee in one of his notebooks). The fusee allowed the diminishing power of the spring to be applied with greater leverage as the watch wound down, by winding a cord of catgut or a minute chain around a spindle of diminishing diameter.

In the early 16th century the clock watch evolved in two directions. Smaller designs, which came to be known as Nuremberg eggs, were originally oval in form and were carried not in a pocket but on a display chain. In the other direction, the clock watch was made larger for setting on a mantel or table and often included a bell or chime to strike the hour and displays of astronomical information. The development of the **balance spring [III]** in the 17th century represented the next notable improvement to the early watches and clocks.

waterwheel

The use of waterwheels for power can be traced back to Roman times, often coupling a bridge over a stream with waterwheels under the bridge and mills above. In 18 B.C. the Roman author Strabo mentioned a water-driven grain mill that had been completed some 45 years previously. Vitruvius, assumed to have lived in the 1st century B.C., spelled out instructions for a vertical undershot waterwheel geared to connect to a vertical axle driving a millstone. Such mills were called Vetruvian wheels. Some scholars have guessed that Vetruvius simply noted an existing machine, possibly imported between Rome and China.

In A.D. 537, Goths besieging Rome cut the **aqueducts [I]** to the city, which fed grain grinding mills. The Roman general Belisarius mounted waterwheels between moored boats, and such barge-mounted mills were found later in both the Middle East and in Europe. In general a single undershot wheel would be equivalent to about one horsepower.

A complex with 16 waterwheels was built from about A.D. 200 to 300 at Barbegal near Arles in southern France, turning 32 flour millstones and producing an estimated 30 tons of flour a day. The complex was built on a hillside; the water flowed in two sluices on each side of the mill structures, turning 8 wheels on each side of the mills. From descriptions, it has been estimated that the mill complex generated the equivalent of about 30 to 32 horsepower, the most powerful mechanical installation of any kind in the world until the development of steam engines. The mills apparently provided all the flour needed for a population of about 10,000.

There are a few indications that through the Middle Ages, engineers developed other uses for waterwheels besides grinding grain. One of the more mysterious is a record, apparently written in A.D. 369, of a water-driven saw for cutting marble on the Ruwar River, a tributary of the Moselle, although it is possible that the document dates from about A.D. 900. The fact that marble is not found in the district throws doubt on the record.

Overshot wheels, like that at Barbegal, require a system to carry the water over the top of the wheel structure. Consequently, the system has certain requirements. It must have a channel to carry the water from upstream to the mill house, which would be most conveniently located near a precipitous drop or fall in the water flow. In addition, there must be a millpond above the mill to provide a steady supply of water. A sluice system must control the flow of water so that just the correct amount fills each bucket on the wheel, as excessive water would simply be wasted, and insufficient water would not provide the optimum turning of the wheel. For all of these reasons, the overshot wheel was expensive to build. The result, however, was aesthetically interesting, with millpond, turning wheel, falling water, and a solid mill house structure. Together, the waterwheel and mill system became a common motif in later landscape art.

It appears that the introduction of waterwheels in Europe for purposes other than grinding grain and as a source of power for industrial work probably dates to between A.D. 980 and 1000.

By about A.D. 1000 there begin to appear many references to fulling mills, for processing fiber, and to waterwheels driving trip-hammers at forges in Germany and Britain. Waterwheels were so prevalent through the late Middle Ages that no accurate count of them can be made, although the Domesday Book, published in 1086, listed more than 5,624 waterwheels in 3,000 English communities. The original London Bridge, with cottages and mills lining a roadway over the Thames River, was built over a system of underbridge waterwheels; other such bridge/mill combinations could be found in France and elsewhere on the Continent. Some waterwheels were constructed to take advantage of rising and falling tides, such as one built between 1066 and 1086 at the entrance to the port of Dover. Where rivers ran too sluggishly over flat land, waterwheels could not be constructed, developing a need for the introduction of the **windmill.**

windmill

The exact location and timing of the invention of the windmill has been a matter of historical interest for decades. It is known that windmills began to appear in Europe in the 12th century A.D., in flat areas where rivers ran too slowly to turn waterwheels for mills. The invention of the windmill that rotated on an axle that was tipped above the line of the horizon to develop a turbine action in its vanes appeared to some scholars to have originated in the East, because of the prior evidence of vertical-, rather than horizontal-axle, windmills in China and Afghanistan as early as the 700s.

Some historians believed that the windmill had its origin in Tibet or China, where wind-driven prayer wheels that rotated on a vertical axis were common, themselves apparently derived from rotating bookcases that held sacred texts. In China in the 8th century A.D., windmills were apparently employed for helping to haul canal boats past canal locks but were not used for grinding grain. In Afghanistan they were used to grind flour. This pattern suggested that the wind-powered prayer wheel of Tibet may have diffused elsewhere in the East, where it was used in canal boat transport, and to the West, where it began to be used in grinding mills.

However, the first written evidence of a vertical-axle windmill, along the lines of the prayer-wheel and Afghan models in Europe, appeared in a notebook of an Italian scholar dated to the period 1438–1450, but there is clear evidence of horizontal-axle windmills as early as the 1170s and 1180s in Britain and France. This distinction suggests that the horizontal-axle windmill was a European rather than an Asian invention.

The first certain case of a European windmill with the horizontal or tipped axle, and vanes pointed into the wind, is in Yorkshire, Britain, in 1185. Within a few years, records show that it was being imitated and had become common. In fact, Christian Crusaders in Syria in the 1190s found the first windmill constructed there, still new. That evidence supports the concept that the windmill came from Europe to the Middle East, not the other way around.

Over the next century, windmills became common throughout northern Europe, where they had several advantages over water-powered mills. The freezing of streams and ponds often brought water-powered mills to a standstill in winter, leading to the construction of some 120 windmills near Ypres in the 13th century. Windmills could also continue to produce power to grind grain during times of military siege.

Together with waterpower, windpower brought all sorts of labor-saving mills to Europe in the 13th and 14th centuries. The impact brought changes to the location of industries, when, for example, areas where waterpower was unavailable soon had mills for tanning or laundering, sawing, crushing ore, running the hammers of a forge, or grindstones for finishing and polishing weapons. Wind-powered mills were used for grinding paint and dye pigments, crushing mash for beer and pulp for paper, and even for polishing precious stones and milling coins. The use of such mechanical force created a form of early Industrial Revolution through Europe in the 14th and 15th centuries and led thinkers to investigate other basic movers, such as the **steam engine [III]**. Windmills became the symbol of Holland, with its lack of falling water, and of La Mancha, Spain, with its high winds. In Paris, the survival of several windmills into the 20th century became a source of local pride. The Moulin Rouge (Red Mill) cabaret is perhaps the most famous.

> The exploratory quality of Western technology announces itself clearly in the twelfth century with the invention of the windmill rotating on an axle tipped slightly above the horizon to secure a turbine action in its vanes.
> —Lynn White Jr., *Medieval Technology and Social Change* (Oxford, Eng.: Oxford University Press, 1962)

wire

In ancient times, wire was formed by forging thin strips of metal and then hammering them into strands. Wire, especially that used in jewelry, has been found from ancient China, Egypt, and the ancient Near East. Gold wire has been found in Egyptian tombs, and a section of **bronze [I]** wire was found in the ruins of Pompeii. The more malleable metals are readily drawn into wire, and gold, one of the softest metals, could be stretched into extremely fine wire for construction of mesh and jewelry. Drawn wire, heated and then pulled to a small diameter, was invented in about A.D. 1000.

The wire would be drawn by hand through a small die, with the diameter limited by the strength of the smith pulling the tongs through the die. In some wire-pulling arrangements the drawer would be seated and would push with his legs and simultaneously pull with his arms.

In about 1350, Rudolf of Nuremberg used waterpower with **waterwheels** to draw wire. The Nuremberg method of drawing wire was at first kept a closely guarded secret. The English wire industry grew after the importation of a group of wire drawers from Saxony in 1564 to a water-powered wire mill known as the Tintern Abbey Works.

Steam-engine use to pull wire developed in about 1800. In the United States, nearly all wire was imported until after the War of 1812, when several wire-pulling establishments were set up near Worcester, Massachusets. Steel wire, useful in the construction of cable and **steel cable suspension bridges [IV]**, was developed by Ichabod Washington in 1831 at his wire mill in Worcester.

In modern practice wire is drawn through a succession of progressively smaller dies until the desired diameter is achieved. A hot rolled metal billet or bar of a standard diameter, as small as ¼ inch, is then mechanically pulled cold through the die, aided by the addition of a lubricant such as tallow, soap, or grease. Some harder alloys, such as those with tungsten or molybdenum, are heated to make them soft enough to be drawn through the dies.

With the growth of the American West, wire, and especially **barbed wire [IV]**, was in great demand for fencing. The wire industry expanded with the development of the **telegraph [IV]**, the **telephone [IV]**, and the **electric light [IV]**.

zero

A mathematical concept so familiar in the modern world that we hardly think of it as invention or discovery, the zero is such a basic intellectual concept that it must rank with the most important discoveries of all time. In Greek and Roman arithmatic and notation, the absence of a zero led to ambiguities and may have hindered the progress of science and engineering. Greek and Roman mathematicians had symbols that could indicate the absence of number, but they did not make use of such symbols in the numbering system.

Placement of a zero in a **decimal number** such as 306 indicates that there are six units, no tens, and three hundreds. In such a usage, the zero is a placeholder to demonstrate that there is nothing in the place. This simple procedure represented a striking innovation, reaching Europe in the 15th century.

Although known to the Hindus and found on a stone inscription made in A.D. 876 at Gwalior, India, the concept of zero was said to be introduced into the Arabic world by Muhammad ibn-Musa al-Khwarizmi, who lived from about 780 to 850. A Persian mathematician born in the town of Khiva (now in Uzbekistan), al-Khwarizmi introduced the zero placeholding system. The word "zero" is derived from the Arabic *sifr,* standing for nothing, and in turn translated from the Hindu word *sunya,* for the concept of a void.

From the Arabic world, the concept migrated to Europe, spreading widely with the diffusion of Arabic learning in the 1400s following the Christian conquest of Spain from the Arabic Moors in that century. The concept had been independently invented by the Maya in Central America before A.D. 300, long before it can be dated to the Hindus.

PART III

THE AGE OF SCIENTIFIC REVOLUTION, 1600 TO 1790

The Age of Scientific Revolution can be dated from the early 1600s through about the end of the 18th century. As the ideas of scientific method were carried into the social and political sphere, the effect became known as the "enlightenment." Textbooks in Western civilization sometimes divide this 200-year period into the 17th century, as the Age of Scientific Revolution, and the 18th century, as the Age of the Enlightenment.

Several features set this period apart from others. In antiquity and in the Middle Ages (periods [parts] I and II in this encyclopedia), it was very rare that an invention was attributed to a single, specific individual. As late as the 14th century, major new developments such as the clock, playing cards, and eyeglasses had been invented anonymously, although later historians have attempted to track down the probable locations of those inventions. However, after the mid-1500s, scientists and inventors, in the Renaissance spirit of individual achievement, sought to ensure that their own names were associated with a specific invention or scientific discovery. The 17th century was characterized by many great disputes over priority as Robert Hooke and Christiaan Huygens fought over who invented the balance spring, Hooke and Isaac Newton over who discovered the laws of motion and gravity, and Newton and Leibniz over the invention of calculus.

In some cases, money was at stake, as states often would reward inventors with a grant of a large pension or single lump-sum payment for the benefit their invention brought to the nation. In other cases, governments would grant a patent guaranteeing the right to exclusively produce the new device and to reap the financial rewards. But frequently, inventors and scientists simply fought for the recognition, realizing that their names would live on in history if their claim to priority were recognized.

The Age of Scientific Revolution was different from earlier eras in another respect. During this period, several key instruments, either to look more closely at the natural universe or to measure more accurately some natural phenomenon, served as tools that led to further scientific discoveries. Historians of science have debated whether the tools produced the discoveries or whether the scientific curiosity drove the development of the tools. Thinking about this chicken-or-egg question is stimulated as we look closely at the invention of the telescope, the microscope, and the mercury barometer and thermometer and at improvements in timekeeping itself, with the pendulum clock, the balance spring, and the chronometer. With all these tools, the scientist and the explorer could better map the universe and begin to achieve sufficient numerical accuracy to be able to establish the fundamental laws by which the universe operated.

Some of the great discoveries were simply improved explanations for long-observed phenomena. The heliocentric solar system, suggested by Copernicus in the 16th century as an explanation for the motions of the planets, was confirmed by the observations of Galileo, and its mechanics were unraveled by Newton. In fact, the idea that the Earth revolved around the Sun had tremendous impact. The idea that the Earth had a "revolutionary" motion meant that in the languages of western Europe, the word "revolutionary" took on its broader meaning of "completely new." Thus, when the French monarchy was overthrown in 1789, the change came to be called the French Revolution. The whole era from about 1600 to 1790 became known as one filled with "revolutionary" ideas, not because they revolved, but because they were as new and startling as the concept of the revolution of the Earth around the Sun.

With the new instruments and with improved laboratory glassware, scientists such as Robert Boyle and Blaise Pascal began to understand the behavior of gases, and other scientists, such as Henry Cavendish, Antoine Lavoisier, and Joseph Priestley, were able to identify oxygen and its role in life.

During this period, the exchange of information among scientists became regularized, with the development of national societies of scientists and the regular publication of papers. Publication became a means of establishing priority, and by the end of the period, the hallmarks of the modern scientific establishment were in place, with what

---

An Era of Science and Invention

Many of the important scientific discoveries and technological inventions of the Age of Scientific Revolution can be attributed to a relatively small handful of people. As can be seen from the following table, which lists 16 of the more prominent individuals born over a period of a century in the order of their birth, in many cases their lives overlapped and the scientists knew each other, worked together, read each other's work, or competed for credit for the same discoveries.

Prominent Scientists in the Age of Scientific Revolution

Scientist	Born–Died	Major Contributions
Galileo Galilei	1564–1642	telescope; pendulum; laws of motion
Johannes Kepler	1571–1630	laws of planetary motion
William Harvey	1578–1657	circulation of blood
René Descartes	1596–1650	optics; mechanistic science
Otto von Guericke	1602–1686	vacuum pump; atmospheric pressure
Evangelista Torricelli	1608–1647	mercury barometer; atmospheric pressure
Robert Boyle	1627–1691	Boyle's law; chemical analysis; litmus paper
Christiaan Huygens	1629–1695	light as waves; pendulum clock; motion
Anton van Leeuwenhoek	1632–1723	microscope; microscopic organisms
Robert Hooke	1635–1703	microscope; law of elasticity; biological cell
Isaac Newton	1642–1727	motion; gravity; optics; calculus
Ole Römer	1644–1710	speed of light
Gottfried von Leibniz	1646–1716	calculus
Denis Papin	1647–1712	pressure cooker; paddleboat
Jakob Bernoulli	1654–1705	calculus; probability
Edmond Halley	1656–1742	orbits of comets

---

later became known as peer review and regular replication of findings to demonstrate the accuracy of a scientist's work. Scientific knowledge came to be formalized, and the methods for presenting the work became better established.

Often scientific societies of the era turned their attention to practical matters. In France, the Paris Academy worked on specific problems assigned by governmental ministries, while in England, following the philosophy of Francis Bacon, scientists hoped to voluntarily address the pressing issues of the day. Two of the most important problems in Britain in the 17th and 18th centuries were approachable with science and engineering. With the depletion of forests for firewood, shipbuilding, and construction, the price of fuel in Britain climbed, and coal became the fuel of choice. As mines went deeper, they tended to flood, and some means of pumping the water out became crucial. This need pressed the development of pumps and led to the steam engine.

The other great practical problem for the expanding British Empire was a more accurate means of determining longitude while at sea to allow better navigation. This issue was so crucial that the British government established the paid position of royal astronomer and built the Greenwich Observatory in 1675. Later, the government offered a huge cash prize of £20,000 for a method of determining longitude that would be accurate. The quest for the prize led to the invention of the chronometer by John Harrison.

The Paris Academy and the Royal Society put scientists in touch with each other and stimulated the flow of information. With both institutions and others devoted to finding practical applications, the Scientific Revolution was as much a revolution in technology as it was in science. Some scientists invented their own tools, while others, hearing of reports from another country, sought to quickly emulate and improve on the tools others had developed.

Several of the major scientists of the period discovered basic laws of nature, and to have a law named after a scientist was perhaps the greatest mark of achievement. For this reason, some findings were called "laws" even before they could be demonstrated to be completely accurate, such as "Bode's law," about the spacing of the planets. However, other laws stood the test of time, such as Pascal's law and, with some modification, Kepler's laws of planetary motion.

The relationships between science and technology, and between discovery and invention, are some of the great issues in the history of human progress. And in this period, a series of developments illuminate different aspects of those relationships. Several of the greatest discoveries were made by the inventors of the very tools with which the discoveries were made. This can be said of Leeuwenhoek and the microscope, Galileo and Newton and their telescopes, and Hooke and his microscope. Some of the great instrument makers were great scientists, and some of the great scientists were great instrument makers. Key to the advancement of science, microscopes, telescopes, vacuum pumps, laboratory glassware, litmus paper, balances, thermometers, and other tools and techniques all came out of scientists' quest for knowledge.

Some simple devices invented in the 18th century had little to do with science but had great impact, including several improvements to textile making such as the flying shuttle and the spinning jenny. Together, these machine improvements, advances in basic motive power, and mechanization of textile manufacturing set the stage for launching the Industrial Revolution. By the end of the 18th century, the beginnings of large capital-intensive industries were in place, preparing the way for many more turbulent developments in the 19th century. These included the vast growth of mill towns and cities, the rise of a capitalist class and a factory working class, and social and political movements reflecting the new social conflicts that came out of the changed conditions.

Hints of the future could be seen by the 1780s and 1790s, with the beginning of coal gas for illumination, air travel, and attempts to harness steam for transportation. Similarly, work in chemistry and electricity led to further breakthroughs that would change the nature of warfare, of manufacturing, and of the everyday life of the consumer in the next century. Some of the household products and systems that we take for granted in the modern age were created in this period, including bleach for clothing and linens, reliable table clocks, carbonated soft drinks, pressure cookers, mass-produced textiles made of cotton or wool, door locks, piped gas for heating or lighting, and flush toilets.

In some of these inventions, the interplay of science and technology was at work when scientific findings were applied to practical, everyday problems. Joseph Bramah developed the flush toilet after a review of the work of Pascal on hydraulics, gaining ideas that soon found fruition in new hydraulic devices such as the hydraulic press. Soda water became popular after scientific experiments by James Priestley on dissolving carbon dioxide in water under pressure. James Watt made some of his improvements to the steam engine from the study of scientific work on expanding gases, although much of his work represented adaptation and improvement of existing mechanical devices.

Despite the focus on great individuals who made contributions, we should remember that several simultaneous inventions demonstrate deeper underlying patterns. The inventions and discoveries, while they can be traced in this period to specific people, would have been invented or discovered even had those specific individuals never been born. The fact that several people came upon similar ideas a few months or a few years apart shows how innovation built on innovation and idea built on prior idea. While the scientists and engineers are rightly honored and remembered for their achievements, their contributions sprang from the juxtaposition of tool, problem, and intellect, reflecting the innovative spirit and ability that had characterized the human race since the Neolithic period.

Chronology: The Age of Scientific Revolution, 1600 to 1790		
Year	**Invention/Discovery**	**Inventor/Discoverer**
1600	pendulum	Galileo Galilei
1608	telescope	Hans Lippershey, 1608; Galileo, 1609
1609	Kepler's laws	Johannes Kepler
1612	thermometer	Galileo et al.
1628	circulation of blood	William Harvey
1635	microscopic organisms	Anton van Leeuwenhoek
1640	Pascal's law	Blaise Pascal
1643	mercury barometer	Evangelista Torricelli
1645	adding machine	Pascal
1647	vacuum pump	Otto von Guericke
1656	pendulum clock	Christiaan Huygens
1658	red corpuscles in blood	Jan Swammerdam
1658	balance spring, clocks	Robert Hooke
1659	Saturn's rings	Huygens
1660	two-lens microscope	Hooke
1662	Boyle's law	Robert Boyle
1665	calculus	Isaac Newton, Gottfried Wilhelm von Leibniz
1666	spectrum of light	Newton, 1666, 1704
1667	cellular structure of plants	Hooke
1668	reflecting telescope	Newton
1672	Cassegrain telescope	Cassegrain
1676	Hooke's law of elasticity	Hooke
1676	universal joint	Hooke
1682	gravitation	Newton
1688	coal gas	John Clayton
1698	steam pump	Thomas Savery
1700	piano	Bartolomeo Cristofori
1712	steam engine	Thomas Newcomen, 1712; James Watt, 1769, 1781–1786
1728	speed of light	Ole Römer
1730	sextant	John Hadley
1730	cobalt	Georg Brandt
1733	flying shuttle	John Kay
1736	chronometer	John Harrison, 1736, 1761
1751	nickel	Axel Cronstedt
1755	magnesium	Joseph Black
1766	hydrogen	Henry Cavendish
1767	spinning jenny	James Hargreaves
1768	textile spinning factory	Richard Arkwright
1769	steam carriage	Nicolas Cugnot

Chronology: The Age of Scientific Revolution, 1600 to 1790 *(continued)*		
Year	Invention/Discovery	Inventor/Discoverer
1771	fluorine	Karl Scheele
1772	nitrogen	Daniel Rutherford
1772	Bode's law	Johann Daniel Titius
1774	chlorine	Karl Scheele
1774	manganese	Johann Gahn
1774	oxygen	Joseph Priestley; independently by Karl Scheele, 1771–1772
1774	boring mill	John Wilkinson
1778	flush toilet	Joseph Bramah
1781	Uranus	William Herschel
1781	molybdenum	Karl Scheele; isolated by Peter Hjelm, 1782
1783	hot-air balloon	Montgolfier brothers
1782	tellurium	Franz Müller
1783	water as a compound	Antoine Lavoisier and Pierre Laplace; Henry Cavendish
1783	tungsten	Juan Elhuyar collaboratively with Fausto Elhuyar
1785	chlorine bleach	Claude Louis Berthollet
1787	Coulomb's law	Charles Augustin de Coulomb
1789	zirconium	Martin Klaproth
1789	uranium	Klaproth
1790	titanium	William Gregor

adding machine

Blaise Pascal (1623–1662) developed a machine in 1642 that could add and subtract numbers with up to 9 digits. According to legend, Pascal, then 19 years old, made the machine to assist his father, a tax collector. A similar legend is associated with Gottfried Wilhelm von Leibniz (1646–1716), whose father also was a tax collector. Leibniz made an improvement to Pascal's *pascaline* by developing a system for stored data, multiplication, and division. Leibniz is better known for his claim to priority of the invention of **calculus.**

Although other machines may have preceded the pascaline, Pascal's machine was patented in 1649 and widely sold through a company he organized. Pascal presented a copy of the machine to Queen Christina of Sweden in 1652, and although he made contributions to knowledge in hydraulics, mathematics, probability theory, and physics, he was most widely known in his own time for the adding machine. Several of the machines have survived into modern times.

Pascal's machine operated with geared wheels linked together so that when a number in the first position was advanced from 9 to 0, a digit in the second column would advance. Similarly, when a number in the first column was reduced from 0 to 9, a digit in the second column would reduce, allowing for subtraction. The odometer on 20th-century **automobiles [IV]** is nothing more than a pascaline connected by linkage to the driveshaft so that the appropriate number of revolutions of the wheels are mechanically transmitted to the dials, adding ¹/₁₀ mile to the growing total for the appropriate number and tripping over the next positions when 9s advance to 0.

Leibniz's stepped calculator, developed in 1673, remained the basic principle behind mechanical adding machines for two centuries. In 1894, William Burroughs, an American inventor, improved on adding machines by introducing a system for advancing the cog- or gearwheels by action of a keyboard.

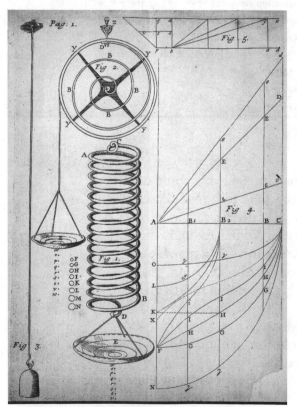

balance spring Robert Hooke measured stress in springs and developed "Hooke's Law," by which the strength of a spring could be calculated. *Library of Congress*

balance spring

The **clock [II]** driven by the action of falling weights was created in the Middle Ages and was improved by the **pendulum** as a regulator of the action in 1656. The Dutch scientist Christiaan Huygens (1629–1695) adapted the pendulum to the clock (following an idea developed by Galileo), and with the addition in 1675 of an anchor escapement mechanism, the pendulum-regulated, weight-driven clock was perfected. However, a **pendulum clock** was not particularly useful aboard ship, where the motion of the ship would disturb the pendulum motion.

In the 1660s, both Huygens and the British scientist Robert Hooke (1635–1703) claimed to have invented the balance spring, suitable both for smaller timepieces and for use aboard ships. Hooke's invention was not manufactured, however, and Huygens' came on the market in 1674, after which Hooke claimed priority of invention. Although the first **watches [II]**, introduced earlier, were several inches in diameter and quite thick, improvements over the 16th and 17th centuries, including the balance spring, reduced their size and improved their accuracy.

A reliable **chronometer** that could help the navigator locate his longitude on the sea was not developed until 1761.

Bernoulli's principle

In 1738, the Swiss natural philosopher and mathematician Daniel Bernoulli (1700–1782) published *Hydrodynamica,* in which he described a variety of aspects of the behavior of fluids. The principle or law that bears his name is that the mechanical energy of a flowing fluid remains constant. That mechanical energy involves the fluid pressure, the gravitational potential energy of elevation, and the kinetic energy of the fluid motion. In effect he described energy conservation for ideal fluids in a steady, streamlined flow. His principle explains why the air flowing over the curved upper surface of an airplane wing moves faster than the air beneath the flat lower surface, creating lift. Bernoulli's principle is employed in the design of heavier-than-air aircraft such as the **airplane** [V] and **helicopter** [VI] and in boat-keel design. Bernoulli's principle can be used to explain how a pitcher can throw a curveball.

Bernoulli also explained that fluid speeds up in constricted pipes, so that the pressure exerted is less where the pipe cross section is smallest. This behavior of fluids is known as the Venturi effect, after the Italian G. B. Venturi (1746–1822), who noted the effects of constricted channels on fluid flow. One could say that Bernoulli's principle explains the Venturi effect.

Daniel Bernoulli was the son of Johann Bernoulli, who together with his brother, Jakob, did much to advance the study of **calculus** and to promote the cause of Leibniz as the inventor of that tool in the dispute with Isaac Newton (1642–1727) over priority. Daniel's father and uncle themselves engaged in several bitter disputes over a variety of scientific issues. According to legend, Johann was so outraged at his son being named as co-prize winner for a study of planetary orbits that Johann thought should be his alone, he physically threw Daniel out of the house in 1735. It was a brilliant but contentious family.

Bode's law

Bode's law is rather unusual in the realm of scientific laws in that it was not discovered by Bode, it is not exactly a law, and even as a principle or relationship, it does not apply very well. A German astronomer, Johann Daniel Titius (1729–1796), first developed the concept in 1766, and it was popularized by Johann Elert Bode (1747–1826) beginning six years later. Some references use the term "Titius-Bode" to describe the concept. The concept or "law" suggests that in the solar system each planet's distance from the Sun follows a regular pattern.

That pattern can be expressed mathematically in several ways. One method is to start with the sequence 0, 3, 6, 12, 24. Add 4 to each of these numbers. Then divide each result by 10, yielding the sequence 0.4, 0.7, 1.0, 1.6, 2.8, and on to 5.2, 10.0, 19.6, 38.8, and 77.2. Assuming that 1.0, the third position, is the unit of measurement of Earth from the Sun—that is, 1 astronomical unit (a.u.)—the other planets are to be found at the other distances. By this calculation, Mars would be found at 1.6 a.u. from the Sun. In fact, it is at 1.5 a.u. However, the calculations were more correct in some cases, as Jupiter is at 5.2 a.u. from the Sun. Venus, the second planet from the Sun, is at 0.7 a.u., exactly in accord with the Bode numbers.

Beyond Mars, one might expect a planet, by this calculation, at 2.8 a.u. In fact the belt of asteroids is at that distance, and the discovery of the large asteroid Ceres in 1801 by the Italian astronomer Giuseppe Piazzi could be taken to fulfill the prediction implicit in the Titius-Bode relationship. More notable was the discovery of **Uranus** in 1781 by William Herschel, at 19.2 a.u., when the Titius-Bode "law" would have placed it at 19.6 a.u. However, the relationship does not apply at all to Neptune and Pluto. So the Titius-Bode law, or Bode's law, is a curiosity in that it suggests a mathematical pattern that is almost, but not quite, a fact and for which there is no physical explanation.

boring mill

The Englishman John Wilkinson invented one of several relatively obscure metalworking tools, the boring mill, in 1774. Combined with Joseph Bramah's **hydraulic press [IV]** of 1795, the metal lathe in 1800 by Henry Maudslay, and Richard Robert's metal planer of 1817, the boring mill made possible the transition from wooden devices to metal machines characteristic of the Industrial Revolution. John Wilkinson's metal boring machine achieved precision in boring the cylinders used in the **steam engine** developed by James Watt (1736–1819). Later, when combined with accurate jigs and guides, such metalworking devices made possible the "American System of Manufactures," which produced **interchangeable parts [IV]**, first for **muskets [II]** at U.S. government arsenals and later for such inventions as the **revolver [IV]** and the **automobile [IV].**

Boyle's law

In 1662, Robert Boyle (1627–1691), an Irish chemist and physicist, formulated the basic relationship between pressure and temperature in gases. Boyle's law states that the volume of a given mass of gas at a constant temperature is inversely proportional to the temperature of the gas. As the gas is heated, its volume increases; when the gas is compressed or the volume decreases, the temperature increases; as a given mass of gas is allowed to expand, its temperature will decrease. The principle lies at the heart of the efficient operation of **steam engines, refrigeration [IV], diesel engines [V],** and many other modern devices.

Boyle made many other contributions to scientific thought. In 1661 he published *The Sceptical Chemist,* in which he argued that the object of chemistry was to find out the composition of substances. He coined the term "analysis," and he introduced litmus paper as a test of acidity after noting that certain vegetable dyes turned color when exposed to acid. He was a founding member of the group that later formed the Royal Society in Britain, one of the first **scientific societies.** He showed the difference between mixtures and compounds, demonstrating that a compound could have very different properties from its component materials. Together with Joseph Priestley (1733–1804) and Antoine Lavoisier (1743–1794), Boyle is regarded as one of the fathers of modern chemistry, leading its transition from alchemy to a science of experimentation and discovery. His approach to chemistry was to investigate the nature of materials in search of knowledge of basic principles, rather than in search for a particular effect, such as transmutation of metals or some cure-all elixir that had characterized alchemy.

Application in the 1760s by James Watt (1736–1819) of Boyle's law improved the operation of the Newcomen steam engine by cooling the steam in a separate condenser.

calculus

One of the greatest disputes in the history of invention and discovery was over the question of who developed calculus first. The two primary contenders for the honor were Isaac Newton (1642–1727) and Gottfried Wilhelm von Leibniz (1646–1716). As Newton worked on the problems of planetary movements and formulated the **laws of motion** and gravity between 1664 and 1666, he developed the binomial theorem, new methods for expansion of infinite series, and his direct and inverse method of calculating "fluxions." He privately circulated his findings, and they were not published until 1711, while a full and detailed explanation of his methods was not published until 1736, nearly a decade after his death.

Meanwhile, Baron Gottfried von Leibniz, a German child prodigy who earned a doctorate in law before age 21, developed a similar mathematical method. In the diplomatic service of the elector of Main and other German states, Leibniz served in Paris and traveled abroad, including a visit to London from January to March 1673. During this period he independently developed a system for calculating series and the progression of curves similar to Newton's, that Leibniz dubbed *calculus integralis*.

With the method that Newton called fluxions and Leibniz called calculus, it became possible to calculate rates of change and to describe curved lines mathematically. As a tool, calculus not only allowed astronomers to apply Newton's laws of motion and gravity to the interactions of the planets, it also soon found uses in a wide variety of fields of engineering and the sciences, allowing complex interactions of forces to be described and predicted. For example, in describing the decreasing pressure of air at increased altitudes, Edmé Mariotte (1620–1684) published a work in 1676 that divided the atmosphere into 4,032 layers, each of which was assumed to be equal in pressure throughout the same altitude. By adding together the height, one could calculate the expected pressure at a particular height. With calculus (with which he was unfamiliar), he would have been able to more accurately describe the decrease in pressure as a function and display it as a continuous curve on a table rather than as discrete and separate numerical values.

Although Leibniz and Newton had built their system on the work of previous mathematicians and natural philosophers, including Pierre de Fermat (1601–1665), René Descartes (1596–1650), Blaise Pascal (1623–1662), and Christiaan Huygens (1629–1695), among others, Newton grew furious when he learned of the work of Leibniz. Newton spread the word that Leibniz had plagiarized his ideas. When a friend of Newton, John Keill, accused Leibniz of plagiarism, Leibniz complained to the British Royal Society, one of the first **scientific societies.** A committee of the society, appointed by Newton, investigated the charges and concluded that Leibniz had received a letter in 1672 from John Collins (1625–1683), a fellow of the society, detailing Newton's methods. The committee concluded that Leibniz had simply rephrased

Newton's method of fluxions. Their report, published in 1712 and titled *Commercium epistolicum* (A Commerce of Letters), in effect convicted Leibniz of plagiarism. Evidence indicates that Newton himself authored the report.

Later historical work, especially an 1852 study conducted by Augustus de Morgan (1806–1871), revealed that the Collins letter did not contain any detailed information on which Leibniz could have based his work. De Morgan concluded that calculus was a true case of independent invention, a judgment that is generally accepted in modern treatments of the subject.

Many other mathematicians contributed to the further refinement of calculus methods, including three born in Switzerland, the brothers Jakob Bernoulli (1654–1705) and Johann Bernoulli (1667–1748) and Leonhard Euler (1707–1783). Euler later used the system to accurately calculate the gravitational relationships among the Earth, the Sun, and the Moon.

carbon dioxide

The Scottish doctor and chemist Joseph Black (1728–1799) discovered the gas carbon dioxide in about 1755. He submitted research on the chemistry of a form of magnesium carbonate for his medical degree. When heating "magnesia alba" and reacting the products with acids, he produced what he called "fixed air," now recognized as carbon dioxide. To an extent, his work foreshadowed the research conducted by Antoine Lavoisier (1743–1794), one of the discoverers of **oxygen.** Joseph Black also made contributions in other areas, including the discovery of latent heat.

Joseph Priestley (1733–1804) developed a method for dissolving carbon dioxide under pressure in water in the 1770s. His technique soon produced a popular fad of drinking carbonated water or "soda," which has persisted in the form of modern soft drinks.

Cartesian mechanistic science

René Descartes (1596–1650) was one of the leading intellectual influences in the Scientific Revolution. His Latinized name, Cartesius, was the source of the term *Cartesian synthesis* applied to a philosophical outlook on the universe that began to dominate universities and **scientific societies,** especially on the Continent of Europe. Descartes was born in Touraine, France, studied in a Jesuit school, and then had a military career. He served in the Dutch, Bavarian, and Austrian militaries, and settled in Stockholm in the last years of his life.

Descartes made very few direct contributions to specific scientific discoveries, although he did apply the principle of **refraction of light,** developed by Willebrord Snell (1591–1626), to an analysis of rainbows, showing how light is refracted twice in each raindrop and explaining the angle at which the observer detected the rainbow. Although he attempted the solution to many other problems, his method of analysis, which was based more on deduction from principles than on induction from empirical evidence, did not lead in his own case to many lasting particular discoveries.

However, Descartes' *Discourse on Method,* written in Holland and published in 1637, was profoundly influential and stimulated the spread of the Scientific Revolution throughout western Europe and the world. In that work, Descartes described all matter as extension or measurable size and all qualities, such as color and weight, as the accidental consequence of the size or relative motion of matter. His method was to require that the mind grasp the mathematical relationships and mechanical interactions of matter. His idea flew in the face of contemporary Scholastic explanations based on handed-down interpretations of ancient Greek physics that held that qualities were inherent in bodies. By the Scholastic interpretation, a cherry was red because redness was inherent in the cherry; by the Cartesian interpretation, a cherry was red because some mathematical relationship of the parts of the cherry accidentally produced the impression of red on the eye. The difference seems very abstract in modern times, but it led to extreme controversies and was at the heart of a different way of looking at the universe. Carried to its logical conclusion, the Cartesian synthesis explained the whole universe in terms of mechanical interactions that could eventually be described mathematically and that were entirely knowable for the human mind. Such a view seemed to many contemporaries to deny a place for God or spirit, suggesting that all the invisible forces of the universe were at the root caused by the mechanical interactions of matter. However, Descartes himself concluded by his own rational analysis that God did exist.

Descartes was influential partly because he wrote in very direct language that could appeal to craftsmen, businessmen, and other people of practical affairs. He suggested that the ordinary mind would be able to grasp science simply using common sense. Rather than claiming to teach his scientific approach, he sought only to describe the method by which his thinking worked, suggesting that mathematics offered certainty and concrete results rather than idle speculation, and that it could be adopted by ordinary people, even those unschooled in universities. His method was to investigate problems by focusing on real objects and on rules that explain their workings, starting with the simple, going to the complex, and keeping careful records. In effect, he described what scientists later called the scientific method, although he was less rigorous in suggesting that experiments be devised to test hypotheses, that became common after the work of Isaac Newton, who developed the **laws of motion and gravity.**

At the heart of why the Cartesian synthesis was revolutionary was the concept that it demanded that the individual be free of preconceptions and traditional concepts but start with physical evidence and proceed logically and mathematically to conclusions. The scientific person should begin with doubts and proceed to pursue truth, using the mind and mathematics, not tradition. Descartes' suggestion that the scientific person could come to the conclusion that God existed without the assistance of tradition angered the church, and his view that the illnesses of the human body could be explained by matter in motion rather than by inherent "humors" angered the medical establishment. For these reasons he found life more comfortable in Holland and Sweden, beyond the reach of the Inquisition.

Some of Descartes' followers, applying his ideas, made numerous scientific contributions. For example, Christiaan Huygens (1629–1695) published in 1659 *The Saturnian System,* a sophisticated analysis of the motions of **Saturn's rings and moons.** On a more popular level, in 1686 Bernard de Fontenelle (1657–1757) published *Conversations on the Plurality of Worlds,* which went through many editions and influenced two generations in France and elsewhere, suggesting that mechanical and mathematical explanations were at the heart of the operation not only of the physical universe but also of society. As head of the Paris Academy of Sciences in the first decades of the 18th century, Fontenelle further advanced the Cartesian approach. Cartesian thinking contributed to a decisive shift in European thought from a religious understanding of nature and of the operation of human institutions to a thoroughly secular understanding of those phenomena.

Cassegrain telescope

Although the Cassegrain construction of **telescopes** is familiar to modern astronomers, virtually nothing is known about the inventor, a Frenchman who may or may not have been a professor of astronomy or medicine at the College of Chartes and who lived from about 1650 to about 1700. His first name also is unknown.

The Cassegrain design represented an improvement over the reflecting telescope developed by Isaac Newton. Cassegrain inserted a small convex mirror in front of the focal point of the reflection cast by a larger concave mirror, reflecting the image directly back through a small hole in the larger concave mirror. The observer was positioned behind the larger mirror. The convenience of the design was that it was considerably shorter than the original telescopes developed by Galileo for equal magnification and that, unlike the Newtonian telescope, it could be pointed directly at the object observed. In the Newtonian model, the observer stood to the side of the telescope and usually used a smaller, externally mounted spotting scope to align the telescope with the target object.

A variation on the Cassegrain was an optical system suggested in 1663 by James Gregory, a professor of mathematics at the University of Edinburgh. Gregory theorized that two concave mirrors could eliminate distortion. Gregory published *Optica Promota,* in which he suggested a primary mirror of parabolic shape with a hole for an eyepiece in the center. In addition, an elliptically curved secondary mirror would reflect light back to the hole and the eyepiece. No "Gregorian" telescope could be built at the time because the precise curves could not be made by grinders of glass of the era. The Gregorian required two concave mirrors with different curves, while the Cassegrain required a concave and a convex mirror.

The Cassegrain was developed about 1672, but it was not until the 1770s that an English optician, Jesse Ramsden, realized that the Cassegrain design had another advantage in that it could be built to reduce the distortion or aberration of the image caused by the curvature of the mirrors. Many small modern telescopes intended for use by amateur astronomers follow the Cassegrain configuration, and a principle similar to that proposed by Gregory has been used in **radio [V]** transmitters and receivers.

cast iron

Although both wrought iron and cast iron were known in antiquity in Asia Minor before 1000 B.C., cast iron was not commonly made or used in Europe until the 17th century A.D. Wrought iron, with less than 0.3 percent of carbon, is worked by heating it to a soft state and then pounding and bending it. By contrast, cast iron, with 2 to 4 percent carbon and other impurities, can be reduced to a molten liquid state and poured into shapes. The Chinese made cast iron in the 6th century B.C., and it was first introduced to western Europe by the 13th century A.D.

Early uses in Europe included cast iron balls for **cannons** [II]. Cast iron soon found uses in grave slabs and firebacks. Cast iron began to be used not only for decorative effects in buildings along with wrought iron but also as a structural member. Although there were some uses of cast iron for structures in China as early as the 11th century A.D., its wide use in Europe for such purposes did not get under way until the mid-18th century.

In 1709, Abraham Darby in Britain introduced the use of coke for iron smelting. The stronger Darby iron could be used in buildings, was less fragile than early cast iron, and was used in 1772 to provide columns to support galleries in St. Anne's Church in Liverpool. From 1777 to 1779, Darby's company cast and built the first cast-iron bridge at the appropriately named town of Ironbridge, in Shropshire.

A malleable cast iron was developed in that period in France, in which a prolonged exposure to heat allowed for working the metal after being cast. Since cast-iron members could be mass-produced, they began to be used to help hold together masonry walls and to brace **chimneys** [II] in urban settings. Later uses included track for **steam railroads** [IV] and framing in **skyscrapers** [IV], later supplanted by **steel** [IV].

cellular structure of plants

In 1665, the English natural philosopher and scientist Robert Hooke (1635–1703) published *Micrographia* (Small Drawings). In the work he described his study of microscopic organisms and noted the discovery of the porous structure of plants. He called the pores "cells" because the boxlike structure he found in cork, among other plant material, reminded him of the cells in a monastery; the name for cells in both plant and animal life derives from Hooke's usage.

Hooke was born on the Isle of Wight, the son of a clergyman. Lacking a fortune, his education was spotty, and although he was admitted to Oxford, he never mastered advanced mathematics. He became a founder of the Royal Society and corresponded with many scientists of his day. His most notable achievements were in the areas of instrument making and experimentation, making contributions in what later generations regard as separate sciences: physics, chemistry, biology, geology, paleontology, astronomy, and optics. He developed the **law of elasticity,** also known as Hooke's law.

> I could exceedingly plainly perceive it be all perforated and porous, much like a Honey-comb, but that the pores of it were not regular.... These pores, or cells were indeed the first microscopical pores I ever saw, and perhaps, that were ever seen, for I had not met with any Writer or Person, that made any mention of them.
> —Robert Hooke on the structure of cork, in *Micrographia* (Small Drawings) (1665).

As a discovery, the cellular nature of plant life was a great contribution, but it simply represented the application of improved instrumentation to observation. Hooke was keenly aware of the interplay between instruments and science and urged his contemporaries to constantly develop instruments as extensions of the human senses. He had become an assistant to Robert Boyle (1627–1691) after entry to Oxford in 1653, precisely because Boyle regarded him as a good mechanic. Hooke had helped build an improved air pump that contributed to Boyle's formulation of his gas law, **Boyle's law.** Because Hooke's background in mathematics was weak, he had difficulty expressing scientific concepts that he discovered. For this reason he engaged in a long and bitter dispute with Isaac Newton, as Hooke had developed the concept that planets were held in orbit by the attraction of the Sun, the gravity force that Newton described with mathematical formulation along with his **laws of motion and gravity.**

Hooke confirmed in 1678 the work of Anton van Leeuwenhoek (1632–1723), who had discovered **microscopic organisms** in the form of protozoa and bacteria. Although remembered for his discovery of the cellular structure of animal and vegetable tissue, Hooke also is well known for his invention of many instruments, including a **mercury barometer,** a hygrometer, an anemometer, and a spiral **balance spring** for clocks, as well as the **universal joint,** the iris diaphragm, and a type of artificial respirator.

Hooke represented a combination of skills particularly adapted to the state of scientific advance in the 17th century, with his focus on using instrumentation to push back the frontiers of knowledge. By combining the abilities of the keen observer with the practical mechanical skills of an instrument maker, he was able to contribute to what later became many different specialized sciences that were in his era subsumed under "natural philosophy."

Charles's law

Jacques Alexandre César Charles (1746–1823), a French physicist, experimented with gases in about 1787 and found that any gas expands by ½₇₃ of its volume at 0 degrees Centigrade for each degree rise in temperature. Stated another way, the volume of a fixed amount of gas is directly proportional to its absolute temperature. Charles wrote of his findings to Joseph Gay-Lussac (1778–1850), who replicated the experiments. Since Gay-Lussac published the results, the law is sometimes referred to as Gay-Lussac's law.

As in many developments, the invention of the technology preceded the discovery of the principle. Charles had flown in a balloon of his own design several years before his conclusions regarding the expansion of gas.

Charles and his brothers had experimented with a gas-filled balloon, using hydrogen, and made their first flight in 1783, only 10 days after the Montgolfier brothers made an ascent in a **hot-air balloon**. On one of his earliest flights Charles reached an altitude of about 10,000 feet. For the next decade, balloonists adhered either to the Charles (hydrogen) or the Montgolfier (hot air) factions.

Charles's law Two competing balloon systems developed almost simultaneously. They were the Charles system, lifted by hydrogen, and the hot-air Montgolfier system. Here a Charles balloon attracts a massive crowd at Versailles. *Library of Congress*

chlorine bleach

The discovery that chlorine could be successfully used to bleach textiles white was made by Claude Louis Berthollet (1748–1822). Chlorine gas had been discovered in 1774 by Swedish chemist Karl Scheele (1742–1786). Berthollet, a leading French chemist, adviser to Napoleon, and colleague of Antoine Lavoisier (1743–1794), developed the technique of using bleach from 1785 to 1787, calling the bleaching liquid *eau de Javel*. In 1799 Charles Tennant, a Scottish chemist, invented a more convenient bleaching powder that, when combined with dilute acid, released the chlorine and made an effective bleach for both textiles and paper. Production of the bleach increased greatly in the 1830s, leading to wider use of cotton.

chronometer

Credit for the invention of the chronometer for timekeeping at sea clearly belongs to the English clockmaker and instrument maker John Harrison (1693–1776), whose fourth chronometer, created in 1760, was so accurate that it became the standard for later timepieces. An accurate clock, set precisely on London time, could be used to determine longitude. The mariner would determine local time at his location by observation of the Sun and, by comparing it to London time, could easily calculate his distance east or west of London. Each hour's difference would be equivalent to 15 degrees of longitude, or each minute of time would represent ¼ degree. The calculation was almost self-evident, as 360 degrees of circumference of Earth, divided by 24 hours, yielded 15 degrees for each hour of difference in Sun time. While it was rather easy to determine "high noon" during a sunny day at sea by use of a **cross-staff [II]** or the exact local time at any hour when the Sun was over the horizon with a **sextant,** the trick was to have a device that would preserve London time aboard the traveling ship, so as to obtain the hours of difference.

In 1714 the British government's Board of Longitude offered a prize of £20,000 for anyone who could invent an instrument capable of determining longitude at sea with an accuracy of ½ degree. Harrison's fourth chronometer, known to history as H-4, maintained accuracy to within 5 seconds on two voyages from Britain to Barbados, representing accuracy in longitude of slightly more than $\frac{1}{60}$ degree, or 1 minute, of longitude, far better than demanded by the ½-degree standard established by the board for the prize.

Harrison solved several problems that had previously made clocks unreliable at sea. One was that changes in temperature affected the length of a metal pendulum, affecting its arc. By constructing pendulums with two different metals whose expansion canceled each other out, he controlled that problem. Another issue was that prior clocks had to be stopped while being rewound, but Harrison solved the issue with an escapement mechanism that allowed the clock to continue working while being wound.

Harrison had to fight to collect the longitude prize, due to the opposition of political enemies. However, he did eventually get

> Mr. Kendall's Watch (which cost £450) exceeded the expectations of its most zealous advocate and by being now and then corrected by lunar observations has been our faithful guide through all vicissitudes of climates.... It would not be doing justice to Mr. Harrison and Mr. Kendall if I did not own that we have received very great assistance from this useful and valuable timepiece.
> —Captain James Cook, entries from log of HMS *Resolution,* 1775

somewhat more than half the prize money. His methods were used to make accurate chronometers for mariners through the next century. However, a rival method of determining longitude, by consulting lunar tables, largely supplanted the pure chronometer method.

circulation of blood

William Harvey (1578–1657), an English physician, is credited with discovering the mechanism by which blood circulates in mammals, including the human race. He published his discovery in *Anatomical exercises concerning the motion of the Heart and of the Blood in Animals,* in 1628. In that work he did not explicitly compare the heart to a pump, although he did so in his lectures and in one small publication.

Harvey's methods, by contrast to some prior work on related questions, were based on direct observation and on dissection of both human and animal cadavers. His views overthrew those of Galen, the ancient Greek whose views on the question had become accepted dogma. Galen believed that the liver changed food into blood and that blood flowed through the veins to the body, which consumed it. Harvey showed that the heart operated as a pump, forcing blood through the arteries to the body, and that it returned through the veins to the heart in a closed circulation system. He also demonstrated the cause of the pulse in the expansion of the arteries following contraction of the heart muscles.

He demonstrated that the pumping chamber, or right ventricle of the heart, sends blood to the lungs, while the left ventricle supplies blood to the rest of the body through the arteries. Among his other findings was a measurement of the quantity of blood in the body at a particular moment and studies of the part played by the egg in mammalian reproduction. He also criticized the concept that illness derived from general changes in humors and instead believed in a more modern notion that it resulted from changes in local tissue.

Harvey's experiments, while contradicting the assumptions of Galen, confirmed the views of several earlier physicians, including the 13th-century Arab Ibn-an-Nafis and the Spaniard Miguel Serveto (Michael Servertus) and the Italians Realdo Colombo and Andrea Cesalpino in the mid-16th century. Each of these doctors had suspected that blood from the veins passed through the heart and thence to the arteries. None, however, had explained the whole system as one of circulation, nor had they traced all the components of the system as clearly as did Harvey.

Harvey at first developed a hypothesis of circulation, which he then later confirmed through experimentation, determining the direction of blood flow. Rather than relying on a nebulous force that moved the blood through the body, Harvey showed the motion as resulting from a specific mechanical action. Like many other discoveries of the Scientific Revolution, those of Harvey derived from careful and structured observation of facts and restating them with clarity.

circulation of water

Edmé Mariotte (1620–1684) was born in France and became a priest, prior of a monastery close to Dijon. He joined the Paris Academy shortly after the establishment of that **scientific society** and made several contributions, including some calculations regarding the pressure of air. He published his work on air pressure in 1676, some 15 years after Boyle. Some references note **Boyle's law** as the *Law of Boyle or Mariotte*. However, in another calculation, Mariotte determined that water circulates through evaporation.

Until Mariotte's work, scientists had imagined that the water springing from the upper reaches of rivers had somehow circulated underground, with springs arising from the interior of Earth due to seepage of seawater or coming from inexhaustible supplies of underground water. The total amount of rain was generally underestimated. Mariotte decided to carefully conduct extensive measurements of rainfall, and he collected data on average yearly quantities of rainfall from the whole region of the Seine River. He compared the rainfall estimates with the known annual quantity of water flowing out of the river, which he determined from the drift of ships in the river and the depth and width of the river. He calculated that the rainfall was about six times greater than needed to account for the river flow.

After making other studies of seepage, rainfall, and river flows, he concluded that springs are fed by rain and snow and, therefore, that the circulation of water on the whole planet takes place on the surface and over it through the atmosphere, not from subterranean passageways, as had been assumed. He also made accurate conclusions about the formation of raindrops in clouds and the evaporation of water vapor from both the ground surface and from bodies of water.

coal gas

Manufactured coal gas or town gas was a result of scientific experimentation in the Age of Scientific Revolution but was not installed as a system for lighting until the first decades of the 19th century.

An English clergyman, John Clayton, experimented in about 1683 with a natural gas spring emerging near coal deposits. He observed that passersby had boiled eggs over a flame from the spring. Digging the coal, Clayton experimented with attempting to fill ox bladders with a gas he distilled from the coal. At the University of Louvain, Jean Pierre Minckelers (1748–1824) experimented in 1785, distilling coal to produce gas with which he illuminated a classroom. He is noted in several works as the inventor of gas lighting.

A French engineer, Philippe Lebon, produced manufactured gas from wood in 1799 and demonstrated a lighting experiment in Paris in 1801. In Britain, William Murdock, an engineer who worked at the factory established by James Watt for the production of the **steam engine,** developed methods of producing and transporting gas distilled from coal, lighting his own house in 1792. Hearing of Lebon's demonstration, Murdock began installing lighting systems in cotton mills as early as 1806. Careful calculations indicated that gas lighting from coal was much less expensive than equivalent light from candles. Murdock investigated the types of coal to determine the best for producing gas and calculated the efficiency of different retort designs.

A German national, Friedrich Albrecht Winzer, moved to Britain and Anglicized his name to Frederic Albert Winsor. Winsor conducted several gas demonstrations in London in 1804. He installed gaslights supplied through tinned iron pipes, and the first London gas-lighting system is usually dated to his 1805 system, established to celebrate the birthday of King George III.

Meanwhile, in the United States, Benjamin Henfrey demonstrated a gas-lighting system in 1802, apparently based on Lebon's system, but Henfrey was unable to raise funds to set up a business. David Melville, a hardware merchant, set up gasworks for lighting cotton mills at Watertown, Massachusetts, and Providence, Rhode Island, from 1813 to 1817. The noted artist Rembrandt Peale, who had grown interested in gas lighting for a museum in Baltimore, worked with a group of financiers to establish the Gas Light Company of Baltimore in 1816, the first commercial gaslight company in the United States. Manufactured gas was installed in pipeline systems in New York (1825), Boston (1829), Louisville (1832), New Orleans (1835), and Philadelphia (1836). A nuisance byproduct of the manufactured gas plants was coal tar as well as other wastes, which, by the 1830s, had become a notorious pollutant of the Thames River in London. News of this fact, as well as the well-known fire hazard represented by gaslights, tended to stand as an obstacle to gas-lighting systems.

By the 1870s, however, as Edison and others experimented with **electric light [IV]** systems, at least fourteen major cities and many smaller communities in the United States were lit by coal gas systems.

condom

The origin of the condom and the term for it are a bit of a mystery, probably compounded by the fact that such items could not be discussed in polite company in the 19th and early 20th centuries. According to one legend, a Dr. Condom sometime during the reign of Charles II (1660–1685) in Britain provided a sheath for contraceptive purposes to the monarch, who wished to limit the number of illegitimate children he might father. No contemporary documented records have surfaced to demonstrate that there was such a doctor in reality. However, the Italian physician Gabriel Fallopio (or Fallopius) (1523–1562), the discoverer of the **Fallopian tubes [II]**, claimed to have invented a linen sheath in 1564 as a prophylactic against syphilis. He may have improved on a Persian device imported to Europe, along with the name "condom," in the Middle Ages. If the term did in fact enter European languages in the 16th century, the mysterious Dr. Condom may in fact be entirely a legend.

References to a sheath made from goat bladder in the Roman writings from the 2nd century A.D. have suggested to some historians that a form of condom was used in that period or even earlier. Condoms made of animal bladders were in common use in the 18th century and often were made available in brothels to limit the spread of disease. The English referred to condoms as "French letters," while the French called them "the British hat" *(la capote anglaise)*. Beginning in the 1840s, condoms were made from vulcanized **rubber [IV]**, and since the 1930s, of latex rubber.

Coulomb's law

The French engineer and physicist Charles Augustin de Coulomb (1736–1806) studied electrostatic and magnetic forces with a newly developed balance he had developed, the torsional balance. He discovered the law that holds that electrostatic and magnetic forces, like gravity, follow an inverse square law—that is, the force between two electric charges is equal to the two charges divided by the square of the distance between the two charges in a vacuum. Coulomb published his findings in 1785.

The force of electrical attraction is known as the Coulomb force or the Coulomb interaction. Along with gravity and magnetic force, the Coulomb force of electrical attraction is a basic physical force. At the atomic level, another natural force, the "strong interaction" or nuclear force, holds atomic nuclei together, even though protons, with positive electrical charges, repel each other.

diving bell

The concept of lowering a weighted inverted tub with air to the sea bottom in relatively shallow waters appears to have been developed in ancient times, with stories from the 4th century B.C. indicating that Alexander was lowered into the Bosporus to observe marine life at some depth. Francis Bacon described a type of metal barrel lowered into the water for the recovery of lost objects. In 1665, a diving bell of this type was used off Britain to recover guns from a sunken ship of the Spanish Armada.

The astronomer Edmond Halley (1656–1742) is credited with the invention in about 1690 of an improved diving bell with a system for replenishing air. His device had a glass window, and he rigged it to be lowered from a spar on a surface ship. Air would be supplied in sealed barrels lowered to the divers. The barrel had a hose attached to a hole in its bottom. When the barrel reached the diving bell, the divers would bring the hose up under the lip of the bell, meanwhile opening a tap at the top of their bell. The water pressure on the outside of the barrel would drive the barrel air into the bell, and the increase in air pressure inside the bell would force some of the spent air from the tap at the top.

Halley developed a salvage operation with his bell and used the proceeds to support his scientific work. He utilized the system in various operations from the 1690s through about 1720.

Fermat's Last Theorem

In 1657, the French mathematician Pierre de Fermat (1601–1665) published a number of problems for other mathematicians to prove, in the form of "theorems." In his most famous, only scribbled as a marginal note, he proposed that it is impossible to have two whole numbers raised to the same power that, when added together, equal a third number raised to the same power, if the power is greater than 2. Stated algebraically,

$$x^n + y^n = z^n \text{ is impossible where } n \text{ is greater than 2.}$$

Fermat had noted the theorem in the margin of a mathematics textbook and claimed he could have proven the theorem if he had more space in which to write out the proof. For more than 300 years no one could prove "Fermat's Last Theorem," and it remained a challenge to mathematicians. However, in 1994, Professor Andrew Wiles of Princeton University published a proof that was accepted as demonstrating that Fermat's Last Theorem was correct.

flush toilet

The water closet, still known as a "WC" in much of the world, was developed by an English watchmaker, Alexander Cummings, who in 1775 patented a valve that could be used to release water from a tank to carry off waste. Three years later, Joseph Bramah (1748–1814), a prolific British inventor who also developed the **hydraulic press [IV]**, improved on the Cummings valve with a hinged and levered arrangement that allowed the tank, mounted high on a wall, to be operated with a pull chain that would be held down until the tank emptied. Thus Bramah is usually credited with the invention of the flush toilet.

However, toilets that emptied to septic cisterns were generally unsatisfactory in urban settings without drain fields, and it was the development of urban sewer systems in the early 19th century that increased the demand for water-flushed toilets. In many settings, sewer and septic tank or cistern systems were slow to replace outdoor privies, which were outhouses constructed directly over a pit that would be refilled with earth when the waste pile approached the surface level of the ground.

According to a widely accepted but ill-documented story, with the development of a sewer system in London in the 1850s, a London plumber, Thomas Crapper, developed a new improvement in 1860. He added an automatic shutoff that allowed the toilet to be flushed with a quick pull on the handle. By etymological legend, his name entered English as slang for the toilet itself.

flying shuttle

Invented by John Kay (1704–1764?) in 1733 in Lancashire, Britain, the flying shuttle was one of three devices that revolutionized textile manufacture, representing a major contribution to the Industrial Revolution. The other two were the **spinning jenny,** invented by James Hargreaves, and the **spinning mule,** invented by Samuel Crompton. All three inventions were resisted by hand weavers as they realized that converting the craft into an industrial process would yield cheaper cloth and make their methods obsolete as manufacturers adapted the methods and equipment. Although Kay patented the "New Machine for Opening and Dressing Wool," Yorkshire manufacturers organized a group to fight his claims for royalty in court. He lost most of his fortune in these lawsuits and moved to France, where he vanished from contact and is presumed to have died in 1764.

In hand weaving, a horizontal weft thread would be passed through vertical warp threads. The weft was attached to a wooden shuttle that would be passed left to right and right to left from hand to hand. In general, handwoven cloth was limited to about a yard in width, representing the distance between the outstretched hands of a seated weaver who would throw the shuttle back and forth. Kay's device simply attached the shuttle to a cord on both ends, allowing the weaver to pull the shuttle through the warp threads without leaving a seated position. Kay's son Robert improved on the flying shuttle by establishing a system of up to four shuttles, each with a different-colored thread, which allowed the weaver to create a variety of colors and textures in the finished product that could be much wider than that made by the single hand-passed shuttle.

Although still a hand process, the flying shuttle allowed for much wider cloth pieces and much faster work as well as for more complex designs, and it stimulated a demand for thread. Thus the flying shuttle is regarded as the device that in turn led to the invention of the spinning jenny, the spinning mule, and eventually the power loom. These devices produced the complete mechanization of the textile industry over the following century. Together the improvements helped launch the Industrial Revolution and helped account for the rise of capital-intensive industry.

Popular histories of the period often draw several moral lessons from the experience of Kay and his flying shuttle. He is presented as a man deprived of a just reward for his major contribution to technological advance. On another level, the resistance of weavers to the innovation is cited as a case of temporary technological unemployment. As the flying shuttle and other devices mechanized the textile industry, greater numbers of people were employed in clothmaking, suggesting that the weavers' protests were shortsighted. However, the shuttle and related devices did eliminate the class of independent weavers, and their complaints about mechanization were indeed justified from a narrow perspective of self-interest.

Glauber's Salt

The first internationally known brand-name patent medicine, Glauber's Salt, was sodium sulfate, discovered in about 1625 by Johan Rudolf Glauber (1604–1670). Also known as the miracle salt *(sal mirabile),* the medicine was sold as a cure for numerous ailments.

Glauber was born in Karlstadt (now in Germany) and settled in Amsterdam. He built a chemical laboratory there, where he experimented with solvents and mineral acids. He described many of his chemical experiments and techniques in *Opera Omnia Chymica,* published from 1651 to 1661. He is regarded as one of the chemists important in the transition from alchemy to modern experimental chemistry, along with Robert Boyle, the discoverer of **Boyle's law.** Glauber may have been the first to support scientific research through revenues from the sale of a legitimate consumer product.

governor

James Watt (1736–1819) regarded the development of a governor in 1788 for a **steam engine** he completed in 1789 as a relatively minor item in the numerous improvements he brought to the design. However, the governor may represent the ancestor of all automatic feedback devices, and in that sense it was a major contribution to technological advancement. Modern engine governors can take many forms, and a familiar modern use can be found in electronic governors that now are provided in automobiles equipped with cruise control.

Watt's design involved two pendulumlike shafts, each with a weighted ball at the end, linked to the engine's revolving flywheel. As the speed of the wheel increased, centrifugal force would swing the weights at the ends of the pendulums outward, lowering a sleeve valve that controlled the flow of steam to the driving cylinder. When the engine slowed, the pendulums would drop and the sleeve move up, admitting more steam. By adjusting the length of the linkage of the pendulums to the sleeve valve, a constant speed could be achieved, or the engine "governed."

Watt apparently derived his idea from a practice common in **windmill** [II] design of the period. In 1787 a British miller, Thomas Mead, had used a similar design of a centrifugal pendulum to regulate the speed of the windmill so that the millstones would not ride up with too high a speed. Originally, similar spinning ball-and-chain systems had been used simply as flywheels in the Renaissance period in some large machines to provide momentum in **crank** [II]-driven devices to ensure a smooth rotary motion. Mead had adapted the ball-and-chain device not to provide a flywheel function, but to create a means of controlling the speed of the turning millstones. Mead's method had been copied in 1788 at a steam-powered mill built by John Renne in London. In correspondence with his financial backer, Matthew Boulton (1728–1829), Watt learned of the use of the millstone governor at Renne's steam-powered mill in 1788, before Watt adapted it to the steam engine. Watt did not attempt to patent the governor, but for some years he kept its design secret by concealing it.

Although the attribution of the invention of an engine governor to Watt is correct, this case is an excellent example of the adaptation of an existing design to an alternate technology, showing that the source of inventive ideas is frequently one of analogy or cross-technology imitation.

hot-air balloon

The brothers Joseph and Étienne Montgolfier are credited with the invention of the hot-air balloon in 1783. However, a Brazilian Jesuit, Bartholomeu Lorenço de Gusmão, who demonstrated a multiballoon airship in 1709 for the Portuguese royal court, preceded them. No one followed up on the Brazilian incident, and air travel can be said to begin in 1783 with the Montgolfier brothers' balloon.

The Montgolfier brothers were papermakers and observed that paper bags would rise over a fire. Working on ever larger bags, the brothers experimented with the use of hydrogen, isolated in 1766 by Henry Cavendish, but they found it difficult to contain in the paper-lined balloons they built. Firing straw and wool to produce heat, they demonstrated an unmanned balloon in June 1783 at Annonay in France. Hearing of the experiment, Professor of Physics Jacques Alexandre Césare Charles (1746–1823) quickly prepared a hydrogen-filled balloon, assuming that was the "Montgolfier air" that had been used. Charles was a respected scientist, later remembered for his discovery of **Charles's Law.**

At a display witnessed by a huge crowd at the future site of the Eiffel Tower, on August 27, 1783, Charles released his balloon, which flew off into the clouds and came down 15 miles away. In September the Montgolfiers held a demonstration at Versailles, with animals aboard, and then on November 21, 1783, the first man to make a balloon ascent, Pilâtre de Rozier, took a risky trip of 5 miles in a balloon of the Montgolfier design. A few years later, he was the first fatality from air flight.

Benjamin Franklin witnessed one of the earliest ascents and immediately realized that ballooning represented a new dimension in human

Means were used, I am told, to prevent the great balloon's rising so high as might endanger its bursting. Several bags of sand were taken on board before the cord that held it down was cut, and the whole weight being then too much to be lifted, such a quantity was discharged as to permit its rising slowly. Thus it would sooner arrive at that region where it would be in equilibrio with the surrounding air, and by discharging more sand afterwards, it might go higher if desired. Between one and two o'clock, all eyes were gratified with seeing it rise majestically from among the trees, and ascend gradually above the buildings, a most beautiful spectacle. When it was about two hundred feet high, the brave adventurers held out and waved a little white pennant, on both sides of their car, to salute the spectators, who returned loud claps of applause. The wind was very little, so that the object though moving to the northward, continued long in view; and it was a great while before the admiring people began to disperse. The persons embarked were Mr. Charles, professor of experimental philosophy, and a zealous promoter of that science; and one of the Messieurs Robert, the very ingenious constructors of the machine.
—Benjamin Franklin to
Sir Charles Banks,
December 1, 1783

travel, predicting the eventual effect on warfare, transportation, and entertainment.

Meanwhile, advocates of the hydrogen system continued to support the Charles design, and the two types of balloons were known as "Montgolfiers" and "Charlières." The competition between the inventors led to rapid improvements in the rigging of passenger-carrying baskets or gondolas, the development of sandbags for ballast, the use of the **mercury barometer** to determine altitude, anchors for landing, and knowledge of air currents. In 1783, Charles made an ascent to about 10,000 feet. In 1785, the first air crossing of the English Channel was achieved, and ballooning became a sport as well as a technological enterprise.

Many histories of aeronautics suggest that the Montgolfier balloon ascent represented the first case of human flight. However, reliable records indicate that the Chinese had used man-carrying kites as a bizarre method of execution in the 6th century A.D. By the 12th century, the Japanese used them to carry men in and out of besieged cities. At the end of the 13th century, Marco Polo reported the use of man-carrying kites by the Chinese.

Nevertheless, the line of progress from the Montgolfier and Charles balloons to the **dirigible** [V] balloon and to experimentation with heavier-than-air aircraft in America and Europe is a direct one, and the hot-air balloon is rightly regarded, as Franklin predicted, as the beginning of a new dimension in human travel.

Kepler's laws

Johannes Kepler (1571–1630) established three laws of planetary motion that, with some modification, have turned out to represent discoveries about the nature of the solar system and celestial mechanics more broadly.

After a career of teaching mathematics and astronomy, Kepler published a study of the orbits of the six known planets, with the title (translated) *The Secret of the Universe,* in which he accepted the **heliocentric solar system [II]** suggested by Copernicus. After publication of Kepler's work, Tycho Brahe (1546–1601) invited Kepler to join him in Prague, and Kepler set about studying the orbit of Mars. After Brahe died, Kepler inherited the careful records of planetary motion that Brahe had assembled.

Continuing his work on the orbit of Mars, Kepler assumed that it must be perfectly circular and that all the planets must move at a regular rate. Using the data Brahe had assembled, Kepler found that the circular, regular-motion model and the data did not conform. Finally concluding that the data were correct and the model wrong, he tried a variety of orbital shapes, finally settling on an elliptical orbit, which he found corresponded very closely with the data.

Kepler then applied his findings from the Mars orbit to the rest of the known solar system and produced three laws of planetary motion, the first two of which he published in 1609, and the third in 1619:

1. The planets move in ellipses, with the Sun at one focus of the ellipse.

2. The line joining the Sun and a planet sweeps out equal areas in equal intervals of time, known as the law of areas.

3. The square of the time of revolution of a planet is proportional to the cube of its mean distance from the Sun. (This principle became known as the harmonic law.)

Kepler's laws and observations led him to believe that a motive force emanated from the Sun and that its force depended on the quantity of matter in the planet, the same problem later addressed by Newton in his law of universal gravitation. Kepler also

"Discovery" always carries an honorific connotation. It is the stamp of approval on a finding of lasting value. Many laws and theories have come and gone in the history of science, but they are not spoken of as discoveries. Kepler is said to have discovered the laws of planetary motion named after him, but not the many other "laws" which he formulated. . . . Theories are especially precarious, as this century profoundly testifies. World views can and often do change. Despite these difficulties, it is still true that to count as a discovery a finding must be of at least relatively permanent value, as shown by its inclusion in the generally accepted body of scientific knowledge.
—Richard J. Blackwell, *Discovery in the Physical Sciences* (Notre Dame, Ind.: University of Notre Dame Press, 1969), pp. 32–33.

believed that the relationship described in his third law suggested a harmony in the universe, the ultimate "music of the spheres." Newton later modified the first law to indicate that orbits can be ellipses, parabolas, or hyperbolas, and the third law was found to apply only to those planets with an elliptical orbit. The laws, as modified by Newton, apply as well to the orbits of satellites around planets and to the mutual orbits of binary stars.

In the discovery of the nature of the solar system, Kepler's three laws, as "amended" by Newton, represented a major step forward in the knowledge of celestial mechanics.

law of elasticity

In 1660, Robert Hooke (1635–1703), while working as assistant to Robert Boyle (1627–1691), discovered the law of elasticity (also known as Hooke's law). The law, however, was not published until 1678. It states that stretching of a solid body is proportional to the force applied to it. Hooke applied this simple principle in the design of the **balance spring,** used in the design of small timepieces such as watches and clocks.

In his dispute with Christiaan Huygens (1629–1695) over the priority of the invention of the balance spring, Hooke pointed out that the reliable use of a coiled spring to provide regulation for timepieces depended on the law of elasticity that he had enunciated. Hooke's law is valid only for small deformations of solid materials subjected to very low strains, but it accurately describes both the force used to compress a spring and the force released from a spring. Hooke was noted for his other disputes over invention and discovery priority, challenging not only Huygens over the balance spring but also Isaac Newton (1642–1727) over the inverse law of gravitation.

Hooke made many other contributions to instrumentation, including improvements to the **vacuum pump** and the **mercury barometer,** and he served for several decades as head of the Royal Society, one of the first **scientific societies.**

laws of motion and gravity

The motions of the planets and the fact that bodies fall to Earth when unsupported have been observed from antiquity. Aristotle had assumed that the speed with which a body fell to Earth was determined by its size. Ptolemy and other astronomers assumed that planets and stars, made of some quintessential element not found on Earth, were supported in space by some form of invisible or crystalline sphere. Galileo deduced that on Earth, bodies of different weights fell with the same speed, and that air resistance accounted for the difference in speed of a dropped feather and a solid object. According to legend he demonstrated this principle by dropping objects from the Leaning Tower of Pisa, but there is no contemporary record of such an experiment.

Isaac Newton (1642–1727) published in 1687 three laws of motion and a law of gravity, a set of principles that together explained both falling objects and the motion of planets in a single set of explanations. His work that spelled out the laws was *Philosophiae naturalis principia mathematica,* usually referred to as *Principia.* The philosophical consequences of demonstrating that events on Earth and in the heavens operated by the same rules were profound.

Newton's three laws of motion and the law of gravity may be stated simply:

First law: A body in a state of rest or of motion in a straight line will remain in that state unless acted upon by an external force. This continuation at motion or at rest Newton called "inertia." Inertia is demonstrated in the collision of a vehicle with a solid object when the passengers and cargo are thrown forward.

Second law: The force on a body is equal to its rate of change in motion times its mass. That is, the acceleration of a body is proportional to the force that acts on it and inversely proportional to the mass of the body. The second law can be stated as

$$F = ma$$

or as

$$F/m = a$$

where F is the force on the body, m is the mass of the body, and a is the rate of acceleration. This principle is demonstrated by the fact that with the same amount of force, a heavier ball cannot be thrown as far or as fast as a lighter one.

Third law: When one body exerts a force on another, the second body exerts an equal force in the opposite direction, known as the principle of action and reaction. This principle is demonstrated in rocket flight or in the trajectory of a toy balloon releasing its air.

Law of gravity: There is an attractive force between objects that is proportional to their masses and inversely proportional to the distance between them. The force of gravity could be expressed with the formula

$$F = (Gm_1 \times Gm_2) \div r^2$$

where F is the force of attraction, m is the mass of the objects, G is the unknown constant of gravitational attraction, and r represents the distance. Thus the force is equal to the gravitational attraction times the first mass times the gravitational attraction times the second mass, divided by the square of the distance between them.

Henry Cavendish (1731–1810) measured the actual numerical value for the attractive force of gravity in 1798, in an experiment known to legend as "weighing the Earth." When expressed mathematically, the force of gravity is an extremely small number. Thus, for example, two objects, each weighing 1 kilogram and suspended in outer space 1 meter apart, would take almost 12 hours to come together based on mutual gravitational attraction alone. The gravitational constant is now expressed as 6.67259×10^{-8} cm³/g sec². An alternate way to express the constant is: .00000000667259 × cm³/g sec².

Newton himself promulgated the idea that he had derived the law of gravity in a moment of inspiration, thus creating the legend that he discovered gravity by observing a falling apple. In fact, he published the concepts in *Principia* in 1687 after nearly a decade of thinking about the problems but suggested that he had worked out the concept immediately after earning his bachelor's degree during the plague year 1666 at home in his garden in a hamlet in Lincolnshire, England. In putting out this story Newton attempted to clarify in his own favor the priority of his ideas over those developed in an exchange in correspondence with Robert Hooke in 1679. Hooke and Newton had been rivals over several issues of priority of publication. These facts account for the variety of dates (1666, 1679, or 1687) often associated with the discovery of the law of gravity.

By eliminating the separation between "celestial mechanics" and "terrestrial mechanics" Newton's laws provided an explanation for how the **heliocentric solar system [II]** worked. Planets would continue in straight lines from their inertial motion, but the gravitational attraction of the Sun pulled them into orbits. The first law of motion, combined with the law of gravity, provided an adequate explanation for the mechanical operation of the solar system. Christiaan Huygens had published in 1673 a study of the **pendulum clock** in which he explored many of the concepts of motion upon which Newton later formulated the complete laws of motion.

The heliocentric solar system model suggested by Copernicus and for which Galileo had argued now made sense. In fact, using the principles and formulas developed by Newton and examining the perturbation of the planet **Uranus,** scientists were able to predict, almost accurately, the location of **Neptune [IV]**.

locks

Robert Barron, who patented a double-acting tumbler system, invented the mortise lock, also known as the Barron lock, in 1778. In the Barron lock, a lever falls by gravity (hence, a tumbler) into a slot in the bolt and prevents it from being moved until the lever is raised out of the slot to exactly the right level by the key. Then the key could be used to slide the bolt. To foil picking, the Barron lock was improved by another British inventor, Jeremiah Chubb, in 1818. Chubb included a "detector"—a spring that would catch and hold any tumbler that had been raised too high. Holding the tumbler up would prevent the bolt from being withdrawn and also would detect that an attempt had been made to pick the lock.

Joseph Bramah (1748–1814), also known as the inventor of the **flush toilet,** made another improvement on the basic mortise lock that Barron developed. Bramah developed a cylindrical key and lock system that he claimed was invincible. In 1784 he offered a 200-guinea reward to anyone who could pick the lock, and the reward went unclaimed for sixty-seven years. A. G. Hobbs, an American lock salesman for another lock company, who took nearly a month to solve the problem, picked it at the Great Exhibition in London in 1851. The prize was immense by contemporary standards, representing more than $50,000 in the early-21st-century equivalent. To make the precisely fitted parts for his lock, Bramah developed a set of machine tools, assisted by Henry Maudslay (1771–1831), whose improvements to the **screw lathe [II]** contributed to the development of **interchangeable parts [IV]**.

Linus Yale, an American, developed a pin-tumbler lock, improved by his son Linus Yale Jr. in 1865. The cylindrical Yale lock, with its small, serrated, flat key, became widespread through the remainder of the 19th and 20th centuries. Linus Yale Jr. offered a prize of $3,000 for anyone who could pick his lock, and that prize went unclaimed.

Although metal and wooden locks had been developed in ancient

Egypt and were made through the Middle Ages, until the development of the tumbler lock by Bramah and the cylinder lock by Yale, all locks had been rather easily picked. The spread of cheap and pickproof effective locks accompanied the growth of the middle classes in the 18th and 19th centuries, allowing a minimal protection of private property without the necessity of mounting a guard or trusted family member to prevent theft.

logarithms

Logarithms are a system of mathematical computation invented in 1614 by the Scottish mathematician John Napier (1550–1617). Observing that multiplication of any number by another can be accomplished by adding their exponents to a common root number, Napier created a list of exponents to the base *e* (2.718 . . .). He published tables that allowed multiplication and division of large numbers by addition or subtraction of the log numbers, in a work titled *Mirfici logarithmorum canonis descriptio* (A Description of the Marvelous Rule of Logarithms). In 1615, after discussions with Henry Briggs, they devised another system, based on the number 10. Napier's original logarithms are known as "natural logarithms," and those to the base 10 are known as "common logarithms."

In 1619, after Napier's death, another published work, whose title in English is *The Construction of the Marvelous Rule of Logarithms,* gave a fuller account of the method he used in building up his original tables. Napier also developed a simple computing device known as "Napier's Bones" and consisting of rods of ivory imprinted with digits, which when arranged in various ways provided shortcut answers to long multiplication problems. Only in use briefly, they were superseded by the development of the slide rule about a decade after his death.

Logarithms and improved tables developed by Briggs and others were widely used by astronomers by the 1630s. They remained in use until the development of **computers** [VI] provided other means of simplifying difficult numerical calculations.

magnetism

Although many of the properties of magnets were well known to prior experimenters, and although magnets had been used in the development of the mariner's **compass** [II] much earlier, the principles of magnetism were codified and described in 1600 by the British physician and physicist William Gilbert (1544–1603). His description of the motion of magnets was clear in pointing out coition (coming together of two magnets or of a magnet and iron), direction (northward-pointing), dip (the angle from the horizon), and rotation. Gilbert attributed magnetism to the soul of the world and assumed all life was governed by magnetism. Despite this somewhat mystical interpretation, he correctly assumed that Earth itself provided a magnetic attraction.

Johannes Kepler (1571–1630) and Galileo (1564–1642) were both impressed with Gilbert's work, with Kepler assuming that some form of magnetic attraction accounted for planetary positions, foreshadowing Newton's discovery of the **laws of motion and gravity.** Often designated in British works as the discoverer of magnetism, Gilbert is probably better recognized as one who analyzed and described magnetic forces.

map of the Moon

Johannes Hevelius (1611–1687) published the first detailed map of the surface of the Moon in 1647. Born in Danzig (Gdansk), now a city in Poland, his German name was Jan Hewel, Latinized to Hevelius. Titled *Selenographia,* the map was based on his own astronomical observations from the roof of his house.

map of the Moon   Johannes Hevelius, a 17th-century beer merchant, built an astronomical observatory on his roof and was the first to map the Moon. *Library of Congress*

Like many 17th-century astronomers who supported themselves financially with another occupation, Hevelius represents an era when no distinction was made between accomplished amateurs and professionals in this science. He was a beer merchant and city councilor. His rooftop observatory, built in 1641, was destroyed by fire in 1679. Many of his observations on star locations and comet paths were gathered and published posthumously by his wife in *Uranographia* (1690).

Mapping the Moon in the mid-17th century represented a clear departure from the pre-Copernican view that celestial bodies were made of completely different materials than terrestrial ones. With designations of mountain ranges and the incorrect designation of flat plains as *mares* or seas, the astronomical community came to accept that Earth, Moon, and the planets were similar in form, an indication of the diminution of the special place in the universe previously assumed to be held by Earth and its human inhabitants.

mercury barometer

The barometer and the principle that air has weight and pressure that varies and can be measured developed through several stages in the 17th century. In 1643 a former assistant of Galileo, Evangelista Torricelli (1608–1647), filled a glass tube with mercury and then placed it in a bowl filled with mercury. Torricelli noted that small changes in the level of mercury in the tube occurred from day to day and correctly surmised that the change in level represented changes in the pressure of the air. Blaise Pascal (1623–1662), who invented the **adding machine** in 1643, turned his attention to the findings of Torricelli. Pascal guessed that air pressure would be lower at higher altitudes and sent his brother-in-law with two mercury barometers of Torricelli's design up a mountain in 1648 to record measurements. His theory was proven correct.

Otto von Guericke (1602–1686), the inventor of the **vacuum pump,** developed a water barometer in 1672. He built a brass tube more than 30 feet high with a small figure at the top, which, when it rose, indicated an increase in pressure, and when it fell, a decrease. The device, on public display, served as the first barometer used to predict weather changes. Robert Boyle (1627–1691) also experimented with the barometer, and in 1655 he introduced the name. Both Boyle and Blaise Pascal developed siphon barometers, in which the bottom of the mercury tube was bent back 180 degrees, eliminating the need for an open bowl of mercury. In 1664 Robert Hooke (1635–1703) invented a linkage system so that changes in barometric pressure could be displayed on a dial. The work

mercury barometer Following up on Torricelli's invention, Blaise Pascal organized an expedition to confirm that at higher altitudes the air pressure was lower. As the men climbed the Puy-de-Dome with one of Torricelli's tubes, they took measurements that confirmed Pascal's guess.
Library of Congress

with barometers by Pascal, Boyle, Hooke, von Guericke, and Torricelli soon led to the application of barometers to weather forecasting, as a drop in air pressure could foretell an incoming storm.

microscope

Whether early magnifying lenses found in ancient Iraq (Assyria) represent the first invention of microscopic lenses has been debated. If so, the origin of microscopes can be traced to the ancient world. There is evidence that craftsmen in ancient Rome used magnifying glasses for fine work. Simple microscopes consisting of a single lens with a magnification up to ten times were in use after the development of **eyeglasses [II]** in about A.D. 1285.

The Dutch draper or linen merchant and optician Anton van Leeuwenhoek (1632–1723) ground a single lens with a magnification of about 500 power in the 1670s. His magnifying lenses were small in diameter and had to be held extremely close to the eye and to the specimen. Leeuwenhoek was the first to discover microscopic life. Quite secretive, Leeuwenhoek did not share his methods widely. The earlier invention of the compound microscope, with two lenses, has remained controversial, variously attributed to Janssen, Galileo, Kepler, and Hooke.

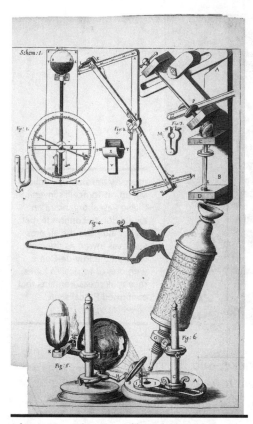

microscope Microscopes like Robert Hooke's with a spirit lamp illumination system expanded the realm of scientific discovery. *Library of Congress*

Zacharias Janssen (1580–1638), according to his son, stumbled on the concept of a two-lens system, the compound microscope, in 1609. Johannes Kepler (1571–1630) described a two-lens system in 1610. Galileo Galilei (1564–1642) used one by 1612. Cornelius Drebbel (1572–1633), another Dutch experimenter, claimed that he invented the compound microscope in 1619. The device was named the "microscope" by John Faber (1574–1629) in 1625. It has been suggested that the microscope resulted from experiments with early **telescopes** in which optical workers and natural scientists tried to examine nearby objects and experimented with improving the lens arrangements by trial and error.

Robert Hooke (1635–1703) made improvements by the 1660s to the device, with which he discovered the **cellular structure of plants.** Hooke's microscope, from which many later models evolved, was mounted in a stand and had a light source reflected by a mirror to the specimen. In the 20th century, the **electron microscope [V]** represented a breakthrough to much higher magnification.

Together with the telescope, the microscope represented the expansion of observation into new realms in the 17th century, leading to a range of new discoveries. Both instruments, invented by a process of experimentation and trial and error, illustrate how sometimes the invention of a technological tool is a necessary precursor to scientific discovery. Like the contributions of the telescope to astronomy, the microscope allowed for new discoveries and observations affecting medicine, embryology, biology, physiology, and zoology and helped establish those fields as separate disciplines.

microscopic organisms

Several experimenters with the **microscope** discovered new findings, including Robert Hooke (1635–1703), who was the first to describe the **cellular structure of plants.** However, Anton von Leeuwenhoek (1632–1723), a Dutch lensmaker, pioneered in the field, achieving much greater magnification than others. Over the more than fifty years following 1672 until his death, he described and drew illustrations of his findings. He described bacteria, sperm, protozoa, and **red corpuscles of blood.** He described the mouthparts of insects and the structure of plants. Many of his discoveries were communicated in letter form to the Royal Society of London.

Leeuwenhoek's microscopes consisted of single lenses, some extremely small. His discovery of protozoa and sperm cells, which he called animacules, contributed to a controversy between ovists and animaculists that lasted through most of the 18th century. The followers of William Harvey (1578–1657), who had discovered the **circulation of blood,** believed that the female ovary produced the entire embryo and were known as *ovists*. The scientists believing that the male sperm was necessary to the production of the fertilized embryo were known as *animaculists*.

orrery

An orrery was an astronomical instrument or demonstration device used to represent the **heliocentric solar system [II].** Usually made of brass and steel, it was driven by clockwork gears that showed the positions and motions of the planets by moving balls about a center sphere representing the Sun. The small spheres representing the planets rested on extended wire arms, making up a moving model that demonstrated numerous effects, including the apparent retrograde motion of the planets when observed from Earth. The first orrery was made before 1719 by George Graham and Thomas Tompion, and was so named because it was presented to Charles Boyle, the earl of Cork and Orrery. The device was sometimes called a lunarium, a tellarium, or a planetarium.

oxygen

The chemist Robert Wilhelm Bunsen (1811–1899), inventor of the **Bunsen burner [IV],** noted in his lectures that the day on which oxygen was discovered was the real birthday of modern chemistry. Even so, that date and its discovery remain obscure because of the more or less simultaneous contributions of several scientists.

Carl Wilhelm Scheele (1742–1786) first isolated oxygen in 1771 or 1772, but he delayed publishing his findings until 1777. The first proven entry by Scheele regarding oxygen in his laboratory notebooks dated to November 1772. Joseph Priestley (1733–1804) was the first to publish his discovery of oxygen, in 1774. However, it was Antoine Lavoisier (1685–1794) who gave oxygen its name, in 1778, and who clarified the role of oxygen in combustion. Lavoisier was mistaken in believing that oxides, when dissolved in water, always generate acids, and the word *oxygen* is derived from two Greek root words meaning "acid generator." In fact, some oxides do form acids when dissolved in water, but oxides of metals dissolve in water to form bases.

Lavoisier conducted experiments in which he determined that what he at first called "eminently respirable air" combined in combustion and in respiration to produce carbon dioxide. Working with Pierre Laplace (1749–1827), Lavoisier showed that combustion and respiration are the same process—one fast, one slow—and he published their finding that **water as a compound** consisted of hydrogen and oxygen in 1783. Lavoisier also pointed out that all substances could exist in three states: solid, liquid, and gaseous.

Lavoisier's explanation of combustion overthrew the existing explanation of fire, which suggested that a substance, phlogiston, was given off when a material was burned. The concept had been developed by the German chemist George Stahl (1660–1743) and had become the accepted explanation. The problem with the phlogiston theory was that in some circumstances phlogiston had to have weight, and in other circumstances it had to have a "negative weight." When certain metals were roasted in the presence of air, they turned into a powdery material, termed a *calx*. In the phlogiston theory, it was assumed that phlogiston had been transferred from the metal to the air. When a calx was heated with charcoal, it was restored to its metallic form. The assumption was that metal had absorbed phlogiston back from burned charcoal. However, the fact that the resulting metal was lighter than the calx would require that the added phlogiston had resulted in a lower weight, requiring phlogiston in this circumstance to have a negative weight. Lavoisier cleared up the matter by showing that a calx was an oxide and that the oxygen was released and combined with the charcoal as the charcoal was burned.

A political liberal and aristocrat, Lavoisier was caught up in the French Revolution. He was a participant in the Fermé Général, a tax-collecting contracting firm, as well as in other government enterprises. During the Terror of the Revolution he was executed for such connections.

papier-mâché

In 1772, Henry Clay, a craftsman in Birmingham, England, improved on an ancient plastic material that had come to Europe by way of Persia (now known as Iran). He used a method of building up paper panels on boards of wood, then impregnating the paper panels with varnish to produce a substance that could be planed, sawed, or doweled. The paper he used was largely made up of cotton fibers, and his resulting papier-mâché was therefore very strong.

Clay held the British patent on the material until 1802, and his process was widely pirated by other manufacturers in the Birmingham area. Through the early 19th century, papier-mâché became so popular that it was used to make furniture including couches, chairs, and tables as well as ornamental items. In the history of technology, papier-mâché is a forerunner of **composite materials [VI]** such as plywood and fiberglass and remains a common material for the construction of masks, decorative items, and fine art.

Pascal's law

Blaise Pascal (1623–1662) was born in Clermont-Ferrand, France, and in his early years met both René Descartes (1596–1650), the founder of **Cartesian mechanistic science,** and the mathematician Pierre de Fermat (1601–1665), known to the history of mathematics for **Fermat's Last Theorem.**

Pascal stated in 1653 the principle that the pressure on a fluid at rest exerts a force perpendicular to the surface and independent of the direction of orientation of the surface. It also stated that additional pressure applied to the fluid would be transmitted equally to every point in the fluid. This set of principles, known as Pascal's law, was later applied by Joseph Bramah (1748–1814) in the invention of the **hydraulic press [IV]** and was reflected in many hydraulic machines for transmitting power, such as **hydraulic jacks [IV], elevators [IV],** brakes, and servomechanisms. Pascal himself was well known in his own time for the invention of the **adding machine.**

Pascal made a number of other contributions in his short life, contributing to the development of probability theory in a series of letters with fellow mathematician Fermat over games of chance. Undergoing a religious conversion, Pascal is also famous for developing "Pascal's wager," which suggests that betting against the existence of God requires accepting infinite odds. As he stated it: "If God does not exist,

Pascal's law Blaise Pascal was famous in his own time for the development of the Pascaline, an adding machine that worked by linked rotors. The principle was used in adding machines well into the 20th century. *Library of Congress*

one stands to lose nothing by believing in him anyway, whereas if he does exist, one stands to lose everything by not believing." This concept and others he included in an unfinished work published in 1670 after his death, *Pensées* (Thoughts).

pendulum

Galileo Galilei (1564–1642) worked out the basic principle of the pendulum, later used as a means of regulating the operation of the **pendulum clock.** Observing a swinging lamp in the cathedral in Pisa, Galileo had noted that it took the pendulum precisely the same amount of time to move in one direction as it did to return. The periodic precision of a pendulum had been mentioned by a 10th-century Arab astronomer, Ibn Yunus, but Galileo did not know of his writings. Galileo suggested that the principle could be applied to a clock. Excluding the effect of the opposition of the air and other accidental factors, he recognized that the period of oscillation is independent of the arc of swing and is proportional to the square root of the length of the pendulum.

Although there are scattered references to the use of pendulums in clocks prior to Galileo's work, he developed a treatise on the subject in

1583. His son Vincenzio Galileo sketched a pendulum clock in 1641, following the concepts of his father. But Christiaan Huygens (1629–1695) was the first to perfect the pendulum clock a few years later and to develop theories of motion based on his study.

pendulum clock

The invention of the **clock** [II] to mark the measurement of the day into hours, and its introduction into European urban life in the 13th and 14th centuries, was both a symptom and a cause of the adoption of new attitudes toward work, time, and daily life.

> Philosophy is written in this grand book, the universe, which stands continually open to our gaze. But the book cannot be understood unless one first learns to comprehend the language and read the letters in which it is composed. It is written in the language of mathematics, and its characters are triangles, circles, and other geometric figures without which it is humanly impossible to understand a single word of it; without these, one wanders about in a dark labyrinth.
> —Galileo Galilei, *The Assayer* (Rome, 1623)

That impulse toward measurement of nature by mechanical means was greatly advanced by the work of Christiaan Huygens (1629–1695).

In 1656, Huygens applied the principle of isochronicity developed by Galileo Galilei (1564–1642) in reference to **pendulums** to a timekeeping device. Doctors had made use of the concept by using a pendulum of thread with a weight to count a patient's pulse rate, known as a *pulsilogia*. Huygens followed up on the concept, combining a pendulum with a weight-driven clock. In his clock, the pendulum would swing exactly once per second, regulating the movement of the clock's hands. Although the force to drive the clock was found either in the weights or in wound spring clockwork, the pendulum provided a timing device to regulate the motion to a steady rate. Within a few years, dozens of clock towers, and then hundreds of smaller clocks throughout Europe, added Huygens' pendulum regulator system.

Often the pendulums would be added to a clock that was already built, leading to some historical mysteries. Since a maker might put the date of construction on the clock face, several 17th-century clocks with pendulums attached to them have face dates preceding Huygens' work. Clock historians have concluded that these few cases represent surviving examples of clocks modified with the addition of a pendulum after Huygens published on the topic.

Huygens' precise clocks, which kept accurate time within a few seconds per day, allowed for far more accurate timing of mechanical, chemical, physical, and meteorological phenomena. However, the pendulum clock could not be adapted for a **watch** [II] or for use aboard a ship, where an accurate **chronometer** would be required. Huygens

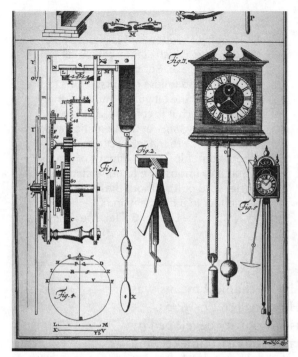

pendulum clock Christiaan Huygens built the first pendulum clock, shown here. His weight-driven clocks could be small enough for home use. *Library of Congress*

developed a **balance spring** for the watch, and in the 1670s he and Robert Hooke (1635–1703) disputed the priority of that invention.

Huygens was born at The Hague, in Holland, into a well-to-do family, and he studied law and mathematics. His invention of the pendulum clock and his description of the patent for it, published in 1657, made him well known. In 1673 he published a thorough study, *Horologium Oscillatorium,* which went far beyond the description of the clock to a study of the mechanics of the pendulum. Using a method reflecting **Cartesian mechanistic science,** Huygens combined mechanical work with mathematical reasoning. To study the motion of the pendulum closely, he included discussion of such principles as the acceleration of gravity, the question of the center of gravity, the moment of inertia, and the laws of centrifugal force. All of these considerations were studied and published about 14 years before Newton did his work on the **laws of motion and gravity.** The exact degree to which Newton drew on the work of Huygens was disputed in their time and remains a matter of discussion today.

piano

The piano was invented by an Italian harpsichord maker of Padua, Bartolomeo Cristofori (1655–1731), in 1700. The first published record of the piano was in 1709, but recently discovered earlier inventories show the 1700 date. Cristofori solved a limitation of the harpsichord, in that the player could not change the timbre or the loudness of the sound produced and could not duplicate a rise or diminishing of volume characteristic of human emotion. As opera became popular in the late 17th century, this limitation of the harpsichord frustrated musicians. The primary difference between the piano and the harpsichord is that the piano connected the keys to hammers that struck the strings, while the harpsichord, based on the harp, plucked the strings.

Cristofori probably made at least 20 pianos before his death, and the word *piano* derived from his name for the device, *gravecembalo col piano e forte* (harpsichord with soft and loud). A surviving Cristofori piano from 1726 has many of the features of later pianos.

Cristofori's design improved on the harpsichord by arranging a different kind of hammer action that allowed the hammer when striking the string to bounce free immediately to its original resting position, with a back check to catch and hold the hammerhead when it fell back. This "escapement" was key to the ability of the piano to be played at different volumes, with a heavier strike to the key producing a harder strike on the strings, without sticking. Pianomakers in Spain, Portugal, and Saxony made improvements over the next century.

The first public piano recital was performed in 1768 by Johann Christian Bach (1735–1782), the eleventh son of Johann Sebastian Bach (1685–1750). As musical styles became more expressive and dramatic, pianomakers made many changes to accommodate the new music, including a wider musical range with more keys, the addition of steel strings, larger soundboards, and the addition of pedals to change timbre by shifting the keys to strike one or two of three parallel strings. A third pedal on some pianos was added to achieve a sustained single note while others were being played.

The development of the piano is a perfect example the interaction between the changed instrument and the changing needs of the musicians and their styles of music. The piano both reflected the demands of the era and provided a means for musicians to create entirely new types of music. The piano keyboard was later replicated on the **music synthesizer [VI]** of the 20th century.

pressure cooker

The pressure cooker was invented in 1679 by Denis Papin (1647–1712), a physicist. Papin was born in France and studied medicine. He worked with Christiaan Huygens (1629–1695), the Dutch scientist who developed the **pendulum clock,** working in Paris after Huygens was appointed to the Paris Academy, one of the earliest **scientific societies.**

In 1675 Papin moved to London, where he worked as secretary to Robert Hooke, who had just been appointed to head the British Royal Society. Papin worked with both Huygens and Robert Boyle on the development of an air pump, and while working in London with Boyle, Papin developed the pressure cooker, or "steam digester." Widely adopted in the 20th century, the cooker worked by closing the pot with a tightly fitted lid with a weighted safety-release valve similar to that designed by Papin. With the higher pressure, the boiling point of water rose considerably, allowing meats to be cooked much more quickly than by boiling at ambient pressure. Papin also developed a hand-propelled paddlewheel boat and proposed the development of the steamboat [IV].

red corpuscles of blood

Knowledge of the nature of blood was advanced by the work of Jan Swammerdam (1637–1680), a Dutch experimenter with the **microscope.** He made many contributions to the study of insects and of the contraction of muscles. In about 1658 he was the first to describe the red corpuscles in blood from a microscopic inspection of the blood of frogs. Like Christiaan Huygens (1629–1695), Anton van Leeuwenhoek (1632–1723), and other natural philosophers in Holland, Swammerdam's access to excellent optics provided tools for new discoveries in the realm of the very small.

Swammerdam studied at Leiden University, completing a degree in medicine at age 30, with a dissertation on human breathing. Between 1667 and 1673 he worked on microscopic studies of insects and other animals. He developed a method of injecting the blood vessels of insects with wax in order to better observe them and studied cell division and muscle contraction in frogs. He concluded, rightly, that insects were not spontaneously generated in waste matter.

Swammerdam's work in other areas included many contributions. He developed minute instruments for the dissection of insects. He described cell division and offered an explanation for penile erection. His explanation for the development of insects through stages of metamorphosis helped in developing the concept of preformation as an explanation of how embryos develop into full individual creatures.

His discovery and description of red corpuscles represents one of many discoveries that derived from the microscope in the 17th through the 20th centuries.

refraction of light

Willebrord van Roijen Snell (1581–1626), a Dutch professor of mechanics and mathematics, discovered about 1621 the law of the refraction of light, also known as Snell's law. It is noted that when a spoon is standing in a glass of water that the handle appears to change angle at the surface of the water. Snell pointed out that the sines of the angles of incidence and refraction always bore the same ratio to each other for a given interface between two media. He called the ratio the refractive index for that interface.

Snell died before he could publish his essay, but the principle was known to René Descartes and to Christiaan Huygens, who both studied refraction. When Descartes published his *Discourse on Method* in 1637, he referred to the law of refraction, without credit to Snell, and it is not known whether he knew of the prior work. Descartes believed light traveled faster in a dense medium, and both he and other contemporaries believed in a particle or corpuscular nature of light. Huygens modified this view with a concept that light traveled in waves through a luminiferous medium or ether. Both Huygens and Descartes sought to reconcile their understanding of light with the question of refraction in different media.

rifled gun barrels

The rifle represented an improvement in accuracy over the **musket** [II]. The term derived from the procedure for cutting spiral grooves in a barrel, leaving ridges that would impart a spin to the bullet, known in German as *rieflen,* developed as early as the 15th century. Although an excellent concept, early rifling was not always successful because if not precisely spiraled, the grooves would catch the bullet, often destroying the weapon. Early successes with rifling came from 1720 to 1770, when German gunmakers in Pennsylvania perfected the technique, producing a marksman's weapon excellent for hunting known as the Kentucky rifle. The barrel was as long as 4 feet, and the shot was often wrapped in a patch of greased cloth or thin leather to ensure a tight fit in the barrel.

Like the musket, the Kentucky rifle was loaded from the muzzle, which was a difficult and often a slow process. Attempts to develop a breech-loading rifle in the early 19th century led to several improvements. In 1848 Christian Sharps invented a single-shot rifle that used a paper cartridge. Closing the breech cut off the edge of the cartridge, exposing the powder to the action of a percussion primer. These weapons were accurate, and those proficient in their use were called "sharpshooters," reflecting the pun on the inventor's name.

A French officer, Claude-Étienne Minié, invented a bullet with a flat base and a pointed nose, a "cylindro-conical" shape, rather than the traditional spherical shape. The base was slightly concave and would expand to tightly fit against the rifling ridges in the barrel. The minié ball was widely used in the American Civil War. Self-contained cartridges with powder in a shell capped by a bullet were introduced by B. Tyler Henry and Frederick Martini in 1860. Following the American Civil War, Henry went to work for the Winchester Company, which produced a repeating rifle.

Rifling of large **cannons** [II] aboard ships improved their accuracy in the late 19th century. In some naval literature, such weapons were referred to as "rifles" or "naval rifles." In the early 20th century, **battleships** [V] were equipped with rifled ordnance with interior bore diameters up to 14 and 16 inches, with experiments in even larger diameters.

Later infantry rifles included the German Mauser (1867), the American Springfield (1903), Enfield (1917), the Garand or M-1 (1934), and the M-16 (1960). The American inventor of the M-1, John C. Garand (1883–1967), began design work on his improved rifle in 1919, but it was not fully demonstrated and cleared for procurement until 1935. The M-1 was the main infantry weapon of the U.S. soldier in World War II and the Korean War. Altogether, some 6 million Garand rifles were manufactured. The Garand fired about three times faster than the 1903 Springfield. Later rifled weapons for the foot soldier included the Soviet Union's Tokarev (1940) and AK-47 (1947). The M-16 and AK-47 can be switched from semiautomatic fire to fully automatic fire, allowing bursts of fire similar to a **machine gun** [IV], and they are known as assault rifles.

Saturn's rings and moons

Like many other astronomical discoveries, finding that Saturn was surrounded by spectacular rings came as a result of close observation with improved instrumentation, rather than through contemplation and reconsideration of a long observed but little understood process of nature that characterized many discoveries. Christiaan Huygens (1629–1695), a Dutch natural philosopher, detected the rings in 1656. Galileo had noted that Saturn had a strange shape in 1610, but with his instruments he could not discern exactly the nature of the phenomenon.

Huygens studied law and mathematics at the University of Leiden and the College of Orange in Holland, and he spent the years 1650 to 1666 at The Hague, working in some isolation from the rest of the growing scientific community. He published a number of works in the field of physics, on light and optics, on the issue of probability, and on geometrical problems. He moved to Paris in 1666, where he met many others working in astronomy, physics, and mathematics. Late in life he visited London and met with Isaac Newton.

At age 25, in The Hague, Huygens took up grinding his own lenses and constructing his own telescopes. He detected Titan, the first observed satellite of Saturn, in 1655, and the rings of the planet in 1656. He published his findings in *System Saturnium* (1659), describing in some detail the phases of the rings. Huygens went on to patent the **pendulum clock,** and he later theorized on the possibility of life on other planets.

Saturn has altogether 18 confirmed satellites, and the 8 largest were discovered by telescopic observation. Titan, discovered by Huygens, is by far the largest, with a diameter of about 3,200 miles, larger than the planet Mercury and only a bit smaller than the planet Mars. The next largest two satellites of Saturn, Iapetus and Rhea, with diameters in the range of 900 miles, were discovered by Giovanni Domenico Cassini (1625–1712) in 1671 and 1672, respectively. Cassini detected two more in 1684, and William Herschel (1738–1822) found two in 1789. An irregularly shaped smaller satellite was detected by telescope in 1848. Six smaller satellites were discovered between 1898 and 1980 by studying photographic plates taken through telescopes, while four more were discovered by the analysis of data from the *Voyager 2* spacecraft. Later analysis has uncovered evidence for at least seven more very small satellites.

The complex nature of Saturn's multiple rings, now designated with letters A through G, was extensively studied in the 20th century. The rings consist of vast numbers of small particles, mostly water ice, that range in size from about ½ inch to about 10 feet in diameter. The rings themselves are quite flat, less than 700 feet thick, but some hundreds of thousands of miles in circumference. Problems in counting the exact number of satellites of Saturn derive from questions of definition as well as issues of detection, for it appears that some satellites up to 30 miles in diameter have swept out gaps in the rings from gravitational attraction but have eluded detection by direct observation.

> In a few short months . . . the moons of Saturn have been transformed. Before November 1980, the ones that were known were no more than dots of light in a telescope. Now they form an array amounting to 17 new worlds.
> —Laurence A. Soderblom and Torrence V. Johnson, "The Moons of Saturn," *Scientific American* 246, no. 1, (January 1982): 101.

scientific societies

In Britain and on the European continent, the formation of scientific and literary societies devoted to discussing the latest findings of scientists contributed to the spread of scientific information and eventually led to several principles that helped science advance in a more organized fashion. With the publication of the transactions of the societies emerged the first scientific journals, and the concept that published work would soon be tested to verify that it could be replicated led to more rigor and to a basic principle of verification.

The first scientific societies in Britain grew up in London and later in smaller English towns as gentlemen sought to learn what was happening in science in the rest of the world. The first British scientific society, in London in 1644, had its origins in a group of young scientists calling themselves the "Philosophical College." They first met weekly at a tavern, and later at Gresham College. After the accession of Charles II to the throne in 1660, the Philosophical College received a charter in that year that formally incorporated the group as "The Royal Society for the Improvement of Natural Knowledge," known informally as the Royal Society. Later, in the provinces of Britain, in Spalding, Liverpool, Northampton, and Edinburgh in the 1730s and 1740s, regional groups would meet to read journals and to discuss their contents, often producing local scientific talent.

The Royal Society was particularly interested in the writings of Francis Bacon (1561–1626), whose 1620 work *Novum Organum* suggested that science should be turned to practical questions. When Robert Hooke (1635–1703), as curator, proposed a set of statutes, he reflected Bacon's concepts that science should influence the practical arts of engines and inventions rather than questions of morals, metaphysics, or politics. Hooke remained curator of the Royal Society from 1662 until his death 41 years later. Hooke was known for his discovery of the **law of elasticity** (Hooke's law) and the **cellular structure of plants,** and for inventions including the spirit level and improved timepieces.

Serving as curator of the society, Hooke published numerous works that reflected his diverse scientific interests. A partial list includes: *An attempt to prove the Motion of the Earth* (1674); *Animadversions on the first part of the Machina coelestis* (1674); *A Description of Helioscopes and some other Instruments made by Robert Hooke* (1676); *Lampas: Or Descriptions of Some Mechanical Improvements* (1677); *Lectures and Collections: Cometa, Microscopium* (1678); and *Of Spring, Explaining the Power of Springing Bodies* (1678).

The Royal Society divided its efforts by establishing eight committees to examine special problems. The largest groups considered mechanical questions, the trades, and agriculture, while smaller groups focused on astronomy, mathematics, and chemistry. The society remained active through the 18th century and into the present.

In Italy, a number of scientific societies came and went. A group called the Academia Secretorum Naturae met in Naples during the 1560s, and in Rome, the Accademia del Lincei was active between 1601 and 1630. The Academia del Cimento met in Florence from 1657 to 1667 and focused on questions of theoretical mechanics, avoiding the study of astronomy because of the condemnation of Galileo by the Inquisition. Germany and Russia hosted short-lived academies, with the one in St. Petersburg, Russia, staffed with the Swiss scientists Daniel and Nicolas Bernoulli and the Swiss mathematician Leonhard Euler.

In France, a lively growth of scientific societies in the provinces and in Paris in the early 1600s set up networks of scientific correspondence and provided meeting places for scientists. Louis XIV's minister Colbert decided that the crown should support such efforts, and in 1666 the Paris Academy of Sciences was founded, consisting of about 20 scientists, all on the royal payroll. They received specific problems sent to them by royal ministers. Although the Paris Academy worked on assigned problems (some quite frivolous, such as making improvements to the royal fountains) and were salaried by the crown, and the British Royal Society was voluntary and set its own agenda, both groups in the 17th century tended to follow the ideas of Bacon, attempting to bring scientific findings to practical use in the everyday arts. From 1699 to 1739, the secretary of the Paris Academy was Bernard de Fontenelle (1657–1757), a popularizer and follower of René Descartes, who had stimulated a new outlook with **Cartesian mechanistic science.** As a consequence, the Paris Academy began to take on more philosophical and social issues.

As the Paris Academy and the Royal Society reported on work through the 18th century, they provided not only forums for the dissemination of scientific findings but also established pathways for the confirmation of scientific findings. In France the trend was to concentrate scientific excellence in Paris, but in Britain, with the flourishing of the provincial societies, independent centers of scientific pursuit grew up in the industrial midlands and in Scotland.

In the British colonies that became the United States, Benjamin Franklin and others formed a group in 1743 that evolved into the American Philosophical Society. In 1780 it received a charter from the Commonwealth of Pennsylvania and, like the Royal Society, represented a voluntary association of people interested in science and other developments. In the early days of the American republic it served as a form of national academy. In 1846 the Smithsonian Institution was established with a bequest of $100,000 from the estate of James Smithson (1765–1829), a British chemist and meteorologist. The Smithsonian took on some of the functions of a national museum. In 1863, during the American Civil War, the U.S. government established the National Academy of Sciences (NAS), which, along the lines of the Paris Academy, served to conduct research at the request of departments of the government. The NAS was expanded in later years, with the addition of the National Research Council in 1916, the National Academy of Engineering in 1964, and the National Institutes of Medicine in 1970.

sextant

The sextant, used as a navigational instrument, was simultaneously invented in 1730 by the Englishman John Hadley (1682–1744) and the British colonial American Thomas Godfrey (1704–1749). Hadley had successfully made a number of **telescopes,** recognized as the best of the era by Edmond Halley (1656–1742), who contributed to the knowledge of **comet paths as orbits [II].**

The sextant consisted of a small telescope mounted on a frame with a sliding scale. The mariner would use the telescope to sight the horizon to ensure that he held the instrument level. He would change the angle of a mirror on the device, reflecting the light of the Sun or another celestial object through a filter to bring the image in line with the horizon. The angle between the celestial object and the horizon could then be read out on a scale. With this information, the latitude of the ship could be determined. A second reading could determine the exact local time by the elevation of the Sun above the horizon at that latitude. If the mariner had an accurate **chronometer** set to show London time (or that of any other known longitude), he could determine the number of degrees east or west of London. The sextant was so named because it allowed a reading of elevations to 60 degrees, or ⅙ of a full circle.

Earlier devices for measuring elevation of celestial objects were less accurate and included the **cross-staff [II]** and the **astrolabe [II].** The invention of the sextant, with its filter and mirror, allowed sighting of the Sun without directly looking at it, a great improvement over the cross-staff, with its risk of eye damage.

size and shape of Earth

The determination of the size of Earth and the fact that it is an oblate spheroid (a sphere flattened slightly at the poles) were discoveries to which many astronomers and geographers contributed. The accurate measurement of the size and the shape of Earth would be essential to accurate mapmaking and navigation. Columbus, like all educated Europeans of his era, assumed Earth to be a sphere, but he mistakenly assumed Earth to be much smaller than it actually is. His mistaken belief in the size of Earth accounted for his notion that he had reached the East Indies when he had sailed across the Atlantic. Compounding several errors, Columbus assumed Earth to be about 29,000 kilometers around at the equator. The actual shape of Earth is 40,075 kilometers in circumference at the equator. Because it is flattened at the poles, the circumference is only 40,007.8 kilometers when measured along any longitude intersecting both poles.

The ancient Greek geographer and mathematician Eratosthenes (276–194 B.C.) developed a method of calculating the size of Earth that was theoretically correct, but due to several errors, his estimate was off by about 15 percent, still much more accurate than the figures used by Columbus. In about 240 B.C. Eratosthenes was appointed chief of the Alexandria **library** [I], where he heard that a well in Syene, far up the Nile in southern Egypt (near the modern Aswan), was so situated that the Sun shone directly into it, illuminating the water at high noon on June 21, the longest day of the year. Eratosthenes knew that no such phenomenon was observable in Alexandria, and he reasoned that Syene must lie at the northern edge of the **latitudinal zone** [I] of the tropics, the latitude now known as the Tropic of Cancer. In the progress of the Sun in its north-to-south passage across the sky, it was at its most northerly point on June 21. If it shone directly into the well at Syene, that would mean that it was at 90° or directly overhead at that location, meaning that that point on the surface of Earth was directly on a line drawn from the Sun to the center of Earth. Eratosthenes calculated that if he could measure the angle of a shadow cast on June 21 in Alexandria, and if he could get an accurate measurement of the distance from Alexandria to Syene, he would know the distance of one defined arc of the 360° of the surface of Earth. From that he could calculate the circumference of Earth, all without leaving the library.

According to travel reports, Syene was due south of Alexandria. If true, that would mean the two points were on the same longitude or meridian. Using a vertical obelisk in Alexandria to define the angle of the shadow, he found that the shadow on June 21 was 7°12′, or about ¹⁄₅₀ of a complete circle. The distance to Syene was reported to be 5,000 stadia or about 578 miles. Multiplying that figure by 50, he determined that the circumference of Earth was 250,000 stadia, or more than

46,250 kilometers. The actual circumference of Earth along a longitude is just over 40,000 kilometers, so his figure was about 15 percent too high.

Although his method was an excellent one, several factors contributed to his error: Syene and Alexandria are not on the same meridian, Syene is not quite on the Tropic of Cancer, and the distance is not exactly 5,000 stadia. But as some of the errors canceled each other out, he came quite close to the actual figure.

In about A.D. 1525, Jean Fernel (1497–1558), a French cartographer, decided that the measurement of the circumference of Earth could be conducted with more accuracy. Using a quadrant, a cruder version of an **astrolabe [II]**, Fernel determined the latitude of Paris. Then Fernel measured the circumference of one of the wheels of a carriage, and as he drove slowly north from Paris he counted the revolutions of the wheel, some 17,024 revolutions, on the way to Amiens. In Amiens he took another measurement of the angle of the Sun to determine the latitude there. The difference was almost exactly 1 degree. He multiplied the distance he had calculated from Paris to Amiens by 360 and came to a value of the circumference that was within 0.1 percent of the correct value. Fernel's figure stood for more than a century, with several refinements.

Using surveying instruments, Willebrord Snell (1581–1626), a Dutch mathematician, developed the basic method of finding distances by triangulation. Using a network of triangles, Snell calculated the distance between the two towns of Alkmaar and Berkgen op Zoom in Holland. Although his calculation was more than 3 percent off, he had introduced a method that, if used accurately, could determine the precise distance between points working from a measured baseline and using angles sighted to distant landmarks such as church steeples and hilltops. In 1633 Richard Norwood, a British mathematician, used a surveying chain invented in 1620 by Edmund Gunter that was exactly 66 feet long, so that 80 chains represented 1 mile. After laboriously measuring by chain the distance from the Tower of London to York, and then measuring the Sun angles, he determined that 1 degree of longitude was 69.2 miles or 110.72 kilometers. By that calculation he came extremely close to the actual circumference of Earth, about as close as Fernel. Jean Picard, a French astronomer, conducted a survey from 1669 to 1670 retracing the route of Fernel, and he calculated that 1 degree was 110.46 kilometers. One degree on a longitude line is actually just over 111.13 kilometers, *on average,* but because of the flattened shape of Earth, it is 110.567 at the equator and 111.9 as one nears a pole.

When Isaac Newton developed the **laws of motion and gravity** in the 17th century, however, a new controversy developed between those believing that Earth was a perfect sphere and those suggesting that it had a slightly irregular shape, either flattened at the poles or flattened at the equator. According to Newton, the centrifugal force of the spinning Earth should lead to the swelling out of Earth at the equator and the flattening of the poles, leading to an oblate spheroid rather than a true sphere. Two French expeditions by members of the French Academy of Sciences—one to Lapland, and one to Peru—in 1742 and 1743, respectively, resulted in accurate measurements that confirmed the oblate shape. In Lapland the expedition headed by Pierre Bouguer found 111.094 kilometers to a degree, while in Peru, Charles Marie de La Condamine found a degree to measure 109.92 kilometers. Later refinements with more advanced instrumentation could only improve on those measurements by minute fractions.

specific and latent heat

Joseph Black (1728–1799) taught chemistry and medicine at Glasgow University, where he met James Watt, the inventor of the **steam engine.** Perhaps stimulated by discussions with Watt, Black conducted experiments on the freezing and melting of water and of water and alcohol mixes. He determined that ice melting in an atmosphere of a fixed temperature and pressure absorbs heat, and that warm water placed in a cold atmosphere of a known fixed temperature releases heat to the atmosphere. In about 1764 he developed the concept of specific latent heat of fusion, the heat required to change a weight of a solid to a liquid with no change of temperature, and the specific latent heat of vaporization, the heat required to change a weight of liquid to vapor with no change of temperature. Later physicists built on these concepts when they developed the **laws of thermodynamics [IV].**

spectrum of light

The discovery that white light is composed of a mixture of colors represents a fundamental advance in human knowledge of the universe. Isaac Newton (1642–1727) first performed a series of experiments related to the topic from 1665 to 1666. He concluded from the work that white light is secondary and heterogeneous and that the colors are primary and homogeneous. The colors of the spectrum, he also concluded, can be quantified. What we now call wavelengths, he referred to as a "degree of Refrangibility."

Newton's simplest and yet most crucial experiment consisted of passing a ray of light through two prisms, which, like the raindrops in a shower on a sunny day, separate light into a rainbow. The first prism would separate the white ray of sunlight into colors. Then, with a slot in a board, one of the colors would be selected and passed through the second prism and then projected to a second board. The red band of the spectrum would not be further separated by the second prism, nor would the colors yellow, blue, or violet. Thus these "primary" colors were inseparable or homogeneous, with a constant angle of refraction from the prism, while the white light represented a combination of the primary colors. Newton called this the *experimentum crucis* or crucial experiment, for it demonstrated the fundamental principle in a clear and irrefutable fashion.

Newton described the experiment in a series of lectures, and he later published them in a work in English, *Opticks,* in 1704. Some of his earlier studies on the topic of light were published the year after his death, in 1728, as *Optical Lectures.* The book *Opticks* had a very logical structure, moving from definitions and axioms through propositions and theorems to experimental proof. At the end of the work he left a series of unanswered questions, or "Queries," for further study. Later scientists often regarded *Opticks* as an appropriate model for publication and reporting of experimental work, and the book is studied as much for its clear expositional technique as for the substance of the findings. Newton held that light behaved as particles, not as waves, while others, such as Christiaan Huygens (1629–1695), argued for a wave theory. Experiments in the 20th century by Max Planck and Albert Einstein led to the conclusion that light consists of **photons [V]**, with characteristics of both particles and waves.

A by-product of Newton's work in optics was the development of the reflecting **telescope,** also known as the Newtonian telescope.

> As in mathematics, so in natural philosophy, the investigation of difficult things by the method of analysis ought ever to precede the method of composition. This analysis consists in making experiments and observations, and in drawing general conclusions from them by induction, and admitting of no objections against the conclusions but such as are taken from experiments or other certain truths. For hypotheses are not to be regarded in experimental philosophy. And although the arguing from experiments and observations by induction be no demonstration of general conclusions, yet it is the best way of arguing which the nature of things admits of, and may be looked upon as so much the stronger by how much the induction is more general. And if no exception occur from phenomena, the conclusion may be pronounced generally.
> —Isaac Newton, *Opticks,* 2nd ed. (London, 1718)

speed of light

The speed of light is often noted as 186,000 miles per second, but the most accurate and modern accepted measurement puts it at 299,792.458 kilometers per second, which is slightly more than 186,000 miles per second. The "discovery" of this speed is actually a question of developing a technique for accurately measuring a constant in nature. The first measurements within 1 percent of the finally agreed-on figure were made during the Age of Scientific Revolution.

Before the 1600s, scientists assumed that light traveled in an instant, supported by a number of astronomical observations. The fact that Earth's shadow passes over the Moon during a lunar eclipse with no apparent lag time between the calculated passage of the Moon out of the shadow and the moment of our perceiving the passage suggested that light traveled the immense distance from the Moon to Earth instantaneously. However, Galileo Galilei (1564–1642) suggested that perhaps the speed was finite but simply too fast to measure by that method or with contemporary instruments. He was right, and his insight might be regarded as the discovery that light had a speed that could be measured, rather than representing an instantaneous event. The discovery or precise measurement of the speed resulted from improvements in technique and equipment.

The Danish astronomer Ole (Olaus) Römer (1644–1710) estimated the speed of light in 1676, accurately predicting a delay in the arrival of light from an eclipse of one of the moons of Jupiter. Römer noticed that the period of revolution of Jupiter's inmost moon, as measured by its eclipse or movement into the shadow of Jupiter, appeared to vary in magnitude, depending on the position of Earth in its own orbit. When Earth was distant from Jupiter, the moon appeared late, but when the Earth was closer to Jupiter in its orbit, the moon appeared earlier. After studying the effect for several years, Römer worked out the idea that light takes some time for its arrival from a distant object. In 1676 he put his calculations before the Paris Academy of Sciences, a newly formed **scientific society**. Römer's ideas first met ridicule, but Isaac Newton (1642–1727) and Christiaan Huygens (1629–1695) supported his findings. Römer concluded that light moved at 214,000 kilometers per second. That calculation was quite a bit off, but he had at least demonstrated that light was not transmitted instantaneously, and he had provided a method for measuring it.

In 1728 the British astronomer James Bradley (1693–1762) studied the apparent changed positions of stars over a 6-month period, resulting from the motion of Earth halfway around its orbit of the Sun. This effect, known as parallax, provided a demonstration of the Copernican conception of the solar system, in which Earth revolves around the Sun. However, Bradley discovered a second effect, or aberration, that derived from the changed angle of the light from the distant star Draconis. Studying the aberration angle and using a good estimate of the size of Earth's orbit, Bradley estimated the speed of light at 301,000 kilometers per second, an extremely accurate estimate, it later turned out. He was 1,208 kilometers per second off, less than 1 percent of the finally agreed-on figure.

Two French scientists, Armand Fizeau (1819–1896) and Léon Foucault (1819–1868), conducted accurate measurements of the speed of light on Earth in 1849 and 1850, respectively. Fizeau developed a device that would reflect a beam of light from a mirror 8 kilometers away from its source. By passing the beam of light through gaps between teeth on the edge of a rapidly moving wheel, he could increase the speed of the wheel until the light was obscured. With this device, Fizeau calculated the speed of light in air at 315,000 kilometers per second. In 1850 Foucault improved the device, using rotating mirrors, and was able to distinguish between the speed of light in air and in water. His estimate for the speed of light in a vacuum was 298,000 kilometers per second. For this reason, Foucault is sometimes noted as the discoverer of the speed of light. His calculation was 1,792 kilometers per second too slow; Bradley's earlier calculation, based on astronomical observation more than a century earlier, had been 1,208 kilometers per second too fast. Later calculations—by Albert Michelson in 1926 and by Gordon-Smith Essen in 1947—approximated the speed in a vacuum to within a few meters per second. In the 1970s and 1980s, scientists using **lasers** [VI] and more accurate measurements of the length of the meter itself resulted in the accepted value of 299,792.458 kilometers per second.

Astronomers use the speed of light to provide a sometimes approximate measurement of distance. Thus the Sun is said to be 8 light-minutes away from Earth (8 × 60 × 299,792.458 kilometers), or 143,900,379 kilometers. The actual mean distance from the Sun to Earth is 149,600,000 kilometers (which itself has become a unit of distance known as 1 astronomical unit or "a.u."), a distance that is slightly more than 8 light-minutes, or 8.31 light-minutes. Other astronomical distances are similarly measured. For example, the nearest star is "about" 4.3 light-years away, and the Milky Way galaxy is "about" 100,000 light-years in diameter. Since such immense numbers when measured in miles or kilometers are extremely difficult to imagine, the light-year or light-minute becomes a convenient, if often inaccurately applied, shorthand way to express astronomical distances.

spinning jenny

After John Kay invented the **flying shuttle,** the demand for yarn began to increase, and in 1761 the British Royal Society of Arts offered prizes for a machine to spin six or more threads at a time, with only one person working the machine. James Hargreaves (1720–1778) invented such a machine, known as the spinning jenny, perhaps as early as 1764, and he patented it in 1770. The term *jenny,* like the term *gin* applied to the **cotton gin [IV]**, may have been derived from the word *engine*. In the jenny, bobbins were placed at the bottom of a frame carrying several spindles. Fiber would be drawn from the bobbins and twisted for a short length, then turned and pressed into place as it was wound on the spindles. The simple linkages allowed a spinner to make several spindles of thread simultaneously. Like Kay, Hargreaves was attacked by angry weavers, but after moving, he formed a company and began to make a profit from his machines. Richard Arkwright (1732–1792) made several improvements with a water-powered jenny or "water frame" in 1769. Arkwright installed both Kay's shuttle and an improved Hargreaves spinning jenny in several cotton mills.

In 1774, Samuel Crompton (1753–1827) developed a hybrid device combining Hargreaves' jenny and Arkwright's water frame that kept the tension steady on the spinning yarn, allowing for a finer and more regular thread. Crompton was too poor to patent his device, and after near riots, in which others sought to discover how he made such fine thread, he publicized the spinning mule, so called because it was a cross between two existing inventions, just as mules represented the crossing of donkeys and horses. In 1812, decades after his invention, Parliament awarded him £5,000, by which point millions of spindles using the Crompton spinning mule were in operation throughout Britain.

When combined, the flying shuttle, the spinning jenny, the Arkwright water frame, and the Crompton spinning mule revolutionized textile manufacture. By moving the manufacture of thread and cloth from a cottage industry into large, capital-intensive factories, these devices are often described as initiating the Industrial Revolution itself. Since the cost of such equipment was so great, companies of investors had to be formed, representing one of the factors that created modern capitalism. No longer did the worker own the tools on which he worked, and the owner of the equipment was in a position to dictate the terms of employment. In this fashion, the development of these key technological improvements can be seen as underlying drivers of much of the 19th century's social and political history.

steam engine

The invention of a **vacuum pump** in 1654 gave rise to a variety of experiments intended to harness the power of a vacuum, leading to a type of steam engine developed by Thomas Savery (1650?–1715), an English military engineer. His 1698 device, which used a boiler and valves to create suction, was a pump for removing water from Cornish mines. Steam would be created in a boiler, and as the steam expanded, it would be released from a valve. Then the valve would be closed, and the boiler cooled, creating a partially evacuated chamber. Then a valve to the lines down into the mine would be opened and the water pulled up. The system worked to a depth of only 25 feet and was never very practical.

Thomas Newcomen (1663–1729) made several improvements to the Savery pump in 1712. He developed a closed cylinder in which a piston was moved by steam pressure. This breakthrough meant that the machine itself could produce motion that could be harnessed through linkages not only to do work but also to open and close valves automatically. Although Newcomen had at first installed valves that were operated by hand, a young boy, Humphrey Potter, developed an automatic system using cords, later replaced by tappet rods suggested by Henry Beighton. Using brass cylinders by 1724 allowed finer tolerances than earlier iron cylinders. The Newcomen engine, linked to a heavy overhead beam, could create a slow, vertical motion that could be harnessed for pumping, running about 10 to 16 strokes per minute.

Although James Watt (1736–1819) is often credited as the inventor of the steam engine, he is more properly regarded as the engineer who made a series of design improvements to the existing devices developed by Newcomen, Potter, and Beighton. In 1765 Watt improved the Newcomen engine by adding a condenser, so that the cylinder did not have to be heated and cooled with each stroke. He also added steam valves and a stuffing box to seal the system. Watt brought a somewhat scientific approach to the problems of the Newcomen engine by carefully identifying several inefficiencies that he sought to address. However, a finer boring system was required to produce a well-sealed cylinder, and a machine designed by John Wilkinson in 1775 allowed for a better steam engine with less wasted heat. Watt also added a flywheel to continue the motion of the engine by momentum, a double-action engine that admitted steam to both sides of the piston in 1782, the steam condenser in 1784, and the **governor** in 1788. Watt's first engines were capable only of linear or vertical motion, but he worked on a method to convert the linear motion to rotary motion. He invented a **crank [II]** and connecting rod system and a geared system for this purpose. After 1794, the crank linkage became widely used. In Britain, the firm of Boulton and Watt maintained a monopoly on steam engine production for 25 years following 1775, and during that time sold more than 500 engines.

An American, Oliver Evans (1755–1819), who developed a high-pressure, noncondensing engine, further improved the steam engine in 1804. Evans built about 50 of his engines, as well as a dredger in 1804 that was probably the first steam-powered land vehicle in the United States. It was amphibious in that it could be moved by paddles in water and by rollers on land. Richard Trevithick (1771–1833) in Britain developed a similar engine and used it to propel the first rail car along a horse-rail system to and from the Penydarren ironworks at Mehthyr Tydfil in Wales in 1804. Both systems used large cylinders to produce steam piped to the pistons. John Stevens (1749–1838), who developed an early **steamboat [IV]**, improved on the boiler system by running a series of pipes through the firebox, allowing the water to recirculate until it became high-pressure steam. Later designs encased the pipes in a large containing case to prevent damage in case of explosion.

The development of many applications for stationary steam engines in the 18th century, and their use early in the 19th century for steamboats, the **steam locomotive [IV]**, and steam-powered carriages led scientists to investigate their behavior and to discover the **laws of thermodynamics [IV]**.

Historians of technology have closely studied the origin of the steam engine because of the central place it held in promoting the Industrial Revolution of the late 18th and the 19th centuries. Like many

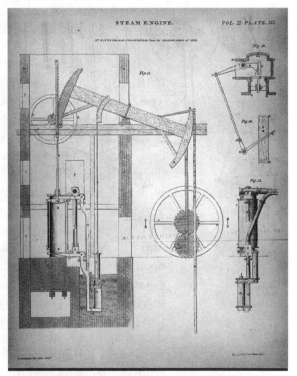

steam engine In the period 1775–1800, the firm of James Watt and Matthew Boulton held a monopoly on the steam engine. This 1822 print shows how the piston action could be transformed both into reciprocal motion for a pump and into rotary motion through a flywheel. *Library of Congress*

other major inventions, it represented the combination of many improvements and refinements, dating from Savery's first inefficient steam-powered vacuum pump in 1698 through the modifications of Newcomen, Watts, and others over the following century.

In the mid-18th century, British scientists and engineers regarded the two most important technical problems of the era as removing water from mines and developing an accurate method of determining longitude at sea. The steam engine solved the first, and the **chronometer** solved the second.

From the point of view of the application of energy to work, the steam engine was the most significant development since the invention of the **waterwheel [II]** and the **windmill [II]** in the late Middle Ages. For a century the steam engine represented such a major force that the period from about 1790 to 1890 is sometimes called the "Age of Steam." Only with the development of the **electric motor and electric generator [IV]**, the **internal combustion engine [IV]**, and the **steam turbine [V]** was the piston-driven steam engine gradually supplanted as the source of motive power for both transportation and stationary applications.

telescope

The first telescopes were developed by Hans Lippershey (1570–1619), a Dutch maker of lenses for **eyeglasses [II]**, in 1608, and the invention is surrounded by several unsupported legends suggesting that the concept was a result of an accidental discovery in his shop. Hearing of the device, Galileo Galilei (1564–1642) made one himself, a simple model with two lenses in a straight tube. The first lens bent the light rays into a focal point near the eye, where a second eyepiece magnified the image. Galileo used the instrument and others with higher magnification to make several astronomical discoveries that helped demonstrate the nature of the **heliocentric solar system [II]**.

Galileo had built a refractor telescope, and to improve the image, a longer tube and larger lenses had to be constructed. In 1668 Isaac Newton (1642–1727) developed a reflecting telescope that used two mirrors that magnified the image and that reflected the image to an eyepiece mounted on the side of the telescope, allowing for a shorter and more compact construction for the same quality and magnification of image. The **Cassegrain telescope**, with a hole in the main mirror, represented an improvement over the Newtonian design. The Cassegrain design was proposed in 1672, but models using the principle were not built for decades. William Herschel (1738–1822) used a reflecting telescope in his discovery of **Uranus** in 1781, although he built an extremely long (40-foot) refractor telescope first.

Refractor telescopes underwent a revival led by amateurs in the 19th century. Alvan Clark, an American astronomer, developed a lensmaking business and built the 36-inch lens for the Lick Observatory on Mount Hamilton in California in 1887. The world's largest refractor was built with funding from Charles Yerkes in 1897, near Chicago. Distortions imposed by the lens led later developers to work on improving mirrors for reflecting telescopes.

Although telescopes made possible many astronomical discoveries, optical telescopes were limited by several factors, including distortions from the lenses or mirrors, as well as by distortions generated by the atmosphere of Earth.

Large 20th-century reflecting telescopes used materials made of Pyrex glass and multiple hexagonal mirrors mounted together to reduce distortion. With the development of the **computer** [VI] it has become possible to reduce distortion further by connecting smaller telescopes, even at locations remote from each other, and electronically combining their images using **charge-coupled devices (CCDs)** [VI].

The mounting of a high-quality optical telescope with CCD capacity aboard a satellite, the Hubble Space Telescope, in 1990 provided a new window on the universe. After a repair mission in 1993 to correct errors in the original installation, the Hubble Space Telescope led to the discovery of thousands of new galaxies, and the receipt of light from the outmost reaches of the universe, whence it had been emitted at the time of the origin of the universe.

Other devices that develop images from radiation, developed in the 20th century, are also called telescopes, or radio telescopes.

Together with the **microscope**, the **mercury barometer**, and the **thermometer and**

You must know, then, that it is nearly two months since news was spread here that in Flanders there had been presented to Count Maurice, a spy-glass, made in such a way that very distant things are made by it to look quite close, that a man two miles away can be distinctly seen. This seemed to me so marvellous an effect that it gave me occasion for thought; and as it appeared to me that it must be founded on the science of perspective, I undertook to think about its fabrication; which I finally found, and so perfectly that one which I made far surpassed the reputation of the Flemish one. And word having reached Venice that I had made one, it is six days since I was called by the Signoria, to which I had to show it together with the entire Senate, to the infinite amazement of all; and there have been numerous gentlemen and senators, who though old, have more than once scaled the stairs of the highest campaniles in Venice to observe at sea sails and vessels so far away, that coming under full sail to port, two hours or more were required before they could be seen without my spy-glass. For in fact the effect of this instrument is to represent an object that is, for example, fifty miles away, as large and near as if it were only five.

—Galileo Galilei, August 29, 1609, to his brother-in-law, describing the effect of his first telescope

temperature scale, the telescope profoundly enriched the tool kit of the scientific community, leading to a host of discoveries over the following centuries. During the era of the Scientific Revolution, scientists using the telescope worked out calculations of the **speed of light,** discovered **Saturn's rings and moons,** calculated **comet paths as orbits [II],** and developed the specialized field of natural philosophy known as astronomy.

thermometer and temperature scale

Daniel Gabriel Fahrenheit (1686–1736) invented the first accurate mercury thermometer in 1724 and developed the first widely used scale of temperature. Fahrenheit was of German ancestry, born in Danzig (Gdansk, now in Poland) and raised in Holland. He did most of his work in Holland, and he is usually identified as Dutch.

Galileo Galilei (1564–1642) and others had developed prior thermometers or "thermoscopes" in 1592 and later. The thermoscopes were glass tubes open at one end, partially filled with water or alcohol-water combinations. These were notoriously inaccurate, as they were subject to changes in air pressure. Duke Ferdinand II (1611–1670) of Tuscany and Cardinal Leopoldo de Medici (1617–1675) both developed in about 1654 a closed glass thermometer, and it was later realized that the phenomenon of changing air pressure had accounted for the variation in temperature readings in the open thermoscopes. Leopoldo's design was later improved and modified to use distilled wine. These "Florentine thermometers" were not very consistent or accurate and were only roughly calibrated. Different scientists working with such thermometers were frustrated in efforts to communicate their experiments, due to the various scales of temperature employed.

In 1695 Guillaume Amontons (1663–1705), a French natural scientist, designed a thermometer using a glass tube filled with compressed air and then topped off with a level of mercury.

In 1708 Fahrenheit developed and suggested the wide use of a scale linked to the melting point of ice and the temperature of the human body. In 1724 he developed a mercury thermometer using purified mercury that would rise or fall in a narrow glass tube, with minimal adherence to the side of the tube, and marked with his scale. Learning of a 16-point scale that had earlier been developed by the Swedish astronomer Ole Römer (1644–1710) that set the temperature of melting ice at 7.5 degrees and the human body at 22.5, Fahrenheit adopted these calibration points, dividing each of Römer's 15 degrees into 4 to create a 60-point spread between the two end points. He later settled on

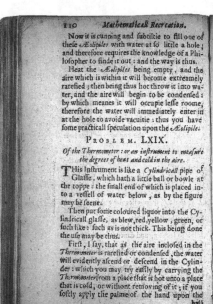
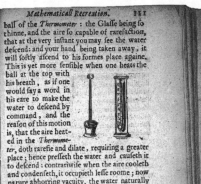

thermometer This 17th-century English translation of a French treatise describes the operation of two types of thermometer: the Italian thermoscope, on the left, and the Dutch two-bulb type. Using such instruments and standardizing measurements, scientists began to exchange more precise experimental data. *Library of Congress*

32 degrees for freezing water, 96 degrees for human temperature, and 212 degrees for boiling water. Later correction of the scale used his freezing and boiling points and adjusted the scale in between so that the normal human temperature was established at 98.6.

Fahrenheit's scale was widely accepted and then partially superseded by the introduction of the 100-point scale suggested by Anders Celsius (1701–1744) in 1741. Celsius proposed that the boiling point of water be set at zero degrees and the freezing point at 100 degrees. After his death, Carolus Linneaus (born Carl von Linné) (1707–1778) and others proposed inverting the scale, and the "centigrade scale" was introduced in about 1747. A century later, in 1848, William Thompson (Lord Kelvin) (1824–1907) proposed a temperature scale that would start at absolute zero but would use the same degree spacing as the centigrade scale. In 1948 the centigrade scale was renamed the Celsius scale in honor of the man who developed it.

Thus the three scales compared as follows:

	Fahrenheit	Celsius	Kelvin
water boiling point	212	100	373
water freezing point	32	0	273
absolute zero	−459	−273	0

thoracic duct

In 1647 Jean Pecquet (1622–1674) discovered the thoracic duct while working on animal dissection. Pecquet was born in Dieppe, France, and studied medicine in Paris and Montpelier.

Pecquet reported his findings in *Experimenta nova anatomica* in 1651. The duct is connected to the lymph system and connects the vessels and organs that transmit nutrients and oxygen to the tissues of the body and carries away waste matter from the tissues. It was later determined that the thoracic system carries lymph, which is the waste fluid, and chyle, the fluid containing nutrients from the digestion of food, to and from the intestinal system into the cardiovascular system. His studies led other investigators to debate and study the nature and purposes of the lymphatic system.

transpiration of water in plants

The understanding of the role of fresh air in human life and of water in the life of plants was greatly advanced by Stephen Hales (1677–1761), an English scientist. Before the discovery of **oxygen,** Hales recognized that *spent air* presented a danger and urged ventilation systems in hospitals, ships, and prisons. He experimented with plants between 1719 and 1725, measuring the flow of water from roots through stalks to leaves, where it evaporated from openings in the leaves. Hales was also credited with the invention of the surgical forceps, and he studied kidney and bladder stones as well.

His work on the transpiration of water in plants was first published in 1727 in *Vegetable Staticks,* and republished with more data in 1733 as *Statical Essays, Containing Haemastaticks, etc.*

universal joint

In 1676, Robert Hooke (1635–1703) described in a published lecture a device he called the "helioscope," an instrument for safely observing the Sun through reflected light. To move the scope, he included what he called a "universal joynt" for transmitting a round motion through a bent or angled pathway. He had some vague ideas that the concept could be applied in a variety of settings, but his intention was to use it in astronomical equipment. **Clocks [II]**, **windmills [II]**, and **waterwheels [II]** had all used gearing that achieved the transmission of power through a right angle. Hooke's addition was to provide a linkage joint in the driving shaft that allowed the angle to be modified from a simple 90° angle to a nearly "universal" number of angles.

Hooke's terminology and joint lay largely dormant for more than a century, finding a few uses in instrumentation. With the development of the **automobile** [IV], the need to transmit rotary motion imparted by the **internal combustion engine** [IV] through a variable angle to the driving axle required the revival of the universal joint and its widespread application.

Uranus

William Herschel (1738–1822), a German-born English astronomer, discovered the planet Uranus on March 13, 1781. In the process of systematically conducting a survey of the heavens with a large telescope of his own design and construction, he visually identified an object that clearly was not a star. He first assumed it was a comet, but later observations by Herschel and others confirmed that it was the seventh planet from the Sun, the first actually "discovered." The other six had all been known since antiquity, being quite visible and noticeable to the naked eye. Once discovered and located, Uranus is actually bright enough to be identified without the aid of a telescope, as it can be seen as a faint dot of light.

Herschel was born in Hannover, Germany, and trained to be a musician. He moved to Britain at age 28 and became an organist in the town of Bath. He grew interested in astronomy but, living on a meager income, could not afford to purchase commercially built telescopes. Building his own instruments led to the discovery of Uranus. King George III allowed him a pension after the discovery, and Herschel then married. William's sister, Caroline Herschel, was known as his assistant, although later examination of the records indicates that she was an accomplished astronomer in her own right. William's only child was John Herschel (1792–1871), who later assisted in founding the Royal Astronomical Society and who was knighted for his work in the fields of astronomy and photography.

William Herschel also discovered the two largest satellites of Uranus, later named Titania and Oberon, in 1787. In addition to Uranus and two of its satellites, Herschel discovered two satellites among **Saturn's rings and moons** and initiated a method of statistical astronomy, assembling star counts for different sectors of the sky. His calculations demonstrated that the Sun itself was in motion.

On at least seventeen different occasions between 1690 and 1781, a number of astronomers, including several of Europe's most eminent observers, had seen a star in positions that we now suppose must have been occupied at the time by Uranus. One of the best observers in this group had actually seen the star on four successive nights in 1769 without noting the motion that could have suggested another identification. Herschel, when he first observed the same object twelve years later, did so with a much-improved telescope of his own manufacture. As a result, he was able to notice an apparent disk-size that was at least unusual for stars. Something was awry, and he therefore postponed identification pending further scrutiny. That scrutiny disclosed Uranus' motion among the stars, and Herschel therefore announced that he had seen a new comet. Only several months later, after fruitless attempts to fit the observed motion into a cometary orbit, did [Anders] Lexell [a Swedish-born astronomer teaching in St. Petersburg, Russia] suggest that the orbit was probably planetary.

—Thomas Kuhn, *The Structure of Scientific Revolutions,* 2nd ed. (Chicago: University of Chicago Press, 1970), pp. 94–95

The British astronomer William Lassell discovered two more satellites of Uranus, Umbriel and Ariel, in 1851. The Dutch-born American astronomer Gerard Kuiper (1905–1973) discovered Miranda, a smaller satellite, in 1948. These five major satellites, named after characters from Shakespeare's *The Tempest* and *A Midsummer Night's Dream,* were soon supplemented by another ten discovered by the spacecraft *Voyager 2,* later given Shakespearean names as well. The larger satellites revealed a number of interesting features, including the largest mountain known in the solar system on Oberon, rising some 12 miles above the surface.

Herschel himself planned to name the new planet Georgium Sidus or Georgian Star, after King George III of England. Others called the planet Herschel. The astronomer Johann Bode (1747–1826) suggested Uranus, the mythological figure who was the father of Saturn, and that name became widely accepted in the 19th century.

Astronomers working on the mechanics of the orbit of Uranus calculated that it was perturbed by the presence of another large but undiscovered planet. These calculations led, in 1846, to the discovery of **Neptune [IV]**.

vacuum pump

Experiments with air, air pressure, and the evacuation of air began to engage scientists in the middle of the 17th century. Galileo Galilei (1564–1642) suggested to Evangelista Torricelli (1608–1647) that he investigate the problem of vacuum. Torricelli noted that mercury would flow only so far out of a tube, and he postulated that it left a vacuum at the top of the tube and was held in place by air pressure. In effect, he had devised the first barometer in 1644. Between 1647 and 1654, Otto von Guericke (1602–1686), the mayor of Magdeburg, Germany, for thirty years (1646–1676), developed the first vacuum pump.

Working with a design similar to a hand-pumped fire extinguisher, Guericke designed flap valves and pumped the air from beer casks and then from hollow copper spheres. At first the copper spheres would collapse, but he made stronger ones that allowed further experiments. He made many observations of the qualities of a vacuum, noting that candles would be extinguished, mice would die, and that a bell rung inside a vacuum transmitted no sound. As a bit of a showman, Guericke held public demonstrations showing that two hemispheres from which air had been pumped could not be pulled apart by two teams of horses. However, when a turncock was opened to admit air, the hemispheres could be easily separated. In another Guericke demonstration, air would be sucked from a bottle into an evacuated sphere, causing the bottle to implode from outside air pressure.

Building on Guericke's work, Robert Boyle (1627–1691) and Robert Hooke (1635–1703) made improvements, allowing a range of experiments on the nature of air in 1659 and later. Boyle had recruited Hooke to work as a mechanic for his laboratory, and asked him to design a vacuum pump along the lines of Guericke's, but one that would be smaller and more useful for laboratory experiments. Hooke's design, publicized by Boyle, was connected to a glass receptacle. An oiled brass plug was fitted into a small hole at the top of a glass chamber as a valve with a turncock to admit air. With the device, Boyle could demonstrate that air has weight and that animals would die in a vacuum. Boyle published a description and illustration of the improved vacuum pump in 1660 in *New Experiments: Physico-Mechanical, Touching the Spring of Air and its Effects* and in 1669 in *Continuation of New Experiments, Physico-Mechanical*. Scientists could readily get glassblowers and mechanics to build the laboratory equipment by following the descriptions.

vernier

A vernier is a precise, two-part scale that is used to measure angles and lengths in very small divisions. A large, stationary scale measures whole numbers, and a smaller, movable scale measures the fractions. A common and familiar modern application is the sliding vernier scale used to indicate weight on a horizontal bar on scales in doctors' offices. Invented in 1631 by Pierre Vernier (1584–1638), a French soldier and engineer, the device represented an improvement to the **astrolabe [II]**, allowing for a fixed and accurate registry of the angle to the Sun or a star.

When the vernier was invented, surveying and other instrumentation did not require the degree of accuracy that the vernier could achieve. It began to catch on and find many applications in **microscopes, theodolites [II]**, and in other instruments in the next century. Although very simple in principle, the vernier contributed to the Scientific Revolution by allowing more precise measurement, and it became part of the tool kit of navigators, surveyors, and, on the laboratory bench, of scientists and engineers over the following centuries.

water as a compound

The fact that water is a compound of hydrogen and oxygen was suggested by Antoine Lavoisier (1685–1794) and Pierre Laplace (1749–1827) in 1783 and demonstrated by the reclusive British scientist Henry Cavendish (1731–1810) in a work he published in 1784. Cavendish had worked on the experiment as early as 1781 but withheld publication until he was certain of his findings.

Cavendish was born in Nice, France, and educated in Britain, dropping out of Peterhouse College of Cambridge in 1753. One of the wealthiest men of his time, he spent much of his time in seclusion in one of several homes in and around London, and according to legend, he was absolutely frightened of women. His demonstration that water could be synthesized out of two elements was only one of several conclusions he drew from his experiments, many of which remained unpublished until his work was studied and edited in the 1840s and 1850s by James Clerk Maxwell (1831–1879).

Cavendish described hydrogen as "inflammable air" and carbon dioxide as "fixed air." He published his findings in a paper, "Experiments on Air," in *Philosophical Transactions* in 1784. In that paper he described the experiment in which he found that water could be synthesized from hydrogen and **oxygen.** He put the two gases in a tube and introduced an electric spark, resulting in an explosion of the hydrogen, the formation of water, and the contraction of the volume of gas. Later he conducted similar experiments, discovering that common air is ⅕ nitrogen. Lavoisier and Laplace had reached similar conclusions in 1783, and the experimental demonstration by Cavendish proved the point.

Cavendish worked on issues of electricity, concluding that electrostatic attractions are functions of the inverse square of the distance between particles and that a charge on a sphere is lodged on its surface, not in its interior, foreshadowing the work of Michael Faraday (1791–1867). Cavendish invented a torsional balance that later bore his name, and using it, in 1798, he worked on discovering the gravitational constant suggested by Newton in his **laws of motion and gravity** and determined the density of Earth. His paper on the topic, "Experiments to Determine the Density of the Earth," published in 1798, established quantitative values of both the gravitational constant and the mass of Earth that were not superseded with more accuracy for a century.

Although Cavendish did not reject the phlogiston theory of combustion, his work in chemistry, electricity, and physics marked this college dropout as one of the most brilliant scientists of his day. By discovering that water was a compound he, together with Lavoisier and others, overthrew the ancient belief that water was an element.

PART IV

THE INDUSTRIAL REVOLUTION, 1791 TO 1890

Many of the major laws of nature that were discovered in the 19th century derived from the observation of already invented devices and their behavior rather than from the use of instruments to observe purely natural events, the process that had characterized the scientific advances of the 17th and 18th centuries. The laws of thermodynamics, the Froude number, Faraday's laws, the Boltzmann constant, principles of electromagnetism, Ampère's law, Ohm's law, and numerous other significant scientific discoveries between 1790 and 1890 are ways of representing the natural laws that govern the operation of human-built machines, electrical devices, pressure chambers, and in the case of Froude, ships and ship models.

The machines were built first; then the principles that governed them were thought through. With the newly discovered principles in place, it was possible to build better machines. As James Clerk Maxwell and others reduced these principles to mathematical formulas, they provided following generations of engineers with new tools with which they could build on the findings in a more organized way, no longer simply relying on cut-and-try methods to improve equipment. By the end of the period, science was beginning to penetrate to an understanding of chemistry and electrical effects, paving the way for the next generation of breakthroughs in the understanding of the building blocks of matter at the atomic level.

In the 17th century, scientists had used instruments to discover and describe natural phenomena. In the 18th century, as technology improved, scientists began to experiment with even better and improved tools such as vacuum pumps and glassware that allowed a deeper understanding of nature. In the 19th century, with better machining and interchangeable parts, a wide variety of useful engines and electrical equipment was produced, at first almost entirely by mechanics who were not trained in the sciences. Although discovery through improved observation equipment and methods continued, the new machinery raised new and interesting issues for scientists. Some scientists turned their talents to explaining the operation of the man-made equipment. Whereas earlier generations had used tools and equipment to broaden the accumulated wisdom of science, now some 19th-century scientists used the scientific method to understand how equipment itself worked. Thus, while science and technology remained separate enterprises and made progress in different ways, the intersections between the two progressing areas were intricate and had been very much studied in recent decades.

The interplay between science and technology, particularly in the fields of electricity and engine building, led to a host of new practical machines that changed communication, transportation, the home, the workplace, and the farm. Although many engineers in the mid- and even late 19th century were self-taught inventors and mechanics with little formal scientific training, by the end of the century, schools of engineering began to include a more scientific curriculum and to lead engineers through the recently formulated principles of electrical theory, thermodynamics, chemistry, and pneumatics, as well as the principles of scale modeling. A self-taught practical innovator, Thomas Edison, regularly employed not only mechanics and craftsmen but also college-trained chemists and engineers in his invention workshop.

Inventions and methods about which visionaries had only dreamed in prior ages were brought into reality during the era of the Industrial Revolution. The concept of interchangeable parts was discussed by French army officers in the 1790s and then pursued by several inventors such as Eli Whitney and Samuel Colt over the next decades. By the 1840s, tools developed in government-owned and -managed U.S. arsenals began to make it a reality. The historical finding that teams of workers at government facilities led the way has somewhat undermined the heroic mythology surrounding the great individualistic and entrepreneurial American inventors.

Other visionary schemes also received government aid and sponsorship. For example, the generals of the French Revolution demanded a means of communicating rapidly across the European continent, establishing a semaphore system that served as a precursor to the electric telegraph. The French government offered a prize for a system of preserving food for troops in the field, and methods of canning meat and vegetables resulted.

In the 18th century, dreamers had visualized the day when steam engines could be used to propel ships and land vehicles. Not until after the improvements of James Watt in the 1780s did the dream come closer to realization. The firm of Boulton and Watt, holding a monopoly for 25 years, produced a wide variety of practical steam engines from 1775 to 1800, when many manufacturers sought ways to employ them in new ways. In the first two decades of the 19th century, steam was widely applied to a variety of transportation systems, from the railway to the river steamboat and including some variations such as road vehicles, farm tractors, and cable cars. With the development of screw propellers for ships, oceangoing steamships became more practical, while steam-driven paddlewheel boats remained efficient and effective on rivers and in harbors. The age of steam power changed the face of Europe, North America, and gradually the rest of the planet.

Together, the railroad and the steamship greatly reduced the time it took to travel long distances. The electric telegraph and the submarine cable by the 1860s knitted together Europe and America with instantaneous communication. With the transcontinental railroad and its parallel telegraph lines completed in 1869, it was possible in 1870 to hear of Berlin news in San Francisco within a day of the event.

By the 1880s, engineering progress was steady and regular, with a host of new inventions coming in clusters. People born in 1815 or 1820, who in their childhood had never seen a railroad engine, were exposed by age 65 to internal combustion engines, the phonograph, the telephone, the lightweight camera, refrigeration, the bicycle, the streetcar, the typewriter, and news of the first automobiles. The term *Industrial Revolution* was appropriate to describe the vast changes that had taken place as skyscrapers began to rise and the shape of life was transformed.

The social consequences of the technological advances were vast and for many people extremely disruptive. By the 1870s, the United States began to develop a single price and market system. Some identical brand-name products, from canned meats to patent medicines, were sold all over the nation. Capital-intensive industries employed workers who owned no tools and had few skills, leading to social inequalities that created new political crises. Agricultural lands opened to the national markets with the railroad net, and new farm equipment increased farmers' productivity. However, increased productivity was not always a blessing to farmers, as prices of basic commodities like cotton, wheat, and corn declined, catching many farmers in a squeeze between fixed mortgage payments and decreasing revenues for their crops.

By the 1880s, farmers and industrial workers represented a force for political insurgency that began to coalesce into new movements that threatened to change the nature of society. In the United States, government regulation to control railroad rates and other large enterprises was being discussed to address the growing power of bankers and railroad magnates.

Similar developments in Europe led to the rise of radical advocates of social change, including Karl Marx and Friedrich Engels, whose writings urged a revolutionary outlook on the new generation of working-class people. The ideas of such men held appeal to those who feared that the new industrial arrangements were destroying human values. Others simply blamed the disruptions on the result of individual immorality, seeing greed for money, ostentation, and power as the forces that changed society for the worse.

The relationship between technological progress and social change was at heart a simple matter. Large and expensive machinery produced better products at lower cost than earlier handcraft methods. To pay for the installation of such machines required accumulations of capital, arranged by groups with access to financial resources or credit. Existing laws protecting individual property from trespass, theft, or unauthorized use by others could serve to protect the position of the owners of the new factory equipment. Even though treated in law in similar ways, however, there were some fundamental differences between personal possessions and factory equipment. Since ownership of the tools was now separated from the men and women who used the tools, the producing workers found themselves in a new position. They were simply hired to use the tools owned by others and were therefore powerless to exert choices as to their hours, their compensation, or other conditions of employment. If they did not like any aspect of a job, as free workers they were free to resign or to stop work. Of course, when they did so, they would go without pay. Like household servants, they had no claim to the tools with which they worked, since they did not own them.

The consequence was the formation of labor unions, which had very few strategies to affect the plight of the workers: either collective work stoppage or political action. The decades from the 1870s through the early 20th century saw one disruptive strike after another and the beginning of the appeal of ideologies and reform strategies that would characterize the labor movement—agrarian reform, populism, socialism, anarchism, and progressive reformism.

Science and technology brought numerous positive social benefits, some tangible and some intangible, despite such disruptions that came with the rise of the factory system, the decline of commodity prices, and the emergence of new huge cities. Improvements in medicine, particularly the use of anesthetics and sterilization of surgical tools, meant that more patients could survive surgery. Travel was not only faster, it also was more comfortable and more available to greater numbers of people at lower prices. Refrigeration and rapid transport of food products meant a more varied diet, less dependent on local supply. The opthalmoscope made the prescription of eyeglasses more accurate and faster. Cheaper and more efficient printing stimulated and was stimulated by increasing literacy. Water supplies, sewer systems, and public lighting improved life in the cities. The sewing machine and the typewriter began to open new employment opportunities for women. All the basic necessities of life declined in real cost: clothing, food, transportation, medicine, and housing. Despite the class struggle stimulated by the Industrial Revolution, technology had brought some undeniable benefits, even to the mass of poor people.

For the middle classes, such as physicians, attorneys, engineers, academics, journalists, and managers, whose employment was less dependent on machinery and tools than on their education, the new technologies provided new sources of entertainment, comfort, and opportunities. By 1890, a person who worked in such a profession could come home, switch on the electric light, make a phone call, go to the refrigerator and take out a cold beer, light a factory-made cigarette with a safety match, then sit down to listen to a phonograph record while reading a newspaper with news of the day from all over the planet. It was a new world, for none of those experiences had been possible 40 years earlier, even to the wealthy.

Some of the developments led visionaries to foretell the technological changes that the next century might bring. Charles Babbage visualized and designed a form of computer. Alexander Graham Bell struggled to build a hearing aid for the deaf. Experiments with the internal combustion engine and with aircraft, submarines, steel ships, and electricity hinted that the dramatic era of progress of the 19th century would continue well into the 20th.

Chronology: The Industrial Revolution, 1791 to 1890		
Year	**Invention/Discovery**	**Inventor/Discoverer**
1792	semaphore	Claude Chappe
1793	cotton gin	Eli Whitney
1794	ball bearings	Philip Vaughan
1795	hydraulic press	Joseph Bramah
1796	battery	Alesandro Volta
1797	parachute	Jacques Garnerin
1801	steam locomotive	Richard Trevithick
1803	steamboat	Robert Livingston and Robert Fulton
1810	canned food	Nicolas Appert
1811	Avogadro's law	Amedeo Avogadro
1812	hydraulic jack	Bramah
1813	bicycle	Karl Drais, 1813; Kirkpatrick MacMillan, 1842
1814	spectroscope	Joseph von Fraunhoter
1816	stethoscope	René Théophile Hyacinthe Laënnec
1816	Stirling engine	Robert Stirling
1820s	galvanometer	André Marie Ampère; Johann S. C. Schweigger
1821	electromagnetism	Hans Christian Oersted; formula for: James Clerk Maxwell, 1873
1821	electric motor	Michael Faraday; Joseph Henry, 1831; Antoine-Hippolyte Pixii, 1832
1821	thermocouple	Thomas Seebeck
1821–1833	Faraday's Laws	Faraday
1824	laws of thermodynamics (formulated, 1852)	Nicolas Sadi Carnot; William Thomson (Lord Kelvin)
1825	Ampère's law	André Marie Ampère
1825	steam railroad	John Stevens
1826	daguerreotype	Louis Jacques Mandé Daguerre
1826	Ohm's law	Georg Simon Ohm
1827	water turbine	Benoît Forneyon
1828	steel-tipped pen	John Mitchell
1829	braille	Louis Braille
1830	matches (safety)	R. C. Bottger, 1848
1830	lawn mower	Edwin Bunting
1830s	streetcar (horse-drawn)	Ernst Werner von Siemens (electric, 1881)
1831	McCormick reaper	Cyrus McCormick
1832	electric generator	Antoine-Hippolyte Pixii
1833	analytic engine	conceived, Charles Babbage
1834	propeller-driven ship	Francis Pettit Smith
1835	revolver	Samuel Colt
1840s	interchangeable parts	Harpers Ferry and Springfield Arsenals
1841	Bunsen burner	Robert Bunsen
1842	anesthetics	Crawford Long
1844	rubber (vulcanized)	Charles Goodyear
1844	telegraph	Samuel Morse
1845	sewing machine	Elias Howe

Chronology: The Industrial Revolution, 1791 to 1890 *(continued)*

Year	Invention/Discovery	Inventor/Discoverer
1845	submarine telegraph cable	Cyrus Field (transatlantic, 1866)
1847	opthalmoscope	conceived, Babbage; Hermann von Helmholtz, developed 1851
1849	pressure gauge	Eugène Bourdon
1850	coal oil	James Young
1851	steel cable suspension bridge	John A. Roebling
1852	elevator	Elisha Graves Otis
1856	steel (Bessemer process)	Henry Bessemer
1859	oil (petroleum) drilling	Edwin Drake
1859	internal combustion engine (two-cycle)	Jean Joseph Lenoir
1860	asphalt paving	unknown
1860s	pasteurization	Louis Pasteur
1864	drive chain	James Slater
1867	dynamite	Alfred Nobel
1868	celluloid	Alexander Parkes
1868	torpedo	Robert Whitehead
1868	typewriter	Christopher Sholes
1869	air brake	George Westinghouse
1869	periodic law of chemical elements	Dmitri Mendeleev
1870s	Boltzmann's constant	Ludwig Boltzmann
1871	Froude's number	William Froude
1873	barbed wire	Henry M. Rose
1873	Maxwell's equations	James Clerk Maxwell
1876	internal combustion engine (four-cycle)	Nikolaus Otto
1876	microphone	Alexander Graham Bell
1876	refrigeration	Karl Paul Gottfried von Linde
1876	telephone	Bell; Elisha Gray
1877	phonograph	Thomas Edison
1878	electric light	Joseph Swan, 1878; Edison, 1879
1880	cigarette (machine-made)	James Albert Bonsack
1886	aluminum by electrolysis	Charles Martin Hall; Paul Louis Toussaint Héroult
1888	pneumatic tire	John Boyd Dunlop
1889	camera (with flexible film)	George Eastman

Elements Discovered, 1790 to 1890

Year	Element	Discoverer
1790	titanium	William Gregor
1794	yttrium	Johan Gadolin
1797	chromium	Louis-Nicolas Vauquelin

Elements Discovered, 1790 to 1890 *(continued)*		
Year	**Element**	**Discoverer**
1798	beryllium	Vauquelin
1801	vanadium	Andres de Rio; disputed claim, Nils Sefström
1801	niobium	Charles Hatchett
1802	tantalum	Anders Ekeberg
1804	cerium	Jöns Berzelius; independently by Martin Klaproth
1804	iridium	Smithson Tennant
1804	osmium	Tennant
1804	palladium	William Wollaston
1804	rhodium	Wollaston
1807	potassium	Humphry Davy
1807	sodium	Davy
1808	barium	Davy
1808	boron	Davy; independently by Joseph Gay-Lussac
1808	calcium	Davy
1808	strontium	Davy
1811	iodine	Bernard Courtois
1817	cadmium	Friedrich Strohmeyer
1817	lithium	Johan Afwedson
1817	selenium	Berzelius
1823	silicon	Berzelius
1824	aluminum	Hans Oersted; independently by Friedrich Wöhler
1826	bromine	Antoine-Jerome Balard
1827	ruthenium	G. W. Osann
1828	thorium	Berzelius
1839	lanthanum	Carl Mosander
1843	erbium	Mosander
1843	terbium	Mosander
1860	cesium	Robert Bunsen, collaboratively with Gustav Kirchhoff
1860	rubidium	Bunsen, collaboratively with Kirchhoff
1861	thallium	William Crookes
1863	indium	Ferdinand Reich, collaboratively with Hieronymus Richter
1868	helium	Pierre Janssen
1875	gallium	Paul Lecoq de Boisbaudran; predicted by Mendeleev, 1873
1876	scandium	Lars Nilson; predicted by Dmitri Mendeleev, 1873
1878	ytterbium	Jean Charles de Marignac
1879	holmium	Per Cleve
1879	samarium	Lecoq de Boisbaudran
1879	thulium	Cleve
1885	neodymium	Carl von Welsbach
1885	praseodymium	von Welsbach
1886	dyprosium	Lecoq de Boisbaudran
1886	gadolinium	Lecoq de Boisbaudran
1886	germanium	Clemens Winkler; predicted by Mendeleev, 1873

air brake

In 1869, George Westinghouse (1846–1914) invented and patented a powerful new braking system, operated by compressed air, for railroads.

Westinghouse conceived of a system that could apply to all of the cars in a train at once rather than requiring brakemen the length of the train to turn braking wheels on a whistle signal from the engineer. After considering several systems, including a steam-driven brake, Westinghouse got the idea for the compressed air system while reading a magazine article about the drilling of the Mont Cenis tunnel in Switzerland. The tunnel, begun in 1857, connected Lyons and Turin. The boring machinery used a compressed air–driven drill that had been designed for mines. A number of compressed air–driven machines had been developed in the 1850s to improve mining efficiency. Air would be compressed aboveground and transported in tubes and hoses or in tanks to the work site to drive disc-cutting and chain-cutting saws for coal.

The Westinghouse compressed air brake had four essential parts. An air pump was driven by the steam engine in the locomotive, compressing air to about 60 to 70 pounds per square inch into a reservoir. A piping system led from the reservoir to a valve that the engineer could control. A line of pipes led from the valve under all the cars, connected by flexible couplings. A set of cylinders and pistons on each car operated the brake shoes. Westinghouse later established a system that would automatically engage the brakes if part of the train were separated from the rest, by reversing the system—that is, the pressure kept the brakes from engaging, and when the pressure was released, the brakes would automatically engage. Westinghouse continued to make improvements on his system, driving the competition out by a superior product.

The new brakes allowed faster railroad speeds and did much to revolutionize rail travel in the late 19th century. After building a fortune in railroad brake manufacture, Westinghouse later went on to develop alternating-current electrical generators.

> My first idea of braking apparatus came to me in this way: a train upon which I was a passenger in 1866 was delayed due to a collision between two freight trains. If the engineers of those trains had some means of applying brakes to *all the wheels of their trains,* the accident might have been avoided.
>
> —George Westinghouse on the air brake, quoted in Mitchell Wilson, *American Science and Invention: A Pictorial History* (New York: Simon & Schuster, 1954)

aluminum

Aluminum as a separate element was isolated by a Danish physicist, Hans Christian Oersted (1777–1851), in 1824 and independently by a

German chemist, Friedrich Wöhler (1799–1882), in 1827. The British chemist Humphry Davy (1778–1829) had isolated an oxide of the metal in 1809 and gave it its name. Later, the British altered the spelling and pronunciation to *aluminium,* adding a syllable.

Oersted's process for isolating aluminum was difficult and expensive, and the American Charles Martin Hall (1863–1914) and the French chemist Paul Louis Toussaint Héroult (1863–1914) discovered a more economical method almost simultaneously in 1886. Both men developed an electrolytic method for separating aluminum from the bauxite ore containing its oxide. Both used direct-current electricity to dissolve aluminum oxides in various types of molten fluorides, and then to electrolytically separate the liquid. Both found that cryolite (sodium aluminum fluoride) was the best material, allowing a low practical boiling point. In Héroult's process, a single large central graphite electrode was immersed in a graphite cell. Hall's method, with two electrodes, was so successful that he reduced the cost of aluminum metal from the equivalent to gold to less than 20 cents a pound. Both men were very young (about 22 years old) when they developed the process. Héroult eventually established a Swiss firm to manufacture aluminum. Hall succeeded in establishing the Pittsburgh Reduction Company, the parent firm of Aluminum Company of America (Alcoa), earning millions of dollars.

With the lowered price of aluminum, it became important in many uses, especially in the bodies of **airplanes** [V] and the structures of **dirigibles** [V]. Alloyed with other metals, it can achieve both lightness and strength. For example, duraluminum, an alloy of aluminum, copper, manganese, and magnesium, was used in the construction of German Zeppelin dirigibles.

Ampère's law

Ampère's law or rule connects electricity with magnetism and was first expressed by French physicist André-Marie Ampère (1775–1836) by about 1825. The rule was later formulated mathematically by James Clerk Maxwell (1831–1879), who credited Ampère with its discovery. Ampère noted that if a person were to travel along a current-carrying wire from positive to negative terminal and were to carry a magnetic compass, the north-pointing compass needle would be directed to the person's left. The magnitude of the current would proportionally deflect the compass needle, an effect first noted by Hans Christian Oersted (1777–1851) in 1819. A device developed by several investigators through the 1820s was later named the **galvanometer** in honor of Luigi

Galvani (1737–1798), who had developed a theory, later discarded, that animal muscles generated electricity. Through the first decades of the 19th century, electrical current was often called galvanic electricity, and the galvanometer measured the flow of that force.

analytic engine

With the prevalence of computers and computerlike devices in the last decades of the 20th century, historians of technology have traced the origin of computer design to many sources. The work of English mathematician Charles Babbage (1792–1871) has been extensively studied because he conceived of many of the aspects of computing devices. A member of several British **scientific societies [III]**, he was elected to the Royal Society in 1816. He began a design for an analytic engine in about 1833 but never completed it. It was a planned device for performing calculations using input from punched cards. The punched-card concept, borrowed from the design of the Jacquard power loom, which had been developed in 1804 by Joseph Marie Jacquard (1752–1834), was later utilized by Herman Hollerith (1860–1929) in the **card punch machine.**

Babbage thought through some basic principles of computing machines, including the need for an input unit; an arithmetical-logical unit, which he called a mill; and a storage facility or store. The latter included what later designers called both memory and output. In addition, he conceived of a control unit that would supervise the flow of operations, extracting information from the mill to the store and providing instructions as to the next operation of the mill. The instructions carried out by the control unit are similar to what later designers called programs.

In 1822 Babbage designed a difference engine or mechanical calculator capable of computing the squares of numbers, working on a commission from the British Admiralty, but he abandoned the project in favor

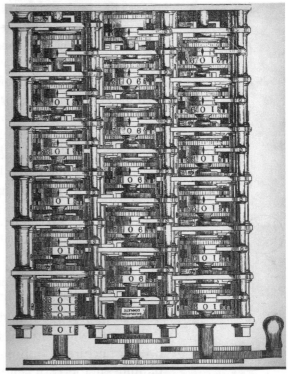

analytic engine Charles Babbage never successfully built his mechanical calculating machine, but he developed concepts such as storage and programming that became the basis for later computers. This 1889 print showed the Babbage system of linkages. *Library of Congress*

of the analytic engine, which would be able to perform several functions. Although he did not complete his project, an assistant, Augusta Ada Byron, the only child of the poet Lord Byron and his wife, Annabella, published his work. Augusta Ada Byron, later countess of Lovelace, was herself a gifted mathematician, and her notes remained a major source for understanding Babbage's work, published as *Sketch of the Analytic Engine*. She translated and published in the 1840s an article written about Babbage's work by an Italian, Luigi Menabrea, who later became prime minister of Italy. Although it was long assumed that Ada, Lady Lovelace, had devised the programming described in her work, recent research has pointed to Babbage as the source.

A simplified and workable version of Babbage's difference engine was constructed in 1854 by a Swedish printer and his son, Georg and Edvard Scheutz. They based their construction on an article published in 1834 by Dionysius Lardner that had described Babbage's early work on the difference engine. The Scheutzes worked for 20 years to complete the project. It was exhibited in 1855 at the Paris Exposition, where it won a gold medal and approval from Babbage himself. In 1991, the British Science Museum completed a Babbage difference engine using materials available in his era and demonstrated that it could raise numbers to the 7th power with accuracy to 30 figures.

Biographers and historians of computing have debated the reasons for Babbage's failure to construct his analytic engine. Some have pointed to his irascible and abrasive personality and others to his approach, which was essentially numeric or digital in an era of analog devices. Many agree that his work was about 70 years ahead of his time and that he might have made more substantive contributions after the 1890s had he lived in that era.

anesthetics

Various types of painkillers had been used in antiquity by doctors during surgical operations. The ancient Greeks used nepenthe, assumed to be a form of opium, and Arabian doctors used opium and henbane. Paracelsus, who had given a start to modern **medicine [II]**, had developed a tincture of opium—that is, a solution in alcohol known as laudanum—in the 16th century. During the 18th century, British surgeons often employed rum to deaden the pain of operations, even including amputations. Since these painkillers were taken by mouth and could not be regulated at all during the operation, they were very approximate in their operation, with the patient often feeling some pain if the dosage was low or possibly risking death if the dosage was too high.

The British chemist Humphry Davy (1778–1829) determined the anesthetic effects of nitrous oxide or laughing gas in 1799, but it did not come into regular medical or dental use for several decades. Both nitrous oxide and ether were used as party entertainments for some years in Britain and the United States in the early 19th century.

The development of modern anesthesia for operations is usually credited to several doctors and dentists in the 1840s. In 1842 Crawford Long, a physician in the American state of Georgia, began to use ether in operations, and in the same year William Clark employed nitrous oxide in a tooth extraction. In 1845 an American dentist, Horace Wells, began to promote the use of nitrous oxide after first using it on himself during a tooth extraction. The use of chloroform was introduced in 1847.

William Thomas Morton, an American dentist who had studied under Horace Wells, made the first successful introduction of anesthetics in a demonstration in 1846. Morton used ether while removing a neck tumor from a patient at Massachusetts General Hospital. Soon, doctors and physicians began to regularly use ether or nitrous oxide to reduce pain in operations. In Britain, Dr. John Snow developed an inhaling device and gave chloroform to Queen Victoria when she delivered her eighth child, Prince Leopold, in 1853.

Many of the early methods of administering anesthetics such as ether, nitrous oxide, and chloroform were quite dangerous, as there were no means to measure or control the dosage. Later advances by anesthesiologists included a system that allowed mixing the gases with oxygen in a rubber bladder, then pumping it to the patient through a tight-fitting rubber mask.

Although coca had been known to South American natives before the discovery of the Americas by Europeans, cocaine was first introduced to Europe in 1860 when Albert Niemann, a German, successfully isolated and named the crystals of cocaine that had the numbing effect. Sigmund Freud used cocaine dosages to reduce dependence on opium and suggested its use for surgery to a colleague in Vienna, Carl Koller (1857–1944). Koller, an eye surgeon, first used cocaine as an anesthetic in 1884, and in 1885, Baltimore surgeon William Stewart Halsted began to use cocaine as a local anesthetic. In 1898 German surgeon August K. G. Bier introduced the concept of spinal injections of anesthetic to block pain, but several negative aftereffects in some cases, including paralysis, gave the method a bad reputation at first. In 1921

Spanish surgeon Fidel Pagés revived the idea of a spinal injection, not directly into the spinal fluid but as an epidural injection, into the area surrounding the dura or spinal fluid.

Other drugs have found their way into the anesthesiologist's tool kit. John Lundy introduced sodium pentothal in the United States in 1933, and a Canadian, Harold Griffith, introduced curare as a muscle relaxant. Other methods have included intense refrigeration, particularly useful in elderly or weakened patients. Acupuncture, involving the placing of needles at specified points in the body, is an ancient Chinese art finding more and more usage in the West.

asphalt paving

The use of natural tars, known as pitch or bitumen, for road surfacing, for sealing brick structures, and for water-sealing **dams and dikes [I]** as well as reservoirs was known to the ancient Babylonians. However, the use of asphalt for modern roads is a 19th-century development.

In the late 18th century, British road engineers set about improving the existing system of dirt roads there, led by two Scots: John Loudon Macadam and Thomas Telford. Macadam's system involved the use of crushed stone, tightly compacted. Heavy steamrollers were employed to pack down the stone. Road surfaces of crushed stone were said to be macadam roads. The development of the railroad in both Britain and the United States tended to retard road construction, and through most of the 19th century, many major cities in the United States could best be reached by rail or water, with no major highway system to connect them. Frequently the networks of roads surrounding and linking cities were simply unpaved dirt tracks, passable only in good weather by coach or wagon.

The use of asphalt was gradually introduced, with asphalt imported to the United States from Pitch Lake in Trinidad and from the Bermudez deposits in Venezuela. The first known asphalt-topped stretch of street was in front of the city hall in Newark, New Jersey, in the 1860s. With the introduction of the **bicycle** and the perfection of the **pneumatic tire,** by the 1890s, political pressure mounted for improved roadways. State and local authorities began to employ water-based concrete and asphalt-based concrete for smooth road surfaces. Asphalt-impregnated crushed stone, known as tarmac for its addition of tar to the macadam method of crushed stone, was first used in that decade.

The emergence of the relatively long-range, gasoline-powered **internal combustion engine**–driven **automobile** increased the demand for roads outside of cities, where stone or concrete-paved streets accommodated the shorter-range electric battery–powered automobile. In the 1920s, the use of gasoline taxes by states in the United States to pay for highway maintenance and construction vastly increased the network of intercity roads, many of which used asphalt-based concrete and tarmac surfaces.

automobile

Of all the inventions of the 19th and 20th centuries, the automobile may perhaps be the one with the most pervasive social impact. The concept of a self-propelled vehicle had been a long-standing technological goal for the human race, and when a series of technical advances made it possible, the automobile was independently developed by at least six separate inventors in the late 19th century.

In about 800 B.C., in the Chou dynasty in China, a steam-powered cart had been described, apparently as a hypothetical or possible device. In 1670, a Jesuit missionary stationed in China built a model cart driven by a steam engine. Nicolas Cugnot (1725–1804) built two steam-powered three-wheeled carts in 1769 and 1771. The first overturned and was wrecked; the second, which never ran, is still on display in Paris.

With the development of the **steam engine** [III] and improvement in its design, several early efforts to develop a steam-propelled vehicle for use on roads, rather than on rails, were successful in Britain and the United States. Richard Trevithick (1771–1833) built a road-capable steam vehicle, the *Puffing Devil,* in 1801. He and others designed and built road **steam locomotives** that ran on regular routes, carrying up to 16 passengers on regular schedules by the 1840s. British rail and horse-carriage interests had legislation passed that effectively put a stop to these ungainly vehicles in 1865, by requiring that road locomotive vehicles be preceded by a man walking and carrying a red flag.

As a propulsion system for a road vehicle, the steam engine presented barriers to developing better automobiles in the era due to the need for fuel and water supply and especially due to the high weight-to-power ratio of the current steam engine designs. In 1860, Jean Joseph Étienne Lenoir (1822–1900) in France developed a two-stroke gas-powered engine, and in 1872, George Brayton in the United States patented another such engine. An Austrian engineer, Siegfried Marcus, developed a petroleum-powered vehicle he drove around Vienna in 1875, but it was outlawed due to the racket it produced.

In 1876, Nikolaus A. Otto (1832–1891) in Germany designed and built an **internal-combustion four-stroke engine,** since known as the Otto-cycle engine, along lines earlier suggested by Alphonse Beau de Rochas (1815–1893). Otto had intended the engines for use as stationary power sources, but their low weight-to-power ratio made them suitable for road vehicles. Karl Benz (1844–1929) and Gottlieb Daimler (1834–1900), who had not met each other, both worked on possibilities of using either two-stroke or four-stroke engines for vehicles in the 1870s, and in 1885 Benz built a working model, which he patented in 1886. Most automobile histories regard the 1886 Benz model as the first automobile. Daimler improved on the Otto engine by substituting a system of electrical ignition for a pilot-light system used by Otto. Several others simultaneously experimented with vehicles propelled with internal-combustion engines, most of which used **coal gas [III]** (also known as town gas) as a fuel. Otto's patent was revoked, and Benz was able to use a four-cycle engine with an electrical ignition system of his own invention in his first three-wheel commercial vehicle, introduced in 1887. Benz introduced a four-wheel model in 1893. Daimler, who had worked for Otto, produced a motorcycle in 1885 and soon began outfitting carriages with motors.

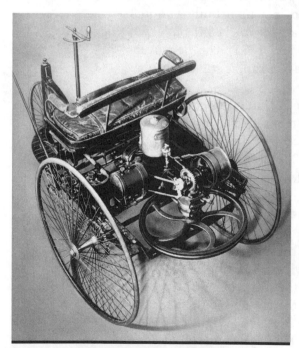

automobile The application of an internal combustion engine to the propulsion of a vehicle occurred to many inventors in the same period. Carl Benz built this version in 1885 and laid the groundwork for the Daimler-Benz corporation, the forerunner of Daimler-Chrysler. *Daimler-Benz Collection, Library of Congress*

The **carburetor [V],** introduced in 1893, made possible the more efficient use of vaporized petroleum fuel, rather than coal gas, eliminating a major technical barrier to automobile development. With the use of liquid naphtha, that fuel was soon named *gasolene* or gasoline, reflecting the heritage of using gas in internal-combustion engines.

The Benz and Daimler firms continued to produce automobiles and merged in 1926 to produce the Mercedes-Benz vehicles. The French company Panhard acquired the Daimler patents and produced vehicles in the 1890s. Meanwhile, in the United States, George Baldwin Selden (1846–1922) patented an automobile in 1879. The two brothers Charles (1861–1938) and James Duryea (1869–1967) produced a working automobile in 1893. In Britain, after a member of Parliament, Evelyn Ellis, purchased a French automobile and drove it through the streets of London in defiance of the red-flag law, Parliament abolished that rule in 1896, allowing the import and later production of automobiles in that country.

The simultaneous or near-simultaneous invention of internal combustion–powered vehicles from 1885 to 1893 by at least six separate inventors demonstrates the nature of technological progress very neatly. The known desire for a road-capable, self-propelled vehicle could not be fulfilled until an engine could be designed light enough to make the concept practical. With the Otto engine, mechanics and engineers in Germany, France, and the United States immediately began designing and building vehicles. The Dunlop **pneumatic tire** developed for the **bicycle** was quickly adapted to the automobiles, making their use on imperfectly paved roads possible.

Several early automobile manufacturers made rapid improvements in design, performance, and manufacturing techniques. Notably, the development of **interchangeable parts** by Ransom Olds in 1901 and Henry Leland, founder of the Cadillac Company, by 1908 made maintenance and repairs a simpler matter. Henry Ford manufactured the Model T in 1908, designing a vehicle that was cheap enough for the workingman to purchase, and Ford constantly worked to bring down the price even more. In 1914 Ford introduced the assembly line, a method already employed in bicycle manufacture, so that during the assembly, the partially completed vehicle would be moved from one workstation to the next. The process allowed even lower prices, and the Model T dropped from its original 1908 price of $850 to $290 by 1925. Cadillac introduced the electric self-starter, developed by Charles F. Kettering, in 1912. The self-starter made driving more accessible to women in the years following.

The social impact of the automobile was more profound in the United States than in Europe, leading to the further development of suburban districts around major cities that had already begun with the advent of carriage routes, **streetcars,** and **subway trains** [V] and, later, to the substitution of highway trucks for railroad distribution of most consumer products and many bulk and industrial products as well. By the 1920s, the introduction of gasoline sales taxes by state governments provided a direct source of revenue for building and improving intercity highways. Automobility, as measured by the lowest number of people per automobile, remained most pronounced in the United States, followed closely by Canada, Australia, and Germany.

Avogadro's law

Amedeo Avogadro (1776–1856) proposed a hypothesis in 1811 that since has been proven and is now known as Avogadro's law. He suggested that equal volumes of all gases, under the same pressure and the same temperature conditions, contain the same number of molecules. Avogadro was born in Turin, Italy, practiced law, and studied physics and mathematics.

His hypothesis was published in French: *Essai d'une manière de determiner les masses relatives des molecules élémentaires des corps, et les proportions selon lequelles elles entrent dans les combinaisons.* At first his work was not accepted, but his theory was later reintroduced, publicized, and demonstrated by Stanislao Cannizzaro (1826–1910), who used it to determine atomic weights of different elements.

Later investigation showed that the number of molecules per gram-mole of a substance is 6.0221367×10^{23}, a number now referred to as Avogadro's number or the Avogadro constant. The volume taken up by 1 gram-mole of any gas is about 22.4 liters at 1 atmosphere of pressure and 0° Celsius.

ball bearings

Although the Italian goldsmith and sculptor Benvenuto Cellini (1500–1571) introduced a form of ball bearing in 1543 to support a statue of Jupiter, the use of ball bearings in a race or groove to reduce friction in a regularly rotating mechanical device was first achieved in the 18th century. An early use was in a **windmill [II]**, built so that the entire structure could rotate around a central post to face the wind.

However, the concept that became essential to modern machinery was a set of radial ball bearings in a pair of rings with raceways cut into them. Philip Vaughan first used that type in 1794. He was an ironmaster of Carmarthen, Wales, who patented the ball bearing system for supporting an axle on a carriage.

Following Vaughan's work, mechanics and engineers made numerous improvements through the 19th century, with a great variety of patents for ball bearings used in applications such as propeller shafts, gun turrets on warships, and **bicycles**. With the coming of the **automobile**, ball bearings became essential to modern machinery and required precise grinding machinery to produce them. They became widely used in turbines, precision instruments, household appliances, and **airplanes [IV]**.

There are three basic parts to a ball bearing. There are two grooved rings or races and the balls with their retention system. The two races are of equal width but have different diameters, with the smaller fitting inside and on the rotating shaft and the larger outside and mounted to the stationary member through which the shaft or axle rotates. The balls fill the space between the two races and run in the grooves of the races. The balls are held loosely in place by a retainer or cage system and rotate with negligible friction. An angular contact bearing has one side of both races cut away so that the balls can be inserted, and such a bearing can take thrust only from the direction of the open side. Often, two angular bearings are mounted together with their open sides facing each other, so that thrust can be taken from both sides. Many ball bearings are prelubricated, with the lubricant sealed inside. Some high-speed ball bearings use air-oil mist rather than liquid lubricants to avoid overheating of the lubricant.

Ball bearings have nearly no friction, even at start-up, which gives them an advantage over sliding bearings, which consist of two smooth rings separated only by a film of oil. Modern ball bearings are made of a high-quality alloy **steel** with small amounts of carbon and chromium.

barbed wire

Barbed wire–making machines were first patented in 1873, although homemade types of barbed wire had been made over the previous six or seven years in the American Midwest. Farmers had sought a cheap form of fencing to keep cattle from straying onto cropland. In Illinois, a natural thorny bush, the Osage orange, provided prickly hedges that worked fairly well, but transplanting and growing the plants was time-consuming. It is possible that the first ideas for barbed wire derived from an attempt to replicate the natural thorns of this plant, an idea given credence by the first appearance of barbed wire in an area in which the Osage orange hedge was already in use. Henry M. Rose developed a type of barbed wire and demonstrated it at the 1873 county fair in Dekalb, Illinois.

Several men who visited the fair came away with ideas for improvements in manufacturing the wire. Joseph Farwell Glidden filed a patent in October, and Jacob Haish filed a separate patent in December 1873. Glidden formed a partnership with a local hardware merchant, Isaac L. Ellwood. Glidden later sold his interest to the Washburn and Moen Manufacturing Company, in Worcester, Massachusetts. Haish and Washburn/Moen engaged in a bitter patent dispute, but Haish eventually profited from leasing barbed wire–making equipment to numerous firms. By the end of the 19th century there were about 400 different types of barbed wire.

In addition to its use in confining or excluding livestock, barbed wire became widely used in warfare, especially during World War I, when it was used to prevent infantry charges against trenches and other fortified positions. With other tools of modern warfare, including the **machine gun,** barbed wire helped account for the three-year deadlock on the Western Front during World War I in which hundreds of thousands of soldiers were killed.

In the 20th century, barbed wire found other uses. Originally intended to fence livestock in or out of a piece of land, barbed wire was an ideal material for the inexpensive construction of compounds in which to detain large numbers of people in prisons and prison camps. Similarly, when mounted in strands atop high, chain-link fencing, it became an effective and relatively inexpensive means to exclude people from secured areas.

battery

There are some indications that batteries and electroplating were known in ancient Iran, in the 3rd century B.C. However, Alessandro Volta (1745–1827) developed the modern battery between 1796 and 1801. Volta's work sprang from a dispute over whether electricity was produced in animal bodies.

In 1780, Luigi Galvani (1737–1798), a professor of medicine in Bologna, observed that a dead frog leg that had been immersed in salt water would jerk when lying on a zinc plate and touched with a steel scalpel. He believed he had discovered animal electricity, and over the next decade, many experimenters sought an explanation for galvanic motion, and his name moved into European languages. However, Volta, a professor of physics with a specialty in electricity at the University of Pavia, believed that the electricity came not from the frog leg but from the metal or cardboard surface on which it was placed. After a series of experiments, Volta built a device in 1800 that produced a steady and large flow of electricity. He connected two bowls of salt water with metal strips made of copper and zinc. Attempting to design a more practical device, he stacked saltwater-impregnated cardboard disks alternately with copper and zinc, and this voltaic pile became a useful battery. Since each disk was a single electric cell, stacking them in groups or a pile that ran in series increased the voltage. The concept that several cells had to be used in groups led to the term battery, by analogy to a battery of several cannons being used to increase firepower in a military engagement. Thus, strictly speaking, a single electric cell is not a battery, although the term *battery* has later been used to describe single- or multiple-cell electrical storage units.

In 1836, John Frederic Daniel (1790–1845), a British chemist and meteorologist, developed an improved battery that could be recharged. The Daniel cell consisted of a zinc electrode in a solution of zinc sulfate and a copper electrode in a solution of copper sulfate, with a porous earthernware diaphragm between the electrodes. Further improvements by a variety of experimenters led to batteries with alkaline, acid, and mercury electrolytes that could serve in many devices to provide small amounts of power as needed.

The lead-acid battery, developed in 1859 by Gaston Planté, was the forerunner of the type used in the **automobile.** The constituents are sulfuric acid and two sets of plates or grids serving as electrodes. Camille Faure patented the grid concept in 1881. One plate is made of pure lead and the other is made of lead dioxide. Each compartment of a multicell automobile battery contains one plate of each metal immersed in the sulfuric acid solution. During the discharge of the battery, the plate materials are converted into lead sulfate. When direct current is run through the battery, it reverses the chemical change, displacing the sulfate from the metals and restoring the acid. Thus the battery can serve as an efficient storage system for electricity, constantly recharged by the **internal-combustion engine** connected to a generator and discharged to operate the **electric motor** starter or other electrical device.

bicycle

The bicycle can trace its ancestry back to a simple device with two wheels in line with a small saddle, propelled by striking the ground with the feet. In about 1690, a Frenchman named de Sivrac had developed this rather primitive type of scooter, known as a *celerifere*. Baron Karl Friedrich Christian Ludwig Drais von Sauerbronn improved it in 1813, with the addition of a steerable front wheel. In Britain they were known as hobbyhorses or dandy horses. Drais demonstrated his running machines, which never caught on in Germany but became popular and even faddish in Britain and the United States. Drais also invented a hand-cranked railway repair car known as a Draisine.

A Scot, Kirkpatrick Macmillan, developed the first pedaled bicycle in 1842, with pedals on the rear wheel. He traveled from Dumfries to Glasgow, some 40 miles, in 2 days. Philipp Heinrich Fischer, a Bavarian mechanic, developed a similar pedaled machine. In France in 1861, a coachmaker named Pierre Michaux was repairing a Drais-type steerable hobbyhorse. Apparently he had never seen one of Macmillan's machines, and Michaux suggested an improvement to the Drais machine, with pedals on the front wheel. The pedal **crank [II]** of these velocipedes was attached directly to the front wheel. In this period, no one understood why it was so easy to maintain balance on a two-wheeled machine while it was in motion but almost impossible when at rest. The gyroscopic effect of the two rotating wheels makes such machines steadier the faster the vehicle moves.

Driven only by directly attached cranks, the front wheel had to be enlarged to achieve greater speed. By the 1880s, the ordinary or boneshaker consisted of a huge front wheel and a smaller trailing wheel. The disparity in size earned them the nickname of penny-farthings, and they became a fad by 1884. The British penny coin had a diameter about three times that of the tiny farthing, or quarter-cent piece. **Drive chains** had been used to drive textile machinery, but a strong and reliable chain small enough to be used for a bicycle went through several improvements, starting in 1868 with one designed by André Guilmet and built by Meyer et Cie. Hans Renold, a Swiss inventor who had moved to Britain, suggested a roller chain similar in design to one appearing in sketches by Leonardo da Vinci. A few other improvements, such as springs for the saddle, ball bearings, and gears and gearshift, came quickly in the early 1880s. In 1885 John Kemp Starley of Coventry, Britain, and the Frenchman G. Juzan separately invented a chain-driven bicycle geared to the rear wheel. The chain-driven bicycle became immediately popular, and with the 1888 invention of the **pneumatic tire** by John Boyd Dunlop, the bicycle industry became a worldwide phenomenon. In Britain the town of Coventry became a bicycle-manufacturing center.

In 1888 there were some 300,000 bicycles worldwide. A century later there were more than 100 million as the device became a cheap and reliable form of transportation, as well as a machine for sport and play. The bicycle can carry at least 10 times its own weight and can travel at speeds up to 6 times that of a running man.

Boltzmann constant

Ludwig Boltzmann (1844–1906), an Austrian physicist, contributed to the evolving **laws of thermodynamics** in several ways. In particular, he established the foundations of a statistical or probabilistic approach to physics by establishing that the approach to entropy, or balance, in an energy system was not an absolute but was a statistically probable effect. For example, when a hot system (such as a bottle of hot water) cools to the surrounding ambient temperature, thereby minutely raising the temperature of the surrounding air, entropy is approached. The effect is a statistical product of minute actions, not an absolute or single effect. He published his ideas in a series of papers in the 1870s, but the principles he suggested were rejected by much of the scientific establishment for several decades.

In the 1890s, the concept of Brownian movement, the random movement of microscopic particles suspended in a liquid, reflected a probabilistic or statistical understanding of physical phenomena very similar to that proposed by Boltzmann. Apparently depressed by the controversy surrounding his findings, Boltzmann committed suicide in 1906. In his honor, physicists named the concept of a constant that relates kinetic energy of a gas atom to temperature the Boltzmann constant, represented by the symbol k. The constant k is the dimension of energy per degree of temperature, a value of 1.380662×10^{23} joules per kelvin (K).

> Hypothesis should be considered merely an intellectual instrument of discovery, which at any time may be relinquished for a better instrument. It should never be spoken of as truth; its highest praise is verisimility. Knowledge can only be acquired by the senses; nature has an archetype in the human imagination; her empire is given only to industry and action, guided and governed by experience.
> —Humphry Davy, *The Collected Works of Sir Humphry Davy*, vol. 8, edited by John Davy, (London: Smith, Elder and Co, 1839–1840) p. 347

braille

Louis Braille (1809–1852) developed the system of braille writing for the blind. He was born with sight but was accidentally blinded in one eye at age 3 and lost the vision in his other eye by age 5. Attending a school for the blind, he learned of a system of embossed symbols, known as sonography or night writing, invented by a French army officer for nighttime communication among soldiers in the field. Braille had already learned a more complex system of embossed type that was very difficult, and he soon began to experiment with a more readily understood system. He settled on a system of a six-dot code, similar to the dots on dominoes and easily detected with the touch of one finger.

Braille encoded not only individual letters but also punctuation, numbers, several common words, some two-letter combinations, and some common endings as well as scientific symbols and musical notations. Since w was not regularly used in French, Braille did not include it in his original system. In 1829, at age 20, Braille published his system.

It was slow to catch on, however, as teachers of the blind were reputed to fear that a class of well-educated blind teachers would replace sighted ones. Two years after Braille's early death from tuberculosis, braille writing received official recognition in France, and a modified system of braille was first used in the United States in 1860. An alternate system, invented in 1845 in Britain by William Moon, partially retained the shape of Roman letters and is still used to some extent, particularly for those who lost sight after learning to read from print.

Frank Hall, who was superintendent of the Illinois School for the Blind, invented a braille typewriter in 1892.

Bunsen burner

The Bunsen burner, a simple gas flame device still widely used in chemical laboratories, was invented in 1841 by Robert Wilhelm Bunsen (1811–1899), a German chemist. The burner consists of a vertical metal tube with a set of holes near the base to admit air and through which a jet of fuel gas is directed. It is ignited at the top opening to produce a steady flame that can be regulated by increasing or decreasing the flow of gas with a simple valve.

Bunsen worked with Gustav Robert Kirchhoff (1824–1887), and together they discovered two new elements in 1860, cesium and rubidium. Bunsen and Kirchhoff developed the burner to heat metal salts to conduct analysis of materials using the **spectroscope,** and the burner soon became a standard and reliable piece of laboratory equipment.

camera

Following the invention of the **daguerreotype** method of photography by Niépce and Daguerre before 1830, a number of experiments and improvements led to the camera that held flexible film. Several claimants to priority in these inventions can be identified.

Progress in the field came rapidly after the 1868 invention of **celluloid.** John Carbutt, an English photographer, claimed to be the first to invent emulsion-coated celluloid films, introducing in 1888 "Carbutt's flexible negative films." At almost the same time, Rev. Hannibal Goodwin applied for an American patent (in May 1887) for a transparent roll film, but his patent was not granted until 1898.

George Eastman (1854–1932) started his firm in 1881, marketing dry photographic plates. In 1884 he patented photographic film made of emulsion smeared on paper. In 1889 he patented a rolled film that had a layer of hardened gelatin on one side of celluloid, developed by H. M. Reichenbach, an Eastman company researcher. After a 12-year lawsuit between the Goodwin interests and Eastman, the courts found in 1914 that the Goodwin application in 1887 was primary, and Eastman settled for $5 million the claims by Ansco, the holder of the Goodwin patent rights.

Eastman introduced the first Kodak camera in 1888. Eastman rapidly acquired patent rights to a variety of specific camera parts, including a small red window for reading the number of exposures on a film roll and a film spool that could be changed in daylight. The original Kodak had a fixed focus lens and could take 100 circular pictures that were 2.5 inches in diameter. Although priced at $25 (the equivalent of several weeks' salary for a well-paid factory worker in the 1890s), Eastman was able to sell more than 100,000 of them in the first two years. Originally Eastman required that the camera be sent to the factory for picture development, reloaded with new film, and returned to the owner with printed pictures. Eastman retained a monopoly on roll film cameras, and for decades the term *Kodak* became synonymous with a small handheld camera of any make. In 1900 Eastman introduced the Brownie, in which the operator could change the film without sending it back to the factory, and which sold for a dollar. The Brownie was named after a popular cartoon character of the era, often featured in advertisements for the camera.

Eastman's improvements had the effect of moving photography from the professional's studio, where it had evolved from the realm of portraiture and art, to the world of the general consumer. By making the camera a household gadget and tourist tool, Eastman led to the popularization of capturing images. Although postcards taken by professionals were still sold at popular tourist destinations, visitors were soon supplementing them with whole albums of personally captured photos.

canned food

The canning of meat and vegetables was the direct result of a government-sponsored search for a means of preserving foods. The French government in 1795 announced a prize of 12,000 francs for such a system. After years of experimentation, Nicolas Appert published in 1810 a treatise, *L'Art de Conserver,* and won the prize for his method of putting cooked meats, fruits, and vegetables in glass containers, which are then boiled long enough to kill bacteria before sealing them.

The glass containers worked well but were fragile when transported.

A British merchant, Peter Durand, who used a tin canister (from which the name tin can was derived), solved the problem in 1810. His wrought iron cans were coated with tin and were soon used by both the British navy and army. Such canned food was particularly useful on sea voyages and exploration. The cans were extremely heavy and sturdy by more modern standards. With the introduction of thin rolled steel, the weight of the can, now formed from three pieces of tin-plated steel, was reduced in the 1850s. A rectangular piece would be bent into a hollow cylinder and welded at the seam, with two disks then welded in place for top and bottom.

However, it took several years before a satisfactory means of opening the cans developed. From the 1820s through the 1860s, soldiers in the field and sailors aboard ships used bayonets, knives, chisels, and even gunshots to get at their rations. Ezra Warner, a Connecticut inventor, developed one of the first successful can openers, in 1858. The first really popular can opener was known as the Bull's Head, introduced in about 1900. In 1870 another Connecticut inventor, William Lyman, invented a can opener with a wheel that allowed a smooth and continuous cut.

card punch machine

The card punch machine and card reader were invented in the 1880s by Herman Hollerith (1860–1929), an American inventor and employee of the U.S. Census Bureau. His system was used for recording information in the 1886 Baltimore census, and then nationally for the 1890 census. In 1896 he organized the Tabulating Machine Company, which eventually grew into the International Business Machines (IBM) Corporation. His machines were a great success in both the United States and Europe, as they cut labor costs in the compilation of statistical data. The machines were known in Britain as *Hollerith machines*.

He designed the card to match the size of the then current dollar bill (about 7.375 inches by 3.25 inches) so that he could incorporate existing money storage equipment in his machinery. Using a typewriter keyboard, the card punch machine would perforate the cardboard, and when fed to a machine, electrical contacts through the holes would close circuits, allowing the counting by electromechanical tabulators. At first many of the procedures, including feeding the selected cards into bins, were done by hand. Later machines included a more mechanized card-feed and allowed for addition and sorting of data.

James Powers, one of the engineers working in the Bureau of the Census, invented an alternative card punch system in 1906. Powers won a commission to develop an automatic card punching machine, and he developed that machine from 1907 to 1909. Powers formed his own company in 1911. Powers' company merged with the Remington typewriter company in 1927, and in 1955 the company merged with Sperry, the manufacturers of the **gyro compass** [V], to form Sperry Rand.

celluloid

Celluloid was the first true plastic invented, although other artificial materials, such as **papier-mâché** [III], had preceded it. Celluloid was first developed in 1856 by a British chemist, Alexander Parkes (1813–1890), a professor of natural science at Birmingham University, who named the material *parkesine*.

Parkes mixed nitrocellulose (also known as guncotton) with camphor, producing a hard, flexible, and transparent material. After attempting to market parkesine, his firm went bankrupt. But in 1868, John Wesley Hyatt (1837–1920), an American inventor, attempted to win a $10,000 prize offered by the New York company of Phelan and Collender for a substitute for ivory to make billiard balls, and he acquired the Parkes patent. Hyatt's brother suggested the name *celluloid* instead of parkesine, and Hyatt began to manufacture the material under that trade name. In Britain, celluloid was sold as Xylonite, a term derived from pyroxylin, the partially nitrated cellulose that was its base. British sources usually credit Parkes with the invention, while American sources stress Hyatt's successful development and marketing of the product.

The plastic was never satisfactory for billiard balls, and Hyatt did not win that prize. However, he soon found other applications for celluloid, including artificial collars, combs, knife handles, baby rattles, piano keys, and Ping-Pong balls. With the invention by George Eastman (1854–1932) of the Kodak **camera,** Eastman used celluloid rather than glass or paper for a rolled photographic film in 1889. Following the invention of **motion pictures** [V] a few years later, the word *celluloid* became synonymous with movies.

cigarette

The Spanish conquistadors discovered that Native Americans made use of tobacco both in the form of rolled leaves, called *cigars* after the Mayan word *s'kar* for smoking, and in the form of ground tobacco rolled in corn husks, resembling the modern cigarette. However, the true cigarette machine-rolled in paper is of much more modern origin.

Legend has it that beggars in Spanish ports would grind discarded cigar butts and roll them in paper in the 16th century, making a paper-rolled smoke a symbol of poverty. In the 18th century, paper-rolled cigarillos became more respectable, and their use spread via Portugal to the Near East and to Russia. The exact spread of paper cigarettes has not been traced, although records indicate their presence in Mexico in the mid-18th century and in France in the early 19th century. The milder Maryland and Virginia tobacco was imported to France for hand-rolled cigarettes in the 1840s.

The ready-made or machine-rolled cigarette seems to be clearly a 19th-century phenomenon, characteristic of the Industrial Age. Cigarette factories employing a hand-rolling method opened in Russia in 1850 and in London in the 1850s. American manufacture of cigarettes in shops began in 1864. In the early 1870s, Greek, Egyptian, and Turkish immigrants manufactured cigarettes in both Britain and the United States. Since cigarettes remained cheaper than cigars, cigarettes increased in popularity during hard times, apparently becoming a fad in the United States following the Panic of 1873.

In 1880, James Albert Bonsack developed a machine that would feed apportioned amounts of cut tobacco into a forming tube where it was shaped into a roll, and a knife would cut the roll into appropriate lengths. James B. Duke improved on the Bonsack machine, and by the mid-1880s, the Duke company in Durham, North Carolina, achieved a daily production of 4 million cigarettes.

Cigarette packaging was automated with a Swedish device previously used to package matches, and adding **cellophane** [V] outer wrap in 1931 helped preserve freshness. Flue-cured tobacco blended with Burley tobacco represented the true American-style cigarette, introduced in 1913 by Richard J. Reynolds.

The fact that cigarette smoking contributes to higher rates of heart disease and lung cancer became obvious in the 1930s. Antismoking campaigns had little success, even when backed by the totalitarian regime in Germany in the 1930s. However, in the United States in the 1950s and 1960s, the requirement for compulsory health warnings and the increasing awareness of the cigarette smoking hazard began to cut into sales. It has been estimated that since the beginning of the widespread use of cigarettes in the 1870s, more people have died prematurely from the effects of cigarette smoking than were killed on all sides in both World War I and World War II.

coal oil

Coal oil and petroleum-derived kerosene were developed after **coal gas** [III], already in use in several European cities by 1808. The invention of coal oil, distilled from coal for use in lamps, was due to a Canadian medical doctor, Abraham Gesner (1797–1868), and a Scottish chemist, James Young (1811–1883).

Notified of oil seepage from a coal mine in Derbyshire in 1847, Young at first sought to make use of the seepage and then to work out a distilling method. Young and a partner, Edward Binney, built the first coal-oil distilling plant at Bathgate, Scotland, and patented the process in England in 1850 and in the United States in 1852. Young called the distillate parafinne oil, which when cooled made a waxy deposit. Eventually his paraffin supplanted beeswax for candles.

By 1855 Young was able to bring the price of coal oil to half that of whale oil. In 1865 he sold his company for a fortune, retiring to pursue scientific work. Meanwhile, Gesner had become intrigued with a pitch lake in Trinidad and had made attempts in the 1840s to distill the pitch. He patented a method for producing illuminating gas from the petroleum tar in 1846. In 1850, at about the same time that Young developed his method in Britain, Gesner worked on distilling heavy asphalt found in Nova Scotia and known as albertite. After several patent disputes with Young, Gesner introduced processes to make kerosene from petroleum rather than coal. Both used sulfuric acid to get disagreeable odors out of the lamp oils. The popularity of coal oil led to the search for ways to obtain the petroleum-based kerosene and the development of **oil drilling.**

The popularity of both coal oil and petroleum-derived kerosene over whale oil was due to the low prices of the new products and also to the introduction in 1856 of a new type of light, known as the Vienna burner, with a flat wick that worked very well with the new fuels. The flat-wick burner in oil lamps became standard and remained virtually unchanged into the 20th and 21st centuries.

cotton gin

The cotton gin to remove seeds from cotton bolls after picking was invented in 1793 by Eli Whitney (1765–1825) and represented a marked improvement over an earlier device invented by Joseph Eve (1760–1835). The Eve machine had been used with West Indies cotton from the 1780s but did not successfully remove the seeds from the long-staple cotton boll grown in the United States. Instead, the Eve machine crushed the seeds, ruining the cotton with oil from the seeds. The term *gin* had been widely used in the West Indies as a colloquial contraction for the word *engine*.

Whitney had traveled to South Carolina, expecting to take up a tutoring position, but stayed on as a guest of Catherine Greene, widow of Revolutionary War general Nathaniel Greene. Staying at the Greene plantation near Charleston, Whitney sought to repay the hospitality of his hostess with various small woodworking projects. Noting his skill and aware of the problem of removing seeds from cotton, she proposed that he apply his ingenuity to the cottonseed-removal problem. After several months in her basement shop he devised a hand-cranked improvement to the Eve machine that removed seeds and, with a brush, cleaned the fibers from iron wire teeth mounted on a revolving cylinder. According to legend and his own later accounts, he used wire instead of sheet iron, finding it available at the plantation, left over from the construction of a birdcage. The brush mechanism may have been inspired by a remark by Mrs. Greene regarding a fireplace hearth brush as a solution to the problem of removing the fiber from the machine teeth.

Although the working model was constructed in secrecy, when it was demonstrated to local planters, they urged Whitney to seek a patent. He did so, working with Phineas Miller, a slave overseer. However, Whitney and Miller found that their patent was widely infringed and copied, as the machine was easy to construct.

Although Whitney and Miller were unable to win lawsuits for patent infringements, several southern state legislatures made them cash grants in reward for their work. Although they received about $90,000 in such awards (equivalent to several million dollars in 21st-century value), the funds were largely consumed in legal fees.

Whitney returned to his home state of Connecticut, where he went on to work on developing a system for the manufacture of

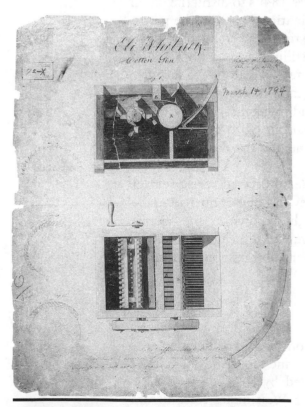

cotton gin The patent for Eli Whitney's cotton gin was filed in 1797. *U.S. Patent Office*

muskets by the use of **interchangeable parts.** Although he is often credited with that concept, he did not successfully achieve the goal.

Meanwhile, the cotton gin revolutionized the production of cotton in the United States, removing a bottleneck in the process. Vast new lands were opened to cotton production, increasing the demand for slave labor. The cotton gin (when driven by mulepower) allowed a single worker to produce seed-free cotton equivalent to the amount produced by 50 workers before the gin. In that sense the cotton gin was a labor-saving device, but ironically it created a labor shortage by making cotton production suddenly very profitable.

Crookes tube

William Crookes (1832–1919), a London chemist and physicist, invented an early type of cathode ray tube, known as the Crookes tube, during researches into radiation inside an evacuated glass tube. Using the **spectroscope,** Crookes discovered the element thallium in 1861. His work on instruments for studying radiation took place between 1875 and 1881. He developed a "radiometer," a device containing a small vane that was black on one side and polished on the other and that would spin when exposed to radiation. Crookes had little formal training, although he went on to be knighted and to receive numerous honors in the field of chemistry. He published very few papers and did not seek patents for the tubes he made, perhaps accounting for the disagreement among sources over the exact dates for their invention.

The Crookes tube was widely copied and used as a form of entertainment or parlor demonstration through the 1880s. The tube could be used to measure the voltage needed to cause an electrical effect to jump from the cathode to the anode in a vacuum. Since there was no conducting material between the cathode and the anode, Crookes had dubbed the transmission a "cathode ray." Reproductions of Crookes tubes from the era continue to be sold as decorative items in the 21st century.

Crookes's cathode ray tubes were used in 1895 by Roentgen in his discovery of **X-rays** [V] and by J. J. Thomson in 1897 in his discovery of the **electron** [V]. As with a great deal of scientific work in the 19th century, the attempt to explain the laws governing the operation of a device led to new theoretical advances.

In 1895 an Austrian physicist, Karl Ferdinand Braun (1850–1918), improved on the Crookes tube by making the anode a ring, allowing the cathode rays to pass through and hit the inside of the glass wall, which he coated with a fluorescent material. The Braun version of the Crookes tube was known as an oscilloscope and used to display the oscillations of an alternating current on a screen. The Braun tube, with its coated surface, was later adapted for the large cathode ray tube used for **television [V]** screens.

daguerreotype

The first processes of capturing an image were invented in France by the collaboration of Joseph Nicéphore Niépce (1765–1833) and Louis Jacques Mandé Daguerre (1781–1851). From 1826 to 1827 Niépce had worked on a process of taking a photograph on a glass plate from nature that required about 8 hours of exposure time, and Daguerre developed a method that took 30 minutes or less. Niépce's first successful photograph has been dated to 1826, taken from an upstairs window in his home in the French countryside. The picture was exhibited in various locations until 1898, when it was lost until 1952, when it was rediscovered and authenticated; it was placed on display at the University of Texas in 2003.

Daguerre worked as a scene painter for the opera, and in 1822 he opened a pictorial exhibition hall in Paris. His diorama used changing lighting effects to create an illusion of reality. Such dioramas, using elaborate constructions with lights shining through painted screens, had become extremely popular over the previous 20 years. Prior dioramas, including one developed by Robert Barker in Edinburgh in about 1793 and one built in Paris in 1804 by the American engineer Robert Fulton, had drawn great audiences, but Daguerre's was far more successful, attracting the attention of Nicéphore Niépce. Daguerre and Niépce cooperated between 1829 and 1833 on improving Niépce's method of capturing images. In describing his procedures, Niépce used phrases later common in photography, including "exposure" and "developer."

Following Niépce's death in 1833, Daguerre continued the work. A latent picture Daguerre had locked away in a cabinet

> M. Niépce, in his endeavor to fix the images which nature offers, without assistance of a draftsman, has made investigations, the results of which are presented by numerous proofs which will substantiate the invention. This invention consists in the automatic reproduction of the image received by the camera obscura. M. Daguerre, to whom he disclosed his invention, fully realizing its value, since the invention is capable of great perfection, offers to join with M. Niépce to achieve this perfection and to gain all possible advantages from this new industry.
>
> —Terms of the agreement signed December 14, 1829, between Louis Jacques Mandé Daguerre and Joseph Nicéphore Niépce that defined their arrangement to invent and improve a photographic process

for a few days developed accidentally. By eliminating one by one various substances in the cabinet, he discovered that fumes from a broken mercury thermometer brought out the picture on the iodized plate, and that with minute mercury exposures, development time could be considerably shortened. He continued working on the process, settling on fixing the image with a salt-water solution. With the discovery of the mercury fume process he obtained consent from Niépce's son to change the name of the firm from Niépce-Daguerre to Daguerre-Niépce. His first successful picture with his improved procedure, a still life, was completed in 1837 and presented to the curator of the Louvre.

The astronomer François Dominique Jean Arago (1786–1853), who served as permanent secretary of the Paris Academy of Sciences from 1830 to 1852, announced Daguerre's methods at a meeting of the academy on January 7, 1839. Daguerre assigned the rights of the daguerreotype process to the French government, and both Daguerre and the heirs of Niépce received annual annuities for the invention.

Daguerre produced a manual describing the process that went through many editions. The method remained the only practical means of photography from 1840 to 1851, when Frederick Scott Archer (1813–1857) developed a wet collodion process of photography. Later progress in image capturing led to portable **cameras.**

drive chains

Leonardo da Vinci (1452–1519) sketched a very modern-looking drive chain in one of his notebooks, but there was no known follow-up on the idea until the 19th century. Iron chains run over cogs were used for lifting buckets and other weights in the 16th century, but they did not have the roller linkage sketched by da Vinci. Only with stronger metals and precision manufacturing could reliable drive chains be built. In 1864 James Slater in Salford, Britain, patented a small drive chain suitable for propelling a **bicycle.** Hans Renold, a Swiss industrialist, later acquired the Slater factory and manufactured an improved roller chain that was even stronger. Similar drive chains were used on later bicycles, on motorcycles, and on some early **automobiles.**

dynamite

Alfred Bernhard Nobel (1833–1896) invented dynamite in 1867. Born in Sweden, Nobel was educated in Russia by private tutors. At age 17 he traveled to France for a year and then to the United States for four years to study chemistry. In Paris he studied under Théophile Jules Pelouze (1807–1867), who had worked on the development of guncotton. In the United States, Nobel worked for a short period under John Ericsson (1803–1889), the Swedish-born developer of the ironclad *Monitor*. The *Monitor* gained fame for defending the Union against the *Virginia* (formerly the *Merrimac*) in an 1862 battle of the American Civil War, the first between two ironclads.

Working in his father's torpedo plant in St. Petersburg in the early 1850s, Nobel experimented with various explosives. His father's company in Russia went bankrupt after the Crimean War (1853–1856), and the family moved back to Sweden. In a small factory near Stockholm, he and his father began large-scale manufacture of nitroglycerine, which had been invented in 1846 by an Italian, Ascanio Sobrero (1812–1888). An explosion in 1864 in the Nobels' Stockholm plant killed Alfred Nobel's younger brother Emil, together with five workers. The accident is often credited with changing the direction of Nobel's work to find a safer explosive. In 1863 Nobel patented a mercury fulminate detonator that made the use of nitroglycerine safer.

Nobel began a series of experiments to determine whether a safer explosive could be made by mixing the oily nitroglycerine, which could be detonated by a shock during manufacture and transport, with some more inert material. After trying a wide variety of substances, he determined that mixing the nitroglycerine with kieselguhr, a type of porous diatomite clay found in Germany, resulted in an explosive that was relatively insensitive to heat and shock but that could be detonated with a percussion cap. He later denied that the discovery of kieselguhr was accidental and insisted it had been uncovered through systematic testing of various substances, including brick dust, paper, and wood. He named the clay-nitroglycerine material dynamite and took out British and American patents for it in 1867 and 1868. Nobel also invented gelignite, a solution of guncotton (nitrocellulose) in nitroglycerine. In 1888 he developed a form of smokeless powder he called ballistite, but he lost a patent fight over this development to the British inventors of cordite.

At first, gunpowder and explosives manufacturers tried to prohibit the manufacture of dynamite. But Nobel's dynamite was more powerful than gunpowder and far safer than nitroglycerine, so it soon proved useful in mining, excavation, and ammunition. Nobel established a chain of manufacturing plants around the world. Together with investments in the Russian petroleum fields of Baku, Nobel made a fortune, which he left to establish a foundation that has granted Nobel Prizes in physics, chemistry, physiology or medicine, literature, and peace. The first prizes were granted in 1901, on the fifth anniversary of his death.

electric light

In popular American legend, Thomas Edison (1847–1931) is credited with the invention of the incandescent lightbulb, with his demonstration of a successful bulb on December 3, 1879. However, Edison's work was only one stage in a much longer process. As early as the 1820s, scientists understood that metals could be heated to incandescence by electric current and sought to construct bulbs from which the air had been evacuated to prevent oxidation of the filament. In 1841 the first patent for an incandescent bulb was granted in Britain to Frederick de Moleyns. A few years later Joseph Swan (1828–1914), a British scientist and inventor, worked with a carbon filament and a vacuum-evacuated bulb.

A more effective **vacuum pump [III]**, invented in 1865 by Herman Sprengel, allowed for better bulbs. Swan displayed a successful carbon-filament incandescent lamp in 1878, fully a year before Edison's success, and Swan illuminated his own house, the first in the world lit by such bulbs. Swan formed a company in 1881. Sometimes both Swan and Edison are credited with the nearly simultaneous invention of the successful lightbulb.

Edison was already well known for other inventions, particularly the **phonograph,** and his research into the electric lamp was financed by the Edison Electric Light Company, a syndicate formed by J. P.

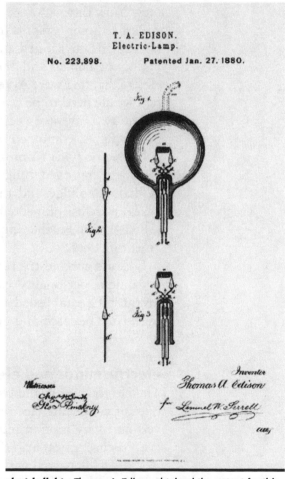

electric light Thomas A. Edison obtained the patent for this electric light in January 1880. *U.S. National Archives and Records Administration*

Morgan and the Vanderbilts. Edison worked with Francis Upton, and together Edison and Upton developed a practical electrical generator. After demonstrating an early bulb in December 1879, Upton and Edison concentrated on a carbon filament using bamboo. During testing of the bulb, William J. Hammer detected a blue glow, later determined to result from the flow of **electrons** [V] from one electrode to another. The effect, however, was not called the Hammer effect but the Edison effect.

Edison proposed systems of distributed electric power from central stations. He constructed a demonstration system in London and the first U.S. power station, operative in September 1882, on Pearl Street in New York City.

Perhaps more important than the bulb itself, the electrical distribution system envisioned and pioneered by Edison made the wide illumination of homes and businesses possible. Edison recognized that if electric lighting were to replace gas lighting, a similar public utility system would need to be created.

The two inventors of the lightbulb, Swan and Edison, after severe patent fights, combined their resources in the 1880s to form the General Electric (GE) Company. Later evolution of the incandescent light included improved tungsten filaments developed at GE in 1910 by William Coolidge and the use of nitrogen and argon as nonreactive gases in bulbs, pioneered in 1913 by GE researcher Irving Langmuir (1881–1957). Further improvements by Langmuir led to the twisted filament in 1934.

Edison's cut-and-try research method, his prolific inventiveness, and his unique personality earned him repute as America's most famous inventor. He established an "invention factory" that many have regarded as the first research-and-development laboratory.

electric motor and electric generator

The electric motor and electric generator are essentially similar devices, one for converting electricity into rotary motion and the other for converting rotary motion into electricity. Two inventors, the British physicist Michael Faraday (1791–1867) and the American Joseph Henry (1797–1878), played parts in both inventions, dating from the 1820s and 1830s. Both were prompted to investigate **electromagnetism** by its discovery by the Danish physicist Hans Christian Oersted (1777–1851) and by the discovery by the French physicist and astronomer D. F. J. Arago (1786–1853) of the relationship between magnetism and electricity.

In 1829 Henry, teaching at Albany Academy, heard that William Sturgeon (1783–1850) had made an electromagnet by winding electric wire around an iron core. Henry set out to make a more powerful magnet and developed the concept of insulating the wires to be able to wind more around the core without them shorting out. By 1831 he was able to demonstrate an electromagnet at Yale University that could lift more than 2,000 pounds. Henry also developed a rudimentary system for using electricity over a 1-mile wire to cause a distant electromagnet to engage. With the addition of an electric relay that he invented in 1835, Henry was able to construct a signaling device that could operate at greater distances, the basis for the electric **telegraph.**

In 1830 both Faraday and Henry discovered the principle of induction, although Faraday published his results before Henry. Faraday had constructed an elementary electric motor in 1821. But in 1831, Henry published a paper explaining the construction of a working electric motor. Thomas Davenport, a Vermont blacksmith, read of the device and patented an electric motor of a different design in 1837. Faraday conceived of a direct-current generator or dynamo in 1831, but he created only a demonstration model.

In 1832 Antoine-Hippolyte Pixii (1808–1835), in France, built a generator that could produce both direct and alternating current, and the generator was demonstrated in London in 1833. The Pixii generator is usually regarded as the first workable electric generator. His device consisted of a permanent horseshoe magnet, rotated with a foot treadle, with a coil of copper wire over each of the poles of the magnet. The two coils were linked and the free ends of the wire connected to terminal points from which a small alternating current could be obtained when the magnet was rotated. André Marie Ampère (1775–1836) suggested a switching device that converted the alternating current to direct current, later known as a commutator.

However, practical applications of both generators and motors did not take off until after the invention of a working dynamo by the Belgian engineer Zénobe-Theophile Gramme (1826–1901) in 1870. Ernst Werner von Siemens (1816–1892) developed an electric-powered vehicle in 1879, and in the United States, Frank J. Sprague installed the first electric **streetcar,** operating on direct current, in Richmond, Virginia, in 1887. The Siemens firm pioneered in the establishment of electric generating stations. With the development of the **electric light** and the creation of power supply systems, developing practical working motors became feasible. A burst of activity over the next decade by many engineers and inventors contributed to the commercialization of electric motive power.

In 1887 the Croatian-born American inventor Nikola Tesla (1856–1943) applied for patents on motors operating on alternating or polyphase current. George Westinghouse, who had developed an alternating-current system for electric lighting, purchased the Tesla patents. At first Edison's firm of General Electric resisted the development of alternating current and continued to install direct-current systems. In the 1890s both the Westinghouse and General Electric firms began constructing electric motors and generators, incorporating a host of improvements made in that decade. In 1895 the first large alternating-current generator, following the Tesla-Westinghouse system, was constructed at Niagara Falls, New York, using a **water turbine** to provide the rotary force. By the turn of the century, General Electric began adopting the alternating-current system for generation and transmission equipment.

electromagnetism

Through the first decades of the 19th century, many experimenters worked with both magnetism and electricity, developing through empirical trial and error such devices as the **electric motor and electric generator** and creating electromagnets.

> An Experiment, like every other event which takes place, is a natural phenomenon; but in a Scientific Experiment the circumstances are so arranged that the relations between a particular set of phenomena may be studied to the best advantage. In designing an Experiment the agents and the phenomena to be studied are marked off from all others and regarded as the Field of Investigation.
> —James Clerk Maxwell, *The Scientific Papers of James Clerk Maxwell*, (Cambridge, England: Cambridge University Press, 1927) p. 505

In 1873 James Clerk Maxwell (1831–1879), a Scottish physicist, stated in mathematical form ideas expressed earlier by Michael Faraday (1791–1867) and André Marie Ampère (1775–1836), in a set of laws known as **Maxwell's equations.** Maxwell postulated that electromagnetic waves propagated through a medium he designated ether, but later work determined that no such medium existed. Maxwell's equations adequately described the behavior of electromagnetic waves, and the equations and findings formed the bases for later advances in **radio** [V].

elevator

Although freight and construction material elevators had existed since the times of the Egyptians, very rarely had they been used to transport passengers. Ropes used to suspend elevators could and often did break, and fear of such accidents prevented the widespread use of elevators to transport people. In most cities, few buildings were built above 5 stories in height, as climbing more flights of stairs was too exhausting for the average person.

The barrier to the development of passenger elevators was not the lift mechanism but the issue of safety. Elisha Graves Otis (1811–1861) essentially solved the safety problem in 1852. A mechanic working for a bedstead company in Yonkers, New York, Otis designed a "safety hoist" that provided a safety catch that would keep the car from falling if the rope or cable broke. He set up a small elevator shop in Yonkers in 1853 and sold his first safety elevator on September 20, 1853, for a freight application. To publicize his invention, Otis set up a dramatic demonstration in May 1854 at an exhibition at the New York City Crystal Palace. There, he built a large open elevator and would have himself raised above a gathered audience. Then he would signal to have the rope cut, and the elevator would safely stop. The publicity stunt worked well, and he soon had more orders.

Otis installed the first all-passenger elevator on March 23, 1857, in the E. V. Haughwout and Company department store in New York. With the simultaneous development of the Bessemer process for the manufacture of **steel,** the way was opened to the construction of higher buildings or **skyscrapers,** and the modern skyline of New York and other cities began to emerge over the next decades. Together, steel construction and the safety elevator changed the landscape, allowing concentration of business, residence, finance, and industry in the small confines of an inner city.

Otis patented an independently controlled steam engine to provide lift in 1861, the year he died. His two sons, Charles and Norton, carried on the work of the firm, establishing the Otis Elevator Company. The company installed the first elevator in the Eiffel Tower, beginning in 1887, and by 1889 it was one of the most popular features of the Paris Universal Exposition of that year.

Hydraulic elevators, working on a different principle, were first built in 1878. Instead of suspending the car or platform from a cable, the elevator was raised or lowered by a hydraulic plunger.

Following the installation of a central electric power system in New York by Thomas Edison to supply power for the **electric light,** and with the development of large-scale **electric motors and electric generators,** the replacement of steam power with electric power became possible. In 1889 the first electric elevator was installed, and automatic systems that made possible elevators without attendants were developed by 1894. By 1949–1950, fully automatic dispatching systems capable of handling a bank of two or more elevators were installed.

Faraday's laws

The British scientist Michael Faraday (1791–1867) developed three laws from 1821 to 1833 that advanced the study and knowledge of electricity and its relationship to magnetism.

Faraday was a chemical assistant under Sir Humphry Davy at the Royal Institution in London, where Faraday learned chemistry and began his own experiments into electricity and electrochemistry.

Faraday experimented in 1821, discovering that when an electrical current was run through a wire that was free to rotate around a magnetic pole, it would do so. In effect, this was the first **electric motor.** After a decade of working on other projects, he returned to electrical problems. He later developed the first **electric generator** by rotating a copper disk with leads from its center and edge between the poles of a permanent magnet. His work with induction led him to develop the first electrical transformer. To explain the generator phenomenon, he assumed that when a conductive material cuts the lines of magnetic force, a current is generated whose strength depends on the power of the magnet, the rate of the cutting of the lines of force, and the conductivity of the material.

In developing his work on electrolysis he created a whole vocabulary to explain electrical phenomena, including the terms *anode, cathode, electrode, ion, ionization, electrolyte,* and *electrolysis.*

Faraday published his findings in three volumes under the overall title *Experimental Researches in Electricity* between 1839 and 1855. His law of induction and his two laws of electrolysis can be briefly stated:

Faraday's law of induction: If a change occurs in the magnetic flux through a loop of wire, the electromotive force induced in the loop is proportional to the rate of change of the flux.

Faraday's first law of electrolysis: The mass of any product liberated in electrolysis is proportional to the quantity of electricity passed through the electrolyte.

Faraday's second law of electrolysis: The masses of products liberated in electrolysis by the same quantity of electricity are in the ratio of their respective chemical equivalents. This concept of the combining power or valence of different substances gave a clue to the existence of electrons.

Faraday's work made possible the further development of electrical systems and the advancement of chemistry. He argued that magnetic, electrical, and gravitational forces pass from one object to another through space. The force of a magnet, he said, did not reside in the magnet but in the field around it.

In 1845, following up on a suggestion by young William Thomson (who later became Lord Kelvin), Faraday experimented to determine whether light could be deflected by an electromagnetic force and discovered that a powerful electromagnet could change the state of polarized light. This is now known as the Faraday effect, or the magneto-optical effect. A lecture he delivered on the topic in 1846 became the basis for the further development of field theory, taken forward by William Thomson and James Clerk Maxwell. Faraday's law of induction became formulated as one of **Maxwell's equations.**

Froude number

In 1871, the British naval architect William Froude (1810–1879) developed a principle for scaling the relationship between models and full-sized ships, based on the mathematical work of the French mathematician Joseph Bertrand (1822–1900) and the French naval architect Ferdinand Reech (1805–1880). The scaling law held that if the speeds of the model and ship were held in proportion to the square root of the ratio of model length and ship length, then the resistance of the model and ship would vary by the cube of the same ratio of model and ship length. With a known model length and a known ship length, and with a measured model resistance, it would be possible to calculate the resistance of the full-scale ship. One factor that interfered with the prediction was the frictional resistance of the ship in the water, and Froude conducted a number of experiments to develop knowledge of the resistance of wood in water to be able to account for and eliminate that variable. The so-called Froude number constitutes the basic relationships, expressed as a ratio, among resistance, velocity, and ship length.

Using his concepts, he visualized a mechanism that allowed for testing the water resistance of different ship hull forms. After experimentation with ship models and wooden planks, Froude built a small water tank at his home in Torquay, near Portsmouth in Britain. Using the mathematical relationship he had established and testing a model, in 1871 he worked out the design for the Royal Navy's *Greyhound,* the first ship whose hull design had ever been pretested. His published report in 1874 showed the value of the predictive method.

The design of the Froude towing tank was emulated later in larger tanks established in the United States, France, Britain, Russia, and elsewhere. A carefully leveled tank of water would be straddled with a pair of rails, on which a carriage could be drawn or propelled. The model would be suspended beneath the carriage, with a spring-loaded device or dynamometer that would measure the resistance of the model as it moved through the water at a known speed. Using Froude's scaling principles, the best ship form could be chosen. Later towing tanks, or model basins, incorporated other features, such as artificially induced waves, greater length, and more precisely leveled track.

Froude's tank was replaced with another, built at Haslar, England, in 1886, by Froude's son. Between 1886 and 1918 more than 500 Admiralty warships were pretested in the Haslar tank, including battleships, cruisers, destroyers, and submarines. In the United States, a model basin constructed in the Washington, D.C., Navy Yard by Captain (later Admiral) David Taylor in 1898 served a similar purpose for the U.S. Navy until 1939, when it was superseded by a much longer and more precise model basin constructed at Carderock, Maryland, near Washington, D.C.

galvanometer

When in 1780 Professor Luigi Galvani (1737–1798) investigated the contractions of a dead frog leg, he thought he had discovered the principle of life itself. He published his work in 1791, in (translated) *Commentary on the Effect of Electricity on Muscular Motion,* creating an international sensation. However, a fellow Italian, Alessandro Volta (1745–1827), the inventor of the **battery,** was able to demonstrate by 1800 that the electricity came not from the frog leg but from an electrochemical reaction of the wet cardboard or metal on which it was placed. Despite the demonstration that Galvani's ideas were incorrect, a device later developed for measuring electricity by a different inventor bore his name.

Studying the "galvanic effect" or electric current, Hans Christian Oersted (1777–1851), a Danish physicist, discovered in 1819 that when a wire through which a current flowed was brought near a compass, the compass needle would swing. In 1820 both the Frenchman André

Ampère and the German physicist Johann S. C. Schweigger made a device to measure the current, and Ampère suggested the name *galvanometer* in honor of Galvani. By carefully controlling the current, a measurement could be made by using the compass needle as an indicator. It was improved by other researchers, including Karl Friedrich Gauss (1777–1855), who from 1832 to 1833 developed an early type of **telegraph.** A French physicist, J. A. d'Arsonval, invented a different type of galvanometer in 1880 by suspending a movable coil between poles of a magnet. A beam of light reflected from a mirror attached to a coil reflected to a scale was used to measure the current. Most later galvanometers are of the d'Arsonval type.

guillotine

During the French Revolution and the Terror that followed, the new government executed tens of thousands of members of the former ruling classes. In 1792 Dr. Joseph Ignace Guillotin (1738–1814) invented an improved version of an ancient execution machine that was vigorously employed during the rule of the Jacobins from 1793 to 1794. The French government and other governments adopted the method of execution widely through the 19th century. A weighted blade, raised between two upright posts, would be released to descend to sever the head of the victim. Dr. Guillotin had intended the device to provide a more humane and sure form of execution than the sometimes inaccurate blow by an axman, hanging on a scaffold, or various forms of torture-execution. Although the guillotine represented a reform, the machine still allowed executions to be public spectacles.

hydraulic jack

Developed by Joseph Bramah (1748–1814) in conjunction with his work on other hydraulic devices that used **Pascal's law [III]**, including the **hydraulic press,** the hydraulic jack soon found several practical applications. Bramah (whose original name had been Joe Brammer) patented a telescoping hydraulic jack in 1812, and it was applied to provide slow lift for heavy loads. Engineers soon used hydraulic jacks as testing devices, and they were used to determine the stress that could be sustained by **chains [I]** in the construction of suspension bridges. In 1858 a total of 18 hydraulic jacks were used to launch the large steamship the *Great Eastern,* later used to lay **submarine cable.**

hydraulic press

The hydraulic press was invented in 1795 by Joseph Bramah (1748–1814), the British inventor and cabinetmaker also known for his development of the **flush toilet [III]**. As he installed toilets, he studied the existing literature on the motion of fluids, learning of **Pascal's law [III]**. Realizing that hydraulic force could be applied to machines, he patented both the hydraulic press and, in 1812, the **hydraulic jack.** His press was first used to compress cotton bales. Together with several late-18th-century inventions, including the **boring mill [III]**, Bramah's inventions helped advance the art of working metal.

interchangeable parts

Although several inventors anticipated the possibility of interchangeable parts, practical implementation of the idea was not achieved until the 1830s at the Harpers Ferry, Virginia, and Springfield, Massachusetts, arsenals of the United States. As early as the 1790s, French army officers had sought to standardize weapons to facilitate field repair through removing undamaged parts from damaged weapons to replace damaged parts in others, a process informally known as cannibalization. Although the concept made excellent sense, the manufacture of machinery in the era could not achieve the degree of standardization required. Mills, lathes, drills, and other metal-shaping devices, although usually powered by falling water and **waterwheels [II]**, were hand-guided with little instrumentation to ensure regularity. Interchangeability has been attributed to both Eli Whitney (1765–1825), the inventor of the **cotton gin,** and Samuel Colt (1814–1862), who developed a reliable **revolver,** but it is more accurate to say that they were both advocates of interchangeability and standardization than that either of them invented the principle.

To make parts for muskets that were truly interchangeable from one weapon to another required that the metal pieces being milled and shaped be held precisely in place by jigs, measured and checked with micrometers that allowed a measurement to be made and fixed precisely. The development of the machine-tool industry at the two U.S. government arsenals meant that by the 1840s, both were making muskets that lived up to the ideal of interchangeable parts. As the concept and the mechanical devices spread, particularly from Springfield to other Massachusetts industries and then more widely, the concept of interchangeability became known as the American System of Manufactures. The clock industry centered in Waltham, Massachusetts, and the manufacture of the **sewing machine** and the **typewriter** all reflected the influence and the transfer of individuals, ideas, and tools from Springfield to the private sector. Later, the principle and the machinery were adapted to the manufacture of the **bicycle** and the **automobile**.

Interchangeability was key to the great increase in industrialization of the 19th century. While first perceived as a means to effect rapid battlefield repair through cannibalization of parts, other advantages soon emerged. If a particular part of a machine was always produced to the same standard, a machine could be repaired simply by ordering a replacement for a damaged or worn part. Before interchangeability, every engine or mechanical device was custommade, and each repair part would have to be made to fit the specific machine. The intro-

> Whereas handicraft, by the very nature of human work, exhibits constant variations and adaptations, and boasts of the fact that no two products are alike, machine work has just the opposite characteristic: it prides itself on the fact that the millionth motor car built to a specific pattern is exactly like the first.
> —Lewis Mumford, *Technics and Civilization* (New York: Harcourt, Brace, 1934)

duction of interchangeability allowed for such equipment as farm machinery, engines, and electrical equipment to remain in service longer, and to be repaired much more economically, simply by the process of remove and replace rather than by machining a new part.

internal-combustion four-stroke engine

The modern four-stroke internal-combustion engine was invented in Germany in 1867 by Nikolaus August Otto (1832–1891) with backing from the businessman Eugen Langen. In 1861 Otto experimented with an earlier **internal-combustion two-stroke engine** design developed by Jean-Joseph Étienne Lenoir. That engine used **coal gas [III]** and was not very efficient. In 1864, with backing from Langen, Otto set up a factory to produce Lenoir-design engines.

The Lenoir design, based on the **laws of thermodynamics** enunciated by Nicolas Sadi Carnot (1796–1832), used an electrical spark to ignite the gas-air mixture.

In 1867 the Otto-Langen company Gasmotorenfabric produced an improved Lenoir engine that compressed the gas prior to ignition. Working on the idea of a compression stroke in 1876, the company produced the first four-stroke or four-cycle internal-combustion engine, soon known as the Otto cycle or Otto silent engine. By 1893 the company had sold more than 50,000 of the new engines.

The design was a great improvement over the Lenoir two-stroke engine, allowing more power for the same amount of fuel. The piston moved through a cylinder, with a shaft connected to a **crank [II]** to convert the lineal motion into rotary motion. The first stroke of the piston would draw down the cylinder, sucking in a fuel-air mixture in the "intake" stroke or cycle through an open valve. The second stroke would compress (and heat) the mixture with closed valves. On the third stroke, the mixture would be ignited and would then expand, and the piston driven back, in the power stroke, still with valves closed. On the fourth stroke, the burned gases would be pushed back out of the cylinder through an opened exhaust valve. Gottlieb Daimler (1834–1900), who worked for the firm, developed a system that allowed more than one cylinder to be mounted to the same crankshaft, starting with a V-shaped configuration of two cylinders. Daimler also pioneered in shifting the fuel from coal-gas to gasoline-air mixtures blended in a **carburetor [V]**, a device developed by Wilhelm Maybach in 1893.

Although the Otto-Langen company made a fortune from the engines and Otto was recognized as the inventor, several others had conceived of a four-stroke cycle before him, including published works by Frenchman Alphonse Beau de Rochas in 1862 and Gustav Schmidt of Vienna in 1861. The Otto patents were invalidated by legal action in 1886, but the company had established its position by that time, selling four-stroke engines for use in the **automobile** and later in the **airplane [V]**.

Along with a few other key inventions in the 19th century, such as **electric motors and electric generators** and systems, the internal-combustion engine had a vast social impact and contributed to the transformation from the Industrial Revolution to the modern Age of Technology.

internal-combustion two-stroke engine

A Belgian-born French inventor, Jean-Joseph Lenoir, invented the first successful two-stroke internal-combustion engine, in 1859. He also constructed a small but noisy car propelled by the engine in 1863. The engine, powered by **coal gas [III]**, was more useful for stationary use in driving printing presses, pumps, and lathes, and by 1865 Lenoir had sold more than 400 of the large, stationary engines. In 1878, two years after the introduction of the Otto cycle four-stroke engine, the British engineer Dougald Clerk introduced another successful two-stroke engine, further improved by Joseph Day in 1891.

A two-stroke engine developed by the American engineer George Brayton was fueled by kerosene, but it was extremely slow. Known as an oil engine, the two-stroke Brayton engine was used as a source for electrical power generation on farms and in other remote locations. Farmers, who often used kerosene for lighting, found the oil engine particularly useful to operate well pumps and as a source of stationary power well into the first decades of the 20th century.

The two-stroke engine combines intake and power in the same stroke, and exhausts in the second stroke. The air-cooled two-stroke engine, with a light and small design descended from the Clerk-Day design, continued to find uses throughout the 20th century in such applications as motorcycles, outboard boat engines, model aircraft propulsion, electric generators, lawn mowers, and other small handheld applications in which light weight and simplicity of design are more crucial than high horsepower or power-to-fuel ratio.

lawn mower

In the 18th and 19th centuries, British estates commonly had greenswards or pastures, naturally mowed by sheep or laboriously hand-scythed when slightly damp. With the growth of suburbs outside of London connected by rail or horse bus travel to the city in the late 18th century, smaller and smaller homes began to emulate the landscaping style of the manorial estates, with green lawns instead of the bricked-in or paved yards typical of the city in the era. The growing market for a mechanized mower occurred to several inventors.

One of the first patents was taken out by Thomas Plucknett in 1805, with a horse-drawn pair of carriage wheels geared to revolve a single large circular blade close to the ground. In 1830 Edwin Budding developed a device that is the forerunner of the modern lawn mower. Pushed by an individual, it had a pair of wheels geared to a rotating cutting blade that threw the cuttings forward into a wooden tray or box. Budding suggested that its use would not only allow country gentlemen to cut their lawns but also provide a healthy form of exercise.

In 1832 an Ipswich company, Ransome, acquired the Budding patent and began to market improved models, some of which used gearing to move the blades in the same forward rotary motion as the wheels and to cast the cuttings to a catch basket at the rear. However, the basic geared Budding design remained the typical hand-pushed mower for a century. A steam-powered mower for large estates was sold in 1893, and by 1900, Ransome had introduced a gasoline-powered internal-combustion engine model, complete with a seat for the operator and a wide swath.

Gang mowers, with multiple cutters towed behind the engine, were introduced in the United States in 1914.

laws of thermodynamics

There are four laws of thermodynamics and several related effects and laws that derive from the laws. The laws were discovered from 1824 to 1852 by several scientists: Nicolas Leonard Sadi Carnot (1796–1832), Rudolf Clausius (1822–1888), James Prescott Joule (1818–1889), William Thomson (Lord Kelvin) (1824–1907), and James Clerk Maxwell (1831–1879). Together the laws of thermodynamics provide a theoretical and mathematical basis for designing engines that convert heat to work, such as internal-combustion and steam engines, and for refrigeration equipment. Implicit in the laws were certain principles, such as the conservation of energy, that contributed to the philosophical basis of 19th-century physical science.

The four laws can be stated fairly simply:

First law: Energy cannot be destroyed; heat and work are interchangeable.

Second law: Heat cannot move from a colder to a hotter body.

Third law: It is impossible to reach absolute zero temperature.

Zeroeth law: Temperature between two contiguous systems tends to equalize or reach entropy.

Various scientists played a part in discovering and stating these laws and the principles that relate to them. However, the impact that these observations had upon the design of early engines is still open to discussion. Considerable practical work in designing steam engines [III] and improving them had taken place well before these laws were stated. That fact is often adduced to support the argument that technology frequently precedes science. The findings and laws developed regarding thermodynamics led to many new devices and discoveries as engineering schools began to incorporate the findings in their curricula, and engineers who designed engines in the 20th century usually had a sound understanding of the laws.

The knowledge that technicians like those at the British firm of Watt and Boulton already designed and improved engines without a theoretical basis for them almost served to affront the Frenchman Sadi Carnot. In 1824 he published *Reflexions sur la puissance motrice du feu* (Reflections on the Motive Power of Fire). He introduced his work with the observation that British engineers had succeeded in making many improvements to steam engines during the Napoleonic Wars without the benefit of science and for the most part without university education. In addition to his published work, he left extensive notes on his researches, in which he reflected the same concern to bring science to bear on engine design. He died at an early age during a cholera epidemic.

Carnot calculated the effects that would take place in a frictionless, idealized steam engine. This notional engine would go through four cycles of conversion of heat to energy. Although an impossible engine to design, the "Carnot engine" or "Carnot cycle" was useful as it imposed a theoretical upper limit to the efficiency of transformation of energy to work through an engine against which the efficiency of actual engines could be compared. The four cycles would be (1) isothermic expansion, (2) adiabatic conversion of motion to work, (3) isothermic compression, and (4) adiabatic conversion of motion back into heat. In this context isothermic refers to a constant temperature; adiabatic means self-contained, with no heat at all either escaping or entering the system. Carnot proposed that heat and work were interchangeable. In the ideal Carnot cycle no heat is lost to friction or to the surrounding environment. Thus 100 percent of heat provided is converted into work, or 100 percent of the work is converted into heat. James Prescott Joule confirmed this postulate and declared it to be the first law of thermodynamics in 1843. The specific principle that energy cannot be destroyed but only transformed from one system to another has sometimes been referred to as the law of conservation of energy.

The second law, that heat cannot move from a colder to a hotter body, developed as a nearly simultaneous discovery by three scientists: Rudolf Clausius; James Prescott Joule; and William Thomson, an Irish physicist. Clausius enunciated the principle in 1850. Joule demonstrated experimentally that energy was not dependent on the materials involved, and a unit of energy, the joule, was named after him, the equivalent of 4.2 calories. William Thomson proposed in 1848 an absolute scale for temperature and suggested in 1851 that Carnot's theory regarding the interchangeability of heat and work was compatible as long as heat could not pass from a colder to a hotter body, the same principle developed by Clausius in 1850. In 1852 Thomson added that mechanical energy tends to dissipate as heat. Thomson was knighted in 1866, and in 1892 he was named as Baron Kelvin, or Lord Kelvin.

To get useful work from a supply of heat it is necessary to have a temperature inequality. However, in nature an inequality of temperature is constantly reduced by conduction of heat and in other ways. But in a

The last decades of scientific development have led us to the recognition of a new universal law of all natural phenomena.... This law is the Law of the Conservation of Force, a term the meaning of which I must first explain. It is not absolutely new; for individual domains of natural phenomena it was enunciated by Newton and Daniel Bernoulli; and Rumford and Humphry Davy have recognised distinct features of its presence in the laws of heat. The possibility that it was of universal application was first stated by Dr. Julius Robert Mayer, a Schwabian physician (now living in Heilbronn), in the year 1842, while almost simultaneously with, and independently of him, James Prescott Joule, an English manufacturer, made a series of important and difficult experiments on the relation of heat to mechanical force, which supplied the chief points in which the comparison of the new theory with experience was still wanting.

—Hermann von Helmholtz,
Lectures 1862. Harvard Classics
(Danbury, Connecticut:
Grolier Enterprises, 1980)

closed or isolated system with irreversible changes, the supply of energy from heat tends to become available for work. Clausius coined the term *entropy* to state the converse of the decreasing availability of energy for work—that is, under conditions in which a system is not isolated, entropy tends to increase. When energy supply is reduced to a minimum (or entropy is at its maximum), no more work can be done by the system. The laws built up by Carnot, Clausius, and Kelvin were expressed algebraically by Hermann Ludwig Ferdinand von Helmholtz (1821–1894) and Josiah Willard Gibbs (1839–1903).

Kelvin declared that absolute zero temperature could not be achieved but only approached. The number approached was 459.67 degrees below zero Fahrenheit, later designated as zero Kelvin or minus 273.15 degrees Celsius. The principle that true absolute zero could never be reached was the third law of thermodynamics.

The "zeroeth" law, proposed by James Clerk Maxwell in the 1850s, stated that two contiguous systems tend to equalize in temperature. By the second law of thermodynamics and the "zeroeth" law, it was assumed that the universe was approaching entropy, the term introduced by Clausius, meaning a state of equilibrium. The principle of conservation of energy implicit in the laws of thermodynamics held that within a system, the amount of energy was constant and could only be transformed from one type to another.

The **diesel engine** [V] and some improvements in steam engine design in the late 19th century could be attributed to the application of the principles of thermodynamics to engine design. Several scholars, however, have indicated that most 19th-century engine improvements derived from work of engineers who were entirely unfamiliar with thermodynamic principles or formulas.

Cooling through expansion of a gas expelled through a nozzle, derived from thermodynamic experiments, allowed for extremely low temperatures and the eventual discovery of the freezing temperature of all gases.

Another, and broader, social consequence of the discovery of the laws of thermodynamics was the spread of the concept that the energy of the universe was being slowly balanced out to entropy, leading to a concern with the conservation of resources. Conservation and wise use of stored energy in the form of fossil fuels was supported by these principles.

Hermann Helmholtz reduced the principle of conservation of energy to a mathematical formula in 1847, and he extended the concept of conservation of energy to biological systems. He suggested that life itself followed the same physical laws as nonliving energy systems. At the philosophical level, Herbert Spencer, among others, struggled with the question of whether the biological conservation of energy allowed for the fact of human initiative or the presumed existence of a human soul—that is, the four laws of thermodynamics did not appear to allow for the engagement in the material universe of sources of energy that were entirely immaterial in nature. By this logic, thought, faith, ideas, or morality could never be sources of physical action.

linotype machine

After the invention of **movable type [II]** and the printing press, the process of setting type by hand meant that printing, although faster than copying manuscripts by hand, was still extremely time-consuming, as each page had to be carefully built up with individual letters in reverse order, making it difficult to check the work until a proof copy was printed to allow the discovery of errors. In the 19th century, with the development of daily newspapers, the typesetting process presented a bottleneck in production that many inventors sought to solve. In 1876 a group of inventors asked Ottmar Mergenthaler (1854–1899), a young mechanic in a Baltimore, Maryland, shop, to assist them in perfecting an automatic typesetting machine. Mergenthaler, the son of a schoolmaster, was born in Württemberg, Germany, and had been apprenticed to a watchmaker. Mergenthaler had immigrated to the United States in 1872.

Mergenthaler experimented with many different machines, finally succeeding in 1884. He began to manufacture a machine in 1886 that solved the problems, later marketed as a "linotype" machine. The operator pressed a key on a keyboard, releasing a matrix with a metal die of a letter from a storage magazine. An air blast would assist in blowing the matrix down a belt to a working line, with small metal wedges inserted for the spaces between words. With the completion of one line of type, the line would be carried off to be cast with molten lead, which quickly hardened. The slug of newly cast type would be deposited in a tray, while the matrices or dies would be returned to the storage magazine. In this fashion the operator could continuously type, and the lines of type would be set automatically.

Newspaper publishers immediately began buying the new machines by the dozens. Later improvements included a machine for punch-cutting the matrices rather than hand-cutting them. Linotype machines remained in use in most major newspapers well into the last decades of the 20th century, when they became replaced by punch-tape systems and **computer [VI]** systems for setting type.

machine gun

The concept of the machine gun, capable of firing small-caliber ammunition in rapid-fire sequence, can be traced to attempts as early as the 14th century, with multiple-barreled weapons. The difficulty of reloading such weapons limited their application.

In 1862 the American Richard Jordan Gatling (1818–1903) patented a hand-cranked rapid-fire weapon with 10 barrels that could fire more than 300 rounds per minute. The cartridges were fed from a hopper. By the 1890s the same weapon had evolved to one driven by an electric motor and capable of firing up to 3,000 rounds per minute.

However, the true machine gun, operated by using the power of the gas from the shell itself, was first developed in 1884 and patented in 1885 by Hiram Stevens Maxim. The development of **smokeless powder,** with a uniform gas emitted after the burning of the propellant, made possible the use of the gas as a source of energy for operating the weapon. Maxim developed one such smokeless powder himself, marketed as cordite. His company was absorbed by Vickers Limited, and later models of the Maxim design were known as Vickers guns. See the table "Early Designs of Machine Guns."

Early Designs of Machine Guns		
Gun	**Inventor(s)**	**Year**
Gatling	Richard Jordan Gatling	1862
Maxim	Hiram Stevens Maxim	1885
Browning	John Moses Browning	1892
Hotchkiss	Laurence Benét and Henri Mercie	1896
Lewis	Isaac Newton Lewis	1911

machine gun Members of the 13th U.S. Infantry Gatling Gun Detachment train with a .30 caliber Model 1895 Gatling Gun in June 1898. *U.S. Army*

Maxim's gun was a single-barrel, belt-fed weapon cooled by water. The British army adopted his weapon in 1889, and many other European countries adopted it. The British used it in the Boer War (1899–1902), and it saw service in the Russo-Japanese War (1904–1905). Maxim, born in the United States, became a British subject, and was knighted in 1901 for his invention of the machine gun. An estimated 90 to 95 percent of the casualties in World War I were inflicted by one or another machine gun.

Maxim also invented the pom-pom gun, a machine gun firing one-pound shells.

matches

Soon after the discovery of phosphorus by Hennig Brand in 1674, the Irish chemist Robert Boyle (1627–1691) and his assistant, Godfrey Haukweicz, attempted to develop a phosphorous match, but without much success.

In the 1830s, "promethian matches" were developed in Britain. These consisted of a roll of paper with a small amount of sulfuric acid that would ignite a sulfur-based mixture when broken open and exposed to air. The first friction matches were developed by a British chemist, John Walker, and were known as "lucifers." The lucifer consisted of a stick dipped in a mixture of sulfide of antimony and chloride of potash that would ignite when drawn over sandpaper. An improvement substituted phosphorus for the antimony, and these matches were known as "Congreves," named after a British rocketmaker. Sold by boxes in the streets by young women, these led to the match girl becoming a fixture of 19th-century London life. Safety matches, which used a slightly less dangerous amorphous phosphorus, were invented in Sweden by R. C. Bottger in 1848 and marketed by 1852.

Due to industrial illnesses from working with phosphorus, its use was banned in many countries after the beginning of the 20th century.

Attempts to develop a mechanical lighter had some success even before the perfection of the match, with an electrically ignited hydrogen lighter by Alessandro Volta (1745–1827) in the 1780s.

Maxwell's equations

James Clerk Maxwell (1831–1879), a Scottish physicist, extended the findings of Michael Faraday and others regarding electric and magnetic fields and established the groundwork for the development of **radio [V]**, **television [V]**, and many other electronic devices.

From 1855 to 1873, Maxwell sought to develop a mathematical expression of the behavior of electric and magnetic fields that the first of **Faraday's laws** had expressed. In his work *Electricity and Magnetism* (1873), Maxwell presented four partial differential equations that explained the relationship between an oscillating electric charge and an electromagnetic field. He also found that the speed of the propagation of an electromagnetic field is approximately the same as the speed of light, and from this he proposed that light is a form of electromagnetic radiation.

Maxwell corresponded with both Faraday and William Thomson (later Lord Kelvin), building on Faraday's early concept of fields that served to carry or mediate magnetic forces. Maxwell formulated as partial differential equations not only the law of induction proposed by Faraday but also **Ampère's law** and **Coulomb's law [III]**. Maxwell found that an extra term, *displacement current,* had to be added to Ampère's law to develop a complete and consistent set of laws that explained all electromagnetic phenomena.

Maxwell assumed that electromagnetic radiation propagated in waves through a medium, ether, and that electromagnetic lines of force were disturbances in the ether. Later work by Albert A. Michelson (1852–1931) and Edward W. Morley (1838–1923) in 1887 demonstrated that the concept of ether could not be supported. However, Maxwell's four equations describing electromagnetic phenomena did not depend on the presence of the ether medium and remained valid. The temporary use in the period 1873 to 1887 of the imaginary concept *ether* as a substance through which radio waves were transmitted caught on in popular culture, and the term was used loosely over the next century to visualize the medium through which **radio broadcasting** [V] is transmitted.

Heinrich Hertz (1857–1894) conducted experiments in 1888 to demonstrate that waves of **electromagnetism,** traveling at the speed of light, could be generated through air and also detected by laboratory equipment. Building on those experiments, Guglielmo Marconi (1874–1927) and other experimenters over the next decade began to recognize that such equipment could be perfected for sending messages at a distance. In this sense, the development of wireless telegraphy and **radio** [V] can be viewed as direct cases of technology as the application of scientific discovery.

McCormick reaper

Cyrus McCormick (1809–1884) developed the harvesting machine that bore his name when he was 22 years old, giving a working public demonstration in 1831. He was able to cut 6 acres of oats in a day. The reaper, drawn by four horses, did the work of six men. The machine did not cut evenly and damaged some of the grain, but McCormick kept improving the reaper, securing a patent in 1834. In effect, the reaper was a much larger version of the **lawn mower** developed at the same time, but designed to harvest the cuttings rather than simply dispose of them.

The reaper used several cutting shears. The horse-drawn, two-wheeled, chariotlike vehicle had a 6-foot axle. Below the axle, a flat iron bar, parallel to the axle and a few inches from the ground, carried steel fingers that separated the grain stalks into bunches. At the base of the fingers, a saw-toothed cutting bar slid sideways, cutting the stalks. The power driving the cutting blade derived from one of the turning wheels.

Despite the workability of the reaper and its obvious laborsaving qualities, it could not be profitably sold due to the high price of iron and a financial panic in which many American farmers went bankrupt.

However, McCormick began selling machines in 1841. After assembling the machines in various shops, McCormick selected Chicago as the site for his own factory. By 1851 McCormick had sold 1,000 reapers, and in 1857 he sold 23,000. He engaged in many patent-infringement suits, which he apparently enjoyed.

The reaper revolutionized American agriculture, allowing the development of large grain farms and paving the way for the change from family farming to larger-scale enterprises in farming, especially where the land was relatively flat and free of obstructing rocks, as in the Midwest and West. His machine helped make Chicago a major wheat exporting port.

The reaper is often traced as one of the causes of the lowering of agricultural prices in the late 19th century and the vast expansion of wheat-producing regions. With the transport of grain by **steam railroad** and **steamboat,** and the emergence of national and international markets, these social effects produced a generation of political unrest and insurgency in the agricultural West linked to the rise of populism and other movements of reform.

microphone

The first microphone was probably developed by the inventor of the **telephone,** Alexander Graham Bell (1847–1922), in 1876. His system worked with a thin disk of metal that vibrated with sound waves, inducing a current in a magnet. The electric current would be transmitted over a wire to another magnet, which would create corresponding vibrations in a metal disk. In effect, the microphone and speaker were identical devices, but the system was too weak to operate effectively.

Thomas Edison sought a more effective means of picking up sound waves, and in 1877 he began working on a system that used a metal diaphragm against a disk of carbon granules. Emile Berliner used a similar system, developed in about 1877 and patented in 1891, known as the first true microphone, called a carbon-button microphone. In 1877 Ernst Werner Siemens introduced a moving coil microphone, in which a coil attached to a diaphragm produced a current when it vibrated next to a magnet. These designs and others all depended on the same principle: The metal diaphragm would vibrate and cause a corresponding change in an electric circuit, either through a magnet, a carbon contact, a coil, or a piezoelectric crystal, among other systems. Coil, crystal, and carbon systems all continued to be improved, finding special uses. Crystal microphones became widely used in public address systems but remained sensitive to shock and rough handling, making them inappropriate for telephones or portable equipment. Simple magnetic or dynamic microphones, similar to the original Bell design, in which the same device can serve as microphone and speaker, have found uses in intercommunication systems.

Neptune

Johann Gottfried Galle discovered the eighth planet of the solar system, Neptune, in 1846. Several astronomers had suspected the existence of an eighth planet when perturbations of the motion of **Uranus [III]** suggested that another planet might be causing deviation from its predicted orbit.

As often happens with discoveries in nature, several scientists almost simultaneously reached similar conclusions. From 1845 to 1846, John Couch Adams in Cambridge and shortly thereafter, Urbain J. Leverrier, the director of the Paris Observatory, at the suggestion of his former professor, Dominique Arago, solved the problem of deriving the mass and orbital elements of the planet thought to be disturbing the orbit of Uranus. The act of discovery was in dispute between the British and the French, eventually resolved in most texts by indicating that the two independently made the calculations. The calculations that led to the prediction of the location of Neptune were regarded as one of the great accomplishments of 19th-century celestial mechanics, utilizing the **laws of motion and gravity [III]** discovered by Isaac Newton. However, Galle, in Berlin, accomplished the actual first sighting of the planet.

Galle (1812–1910) was born in Wittenberg, Germany, and educated at Berlin University. He taught mathematics and became an assistant astronomer at the Berlin Observatory in 1835. In 1838 he discovered one of the rings of Saturn, and he also discovered several comets. In 1846 Leverrier requested that Galle search for a new planet in a specified location.

On September 23, 1846, Galle and his colleague Heinrich d'Arrest located the planet astronomically within an hour of beginning the experiment, comparing their sightings to a printed star map and using Leverrier's calculations. The planet was only 55 minutes of arc (less than 1 degree) from the predicted location. Within a few months, William Lassell (1799–1880) discovered visually that Neptune had a large satellite, Triton, later determined to have a diameter of

Neptune With its moon Triton in the foreground, Neptune was photographed by the unmanned spacecraft *Voyager 2* in 1989. Neptune was originally discovered by several astronomers nearly simultaneously in 1846. *National Aeronautics and Space Administration*

1,350 kilometers. More than a century later, in 1949, Gerard Kuiper (1905–1973) discovered a second satellite, Nereid, with an eccentric orbit.

The space-reconnaissance vehicle *Voyager 2* conducted a flyby of Neptune in 1989 and returned signals showing the existence of six additional quite minute satellites of Neptune. Later these moons were given classical names: Naiad, Thalassa, Despina, Galatea, Larisa, and Proteus. By contrast to Triton, the other seven moons are all quite small. In addition, Neptune has a number of rings discovered by *Voyager*. The rings were named after astronomers: Adams; Leverrier (with two extensions called Lassell and Arago); and the innermost broad ring, named Galle. Fortunately, the rings provided enough naming opportunities to honor all the major participants in the planet's discovery.

Galle's visual confirmation of the prediction of Leverrier (and although Galle did not know of it, the prediction of Adams) served as an often-cited case of the predictive nature of scientific laws.

Ohm's law

The basic relationships among current, resistance, and electric potential or voltage were established in 1826 and published in 1827 by Georg Simon Ohm (1789–1854). Born to a humble family in Bavaria, Ohm studied mathematics and physics under the guidance of his father, and after some college training, entered teaching at the secondary and college levels. He had developed several experiments for his own edification in the laboratory of the Jesuit College of Cologne, where he taught. To achieve his ambition of a university appointment, he decided to publish his findings. In *Die galvanische Kette, mathematisch bearbeitet* (The Galvanic Circuit Investigated Mathematically) he expressed the relationship he deduced experimentally in a set of algebraic formulas:

$$V/I = R$$

$$I = V/R$$

$$V = IR.$$

Here V equals potential electric difference measured in volts, I represents the current measured in amperes, and R equals a quantity of resistance now named in his honor as ohms. Expressed another way, when 1 ampere of current through a conductor results in a drop of 1 volt, the resistance of the conductor is 1 ohm.

At first Ohm's findings were not well received, as most workers in the field of electricity did not think mathematically. However, his work began to be accepted, and in 1841 he was awarded the Copley Medal of the Royal Society of London. In 1849 he became collection curator at the University of Munich, where he began lecturing; and in 1852, two years before his death, he was accepted as professor of physics at that university.

Ohm's law adequately explains the behavior of direct-current circuits, but it needed to be modified slightly to address alternating-current circuits. In alternating current, resistance is complicated by the reverse effect, reactance, and together resistance and reactance are considered impedance. Thus, if impedance is designated Z, Ohm's law can be restated for alternating current as $V/I = Z$.

Like some of the other major laws of nature that were discovered in the 19th century, Ohm's law derived from the observation of already invented devices and their behavior rather than from the use of instruments to observe purely natural events, which had characterized the scientific advances of the 17th and 18th centuries.

oil drilling

In the 1850s, the development of **coal oil** for lighting had begun to supplant whale oil, and lamps were designed for the refined product. Several entrepreneurs recognized that lighting oil distilled from petroleum would offer a good substitute if it could be provided in quantity. In 1857 A. C. Ferris began searching for Seneca Oil (so named for the Native American tribe who had used it as a medicine) by digging holes in oil-bearing lands in western Pennsylvania. Meanwhile, Samuel Kier had found oil in his salt wells near the Allegheny River and began selling it as a substitute for coal oil.

Learning of the Kier business, George H. Bissell and Jonathan G. Eveleth decided to finance a project to drill wells especially for oil, instead of oil as a by-product of brine wells. They took Colonel E. L. Drake in as a partner and as superintendent of the fieldwork. In 1858 Drake took a team to Titusville on Oil Creek in western Pennsylvania. The method employed, as in brine wells, was to hoist a chisel-shaped bit into a wooden derrick structure and drop it into the ground. As the rock and earth were chopped, the hole would be washed clean with surface water. With a steam-powered rig, the bit could be repeatedly forced to chop into rock and earth.

However, when water was struck, it became difficult to clear the debris. Colonel Drake invented a method of dropping the bit through an iron pipe, lowering the pipe into the hole, and then attaching another length to the top of the pipe as it lowered into the ground. Drake found some small quantities of oil. The next year, Drake returned with "Billy" Smith, and on May 28, 1859, the Drake-Smith team hit oil strata at about 70 feet. Using hand pumps, they removed 8 barrels, and within 2 months, the well yielded more than 20 barrels a day. Oil drilling is often simply attributed to Colonel Drake, and the invention dated to 1859.

Within a year, an oil rush began in western Pennsylvania. Hundreds of wells pumped oil through local pipelines to underground tanks and vats, where it then would be barreled for shipment to refineries. Companies that had refined coal oil got into the petroleum refining business, thermal cracking the crude oil into naphtha (for cleaning), heavier oils or kerosenes used for lighting and heating, and even heavier greases for lubrication and asphalt for paving. By the late 1860s the petroleum business had rapidly expanded, and petroleum-derived kerosene tended to replace both coal oil and whale oil for lighting. Not until after 1900 did a large market develop for naphtha or gasoline as a fuel for **internal-combustion engines.**

The chisel bit was superseded by Howard R. Hughes with the invention in 1921 of a gear-toothed rotary drill bit that could cut through rock, allowing much deeper holes, since the bit did not require as frequent replacement as the chisel bit. With the worldwide expansion of oil drilling in the decades following Hughes's invention, he earned a vast fortune from his patented drill bit.

opthalmoscope

The opthalmoscope is a machine for shining a light through the pupil of the eye to examine the retina. The scope has a set of **lenses [II]** that allow the examiner to bring the retina into clear focus. By changing the lens required to achieve focus, the examining physician can determine the curvature of the natural lens in the eye and by this means determine the degree of short-sightedness or long-sightedness of the patient. Furthermore, the opthalmoscope allows the physician to detect certain diseases, such as diabetes or high blood pressure.

A version of the opthalmoscope was devised in 1847 by Charles Babbage (1792–1871), who invented a notional computer known as an **analytic engine.** However, the concept, like that of his computer, lay fallow, and the opthalmoscope was independently invented in 1851 by Hermann von Helmholtz (1821–1894). Helmholtz worked in a variety of scientific fields, including the nature of sound, and was the first to explain the workings of the inner ear. His ideas regarding the conservation of energy were incorporated in the **laws of thermodynamics.**

In his work on the physiology of vision, in 1855 Helmholtz also developed the ophthalometer, which measures the curvature of the eyeball. His study of the perception of color led to an explanation of color blindness.

parachute

Although Leonardo da Vinci (1452–1519) had sketched a conceptual parachute in about 1485, the first known practical use of a parachute by a human descending from an aircraft was in 1797. Jacques Garnerin, experimenting with a hydrogen-filled balloon of the Charles design, jumped from about 3,000 feet, hanging in a bucket under a parachute designed much like a ribbed **umbrella** [I]. The parachute swung from side to side as the air escaped, making Garnerin extremely sick, but he landed safely. Public demonstrations of **hot-air balloons** [III] began to include a performance of a parachute jump. Improvements to the parachute over the next decades included vents to allow more stable descent and folded rather than ribbed systems. In these descents, a line attached to the balloon would pull open the parachute.

The first jump from an **airplane** [V] was in 1912 in St. Louis, using a line attached to the plane to open the chute, similar to those used in prior ballooning demonstrations. The first rip-cord parachute, known as the Stevens Life Pack, was demonstrated a few months later in 1912, but it was not until 1919 that a regular rip-cord pack for military use was designed and tested.

pasteurization

Named for the inventor of the process, French chemist and microbiologist Louis Pasteur (1822–1895), pasteurization was originally conceived as a method for preventing the souring of wine and beer. Pasteur believed that during the fermentation process, wine and beer were contaminated by bacteria that in turn produced lactic acid, leading to souring. He also proposed, correctly, that the fermentation process was caused by yeast, not as a catalyst, as had been assumed, but as a living organism. His germ theory of fermentation was first proposed in 1857. He argued that microbes did not spontaneously generate in foods or fermenting mixtures but resulted from the growth of preexisting organisms. His antibacterial process required heating the brew at a low temperature for an extended period of time to destroy the bacteria, and he recommended that heat be used to sterilize a wide variety of food products. Despite his development of the process in the 1860s, pasteurization of neither beer nor wine became established in Europe as connoisseurs claimed it destroyed the flavor as well as any bacteria.

Pasteur's claim to fame in his own time was established not by the process for pasteurizing foods or drinks but by his work on vaccination against diseases, particularly rabies. In 1882 he developed a vaccine against rabies as well as a treatment for humans who had been bitten by rabid animals.

Pasteurization of milk was advocated by Alice Catherine Davis (1881–1975), a scientist working for the Dairy Division of the Bureau of Animal Industry in the United States. She discovered several bacteria in milk and theorized that the bacteria came not from contamination in shipping but from the cows themselves. Brucellosis, or undulant fever, was a disease traced to such bacteria. Pasteurization of milk was controversial in the 1920s, but by the 1930s it became standard practice in the United States. The process requires heating milk to about 145°F for a half hour or to a higher temperature for a shorter period. Later development of ultrahigh-temperature pasteurization required flash heating to about 300°F for a second or two, after which the milk, if hermetically sealed, may be stored for months without refrigeration. Some foods have been sterilized or pasteurized with dosages of beta or gamma radiation.

periodic law

Dmitri Mendeleev (1834–1907) developed the periodic law of elements in 1869. A colleague, Professor Menshutken, made a formal presentation of Mendeleev's ideas to the Russian Chemical Society in the same year, due to Mendeleev's temporary illness. Mendeleev's work had been preceded by that of John Alexander Reina Newlands, who in 1865 presented a "law of octaves," but Mendeleev's principles had more data and went further in arranging the elements.

Mendeleev (also spelled in English as Mendeleyev) was raised in a large family of children by his widowed mother, who earned a living managing a family glass factory. After a series of family tragedies, including the loss of the factory to fire, Mendeleev entered St. Petersburg University and became a teacher of chemistry. He was the first to understand that all chemical elements are members of a single related and ordered system. He predicted that several elements, then unknown, would be discovered, and predicted their qualities. Three of them were

> I am convinced that I have found a precise, practical solution of the arduous problem which I proposed to myself—that of a process of manufacture, independent of season and locality, which should obviate the necessity of having recourse to the costly methods of cooling employed in existing processes, and at the same time secure the preservation of its products for any length of time. These new studies are based on the same principles which guided me in my researches on wine, vinegar, and the silkworm disease—principles, the applications of which are practically unlimited. The etiology of contagious diseases may, perhaps, receive from them an unexpected light. I need not hazard any prediction concerning the advantages likely to accrue to the brewing industry from the adoption of such a process of brewing as my study of the subject has enabled me to devise, and from an application of the novel facts upon which this process is founded. Time is the best appraiser of scientific work, and I am not unaware that an industrial discovery rarely produces all its fruit in the hands of its first inventor.
> —Louis Pasteur on the development of pasteurization, Paris, June 1, 1879

discovered in his lifetime: gallium, scandium, and germanium. His accomplishment of establishing the periodic table at age 35 established his reputation in Russia and throughout the world.

In 1868 he published a text, *Principles of Chemistry,* which became widely used. He was known as a popular teacher; his humble background and his popularity with a wider public made him a Russian national hero by the 1890s. Although the periodic table has been refined and filled in over the decades as a result of empirical research, several of his principles are still reflected in the current versions of the table. He stated that the "properties of elements are in periodic dependence on their atomic weight." Whether the table is presented horizontally, as he first arranged it, or vertically, in its best-known version, or in a spiral, the groupings reflect his thinking.

The other principles announced in 1869 have also stood the test of time: Elements similar in chemical properties have atomic weights that are nearly the same or that increase regularly; the groups of the elements correspond to their valences; the most widely diffused elements are the lightest; the magnitude of atomic weight determines the character of the element; elements will be discovered in analogous spots, predicted by gaps in atomic weight; atomic weights of some elements can be predicted by contiguous elements; and certain characteristics of elements can be foretold from their atomic weights.

The first of the elements predicted by Mendeleev, "eka-aluminium," was discovered in 1875 by Lecoq de Boisbaudran, and he named it gallium. That discovery squelched critics who had scoffed at Mendeleev's system, and his ideas were taken much more seriously after that. In 1882 Mendeleev won the Davy Medal from the Royal Society in Britain, and in 1905 he won the Copley Medal. A number of universities granted him honorary degrees, and the Russian government selected him for the post of director of the Bureau of Weights and Measures. He made several contributions in the area of armaments—for example, determining the ideal formula for **smokeless powder.**

phonograph

In 1877 Thomas Edison, experimenting with a **telephone** receiver, thought of a means of recording the voice by using a vibrating needle to impress varying indentations on a piece of foil. Turning the concept over to a Swiss mechanic in his establishment, John Kruesi, with instructions to "make this," the device, which Edison named the phonograph, was invented. Edison's phonograph used a tinfoil surface mounted on a cylinder as the recording surface. A thin diaphragm picked up sound, and a stylus transmitted the vibrations to the tinfoil, progressing along the cylinder as it moved through a lead screw. An identical screw, stylus, and diaphragm played the sound back on the other side of the cylinder. On December 6, 1877, Edison recorded "Mary Had a Little Lamb," the first known recording. This original device was sold for a few years as a novelty, as Edison brought the attention of his group to the development of the **electric light.** Although Charles Cross (1842–1888) had planned a sound-recording device, Edison was the first to develop and patent it. Edison received the patent on February 17, 1878.

Alexander Graham Bell, working with his brother Chichester A. Bell and with C. S. Tainter, began in 1881 to seek improvements to the phonograph. In 1886 Bell patented a wax cylinder method that became employed as a dictation machine, dubbed the graphophone. Music could be recorded by having a musical group perform a short piece in front of ten recording instruments, and then, after reloading the machines, in front of another ten. This laborious process could manufacture a few cylindrical records. Both the Edison and Bell cylinders impressed the vibrations in a vertical direction in the grooves, in a hill-and-dale pattern.

In 1892 Emile Berliner patented in Germany an improved process, using discs instead of cylinders, and he developed a method by which a steel negative could be prepared from a zinc positive recording. Using the negative, hundreds of hard rubber or shellac records could be produced. Berliner's method impressed the grooves from side to side. Berliner began manufacturing the instrument under the name gramophone, and it entered the American market in 1893.

In 1897 Eldridge Johnson, of Camden, New Jersey, developed improvements to the Berliner record system, using a method of engraving the discs rather than etching them. Johnson founded the Victor Talking Machine Company in 1901 and soon produced millions of records from studios established in Camden. One of Johnson's first successes was a series of recordings by Enrico Caruso that brought fame both to the phonograph and Caruso. The Berliner/Johnson discs standardized at a speed of 78 revolutions per minute (rpm) in 1913 and were limited to about 4 to 5 minutes of playing time.

Further improvements to recording systems came with the development in 1948 by Emile Goldmark at Columbia Broadcasting System of long-playing (LP) records that could record about 20 minutes of sound, more suitable for longer pieces of music. By contrast to the 78-rpm record with about 80 grooves to the inch, the LP had some 250 grooves per inch. In competition, Radio Corporation of America introduced a

45-rpm disc that could play about 8 minutes per side. Stereophonic recording (with two tracks, one hill-and-dale, the other lateral) was introduced in 1958, with compact disc and **laser disc [VI]** technologies supplanting the earlier recording systems in the 1980s and 1990s, respectively.

pneumatic tire

The development of the air-filled tubular tire made possible a smoother-riding **bicycle** and advanced the acceptance of the **automobile**. A British inventor, R. W. Thomson, who suggested leather casing filled with either horsehair or air, patented the concept in 1845. In 1888 a Scottish veterinarian, John Boyd Dunlop, resident in Belfast, Ireland, independently invented such a tire for his son, who asked for something to cushion his tricycle on cobbled streets. Taking a garden hose, Dunlop cut two proper lengths and wrapped them around the rear wheels of the tricycle, sealing the ends with glue. He inserted a valve for inflation and wrapped the tire to the wheel. The resulting assembly was called a mummy or pudding tire, from its wound bindings.

A year later a bicycle racing enthusiast noted the tricycle tire and realized its value, leading to newspaper publicity of the concept. Dunlop patented the new tire, and Harvey DuCros, an Irish financier, worked with Dunlop to establish the Dunlop Rubber Company. Later, a dished rim was patented in 1890, and further improvements, such as embedding wire in an outer casing, improved the concept. The Dunlop tire helped spread bicycling from a relatively small sporting populace of 300,000 worldwide to tens of millions of customers.

pressure gauge

In 1849, Eugène Bourdon (1808–1884), a French scientist, invented a type of pressure gauge that allowed pressure of a fluid, either liquid or gas, to be measured on a circular dial. A flattened tube is bent in the curved shape of a C, with one end connected to the fluid being measured and mounted on a fixed point. The other end is sealed and free to move in an uncurling direction as the pressure increases. The moving end is connected with a geared linkage to a pointer that moves as the pressure changes. Bourdon patented the gauge in 1850, and it found wide application in **steam engines [III]** and in pressurized hydraulic and oil systems in a variety of engines. He personally built more than 200 small steam engines for demonstration purposes. As a simple analog device, the Bourdon gauge remained the basic means of reflecting oil pressure in engine systems used in the **automobile** throughout the 20th century.

propeller-driven ship

The idea of propelling ships with a screw propeller rather than with paddles or **paddlewheels [II]** had been developed as early as 1752, when the Swiss physicist Daniel Bernoulli (1700–1782) suggested it. However, for the principle to work, an efficient engine would be required, and the concept remained an unrealized idea for several decades. John Fitch, an American inventor who developed a paddle system using a **steam engine [III]**, toyed with the concept of a screw propeller in the 1790s, as did several other experimenters. The principle relied on an ancient device, **Archimedes' screw [I]**, which was a spiral made of wood or metal fixed around an axle and surrounded by a tube. With one end immersed in water, when the screw was turned, water would be pumped out of the other end. The notion was that if such a screw could be mounted horizontally beneath or behind a ship it would propel the vessel. However, unless turned rapidly, the screw would simply slip through the water without imparting the desired force to the ship.

propeller driven ship
The modern ship propeller derives from an 1834 discovery that a partial, single turn of a screw is effective as a propulsion device. *U.S. Coast Guard*

In 1828 an Austrian forest manager, Joseph Ressel, developed an experimental ship at Trieste, a port on the Adriatic Sea, using a 6-horsepower steam engine with a screw mounted at the stern. The *Civetta* achieved a speed more than 7 knots, but after a boiler accident, the ship was never tried again. In 1834, Francis Pettit Smith, a British gentleman farmer, experimented with a wooden screw propeller. He built a 237-ton ship, the *Archimedes,* and hoped to promote the idea with the British Admiralty. At ship trials in 1838 the ship performed well, achieving a speed of more than 4 knots. However, the turn-and-a-half propeller broke during the trials, shearing off a half-turn. Expecting to limp back to port on the remaining single turn of the screw, Smith was astounded to find that the ship moved faster with the damaged propeller than with the complete one. This accidental discovery led to further development, and Smith soon achieved speeds with the *Archimedes* of up to 13 knots, and on long voyages the ship could average 11 knots.

Ship innovators soon developed the idea further, with Isambard Kingdom Brunel (1806–1859) building the propeller-driven 3,000-ton *Great Britain,* which steamed in 1844 to the United States in a record 14½ days. Cyrus Field later used the *Great Britain* to lay **submarine cable.** The Swedish-born John Ericsson (1803–1889) patented the concept of screw propulsion in 1838, and in the United States in 1849 he built the *Princeton,* the first metal-hulled, screw-propelled warship. The engines were mounted below the waterline for further protection. Navies soon recognized the superiority of propellers over paddlewheels, since the fully submerged propeller was a more difficult target for enemy gunfire than the paddlewheel, mounted above the waterline. The term *screw propeller* survived in common usage well into the late 19th century, long after the characteristic screw shape of the early experiments had been abandoned in favor of the more efficient single-turn propeller, with blades mounted around an axle. Ericsson continued to build propeller-driven ships for the U.S. Navy, earning a place in history with the Civil War *Monitor.*

Propeller developments later in the 19th century and in the 20th century included variable-pitch propellers and scientific studies of propeller shape to reduce cavitation, the tendency of a high-speed propeller to create cavities in the water that impede its efficiency and damage the blade metal. Other work included reducing the noise emitted by spinning propellers, as part of the effort to silence a ship to make it less detectable during warfare.

refrigeration

The ancient Chinese, Greeks, and Romans used snow and ice carefully stored in underground chambers to cool drinks and to help preserve food. However, the use of ice in many climates depended upon harvesting it in the winter and storing it for use in the summer. Ice harvesting from lakes in New England in the 18th and 19th centuries became a major local industry for some communities.

In the 19th century, many inventors developed icemaking machines and refrigeration systems based on the principle that when a liquid changes to a gas, it absorbs heat. By compressing the gas to a liquid and then releasing the pressure to let it absorb heat as it expanded in a self-contained cycle, the air or other fluid surrounding the device can be reduced in temperature. With the principle widely understood by the mid-18th century, a number of prototype refrigeration machines were invented, including one in Scotland by William Cullen in 1758 that used ether and another by American inventor Oliver Evans (1755–1819) in 1805 that also used ether. Evans visualized but never produced refrigerated drinking fountains. Jacob Perkins, an American resident in Britain, patented another type of ether-compression refrigeration machine in 1834. A doctor in Florida, John Gorrie, patented an icemaking machine using compressed air in 1851.

The concept of using ammonia as the compression refrigerant and a water jacket as the heat absorbent that could be circulated to the area to be cooled was developed by Ferdinand Carré in France in 1859. At a London exhibition in 1862, Carré demonstrated an icemaking machine that amazed visitors by continuously producing enormous blocks of ice. Another system, using sulfur dioxide, was installed in London in 1876 to build an indoor skating rink, soon followed by others, at Manchester and elsewhere.

Best known and most often cited as *the* inventor of modern refrigeration was the Bavarian Karl Paul Gottfried von Linde (1842–1934), who in 1876 developed an ammonia system for brewery refrigeration. After some modification, Linde manufactured small refrigeration units for home use. By 1891 he had sold more than 12,000 refrigerators. Later home refrigerators used a variation on the Linde design. Linde went on to develop a continuous process for reduction of pure oxygen and nitrogen from air, in commercial amounts.

Even though household refrigerators were expensive, they caught on quickly. The Kelvinator company began producing them in 1916, and the Guardian company, a forerunner of Frigidaire, began in 1917. In

1920 there were an estimated 20,000 refrigerators in the United States. By 1936 there were 2 million.

One problem with ammonia refrigerators was that when they developed a leak, the gas was irritating and slightly corrosive. In 1930 Thomas Midgley (1889–1944) discovered a halogenated hydrocarbon or fluorocarbon that was trade-named **Freon [V]** by the du Pont company. From the late 1930s until the 1990s, Freon was widely used as a refrigerant. However, it was discovered that escaping Freon rises to the upper atmosphere, where it yields chlorine that has the effect of breaking down the ozone layer. Since the ozone layer of the atmosphere provides protection from harmful ultraviolet radiation, other refrigerants were widely substituted for Freon in the last decade of the 20th century.

revolver

Samuel Colt (1814–1862) invented the first successful single-barreled pistol with revolving cartridge chambers, and he patented it, first in Britain and France in 1835, and then in the United States in 1836. A separate cap and ball would be loaded into each chamber from the front. Colt's pistol worked on the single-action principle—that is, a single action would cock the hammer, bringing the next charge into line, and then a pull on the trigger would release the percussion hammer.

Colt had been sent to sea at an early age, and on his travels he observed several repeating pistols. Noting the mechanism of the ship's wheel that locked in position, he conceived the idea of a revolving set of cartridge chambers and carved several wooden models of his idea. On his return to the United States in 1832, at age 18, he set up a factory in Paterson, New Jersey, to manufacture his new weapon. He had some initial success in sales to the army for use in the Seminole War and to the Texas Rangers. However, his company went bankrupt in 1842.

Interest in the revolver picked up again during the Mexican War, and Colt worked through Eli Whitney's plant in Whitneyville, Connecticut, to begin fulfilling an army order for 1,000 pistols. Whitney had earlier earned fame as the inventor of the **cotton gin.** In 1847 Colt opened his own plant, the Colt Patent Arms factory, in Hartford, Connecticut. By 1855 Colt had the largest arms factory in the world. He died in 1862, but his firm continued to manufacture weapons, expanding further with sales during the American Civil War. The principle of **interchangeable parts** has often been incorrectly attributed to either Whitney or Colt. Both men sought to achieve interchangeable parts for their weapons, but it was the development of machine tools at Harpers Ferry

Arsenal and at the Springfield Arsenal that allowed both manufacturers to begin to achieve this goal.

When Colt's American patent expired in 1857, Horace Smith and Daniel Wesson introduced a new revolver with copper cartridges that could be loaded easily into pistol chambers from the rear. On the expiration of the Smith & Wesson patent in 1872, the double action was introduced to the Colt pistol, simplifying the mechanism. A double-action revolver could either be operated as Colt's original, by manually cocking the hammer that would also rotate the cylinder, or it could be fired simply by pulling the trigger, which at once moved the chamber, cocked the hammer, and released it to fire the chambered cartridge.

Colt's Peacemaker model, introduced in 1873, became the famous weapon that "won the West."

rubber

When Europeans first made contact with the Americas following the explorations of Columbus, they found that the Native Americans made use of the hardened latex sap from trees, particularly the *hevea* tree. In addition to bouncing balls, they used natural rubber gum to chew, and they made numbers of practical objects such as bowls and waterproofed baskets. Samples were brought back to Europe, most notably by François Fresnau, who had lived in Brazil. Charles Marie de la Condamine (who led an expedition to determine the **size and shape of Earth [III]**), donated samples from Fresnau's expedition to the French Academy in 1751, with a report on the process by which natives in the Amazon collected the sap and dried it over fires. The Quecha word for the material, meaning "wood-tears," was *caoutchouc*. Many European languages modified the Quecha word to refer to the material.

Joseph Priestley (1733–1804), the British chemist noted for the discovery of **oxygen [III]**, coined the term *rubber* after noting that a small block of the material could rub out pencil marks. In English the material was often called *India rubber*, meaning it came from the West Indies.

However, when natural rubber was shipped to Europe, it tended to harden. Various solvents, such as turpentine, could soften the material. In about 1820, Scottish manufacturer Charles Macintosh (1766–1843) developed a process of laminating fabric sheets with a solution of rubber and naphtha, itself derived from coal. Macintosh soon produced raincoats and galoshes from these natural rubber cements. By 1823 Macintosh had developed a thriving rainwear business, with international exports, and his name became a generic term for a waterproof coat. Macintosh's associate Thomas Hancock (1786–1865) established a small research laboratory and developed several machines to shape and mold rubber under heat, taking out 17 patents from 1820 to 1847. Elastic bands, garters, and many other uses developed for rubber in this period.

A tree grown in Malaya, the *getah,* yielded a gum that was a form of rubber, known in Europe as gutta-percha, often used in the mid-19th century as a form of electrical insulation.

However, rubber continued to present problems, in that it would harden and crack in cold weather and grow sticky in hot weather. In the United States in 1839, Charles Goodyear (1800–1860), a hardware merchant, developed a process of mixing rubber with sulfur and heating it that resulted in a more stable yet still elastic substance. He had experimented with a variety of materials and processes, discovering the method more or less by accident, working on a kitchen stove. He called the stabilization process vulcanization and the new product vulcanized rubber. Goodyear obtained a patent on vulcanization in 1844. However, Thomas Hancock heard of Goodyear's work and took out a British patent in 1843, preventing Goodyear from exploiting his method in Britain. Struggling against patent infringement in the United States and Europe, Goodyear died penniless in 1860. Both Hancock and Goodyear published accounts of their discovery of the vulcanization process.

With vulcanization making rubber more practical for multiple uses, European nations soon established rubber plantations in tropical colonies, particularly in the Dutch East Indies and Malaya, while the American-owned Firestone company later established plantations in the West African nation of Liberia. Hard rubber tires for **bicycles** led to the development of the **pneumatic tire,** at first for bicycles and then for the **automobile.**

semaphore

The signaling semaphore was invented in 1792 and was the first effective means of communicating over long distances. Decades before the invention of the electric **telegraph,** the semaphore was called a telegraphic system. The invention was the creation of Claude Chappe, a supporter of the French Revolution. He and his brother had invented a system of communicating, using a post with a crossbar. On each end of the crossbar, a pivoting bar could be changed to different angles. Chappe had worked out a code in which the position of the pivoting arms could be read as particular common words or letters. He proposed the system to the French Revolutionary Council, which appointed Chappe as the *ingénieur télégraphie*. He established a line of stations across France in 1794, at intervals of about 10 miles. At each post, an operator with a telescope would pick up messages and pass them along.

Ultimately, the semaphore system covered some 3,000 miles. The British Admiralty imitated the system and set up Telegraph Hill stations between London and English Channel ports. The semaphore network also was emulated in the British colonies in North America. Although the system was quite efficient, allowing a message to be transmitted over several hundred miles in a few minutes, it was expensive to maintain, requiring attentive observers along the whole line of signals. Furthermore, it could not operate at night or in rain or fog. The concept of the semaphore network, however, is the ancestor of modern communication networks, including not only the telegraph but also the **telephone** and the **Internet [VI]**.

The system of signals was adopted for visual signaling with flags from ship to ship, still used when radio silence must be maintained. The semaphore also is the direct ancestor of signals used in **steam railroad** systems throughout the world.

sewing machine

Several attempts at developing a machine that could sew were made between 1790 and the 1830s. In Britain, Thomas Saint developed plans for a machine for sewing leather but never built one. In the 1820s a French tailor, Barthélemy Thimonnier, developed a sewing machine and sold a number, but indignant tailors broke into the shops that had installed them and destroyed the machines. In the United States a prolific inventor, Walter Hunt, invented a sewing machine in 1832, but it had the defect that it could sew in only a straight line.

Success eventually came to an American inventor, Elias Howe (1827–1867), after many struggles. He worked on a design in 1843, and held a demonstration in 1845 in Boston that showed that one person at his machine could outstrip the work of five seamstresses. He patented the machine in 1846. The device had a horizontal needle that would penetrate two layers of cloth, carrying a thread through a plate behind which a shuttle would interweave a cross-stitch. The seam had the virtue that if it was broken in one spot it would not unravel. After an unsuccessful venture in attempting to sell his machine in Britain, he returned without funds to the United States in 1849, only to discover many imitators infringing his patent. Embittered by the death of his wife from lack of medical care, he plunged into several lawsuits. Eventually the courts sustained his cases, and soon he agreed to license the patent to a patent-pool group established by one of the infringers, Isaac Merritt Singer. The Singer machine had a vertical needle, a foot treadle to provide power, and a convenient arrangement for turning a seam. Despite its differences from the Howe machine, the courts ruled that Singer had infringed the Howe patent. Historians have tended to credit both Howe and Singer as inventors of the machine.

Singer arranged back royalties to Howe, and future royalties to Howe on new machines. By 1859 Howe was earning more than $400,000 a year in royalties. Singer retired from the business in 1863, with a fortune estimated at $13 million. Considering the 19th-century purchasing power of the dollar, both of these men had become immensely wealthy. The Singer sewing machine was featured as a front-page story in the *Scientific American* of November 1, 1851.

The machine revolutionized the garmentmaking industry, making the preparation of precut and presized clothing on a massive scale possible.

skyscraper

The development of the skyscraper was made possible by several mid-19th-century developments, most notably the invention of the **elevator** by Elisha Graves Otis in 1852 and the development of **steel** manufacturing by Henry Bessemer in 1856.

Although these technologies made it feasible to construct a skyscraper, the first tall building partially utilizing a steel frame construction was not built until 1884–1885. The rising price of urban land, rather than the technological possibility, drove the desire for taller buildings. Architect William Le Baron Jenney (1832–1907) designed the 10-story Home Insurance Company Building in Chicago, often described as the "first skyscraper." However, this building had square cast-iron wall columns, wrought iron beams to the sixth floor, and steel beams above the sixth floor. While the beams and girders were fastened together and bolted to the columns to form a cagelike framework that constituted the main load-bearing structure, later

> When builders began erecting exceptionally tall buildings supported by a light iron or steel frame . . . the problem of providing adequate resistance to the wind proved difficult to solve. Earlier forms of construction gave only limited guidance. In fact many of the devices and techniques adopted by the builders of skyscrapers are derived not from earlier building practice but from methods devised for the construction of long-span bridges.
>
> —Carl W. Condit, "The Wind Bracing of Buildings," *Scientific American* 230, no. 2 (February 1974): 92

skyscrapers took the evolution further. The 1889 Leiter Building in Chicago, designed by Jenney, was the first in which there were no self-supporting walls. The 16-story Manhattan Building and the Fair Building, both built in 1891, had wind bracing. By 1895, Jenney and other Chicago architects had perfected the concept of a wind-braced steel structure with "curtain walls"—that is, walls that hung from the frame, rather than walls that served to bear the load above.

A number of other Chicago architects who studied under Jenney carried on his work, including most notably Louis Sullivan and William Holabird.

A building boom between the 1890s and the 1920s changed the look of American cities, with many skyscrapers incorporating Art Deco features, such as the Chrysler Building and the Empire State Building in New York City. Building codes required that tall buildings be "stepped" back from the street, to allow light and to prevent the streets from becoming dismal canyons surrounded by sheer walls.

Few skyscrapers were built during the 1930s and 1940s, but in the post–World War II era, a new design began to emerge, characterized by plain glass curtain walls and featureless exteriors. This ultramodern style was followed after the 1970s by a "postmodern" style in which some decor, color, and the arrangement of open plazas at the base of skyscrapers contributed to a less mechanical or impersonal look.

smokeless powder

The first smokeless powder for ammunition was introduced in Austria in about 1852 and used extensively in hunting. It was made from a form of cellulose, such as found in cotton or wood that had been processed with nitric acid. Known as guncotton, the new compound, nitrocellulose, was a form of plastic, which when burned behaved very much like black powder or **gunpowder** [II].

In 1888 Alfred Nobel (1833–1896), the inventor of **dynamite,** developed a form of advanced smokeless powder using nitroglycerin. Hiram Maxim (1840–1916) developed a smokeless powder using cotton fiber in the same year, with beeswax or petroleum jelly as a stabilizer. It could be extruded in small cords, and the British called it cordite. Both the double-base powder that included nitroglycerin, or the single-base powder, with nitrocellulose alone, included new organic compounds and differed from earlier black powder mixtures of sulfur, carbon, and saltpeter.

The Nobel formula was about 40 percent nitroglycerin and 60 percent nitrocellulose, while the Maxim cordite was about 15 percent nitroglycerin. Both double-base smokeless powders were some 15 percent more powerful than earlier brown and black powders and burned with a more uniform pressure. Both powders achieved a higher velocity for the projectile. Maxim won a patent lawsuit in Britain that determined that his powder was sufficiently different from Nobel's to entitle Maxim to manufacture it without royalty to Nobel.

By the 1890s, scientists at the U.S. Navy's research facility at Newport, Rhode Island, sought to find another formula, working on a pure nitrocellulose formula, regarding the double-base powders with nitroglycerin as too dangerous for shipboard use. In 1896 Dmitri Mendeleev, who had developed the **periodic law** of the elements, published results from earlier work determining that nitrocellulose with about 12.44 percent nitrogen was the chemical ideal. Russia adopted that formula in the early 1890s. American and European navies continued experiments, settling on ranges of nitrogen between 12.45 and 12.80 percent in their product.

Smokeless powder when used aboard naval vessels had several advantages, primary among which was the fact that without a cloud of smoke near the vessel to obscure repeated firings by sight, it became important to develop rapid-fire weapons. Accordingly, the navies of the world entered a period of intense innovation, attempting to improve the breech-loading mechanisms of artillery pieces so as to increase the speed with which they could be fired and reloaded. American rapid-fire deck guns were particularly effective in the Spanish-American War of 1898.

spectroscope

After Isaac Newton (1642–1727) analyzed the **spectrum of light [III]** in 1666, experimenters continued to study color and light, following up on the lines of inquiry he established. In 1814 the German physicist Joseph von Fraunhofer (1787–1826) discovered dark lines in the solar spectrum that have since been known as Fraunhofer lines. He found 574 lines between the red and violet ends of the spectrum and correctly identified and accurately mapped them. Fraunhofer had learned optics working in his father's glass workshop at age 10, and at age 19 he joined the optical shop of the Munich Philosophical Instrument Company.

Fraunhofer determined the refractive indices and dispersion powers of different kinds of optical glass, and he used a diffraction grating to achieve the prism effect to separate white light into a spectrum. In 1859 Gustav Robert Kirchhoff (1824–1887) showed that all elements, when heated to incandescence, produce individual characteristic spectra, providing the explanation for the Fraunhofer lines. Kirchhoff concluded that the lines are due to the absorption of light by specific elements in the Sun's atmosphere. The use of the Fraunhofer spectroscope as a tool of astronomical spectroscopy can be traced to Kirchhoff's 1859 work. Kirchhoff had initiated the method of chemical analysis that can be used to determine the chemical nature of distant stars. Robert Bunsen (1811–1899), the inventor of the **Bunsen burner,** worked with Kirchhoff, using the spectroscope to discover two new elements, cesium and rubidium, in 1860.

steam locomotive

The application of the **steam engine [III]** to ground transportation on rails can be traced to several contemporaneous developments in the late 18th century. With the improvements made by James Watt (1736–1819) to the steam engine in the 1780s, various inventors conceived of the idea of using it for transportation both in **steamboats** and on **steam railroads.** Railways were already in place with horse- or mule-drawn carts in and around mines for transport of heavy ore.

In 1796 Richard Trevithick (1771–1883) in Britain developed an improved steam engine, and in 1801 he used it to pull a passenger car. In 1804 he developed an engine to move carts along a horse-rail system to and from the Penydarren ironworks at Mehthyr Tydfil in Wales. That system ran about 9½ miles and could carry 10 tons of iron at about 5 miles per hour. To popularize his concept, Trevithick set up a circular track at Euston Square in London and offered rides to the public. Although Trevithick went on to build off-road vehicles that were predecessors of the **automobile,** he never received much credit for the locomotive. Between 1804 and the 1820s, many British mines installed a variety of steam-powered rail systems for local hauling, but none was intended as a public passenger-carrying line.

George Stephenson (1781–1848) made a number of improvements to steam engines and mining-rail systems. In 1814 he developed the Blucher, a steam engine with flanged wheels that kept it on the rails more efficiently. In 1825 he opened a commercial railroad line between two towns about 30 miles apart in England, with the engine Locomotion I. Several experimental engines competed in 1829 at trials held at Rainhill, outside London. Robert Stephenson (1803–1859), the son of George, designed the Rocket, named the winner of the Rainhill trials. In 1830 the Liverpool-Manchester Railway opened as the first commercial, passenger-carrying line. Robert Stephenson introduced several improvements that were prototypes for later locomotives, including a multiple fire-tube boiler; a drafted firebox to increase the heat under the boiler by exhausting steam through the firebox smokestack; connecting the steam piston to the wheels by connecting rods rather than gears; and flanged wheels to grip the tracks.

Many improvements to the steam locomotive were developed in the United States. American mechanics introduced such features as lighter locomotives and a set of small wheels to constitute a leading truck that would allow the locomotive to follow curved or irregular track. Other American features included a screened wood-burning smokestack, a cowcatcher, an enclosed cab, a bell, and a headlight. Improvements to the suspension and the development of an 8-wheeled configuration with 4 large driving wheels and 4 smaller wheels in front produced the American Standard locomotive, which remained the basic design from about 1840 to about 1890. Locomotives were later designated by their wheel configurations, with a number for the leading wheels, the driving wheels, and the trailing wheels. Thus the American Standard was known as a 4-4-0 configuration, whereas a style developed in 1866, the Consolidation Type, had a 2-8-4 pattern, was widely adopted, and was still produced as late as 1949.

steam railroad

The first railroad promoters had to be very explicit to call their concept a "steam railroad," since many dirt roads had been improved with wooden rails, sometimes topped with iron, along which horses could draw carriages. Thus a steam railroad represented a new departure, and eventually the term *railroad* came to refer only to a tracked system over which there was no horse-drawn traffic but only engine-pulled trains of cars.

The concept occurred to several developers about the same time. John Stevens (1749–1831?), an American who had competed with Robert Livingston over the development of **steamboat** travel, received from New Jersey the first railroad charter in the United States. While he raised capital for his first line, the Camden and Amboy Rail-Road and Transportation Company, to link the New York and Philadelphia regions, other entrepreneurs started steam railroad lines in South Carolina, Pennsylvania, and Ohio.

To raise funds, Stevens built his own engine and ran it on a track built on the lawn of his estate in Hoboken, New Jersey, in 1825. The next **steam engine [III]** to run on tracks in the United States was an engine imported from Britain, the Stourbridge Lion, operated by Horatio Allen in 1829 on a track from a coal mine to a canal in northeastern Pennsylvania. However, the British engine was too heavy for the track, and American promoters began to design lighter engines. Allen moved from the Delaware and Hudson Company to work with the South Carolina Canal and Railway Company. There he designed and had built at the cannon works at the West Point Foundry an engine called The Best Friend of Charleston. For another engine he built in 1831, the South Carolina, Allen designed two pivoting trucks that allowed the engine to follow curves, one in the front and one in the rear of the engine. John B. Jervis also invented a single, front-mounted pivoting truck, in 1831.

In 1836, a Philadelphian named Henry Campbell patented a style of engine that came to be called the American Standard type, consisting of two pairs of driving wheels and a single swiveling truck on front to follow rail curves. From the early 1830s to the 1850s, American steam engines for railroads went through rapid development, increasing in speed, power, and durability.

As a system, the railroad incorporated dozens of separate improvements and inventions, including elements of carriage and coach design, the track and its support system, tunnels, and other auxiliary equipment. John Stevens's son Robert invented the T-rail in 1830. He also invented the hook-headed spike to attach the rails to wooden ties. In general, American railroads had more curves per mile and were built much more rapidly and carelessly than British lines.

To control traffic and reduce the likelihood of accidents, efficient communication over the line was required. Railroads quickly employed both the **semaphore** system developed in France and the electric **telegraph** to manage traffic, stimulating the creation of a telegraph network linking many cities in the United States well before the Civil War.

The rapid development of the steam railroad in the United States had several far-reaching social consequences. Although the extensive Mississippi and Ohio River system of the interior of the United States provided a natural highway for commerce, the railroad soon made possible the opening of regions far from water transport. By crossing the United States with a transcontinental connection in 1869, the nation was united not only politically by the Union victory in the Civil War but by transportation as well. By the last decades of the 19th century, a national price and market system began to emerge in which products manufactured in Chicago and St. Louis, such as butchered meat and milled wheat flour, as well as products manufactured in Pennsylvania and New England, such as textiles, were regularly traded on a nation-wide basis. With such exchange came a lowering of prices and the elimination of inefficient and more expensive producers from competition.

The demand that the railroad created for reliable **steel** drove the growth of that industry. As late-19th- and early-20th-century economists examined the history of industrial development, many concluded that the railroad was the key industry that stimulated others. For this reason, political leaders in countries such as Argentina, Chile, Mexico, and Russia sought to construct extensive rail nets on the assumption that they would launch their countries into modernity and prosperity. Another social consequence of the railroad in the United States was the need for some form of regulation of commerce on a national basis, leading to the formation of the Interstate Commerce Commission in 1887, the first federal regulatory body in the nation.

steam turbine

A steam turbine consists of a wheel that is turned by injecting steam to blades on its periphery, in an enclosed system. The concept had been suggested and illustrated as early as 1629 by an Italian, Giovanni Branca, in his book *Le Machine*. William Avery developed an early application in 1831 in the United States, but his machines were subject to breakdown and operated inefficiently.

The Swedish inventor Gustave De Laval developed an effective steam turbine in 1883, and he shares credit as the inventor of the modern steam turbine. He utilized small turbines to drive a device that separated cream from milk by centrifugal force, operating at extremely high rates—as many as 30,000 revolutions per minute. Those rates were too high for practical transportation uses, although De Laval went on to produce steam turbines for a wide variety of stationary applications, and his company later produced gas turbine engines as well.

In 1884 Charles Parsons (1854–1931), a British engineer, patented a steam turbine similar to De Laval's design, with several sets of vanes mounted on a rotor. It was Parsons' improvements that made the steam turbine practical for many heavy-duty applications calling for slower rates of revolution and more power. Stationary blades forced the steam against the rotating blades, using series of blades to step down the speed and pressure. His turbines, which also rotated at extremely high rates, could be used to drive electrical generators. Parsons developed a steam turbine–propelled ship, the *Turbina,* which achieved record speeds of up to 34 knots in 1897. Parsons turbines were later installed in the liners *Lusitania* and *Mauritania*. In the United States, Charles Curtis perfected the large-scale steam turbine and sold the rights to the design in 1902 to General Electric, which used them to construct power-generating plants.

steamboat

Several individuals can be credited with the invention of the steamboat, but all of the efforts in the period from 1690 to the 1780s saw very limited success. Among these was a steamboat using a Newcomen **steam engine [III]** designed and built by a Frenchman, Denis Papin, in 1707. In about 1737, Jonathan Hulls in Britain patented a steamboat, but his design would not have been workable.

It was not until James Watt (1736–1819) improved the steam engine in the 1780s that reliability and power-to-weight ratios became practical for shipboard use. In 1783 a French marquis, Claude de Jouffroy d'Abbans, installed on a riverboat a steam engine built by James Watt. None of these efforts produced lasting results. In the United States, early experiments by James Rumsey and John Fitch in 1787 proved interesting but did not attract investors. The Fitch steamboat had a system of vertical oars, and he operated it as a ferry on the Delaware River out of Philadelphia.

A sustained success came out of the teamwork of Robert Livingston (1746–1813) and Robert Fulton (1765–1845). Livingston concentrated on securing monopolies on routes and arranging financing, while Fulton worked on engine and propulsion designs. In 1803 they demonstrated in France a design along the lines of that of Fitch. In the United States from 1807 to 1808 Fulton built and launched the *North River Steamboat of Clermont* (known to history as the *Clermont*), followed by more than a dozen other designs from 1808 to 1814, using paddlewheels rather than the oar system. As a result, most American treatments of the subject credit Robert Fulton as the inventor of the steamboat, with 1808 as the year of its introduction, despite the claims of nearly a dozen predecessors in Europe and the United States.

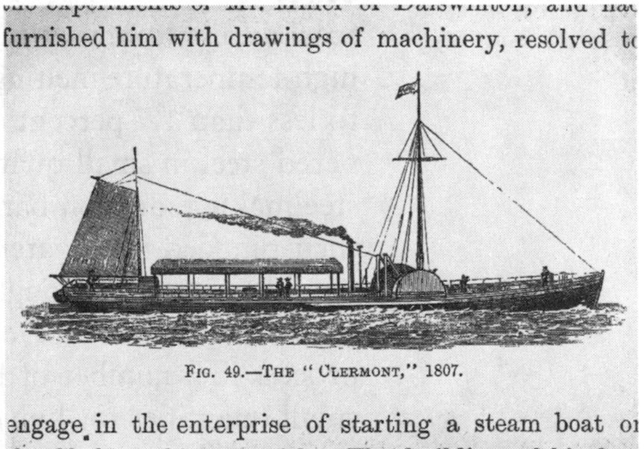
furnished him with drawings of machinery, resolved t

Fig. 49.—The "Clermont," 1807.

engage in the enterprise of starting a steam boat o
the Clyde at his own risk. The building of his boa
was commenced in October, 1811, and it was launche

steamboat In the period 1790–1810, numerous inventors experimented with steam propulsion of ships. Robert Fulton and Robert Livingston established a partnership and made many successful voyages with the side-wheeled *North River Steamboat of Clermont* in 1807 and 1808.
Library of Congress

In 1811 Fulton launched the *New Orleans* in Pittsburgh. On its voyage downriver that year it encountered a massive earthquake centered at New Madrid, below the point where the Ohio and Mississippi Rivers join. Later the vessel began regular service between New Orleans and Natchez, Mississippi. Steamboats with sidewheels and stern paddlewheels proved well suited to the developing United States, where long, navigable rivers but few roads connected vast stretches of territory. Simultaneously with the developing of the **steam railroad,** the steamboat provided a means for carrying passengers and cargo throughout the burgeoning Mississippi River valley in the mid-19th century. In 1814 New Orleans had about 20 arrivals of steamboats, but by 1834 there were more than 1,200 per year. Fulton steam ferries also became valuable as river crossing vessels at Philadelphia, New York, and Boston. In New York City the street across lower Manhattan connecting two ferry terminals was named Fulton Street.

As a means of ocean travel, however, the paddlewheel-driven steamboat was relatively unreliable, and steam power did not fully supersede sail power for ocean travel during the 19th century. The **propeller-driven ship** began to replace paddlewheel and sail-and-paddlewheel power for oceangoing vessels and for warships in the 1850s and 1860s.

steel

Steel differs from iron in that steel's carbon amount is reduced through high-temperature melting and oxidizing of the carbon until it is reduced to less than 1.7 percent of the metal content. The manufacture of tempered steel in small quantities dates back to the Middle Ages. In early steelmaking, an iron bar would be repeatedly heated in a coal fire and then plunged into water or blood. The repeated contact with the air would eventually slough off oxidized carbon. Tempered steel, with traces of carbon and other minerals, was manufactured by such a process in a number of European centers, such as in Toledo, Spain, in small quantities for knives, ax blades, and swords. However, the 19th-century advances in steelmaking made possible production of rails, beams, and sheets of steel that had a major impact on advancing the Industrial Revolution, affecting the daily life of millions. Later invention of alloys of steel developed new metals with different characteristics. Steel with about 11 percent chromium allowed for stainless or rustless steel, while alloys with tungsten produced much stronger metal.

In 1856 the English engineer Henry Bessemer (1813–1898) invented a blast furnace system in which raw iron was placed in a container or converter lined with heat-resistant bricks. Bessemer opened a plant in Sheffield, England, to which he imported phosphorus-free iron ore from Sweden and began to produce high-quality steel in 1860. In Bessemer's system high-pressure air was streamed from below into a large vessel containing molten iron to burn out impurities such as carbon. After the carbon burned, carbon ash (or slag) sank to the bottom of the container, which would then be tipped to pour off the purified molten metal from the top. In a variation on the method, in 1864, German-born William and Friedrich Siemens, working in Britain, developed the open-hearth method, in which iron was blasted from above by pre-heated air and gas. A year later, Pierre and Emile Martin in France made several improvements in the method. In 1948 the process was improved in the basic oxygen method, by inserting a tube or lance from above into a converter and blasting the molten metal with oxygen rather than air. The Bessemer, Siemens-Martin, and basic oxygen methods are all similar in that they permit large-scale production of steel through the use of forced gas (either air or oxygen) to speed up and increase the combustion of carbon in the molten metal.

With the development and spread of the Bessemer process, it was possible to replace iron railroad tracks with steel rails and to begin the construction of steel-frame buildings and bridges. Large-scale steel production stimulated **steam railroad** construction in the United States and around the world in the last half of the 19th century. Following the Civil War in the United States, constructing buildings taller than 5 or 6 stories was made possible by the supply of steel girders that could be riveted together. With solid foundation into rock in locations such as Manhattan, the invention of steel girders, together with the **elevator**, made the clustering of **skyscrapers** characteristic of modern cities possible. Sheet steel allowed for the construction of much stronger boilers for **steam engines [III]**, as well as finding many other applications, such as railroad cars and eventually **automobile** bodies.

steel cable suspension bridge

The suspension bridge using rope was invented in ancient China and apparently independently in the Inca Empire in Peru, but the adaptation of steel cable to the principle of a suspension bridge is a triumph of the Industrial Revolution in the 19th century. The major pioneer in the work was the American engineer John A. Roebling, who built a suspension bridge in 1851 over the Niagara River gorge and another over the Ohio River at Cincinnati in 1867, later named for him. His most famous work was the Brooklyn Bridge.

Roebling began planning a bridge over New York's East River in the early 1850s. In 1867 he was named chief engineer for the Great East River Bridge or Brooklyn Bridge, connecting Manhattan with Brooklyn. In 1869 he contracted tetanus from an accident that crushed his foot, and his son, Washington Augustus Roebling, supervised completion of the bridge. The bridge opened for traffic in 1883. It was such a striking achievement that many contemporary commentaries on 19th-century technology viewed the Brooklyn Bridge as one of the wonders of the age.

The bridge combined three unique features, later emulated in numerous 20th-century bridges: underwater piers for foundations, stone towers, and steel suspension cables. Roebling was the first to advocate the use of steel cables rather than iron. In 1964 the Brooklyn Bridge was declared a national historic landmark.

steel-tipped pen

Writing with ink requires a readily held pointed instrument, and papyrus reeds apparently were used by Egyptians and fine brushes by the ancient Chinese. The use of bird feathers or quills has been dated in Europe to the 7th century A.D. and may have been prevalent earlier. Penknives were small knives used to sharpen quills. Fountain or reservoir pens were introduced in the 17th century but had many defects, including leaking, clogging, and problems of refilling. Handcrafted pen tips of silver, gold, and brass were in wide use by the 17th century.

In the 19th century, machinemade steel pen tips were introduced. In 1828 John Mitchell in Birmingham, England, produced a steel pen tip, and another design was manufactured in France in the same year. In 1830 another British inventor, James Perry, redesigned the pen tip. Lewis Edson Waterman, an American salesman, invented a practical version of the reservoir pen in 1884 that used capillary action and successfully went into business making them. According to legend, he was inspired to invent the reservoir pen when a fountain pen leaked all over a contract with a client that was about to be signed. His first pens were filled by a plunger or eyedropper, and later improvements by 1913 led to a lever that compressed a rubber reservoir and released it, to suck in the ink. The **ballpoint pen** [VI], invented by the Hungarian brothers Ladislas and Georg Biro during World War II, represented a further improvement.

stethoscope

In 1816 the French physician René Théophile Hyacinthe Laënnec (1781–1826) invented the stethoscope. In 1819 he publicized the tool and the method of diagnosis in a book, *Traité de l'auscultation médiaté,* which was widely adopted. According to his account, he thought of the method while observing children playing by tapping on a long piece of wood and listening to the sound at each end. He experimented with paper tubes, and then in his own woodshop, he developed a tubular wooden stethoscope. Later improvements led to flexible binaural tubes.

Laënnec served as a physician to Cardinal Fesch, an uncle of Napoleon, and was chief physician at Necker Hospital before retiring in 1818. He was briefly professor of medicine at the Collège de France in 1822, but he died in 1826 of heart failure.

The stethoscope, with its extremely simple principle, revolutionized the diagnosis of heart ailments and soon became a standard part of the medical tool kit.

Stirling engine

A Scottish clergyman, Robert Stirling (1790–1878), invented the Stirling engine in 1816. The Stirling engine design is an external-combustion engine, as distinct from the later **internal-combustion engines** and the **diesel engine** [V]. Stirling's concept was much simpler. Wishing to develop an engine that would not require a boiler, since boiler explosions in **steam engines** [III] were common causes of fatalities, he conceived of an engine that would use heat to alternately heat, compress, and cool a gas such as air.

The Stirling engine consisted of a piston that would move up and down in a cylinder. External heat would be applied to the upper wall of the cylinder, causing the fluid or gas there to expand, driving the piston down. Beneath the piston, colder fluid would be driven through connecting tubes back to the top of the cylinder, to be heated and expand again.

Versions of the Stirling engine were constructed and used in a variety of applications from about 1820 to 1920 but fell out of favor. Nevertheless, since they held the potential for extremely low emission of exhaust gas, the concept was revived, particularly by the N. V. Philips Company of the Netherlands. In the late 20th century, interest in the Stirling engine as a possible improvement for automotive transportation grew. Engineering improvements brought its weight per horsepower in the range of regular gasoline internal-combustion engines, but with vastly reduced environmental emissions. However, the engine never went into full-scale production in the United States.

streetcars

Streetcars or trolleys had their origins in horse-drawn street railway trams that were introduced in the United States and European countries in the 1830s. A horse-drawn carriage on rails could carry twice as many people for the same horsepower as the prior "omnibuses," or carriages open to everyone (who paid the fare) that traveled on dirt or paved roads. With rails, the ride was smoother and friction was reduced. The first horse-drawn tram to replace omnibuses in the United States was introduced in New York City in 1832, soon followed in many smaller cities and towns throughout the nation.

Similar routes became popular in Paris in the 1850s and in Britain in the 1860s. In the 1870s, several cities experimented with steam-driven systems, but in general, the **steam engine [III]** was too heavy for the light tramlines installed. One solution that caught on, particularly in hilly cities such as Seattle and San Francisco, was the cable car, devised by Andrew S. Hallidie. The San Francisco system was inaugurated in 1873, and parts of it survived into the 21st century. The bell-ringing system to communicate between the ticket taker and the gripper was modeled on similar systems used to communicate between captains and engine-room personnel on early steamboats. The cable cars operated with a central steam-powered engine pulling a cable underneath the tracks, to which the cars would attach and travel at steady speeds, stopping by releasing the cable and using friction brakes.

As these systems were installed, the first electric-powered street railways were built, the first in a suburb of Berlin by Ernst Werner von Siemens (1812–1892), demonstrated in 1879 and introduced in 1881. The Siemens system at first used one metal rail to supply the electric current and the other rail to return the current. These systems soon proved dangerous, especially when a crossing vehicle or a pedestrian touched both rails simultaneously. Siemens soon substituted a system with a fish rod–type arrangement that would draw current from an overhead line. This *troller* gave the electric streetcar its name as *trolley*.

Such systems soon proliferated in American and European cities, replacing the older horse-drawn trams rapidly. In the United States, Frank J. Sprague (1857–1934) developed a swiveling trolley pole that gave the streetcar more flexibility in rounding corners, installing his first system in Richmond, Virginia, in 1888. Systems utilizing the Sprague device proliferated in the first decade of the 20th century. By 1917 there were almost 45,000 miles of streetcar track in the United States, and passenger carriage on these lines peaked from 1919 to 1920. However, with the advent and growing popularity of automobiles, the streetcar, with its fixed route and constant stops, proved a bottleneck in urban traffic, and usage began to decline. Some streetcar lines supported the building of entertainment parks on the outskirts of cities to draw travelers, and in many cities, the existence of streetcar lines running out to country locations stimulated the development of streetcar suburbs, forerunners of the automobile commute–based bourgeois utopias of later suburban sprawl. As with other forms of early-20th-century transportation, the streetcar became an icon of urban nostalgia.

submarine cable

The first international underwater **telegraph** cable was laid in 1845 across the Strait of Dover, between Britain and France. The first cable was insulated with raw **rubber** (gutta-percha). A second cable the same year used a copper core and had thicker insulation. However, the first underwater cable had been laid earlier, by Samuel Morse (1791–1872), between Governor's Island and Castle Garden, New York, in 1842. A later Morse cable connected the 13 miles between the Nantucket and Martha's Vineyard islands off Massachusetts.

In 1854 Cyrus Field, who had successfully retired as a paper merchant at age 35, set up the New York, Newfoundland, and London Electric Telegraph Company, with plans to lay a cable across the Atlantic. The first link, from Newfoundland to the mainland of North America, was successful but had cost more than $1 million. Field then organized in Britain the Atlantic Telegraph Company. The first transatlantic cable, laid in 1855 under the supervision of the noted physicist William Thomson (Lord Kelvin) (1824–1907), worked only a few weeks before failing. However, Thomson supervised the laying of a more durable cable in 1866, using the large steamship *Great Eastern*. After several failures, the design of grappling instruments to recover a broken cable from the ocean floor captured public imagination as a great success. Cyrus Field was on hand to send the first message from Newfoundland to Britain announcing the linkup. Transatlantic communication was regularly under way after July 27, 1866.

telegraph

Although the American artist Samuel Morse (1791–1872) is usually credited with the invention of the telegraph, that invention, like many others, was a case of near-simultaneous work in several countries, and several different pathways to the solution of the same problem. Although Morse conceived the idea as early as 1832, the first working line using principles he developed was not built until 1844, connecting Baltimore and Washington, D.C.

The problem of communicating information over a distance had been addressed with many primitive systems, including visual signals such as smoke or light from relay stations among Native Americans, to drums audible at distances of several miles in Africa and the Brazilian forest. During the French Revolution, Claude Chapp introduced a system of **semaphores,** mounted 6 to 10 miles apart on hilltops, to relay signals. Such a system had been suggested more than a century previously by Robert Hooke (1635–1703).

The Chapp system, once installed, was quite practical. Signals from Paris to Lille, a distance of 130 miles, could be read within 2 minutes. Napoleon made great use of information relayed by the semaphore system. In 1809 he learned of the occupation of Munich by the Austrian army on the same day it happened, and he marched troops within 6 days in a successful counterattack for his Bavarian ally. The king of Bavaria, impressed, asked Thomas von Sömmering to improve on the Chapp semaphore telegraph.

Sömmering began work on an electric telegraph system, using a voltaic **battery** system, and constructed a working model. Improvements to the electric telegraph system followed fairly rapidly after the discovery in 1819 by a Danish physics professor, Hans Christian Oersted (1777–1851), that an electric current could be used to deflect a needle. Using that principle and inspired by the Sömmering device, Carl Friedrich Gauss (1777–1855), in Göttingen, Germany, set up a short telegraph line from the university there to an astronomical observatory 2 miles distant in 1833. That system was used to transmit technical and scientific information, using a code to interpret minute deflections in an iron bar.

Gauss did not attempt to make a practical commercial system, but he did anticipate that such a system could be established. A Gauss student, August Steinheil, improved on the Gauss system with a device that would mark printed dots on a paper band. The Steinheil system was established in Munich in 1837. A Russian diplomat, Baron Paul Schilling, was impressed with the early Steinheil system and worked on improvements to it, demonstrating his own system at a scientific meeting in Bonn in 1835. William Cooke (1806–1879) took the concept to Britain, where he worked with Charles Wheatstone (1802–1875) to develop systems, the first of which Cooke and Wheatstone patented in 1837. Although Wheatstone and Cooke sometimes disputed the primacy of their invention, they cooperated and installed an electric telegraph with a single needle-reading device along the railway line from London's Paddington station 19 miles to Slough, in 1844. That system was patented in 1845.

All of these developments preceded the development by Samuel Morse, in the United States, of a different system. Morse was a fairly successful portrait painter, but after the death of his wife, he temporarily gave up painting and traveled to Europe to recover his spirits in 1832. In France he witnessed a demonstration of the electromagnet by Professor André Marie Ampère (1775–1836), and aboard ship on his return to the United States, he speculated on the possibility of using such a device as a telegraph. By the time of Morse's first concept of the telegraph in 1832, a working model had already been constructed by Joseph Henry (1797–1878) in the United States.

Morse took a position teaching art at New York City University and there began experiments with magnets and wires. He had no previous experience or training in the emerging field of electricity, and his early models were quite primitive. Nevertheless, he succeeded in establishing a system that would transmit a signal a short distance. In the United States, Joseph Henry had worked with electromagnets and early **electric motors and electric generators.** Morse learned of Henry's development of an electrical relay in 1835 that would take a weak signal and then resend it along a line. Henry invented the relay to improve on his own version of a telegraph, developed in 1831. With relays, the 1- or 2-mile limit to signals encountered by Morse could be indefinitely increased.

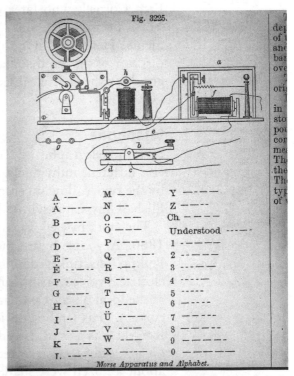

telegraph Samuel Morse was credited with the invention of the telegraph by Americans, but the invention came out of near-simultaneous work in several countries. This early version of his code included vowels with umlauts as well as a symbol for "ch." *Library of Congress*

Morse worked with a student of his, Alfred Vail, and on September 4, 1837, Morse successfully transmitted a message, using an existing U.S. Navy code. He and Vail developed an improved code that relied on long and short signals—the dot-and-dash system. Assigning the simplest codes to the most frequently used letters of the alphabet in English, his Morse code was perfected by January 1838.

However, it took Morse and Vail several years to get a commercial line established. Hoping to receive support, Morse had a bill introduced into the U.S. Congress to provide financing. The bill finally was enacted in 1843, and the Washington-to-Baltimore line was opened in May 1844, about the same time as the Paddington-to-Slough line in Britain that used the Wheatstone-Cooke system.

Soon attempts were made to lay a **submarine cable,** with a successful connection across the English Channel in 1851 and a temporary connection from Britain to the United States later in the decade. Morse grew wealthy from his invention, as did his supplier of copper wire, Ezra Cornell, who later endowed Cornell University in Ithaca, New York. Both Cooke and Wheatstone were knighted for their achievements. Henry did not follow up on his telegraph work but became director of the Smithsonian Institution and fostered the use of the telegraph by the U.S. Weather Bureau.

telephone

Like the **telegraph,** the telephone was simultaneously invented by several different individuals. In this case, the filing of patent information by the two leading inventors, Alexander Graham Bell (1847–1922) and Elisha Gray, came within hours of each other.

The work by Bell was inspired by an earlier invention, dating back to 1860. A young high school physics teacher in Frankfurt, Germany, Philipp Reis, tinkered in the school workshop, attempting to develop an electrical voice transmission system. He gave a lecture in October 1861 at the local physics association on "telephony by means of the galvanic current." His sending device was a bladder over a wooden cask, and the receiver was a needle inside a violin body. After publication of a report on his lecture, he received requests for copies of the instrument, and about a dozen were made in the shop of a local Frankfurt workman. There remains a dispute as to whether the Reis device could transmit a clear version of the human voice. The Reis telephone was described in a popular German magazine, and one copy of the device was on display at Edinburgh University. The Scottish-born Alexander Graham Bell was studying at Edinburgh in 1862, and he became enthralled with the concept that such a device could be used to assist the hearing-impaired. Bell visited Charles Wheatstone (1802–1875), one of the inventors of the **telegraph,** and discussed technical advances that might be applied to the telephone.

Moving to Boston, Bell engaged in training teachers of the deaf, and his future father-in-law, Gardiner Hubbard (1822–1897), helped fund his experiments with equipment. In 1874 Bell engaged an assistant, Thomas Watson, and the two quietly tried to perfect improvements to the Reis device. In 1875, at the Smithsonian Institution, Bell visited Joseph Henry (1797–1878), another inventor of the telegraph, who encouraged him to continue his work.

Hubbard filed Bell's application for a patent for the telephone on February 14, 1876 (issued March 7, 1876). Two hours after Hubbard filed the Bell application, Elisha Gray filed papers with the patent office indicating he had already begun work on such a device and intended to file a patent, only then discovering that the Bell documents had already been deposited. On March 10, 1876, Bell succeeded in transmitting from the attic to the ground floor of his home a line that later became famous, "Mr. Watson, please come here, I want you." Bell displayed his invention in 1876 at the Centennial Exposition in Philadelphia, where it drew the attention of the emperor of Brazil, Dom Pedro II, and then the attention of a wider public.

The Western Union Telegraph Company purchased Gray's patents and then hired Thomas Edison (1847–1931) in 1876 to work out improvements to the telephone that would not require infringing on the Bell patent. Included in Bell's patent suits was a dispute with Emile Berliner, who had improved on the original Reis device with a carbon microphone and a transformer, patented in 1880. Western Union and Bell resolved their patent differences in 1879, but a continuing set of patent fights with others were not resolved in Bell's favor until 1888.

Scientific American published a front-page feature on the telephone in October 1877, and the director of the German Post Office, Heinrich Stephan, sought to acquire a telephone set. He soon obtained one from the manager of the London Telegraph Office and installed sets in Berlin and the suburb of Potsdam. There Ernst Werner Siemens (1816–1892) observed the system and recognized an opportunity, since Bell had not secured a German patent. The Siemens company soon mass-produced telephones, and they spread rapidly in Germany.

The first telephone exchange was set up in New Haven, Connecticut, in 1878, with others soon established in London, Manchester, and Liverpool. In 1889, Almon Strowger patented a dial system for connecting phones that could replace operators, but such systems were slow to be installed, as the expense of replacing existing manually operated switchboards stood in the way. Several cities in the United States led the way with dial systems, including La Porte, Indiana, in 1892, and San Francisco after the 1906 earthquake destroyed the earlier systems. Siemens installed Europe's first automatic exchange in 1909, in Munich.

Bell became a U.S. citizen in 1882. He left a large portion of his fortune to promote the teaching of the deaf and was a founder of the National Geographic Society and its president from 1898 to 1903, succeeding his father-in-law in that position.

The telephone revolutionized communication, and by the 1890s and the first decade of the 20th century it had become essential to the conduct of business in many major cities of the world.

thermocouple

In 1821 the German physicist Thomas Seebeck (1770–1831) discovered that when a loop of metal is constructed of two different metals and heated, a small electric current is generated in the direction of the heat flow. Later called the Seebeck effect, this discovery led to a number of devices, primarily the thermocouple.

A thermocouple can be used to measure heat very precisely by measuring the electric current generated from the bimetallic strip. Although various metals and alloys can be used, the most accurate thermocouples for measuring temperature were made from platinum and platinum alloys. One convenient aspect of a thermocouple is that the temperature of a vessel or object can be read at a distance by running an electrical wiring system from the thermocouple to the analog or digital gauge.

During World War II, a thermocouple heated by a small kerosene lamp was used to provide power for a field radio.

During the age of space travel, the Seebeck effect was used to make **radio-thermal generators** [VI] or RTGs, which could provide power aboard long-distance-traveling space probes. A radioactive **isotope** [V] with a long decay period would be used to generate heat, and a bimetallic loop would generate electricity to power radios and other equipment aboard the spacecraft.

torpedo

The term *torpedo* derived from the name for a type of electric fish and was first applied to what are now called mines—explosive devices moored or free-floating that explode on contact with a ship. In the 1860s, spar torpedoes were introduced, including one attached to the **submarine** [V] *Hunley,* operated by Confederate seamen who sank the USS *Housatonic.* All the attacking sailors perished in the engagement.

In Austria, a Captain Luppis introduced the concept of a self-propelled boat that would be guided by long lines and carry a charge. The system proved unworkable, and he sought improvements to the concept from Robert Whitehead (1823–1905), a British manager of a marine engine factory in Fiume, Austria (now a port in Croatia). Whitehead improved the concept, developing an automobile torpedo—one that was self-driving. A compressed-air engine that could travel a few hundred yards drove his weapon. He continued to work on the device, and by 1868 he was able to demonstrate a successful torpedo that had a range of 1,000 yards and that would travel at 7 knots. He gradually increased the underwater speed, achieving 29 knots at ranges of up to 1,000 yards. The British employed a Whitehead torpedo in action in 1877. In the United States, John Adams Howell (1840–1918) developed a self-steering model of the torpedo in 1885 and was the first to introduce a balance system using a gyroscope in torpedoes, later emulated by Whitehead in 1896.

Self-propelled or automobile torpedoes saw continual improvement through the 20th century, with the development of air-dropped torpedoes in the 1930s (successfully used by the Japanese against moored battleships at Pearl Harbor in 1941) and acoustic-homing torpedoes in the 1940s. Later improvements included longer range and faster torpedoes.

typewriter

The typewriter was invented by Christopher Latham Sholes (1819–1890), an American newspaper editor who had served briefly in the Wisconsin legislature. Although several other inventors had attempted to develop a writing machine, Sholes's 1868 patent is generally recognized as the starting point for the development of the typewriter. Earlier attempts possibly included a writing machine patented in Britain in 1714 by an engineer named Henry Mill. No drawings or plans for that device survive, and it may have been more in the nature of a stencil. In 1829 William A. Burt of Detroit patented a printing machine that required that the letters be rotated by hand and that the keys be inked by hand. In 1845, Charles Thurber developed a similar rotating-alphabet printer, which was a bit faster. An 1850 patent issued to John Fairbank included paper on a continuous roll, and another 1850 patent was issued to Oliver T. Eddy for a machine that used an inked ribbon. However, Sholes's patent and later improvements to it were the ancestors of the 20th-century typewriter.

Sholes worked with a machinist, Samuel W. Soulé, and together they patented a page-numbering machine in 1864. Another machinist, Carlos Glidden, noted an article in *Scientific American* about an effort to develop a writing machine in Britain and brought it to the attention of Sholes and Soulé. Sholes and the two machinists developed their own writing machine, working on improvements and patenting it in 1868. After seeking funding from several sources, including Thomas Edison (1847–1931), who turned him down, Sholes secured backing from another newspaper publisher, James Densmore. Densmore suggested improvements, and Soulé and Glidden backed out of the business. After several further improvements to make the machine smaller, lighter, and more workable, Densmore negotiated a manufacturing agreement with the Remington Arms Company in 1873. To prevent the keys from sticking, Sholes developed a keyboard that spaced the most commonly used letters apart from each other, and his *qwerty* keyboard became a long-lasting convention.

Remington had difficulty marketing the device, but after he gave away a number of the machines, the typewriter began to catch on in the late 1870s and 1880s. The Sholes machine typed the letters on the underside of the roll or platen, so the resulting text was not visible until rolled out, a so-called blind typewriter. In 1892 Rev. Thomas Oliver invented an improved machine with slanted keys that struck to a common visible printing point. In 1893 Franz X. Wagner patented another machine that, like the Oliver machine, typed the letters at a visible point in front of the operator. John T. Underwood purchased the Wagner patents and formed the Underwood Typewriter Company in 1895. Several machines through the 1890s had two keyboards, with separate upper- and lowercase alphabets. In 1904 L. C. Smith Company introduced a machine that allowed for shifting between lowercase and capital letters using a single keyboard. Later improvements included electric typewriters introduced in the 1920s by the Woodstock company and a design adapted by International Business Machines that moved a ball with letter typefaces across stationary paper on each line.

One of the social impacts of the typewriter was the opening of a new class of jobs to women in the late 19th century. At that time, secretaries were nearly all men, but they disdained the newly invented mechanical writing machinery, perhaps regarding it as insulting to their abilities. Women were hired as operators of typewriters, and for a period, a typewriter operator was also known as a *typewriter.* By the first decade of the 20th century, increasing numbers of women found employment in this new role, partially contributing to the political and social emancipation of women over the following decades.

water turbine

The water turbine represented a great improvement over the **waterwheel** [II]. The concept of a turbine or enclosed waterwheel with greater efficiency than either a **steam engine** [III] or a waterwheel was under serious discussion in the late 18th and early 19th centuries as a source of power for the growing Industrial Revolution. In France, Claude Burdin, a professor in the School of Mines at St. Étienne in Loire, proposed the concept, and developed the term *turbine* from the Latin *turbo,* for whirlpool.

Benoît Fourneyron (1802–1867) studied under Burdin. Fourneyron began work on the concept in the early 1820s and by 1827 demonstrated the first practical turbine, designated as an outward-flow turbine. He continued to make improvements on the efficiency of the turbine, and in 1833 he won an award of 6,000 francs from the French Society for the Encouragement of Industry. The water passed through guide passages in a wheel, striking fixed wheel vanes in an outer ring and escaping out the periphery of the wheel, driving the mechanism at more than 2,000 revolutions per minute. Fourneyron installed more than 1,000 turbines around the world, including applications in the United States and Europe. He developed a number of other inventions, including a rolling mill, mining equipment, and a system for lubricating bearings within turbines.

With the development of electric power systems in the 20th century, following the development of the **electric light,** water turbines along the Fourneyron design lines became essential to the conversion of the force of falling water into electricity. When attached to a generator, the turbine was key to the production of hydroelectric power.

PART V

THE ELECTRICAL AGE, 1891 TO 1934

The 1890s and the first decades of the 20th century saw the introduction of some extremely popular new consumer items. A whole list of products changed the day-to-day life of the average person: safety razors, aspirin, thermos bottles, electric blankets, breakfast foods, cellophane, and rayon. To this array were added a constant flood of improvements to earlier inventions that also found their way into daily life, such as the camera, the typewriter, the phonograph, the refrigerator, and most of all, the automobile. The automobile, a novelty in 1895, was widespread by 1910, and it led to gas stations, traffic lights, and highway systems. The social impact, particularly in the United States, was profound and dramatic. By the 1930s, the automobile had changed the commercial landscape, with motels and highway cabins, roadside diners, fruit stands, drive-in theaters, and the beginnings of coast-to-coast highway travel.

As the generation born in the 1880s and 1890s grew to adulthood, they were stunned by the rapid changes, particularly impressed by the automobile, the radio, the airplane, and plastics. Like their grandparents, who had witnessed profound 19th-century transformations brought on by the Industrial Revolution, the first 20th-century generation frequently remarked on the positives and negatives of technical progress.

Life was more comfortable but faster. Cities became gradually cleared of the flood of horse manure, but soon the streets were choked with cars and their fumes. Automobile accidents began to claim hundreds, then thousands, of victims. Information flowed more quickly across the cable lines and later over the radio, but much of the news was itself depressing, bringing the horrors of war, political tyranny, revolution, and natural disaster to the morning breakfast table. While eating corn flakes or shredded wheat, the average citizen learned of the Japanese defeat of Russia in 1905, the progress of a naval armaments race between Britain and Germany unlike any the world had ever seen, and then the horrors of trench warfare in World War I. Even before radio was introduced, phone companies experimented with using telephone lines to carry news, speeches, and entertainment over the networks as a type of broadcast system.

The frontiers of scientific discovery moved into new areas of physics and chemistry in the 1890s and the first years of the 20th century, opening some new and startling possibilities. While Einstein's theory of relativity represented a theoretical explanation for the relationship of space and time, rather than the discovery of new laws of nature, related findings such as Planck's constant, the Heisenberg principle, and the nature of the atom and its subatomic particles suggested that some of the certainties of Newtonian physics were no longer valid.

As in the past, the pathway between invented technological tool and discovered scientific principle was a two-way street. The Crookes tube, which had tested the effect of electrical leads in a glass-enclosed vacuum in the 1880s, spun off into a number of parlor-demonstration devices. Then suddenly, in the 1890s, a burst of scientific discoveries poured out of experimentation with the Crookes tube toy-turned-tool. Within a decade, using cathode ray tube variations on Crookes's invention, scientists discovered radioactivity, radium, polonium, X-rays, and the electron. Such new discoveries in the realm of physics, including a deeper understanding of the nature of isotopes, the electron, and the nucleus of the atom, began to pave the way for the Atomic and Electronic Ages that would follow in the later years of the 20th century. And perhaps even more important for the average person, the original Crookes cathode ray tube, with some modification, provided the ideal television screen, already in experimental commercial use by the early 1930s.

The short period from the 1890s to the 1930s was one of great inventiveness, but many of the inventions that had the greatest impact on daily life were improvements over existing machines (as with electrical lighting and internal-combustion engines) or simply the ingenious creations of people with good ideas, such as the work that led to the zipper, the brassiere, and the parking meter. Such useful, everyday items could have been made decades earlier with existing tools and materials, but they had to await the matching of need, idea, and the perseverance of individual inventors.

The times had changed, and new terms to define the era seemed needed. In many ways this was an Electrical Age, with the wider spread of such electric-based inventions as the telephone, the phonograph, the electric light, the electric refrigerator, and other household appliances. As Edison had realized, a network of electrical supply systems was needed to make electric light viable, and then with the networks, vast new markets for other devices quickly followed for the office, home, and factory. Many of the new products were already available in quite workable form in 1890, but with readily available electric power, their manufacture and sale became big business, with General Electric and Westinghouse leading the way in the United States and Siemens in Germany.

New weapons and ships and vehicles to carry them gave the wars of the early 20th century a different technological character than those of the 19th century, as war moved off the surface of the planet above to the air and beneath to the sea. Navies adopted the submarine, destroyer, and battleship. Radio and sonar brought early electronic devices to warfare. With older inventions, such as the machine gun and barbed wire, World War I (1914–1918) was a chaos of human slaughter. The tank, dirigible, and airplane were used in that war to devastating effect.

With the establishment of electrical distribution systems in the 1890s, a wide variety of household appliances soon reached the market, as this 1907 advertisement attests. *Library of Congress*

The new technologies of war struck deep in social impact. World War I nearly destroyed a whole generation of Europeans, leaving the survivors as the "Lost Generation," disillusioned, bitter, and alienated. Out of the alienation came a burst of creativity and modernism in art, literature, and music.

Ongoing technological changes gave the cultural products of the Lost Generation specific qualities. With the independence and freedom to travel provided by the automobile, relations between men and women altered in obvious and subtle ways. Religious leaders were shocked that the younger generation saw the automobile as a mobile bedroom. Employment of women in office and sales work and their

Chronology: The Electrical Age, 1891 to 1934

Year	Invention/Discovery	Inventor/Discoverer
1890	subway trains	multiple inventors
1890s	radio/wireless	Guglielmo Marconi
1890s	destroyer	multiple inventors
1890s	submarine	Simon Lake 1897; John Holland 1898
1892	escalator	Charles D. Seeberger; Elisha Graves Otis 1899
1892	reinforced concrete (earlier patents)	François Hennebique
1892	thermos bottle	James Dewar
1893	carburetor	Wilhelm Maybach
1893	diesel engine	Rudolf Christian Karl Diesel
1894	breakfast cereal	John Kellogg and William Keith Kellogg
1894	argon	Lord Rayleigh and William Ramsay
1895	motion picture	Lumiere brothers and Thomas Edison
1895	X-rays	Wilhelm Roentgen
1895	X-ray machine	multiple inventors
1896	Lorentz transformation	Hendrick Antoon Lorentz
1896	radioactivity	Antoine Becquerel
1897	electron	J. J. Thomson
1898	krypton	William Ramsay and Morris Travers
1898	neon	Ramsay and Travers
1898	polonium	Marie Curie and Pierre Curie
1898	radium	Marie Curie
1898	tape recording	Valdemar Poulsen
1898	xenon	Ramsay with Travers
1899	actinium	André Debierne in the Curie laboratory
1899	aspirin	Felix Hoffmann
1900	dirigible balloon	Ferdinand Zeppelin
1900	photon	Max Planck
1900	Planck's constant	Planck
1900	radon	Friedrich E. Dorn; atomic weight by William Ramsay, 1904
1901	instant coffee	Sartori Kato
1902	ionosphere	Oliver Heaviside
1902	photoelectric cell	Arthur Korn
1902	safety razor	King Camp Gillette
1903	airplane	Orville and Wilbur Wright
1903	oxyacetylene welding	Edmond Fouché and Charles Picard
1904	caterpillar track	Benjamin Holt
1904	diode	John Ambrose Fleming
1904	fax machine	Arthur Korn
1904	rayon (term, 1889)	Samuel Courtauld
1906	Bakelite	Leo H. Baekland

Chronology: The Electrical Age, 1891 to 1934 *(continued)*		
Year	**Invention/Discovery**	**Inventor/Discoverer**
1906	radio transmission of voice (first broadcasting station, 1920)	Robert von Lieben
1906	triode valve (tube)	Lee de Forest
1906	vitamins (named by 1919)	Frederick Gowland Hopkins
1906	battleship	multiple inventors
1907	gyrocompass	Elmer Sperry
1909	electric charge of electron	Robert Millikan
1911	superconductivity	Heik Kamerlingh-Onnes
1912	stainless steel	Harry Brearley and Elwood Haynes
1912	cellophane	Jacques Brandenberger
1912	electric blanket	Sidney I. Russell
1912	loudspeaker	de Forest and Edwin Armstrong
1913	zipper	Witcomb L. Judson
1914	traffic lights	multiple inventors
1915	catalytic cracking of petroleum	Eugène Houdry
1916	brassiere	Mary Phelps Jacob
1916	tank	Ernest Dunlop Swinton
1918	sonar	Paul Langeuin and Constantin Chilowski
1921	lie detector	John Larson
1922	aircraft carrier	multiple inventors
1922	insulin	Frederick Sanger
1922	isotopes	J. J. Thomson
1923	hearing aid	Marconi Company
1923	television	Vladimir Zworykin
1926	liquid-fueled rocket	Robert Goddard
1927	uncertainty principle	Werner Heisenberg
1928	Geiger counter	Hans Geiger
1928	Raman spectrography	Chandrasekhara Venkata Raman
1930	deep freezing (foods)	Clarence Birdseye
1930	Pluto	Clyde W. Tombaugh
1930	freon	Thomas Midgley
1932	parking meter	Carl C. Magee
1932	neutron	James Chadwick
1933	electron microscope	Ernst Ruska

enfranchisement in Britain and the United States signaled a liberation, represented by changes in clothing and hairstyles and noticeably different behavior and manners. Well-brought-up young ladies were now seen smoking cigarettes in public, driving their own cars, and wearing types of clothing that would have shocked their mothers two or three decades earlier.

Technology had played a role in bringing those social changes and would continue to shape the ways in which the changes were felt. The Jazz Age denoted the rise of a new music and a new culture that went with it. In the 1920s, a young generation of people of all races began to identify with the music of the alienated or marginalized population of African Americans. The blues, deeply rooted in black culture, moved into the mainstream, not by accident, but because the music resonated with the national and international mood of youth. The phonograph record, the radio, and motion pictures all caught the changing tempo of the times and reflected it from the masses to the creative elites and back to the masses. Music moved out of the parlor and the amateur gathering as the radio and record industries created the 20th-century phenomenon of professional popular music.

Politics and international affairs also were reshaped by technology. The great ideological conflicts between the flexible democratic and capitalist systems of North America and western Europe and the authoritarian Fascist and Communist regimes that began to take shape in the Soviet Union, Italy, and Germany were at first fought with the tools and instruments of communication media before being fought with the arms of war. Thus the cheap newspaper, set on linotype machines, with telephoto (or facsimile) pictures from all over the world, carried the news, advertising, and propaganda of the day, as did the motion picture and radio broadcasting. Mass politics still incorporated some earlier methods of arousing popular support, such as the parade, the torchlight gathering, and the public address to huge crowds, but all such events were made more dramatic with the new technologies of floodlighting, recorded music, and the amplified sound of loudspeakers. And when filmed or broadcast, a political rally in Nuremberg or Rome could impact millions, not just thousands. Some of the great demagogues of the era twisted such techniques to sinister purposes: Benito Mussolini in Italy and Adolf Hitler in Germany built their followings with loudspeakers, mass meetings, and radio and through leaflets, newspapers, and cheaply printed books. In the United States, master politicians such as Franklin Roosevelt and religious leaders such as Billy Sunday and Father Charles Coughlin became adept at using the radio to convey their messages.

Some of the popular heroes of the day were men and women who contributed to the new technologies or used them in new and striking ways. Thomas Edison, Henry Ford, and the Wright brothers became heroes of invention; Madame Curie and Albert Einstein earned fame as scientific geniuses; Charles Lindbergh and Amelia Earhart, as aviation greats. Radio, film, and recorded music created flocks of new media celebrities, sports and film idols, crooners and divas, including Jack Benny, Babe Ruth, Charlie Chaplin, Rudolph Valentino, Enrico Caruso, and Bessie Smith.

Despite such bright diversions and mass impacts, technology held a darker threat. By the 1930s, scientists could foresee that the weapons of the future, when they tapped into the energy of the atom, could be even more terrible than those of World War I. The Atomic Age was about to be born in the mid-1930s, with the discovery of the neutron and further experimentation with radiation. The science fiction and the science predictions of the 1930s became the science fact and the technological realities of the next decades.

aircraft carrier

No sooner had heavier-than-air **airplanes** been developed than naval officers began to consider the advantages that would derive from ship-launched planes. The first successful launch of an aircraft from aboard a ship was on November 14, 1910, with a test flight from a temporary wooden deck installed on a U.S. Navy cruiser. A few months later, the same pilot, Eugene Ely, landed a plane aboard the cruiser *Pennsylvania*. In 1918 the British HMS *Furious* was the first aircraft carrier to launch planes, a few Sopwith Camels, into military action. In 1919 the U.S. Navy converted the coal-supply vessel *Jupiter* into an experimental craft designed to carry and launch aircraft. Renamed *Langley*, it was commissioned March 20, 1922.

The Japanese built the first aircraft carrier from the keel up, *Hosho*, in 1922, and the first U.S. aircraft carrier built for the purpose was *Ranger*, commissioned in 1934. The destruction of **battleships** in Pearl Harbor by Japanese carrier-launched aircraft increased U.S. reliance on aircraft carriers,

aircraft carrier   First developed at the end of World War I, aircraft carriers evolved over the next decades into the largest warships afloat. Here the nuclear powered U.S.S. *Enterprise* launches A-4 Skyhawk bombers. *National Archives and Records Administration*

particularly in the Pacific Theater of World War II. During the war the United States launched 23 fleet carriers of the Essex class, displacing about 36,000 tons, with a top speed of 33 knots.

After World War II, the introduction of **jet engine aircraft [VI]**, with requirements for longer landing decks, led to the construction of super-carriers. The British introduced the concept of the angled or canted deck, together with a steam-powered catapult system for launching aircraft. Four carriers of the Forrestal class, built between 1951 and 1959, had decks more than 1,000 feet long. The *Enterprise,* launched in 1960 and commissioned in 1961, was powered by eight **nuclear reactors [VI]**. In addition to nine carriers of the Enterprise class, the United States built six more of the Nimitz class, each powered by two nuclear reactors.

Other varieties of carriers have been introduced, including a smaller version, known as escort carriers, used in World War II in anti**submarine** warfare. The United States built 79 escort carriers in World War II, while the nations of the British Commonwealth together had more than 40. An escort carrier was generally slower than a fleet carrier, with a top speed of less than 20 knots. Several light carriers were built on hulls originally designed for cruisers and could achieve generally higher speeds than fleet carriers.

airplane

The invention of the airplane or, as it was originally known, the *aeroplane,* like that of the **automobile [IV]**, represented the culmination of a human technological dream of long standing. Leonardo da Vinci (1452–1519) had worked on plans of an aircraft with flapping wings, producing at least 500 sketches between 1486 and 1490. The technology of developing such a large-scale *ornithopter* capable of carrying a pilot was not achieved by the beginning of the 21st century. In 1804, George Cayley (1773–1857), a British baron living in Yorkshire, developed and flew a fixed-wing glider with a wingspan of more than 3 feet. Otto Lilienthal (1848–1896), a German mechanical engineer, carried out numerous successful manned glider flights between 1891 and 1896, dying on August 10, 1896, from a crash in one of them.

Samuel Pierpont Langley (1834–1906), secretary of the Smithsonian Institution in Washington, worked on powered flight and successfully flew a steam-powered but unmanned craft he called an *aerodrome.* Other experiments included a steam-powered aircraft built by Hiram Maxim (1840–1916), the inventor of the **machine gun [IV]**, and another steam-powered craft that flew some 900 feet, built by a French engineer, Clé-mont Ader (1841–1926), both in the 1890s.

Wilbur (1867–1912) and Orville (1871–1948) Wright began work in 1896 on the problem of powered flight, studying the existing literature. They built a number of models they tested from 1900 to 1901 at Kitty Hawk and Kill Devil Hills, North Carolina. Finding the results unsatisfactory, they decided to develop their own data and built a small wind tunnel to conduct aerodynamic tests of wings. They developed the wing, which they called an *aeroplane,* the term later used to identify all heavier-than-air winged aircraft. They filed the patent application for the design of their *Flying Machine* on March 23, 1903, well before their first flight, and the patent was issued May 22, 1906.

The wing had a curved upper surface, and relying on **Bernoulli's principle [III]**, the forward motion of the wing through a stream of air provided lift. Two propellers mounted to the rear of the wing, and turning in opposite directions to prevent torque twisting the aircraft in flight, provided the forward motion. They built their own lightweight (152 pounds) gasoline-powered **internal-combustion engine [IV]**, rated at 12 horsepower, for propulsion. The design they developed, with two wings

airplane At Kitty Hawk, the Wright brothers experimented with air flight in 1903 but did not publicize their work for several years. *Library of Congress*

Be it known that we, Orville Wright and Wilbur Wright, citizens of the United States, residing in the city of Dayton, county of Montgomery, and State of Ohio, have invented certain new and useful improvements in Flying-Machines, of which the following is a specification.... Our invention relates to that class of flying machines in which the weight is sustained by the reactions resulting when one or more aeroplanes are moved through the air edge-wise at a small angle of incidence, either by the application of mechanical power or by the utilization of the force of gravity.

—Wright brothers on the aeroplane. As they used the term, *aeroplane* referred to the wing of the flying machine. From patent application 821,393, Flying Machine. Filed March 23, 1903.

mounted one above the other and connected with struts and wires—a biplane—resulted in the 1903 *Wright Flyer.* Their first successful flight was December 17, 1903. The Wright brothers continued to experiment in relative secrecy and finally began public demonstrations in 1908. A great many experimenters followed, some hoping to compete for a prize of £1,000 offered by the London newspaper *Daily Mail* for the first successful flight across the English Channel. Louis Blériot collected it with his flight of July 25, 1909, in a single-wing plane of his own design.

A bitter dispute over the patents of the Wright design took place, with Langley insisting that his designs represented priority of invention. As a consequence, the Wright brothers never grew wealthy from their invention, but they were widely hailed as the pioneers of manned, heavier-than-air flight.

Between 1903 and 1927 the biplane design dominated airplane manufacture. Studies revealed that the struts and wires, together with the relatively stubby wings, produced drag. A longer single wing, together with other improvements, such as streamlining, reduced air resistance. In particular, in 1928, the National Advisory Committee for Aeronautics (NACA) in the United States introduced the concept of enclosing the engine in cowling and providing a streamlined shape, significantly reducing air resistance. NACA cowling and other measures improved air design through the 1930s. It became apparent to aircraft designers in that decade that the airframe no longer provided the limitation on upper speed, and that with piston engines, aircraft would reach an upper speed limit of about 400 to 450 miles per hour. To achieve higher speeds, a new propulsion system would be required, and investigators in Germany, Britain, and Austria simultaneously worked on rocket and **jet engine aircraft [VI]** to produce the next breakthrough.

The impact of the airplane was most immediately felt in the nature of warfare. Aircraft used to scout enemy positions, to provide for artillery spotting, and to drop bombs on enemy positions soon encountered improved aircraft designed as fighters. Between the two world wars, the development of long-range bomber aircraft and high-speed fighters presaged the major role that such craft would have in future warfare. Air transportation of mail, cargo, and passengers also began in the interwar era, with Pan American World Airways pioneering in the development of transoceanic flights of large amphibious *clippers* that could alight on the surface of harbors.

aspirin

Nineteenth-century chemists knew that salicylic acid could be used to reduce pain, but the acid was harsh, burned the throat, and caused stomach distress. In 1853, Charles F. Gerhardt, a French chemist, was able to synthesize acetylsalicylic acid. In the search for **anesthetics [IV]** during the mid-19th century, the acid was regarded as a minor but possible alternative.

However, the true invention of aspirin as a medicine is credited to Felix Hoffmann, who in 1897 developed an improved method of synthesizing the drug. Hoffmann developed the process while working for Friedrich Bayer and Company. The first patent was issued to that company, of Elberfeld, Germany, in March 1899, for the synthetic acetylsalicylic acid. Bayer began to market acetylsalicylic acid in 1899 under the trade name *Aspirin*. Hermann Dreser recommended its use in medicine, and it immediately became the best-selling and most genuinely beneficial of the many "patent medicines" on the market.

At the end of World War I, as part of the reparations agreements by Germany with the Allies, the trade name passed into common usage, and in both Britain and the United States, the name *aspirin* became the generic name for the drug. However, Bayer retained the trade name in many countries. A dentist in Wisconsin invented a variant of the product, *Anacin,* in 1918.

Aspirin has several effects that are not well understood. It is able to reduce fever by acting on the hypothalamus in the central nervous system of the brain. It apparently reduces the temperature by increasing the body's blood flow and increasing the process of perspiration. It also has an effect in the relief of pain, with a selective depressant effect on part of the brain. It has also proven effective in reducing intense muscle contractions during labor by interfering with the production of naturally produced body chemicals known as prostaglandins. In addition, the daily use of aspirin in small quantities has been suggested to reduce the risk of heart attacks and strokes. A vast quantity of aspirin is sold worldwide, with more than 30 million pounds consumed annually in the United States alone, usually for the relief of minor pains and headaches. As it is relatively harmless, and at the same time quite effective, it has great advantages over more narcotic and habit-forming painkillers such as morphine.

Aspirin is often held out as an example of a medicine or drug whose effects and benefits are well known but for which detailed scientific explanations have been extremely difficult to establish.

Bakelite

Often called the first commercial plastic, Bakelite was invented in 1904 but had been preceded by **celluloid** [IV] in 1868. Bakelite was the product of a Belgian-born chemist who became a U.S. citizen, Leo Hendrik Baekeland (1863–1944), who developed the plastic from formaldehyde and phenol. Bakelite was the first "thermosetting" plastic, one that after having been cooled was extremely hard and resistant to high temperatures. He developed a three-stage process involving the use of bases rather than acids to cure the product and using extremely high temperatures to mold and set it. Thus the name for the product represented a pun on both his name and the process for manufacture. He published his findings in two supplements to *Scientific American* in November 1909. The new plastic soon found many uses, particularly for electrical insulators.

Baekeland had previously developed a type of photographic paper in 1893, and he sold the rights to the paper in 1899 to the Kodak Company for $1 million. He patented Bakelite plastic in 1909 and then founded the Bakelite Corporation in 1910 to produce the powder from which the plastic could be manufactured. He used his own funds and those of a few friends to launch the company, and it later became a part of the Union Carbide and Carbon Company.

battleship

The term *battleship* derives from the 18th- and 19th-century concept of a "line of battle ship" or "ship of the line," referring to the European naval strategy of lining major ships so as to present their sides to the enemy, allowing the most effective deployment of side-mounted **cannons** [II]. In the mid- and late 19th century, European navies experimented with ironclad, wooden-hulled ships with **steam engines** [III] and sail power. The first ironclad was the French oceangoing *Gloire,* launched in 1859 and soon followed in the United States by many experiments with iron cladding of wooden hulls. The first fully iron ship was the British *Monarch,* launched in 1869.

As heavier warships evolved through the period, navies introduced steel ships by the 1880s. Mounting guns in armored side sponsons or on rotating armored turrets on the deck changed tactics. With forward and astern firing weapons, ships no longer had to present their sides as targets to the enemy, and the ship-of-the-line concept began to vanish. The term *battleship* became more generic. Although small by later standards, the battleships of the 1890s tended to be all-steel and all-steam (that is, steel-hulled and driven solely by steam engines, with no sails), mounting a mixed armament of large-caliber weapons firing shells up to 10 inches in diameter. Such ships usually carried smaller armament as well, with barrels of 5 inches or so in diameter to clear the decks of enemy ships and even smaller weapons to fend off patrol boats and torpedo boats. Smaller armored vessels, dedicated to the destruction of torpedo boats, were known as *torpedo-boat destroyers,* the origin of the small, fast **destroyer** that later proved effective in anti**submarine** warfare. Faster, heavily armed, but lighter and less armored ships were classed as cruisers.

However, in 1906, with the British introduction of the steam turbine–powered *Dreadnaught,* the term *battleship* began to be restricted to the heaviest class of armored ships that were armed only with heavy, long-range guns. That development established a new definition of *battleship* that continued through the rest of the 20th century. In fact, other navies

battleship The USS *Wisconsin* being nudged by tug to the Nauticus Museum in Norfolk, Virginia, in December 2000. The 20th-century battleship was introduced with the British *Dreadnaught* in 1906. Vulnerable to attack from the air, battleships were used sparingly after World War II and fully retired from the U.S. Navy in the 1990s. *U.S. Department of Defense* sometimes referred to their battleships in the next decade as *dreadnaughts*. The original *Dreadnaught* was equipped with ten 12-inch guns. Smaller, heavily armed, but faster battleships were sometimes designated *battle cruisers*.

In 1922, the five major naval powers signed a treaty at the Washington Conference that restricted battleships to 35,000 tons. Over the next decade, fast battleships with speeds of more than 30 knots, with heavy armament, and limited to 35,000 tons dominated the class. However, with the rising international tensions of the 1930s, the Washington treaty limits were abandoned. Germany built two ships of about 50,000 tons (the *Bismarck* and the *Tirpitz*), while the United States built four of the 58,000-ton Iowa class. That class of ships mounted 16-inch guns, and the ships were 887 feet long. The Japanese *Yamoto* and *Musashi* were 72,000 tons, and although two more of that class were under construction during World War II, one was converted to an aircraft carrier and the other was utilized for spare parts. The Japanese also introduced 18-inch weapons in the Yamoto class.

With the sinking of two American battleships and severe damage to three others by air attack at Pearl Harbor and the sinking of a British battleship and battle cruiser off Malaya in December 1941, the vulnerability of large ships to air attack was clearly demonstrated. Within three days, USS *Arizona* and *Oklahoma* and HMS *Prince of Wales* and *Repulse* were sunk. By necessity, the United States began to fight the Pacific portion of World War II with greater reliance on **aircraft carriers**. Although the air wing of the U.S. Navy claimed that the day of the battleship had passed, battleships saw action in the Korean War, from 1950 to 1953, and the *New Jersey* saw action in the Vietnam conflict in the 1960s. Battleships were not fully retired from the U.S. Navy until the 1990s.

brassiere

The origins of the brassiere and the word itself for the garment are not entirely clear. Since ancient times, women have used various clothing arrangements to support the breasts. A "patent bust improver" was advertised in Britain in 1902. *Vogue* magazine carried an advertisement for a brassiere in 1907.

The first person to patent the brassiere was Mary Phelps Jacob. According to her own account, she became disgusted with dressing in the complex and restraining **corset** [II] of the period and worked with her maid to make a simpler undergarment out of two handkerchiefs and some ribbon. She then showed her garment to a few friends, who asked for copies. She patented the "backless brassiere" in 1914 and went on to produce several hundred. However, she lost interest and sold the patent to the Warner Brothers Corset Company for $15,000. That firm made more than $15 million from sales over the next few years and remained a major producer of the garment into the 21st century. The Warner company introduced the system of cup sizes—A, B, C, and D—in the 1930s. The word *bra* first appears in print in 1937.

The term itself is something of a mystery. One theory is that the name derived from the French word *bras,* for arm. With the introduction of similar products in Europe and the United States, it appears that several inventors were working on an alternative to the corset in the first decade of the 20th century.

The brassiere was widely adopted in the 1920s, and along with the public smoking of **cigarettes [IV]**, and driving of the **automobile [IV]**, wearing brassieres rather than corsets became a symbol of the liberation of women in that decade.

breakfast cereal

Three different breakfast cereals were invented between 1893 and 1903, all growing out of a health-food movement that swept the United States at the time. The first was developed in 1893 by Henry D. Perky, an inventor in Denver, Colorado, who developed a machine to shred wheat and make it into pillow-shaped biscuits.

As part of the interest in healthier foods, Dr. John Harvey Kellogg (1852–1943) conducted many experiments with a variety of items after noting as a medical student that a light breakfast of fruit and grains left him more energized than the typical high-fat breakfasts of the day. In 1876 he took charge of the Adventist Health Reform Institute in Battle Creek, Michigan, and renamed it the Battle Creek Sanitarium, purposefully changing the spelling from the established *sanitorium*. He invented Granola (now a generic term) in 1877 and a substitute coffee and experimented with peanut butter and various meat substitutes. Kellogg was assisted by his brother Will Keith Kellogg (1860–1951) as both administrator and food experimenter.

In 1894 the Kellogg brothers produced flaked wheat, which they served at the sanitarium, and soon patients began to request the product after returning home. By 1898 the Kelloggs were producing a flaked wheat product. However, the two brothers split up, and in 1903 Will Kellogg established the Kellogg Toasted Corn Flake Company, which added malt, salt, and sugar to the corn.

One of the patients at the Kellogg sanitarium was Charles W. Post (1854–1914). Post had stayed at the sanitarium in 1891 and suggested that he help promote the coffee substitute. When the Kelloggs did not support him in that business, he established his own firm in 1895 and produced "Postum Cereal Food Coffee," a product that was an immediate success. In 1897 Post developed a bread stick made of wheat and malted barley. Believing that grape sugar was produced in the product and noting its nutty flavor, he dubbed the product Grape-Nuts. Thus, by 1903, shredded wheat, Grape-Nuts, and cornflakes were all available as food products. Post's company eventually grew into the General Foods Corporation.

From 1903 to 1913, dozens of other breakfast foods came on the market, most of which did not survive. In 1904 Post introduced a wheat-flake product he first dubbed "Elijah's Manna," but when religious leaders objected, he changed the name to Post Toasties.

Four basic types of dry or ready-to-eat breakfast cereals emerged from this burst of enterprise: flaked, puffed, shredded, or granular. Cooked cereals that had been on the market earlier included rolled oatmeal and wheat farina. Consumption of all types of dry cereal vastly increased in the 1950s as the manufacturers launched advertising campaigns to encourage mothers of the baby-boom generation born after World War II to serve cereal to children. As sugars and other flavorings were added to the cereals, numerous small competitors and the major companies began to revive the origins of the dry cereal with an emphasis on health benefits, fewer additives, and natural flavors.

carburetor

The carburetor to mix liquid fuel with air for the **internal-combustion engine [IV]**, making possible the use of liquids such as kerosene or gasoline (naphtha) rather than gases such as **coal gas [III]** as the fuel, was invented in about 1893 by Wilhelm Maybach (1846–1929). His first design was quite simple, with two screws, one of which would control the amount of air, and the second, the amount of liquid. The problem with the design was that only one richness of fuel could be set for the engine until the operator stopped the car and readjusted the machine. Carl Benz solved that problem in 1893 with the development of a butterfly valve below the spray nozzle in the throat or intake of the carburetor that would allow constant variation in the mixture. The throat was designed with restriction, relying on the Venturi effect (explained by the **Bernoulli principle [III]**) to create low pressure to draw the fuel through a jet. Another innovation was the introduction of a closing valve or choke above the spray nozzle to reduce the air supply, allowing a temporary extremely rich mixture for starting. With modification, the choke could be linked by a heat-sensitive system to automatically allow a leaner mixture (with a higher proportion of air) as the engine warmed up, although in early models, the choke was adjusted by hand.

Prior to Maybach's carburetor, **automobile** [IV] engines designed by Benz and Daimler had used engines that ran on illuminating gas or coal gas. With the carburetor design and the use of more readily handled liquid fuel, sales of automobiles began to increase and the market for petroleum products quickly altered. Whereas naphtha or gasoline had been used primarily as a cleaning fluid, it suddenly became the fuel of choice, and **catalytic cracking of petroleum** to increase the produced ratio of gasoline from crude oil became ever more crucial in the early years of the 20th century.

Maybach's carburetor was widely copied in other engines, and the Daimler company brought patent suits against infringers. However, a British court ruled that a gasoline-powered tricycle developed in 1884 by Edward Butler preceded the spray-fed carburetor principle and that Maybach's was not the "master patent" in the field. Butler's original "petrocycle" was never widely marketed, and the Maybach carburetor, with improvements, became the leading design.

catalytic cracking of petroleum

When crude oil is obtained from its underground reservoirs, it represents a mixture of a great number of hydrocarbon compounds. Even before the first successful **oil drilling** [IV] in the United States in 1859 in western Pennsylvania, chemists recognized that through a simple process of distilling, various hydrocarbon mixes could be boiled off from the crude oil at different temperatures. This process of thermal cracking could separate "high end" distillates with lower boiling points, such as naphtha, gasoline, and kerosene, from "low end" products such as heating oil, heavier oils for lubrication, and asphalt. With the development of the **internal-combustion engine** [IV] relying on the Otto cycle and the **carburetor,** the demand for gasoline vastly increased from 1895 to 1910. Using thermal cracking, the mix of hydrocarbon compounds that together represented *gasoline* generally came to about 20 percent of the total yield from crude oil. During and after World War I, the search for any method that would increase that proportion intensified.

catalytic cracking of petroleum This depiction of a modern petroleum refinery was produced by a computer-drawing program. *National Aeronautics and Space Administration*

In 1915, A. M. McAfee developed a method that used anhydrous aluminum chloride as a catalyst in refining, and the Gulf Refining Company installed it in 1916 in Texas. However, the catalyst remained workable for only a few hours, and no convenient way to regenerate the catalyst was found. In Germany and the United States several researchers looked for a method of using metal oxides as catalysts. Eugène Houdry, a French engineer, examined a process developed by E. A. Proudhomme, a pharmacist from Nice, to make gasoline from coal. Houdry established a laboratory and worked with Proudhomme to perfect the method and to study catalysts. Houdry selected a catalyst made from activated clay, which had been used in bleaching lubricating oil. His method held promise for increasing gasoline production.

After several unsuccessful attempts to sell his equipment to various refining companies, Houdry eventually negotiated with Sun Oil Company, which licensed the patents and agreed to finance further work. In 1935 Sun Oil Company worked with Socony (Mobil), and the latter company began using the Houdry process to manufacture higher-octane aviation gasoline. Standard Oil Company later improved on the Houdry process, setting up a fluidized bed system of exposing the catalyst, which allowed for continuous operation.

During World War II the demand for high-octane gasoline, rated at 100 octane, was so great that the U.S. petroleum administrator for war (PAW) set up a system of priorities and distribution so that all refineries in the United States had to share additives and methods. Some crude oils (including most California crudes) were rich in higher-octane petroleum. Some refineries had developed methods for enhancing octane. Furthermore, the PAW set a policy allowing refineries to add more tetraethyl lead to gasoline (temporarily up to 6 cubic centimeters per gallon) to increase its octane rating. Through all of these methods, an increased supply of *av-gas* or aviation gasoline was achieved for the duration of the war. In the postwar years, automobile companies began to produce more engines that required high-octane gasolines.

caterpillar track

The concept of a vehicle that would lay its own continuous track was suggested as early as the 18th century. In 1770 a British inventor, Richard Edgworth, patented a "portable railway." Various special-purpose vehicles using self-laying tracks included a 19th-century log-hauling machine in Maine and an 1888 system with tracks designed by F. W. Batter.

However, Benjamin Holt designed the first successful commercial vehicle using a self-laying track system in 1904 and sold his first Holt Tractor in 1906. Meanwhile, a British design, using lubricated links for the track, was designed by David Roberts, and sold by Ruston Hornsby and Sons in 1904. The Hornsby company demonstrated the system for the British War Office in 1907. Both the Holt and Roberts tracks consisted of a continuous loop of metal plates linked by hinges, inside which wheels supported the vehicle. The loop of track was propelled forward by a geared drive against the inside of the loop.

The Roberts-Hornsby patents were sold to the Holt company in 1912. Meanwhile, the Holt company continued to manufacture its "caterpillar"-tracked tractors, and they appealed as the basis for the **tank** during World War I. Holt continued to sell caterpillar-tracked farm tractors, with about 500,000 in use by the mid-1920s. In 1925 Benjamin Holt's company merged with the C. L. Best Tractor Company to form the Caterpillar Tractor Company (known after 1986 as Caterpillar, Incorporated). One of Caterpillar's most successful early sales was some $18 million worth of construction equipment to the Soviet

Union used in building hydroelectric dams and other major construction projects in the early 1930s.

cellophane

Cellophane was patented in 1912 by a Swiss-born French chemist, Jacques E. Brandenberger, using cellulose softened with glycerol. He had worked since about 1900 to find a coating for tablecloths to make cotton more readily cleaned. By 1908 he tried working with a form of cellulose treated with sodium hydroxide in water; the combination had been developed by Frederick Cross and Edward John Bevan. After experimentation with processes to make the material more flexible and thinner, he finally concluded that the thin sheet material he developed would be more readily marketed as a separate product. Later court decisions over Brandenberger's patents concluded that his complete process, including the use of glycerol and a slotted machine to produce thin sheets, entitled him to credit for the invention.

Since the material was impermeable to dry gases, it found immediate uses in packaging and in eyepieces for gas masks during World War I. The largest French manufacturer of **rayon,** Comptoir de Textiles Artificiels, agreed to finance Brandenberger's work and formed a special company, La Cellophane, for its production. In 1923, the American firm of du Pont, which had worked with Comptoir de Textiles Artificiels in production of rayon, set up a separate firm, du Pont Cellophane Company. The French firm transferred the patents, manufacturing processes, and know-how to the American company, which began producing cellophane in 1924. In 1926 two du Pont researchers, William Hale Charch and Karl Edwin Prindle, developed a moistureproof cellophane. Du Pont patented the process and then licensed the production to Sylvania Industrial Corporation, where production of the new cellophane began in 1933.

deep-freezing food

Frozen food had been used for centuries as a means of preservation, and in the Arctic, native peoples have been aware that frozen meat will keep for months if not thawed. In Britain in 1842, H. Benjamin developed a method of freezing food in salt brine, and in 1861, frozen fish were introduced in the United States. With the development of **refrigeration** [IV] in the 19th century, freezing of foods for shipment became common, with meat-freezing plants in Australia, Argentina, New Zealand, and the United States for local and international shipment.

However, deep-freezing of food for retail sale is a 20th-century development. Clarence Birdseye observed Eskimo freezing of foods in Labrador in 1923, learning that quick-freezing of food reduces the amount of ice crystals in the food, allowing the thawed food to better retain its original flavor and color. Birdseye experimented with various methods, freezing rabbit meat and fish in a refrigeration plant in New Jersey by 1924. By 1930 to 1933, he began to sell "quick frozen foods," so called to distinguish them from foods that had been frozen in cold storage. He improved the process with machines to run the food packages through freezing units on hollow conveyor belts and continued to research different methods for different foods. By regulating the speed of the belts through the freezing units, the amount of time exposed to low temperature could be varied precisely. During World War II, shortages of fresh meats and vegetables led to widespread consumer acceptance of frozen food. By 1945, airlines began to experiment with whole frozen dinners, and by 1954 the so-called TV dinner was introduced.

destroyer

Originally built in the 1890s to counter the automobile **torpedo** [IV] that could be launched from small patrol boats, the fast torpedo-boat destroyer was unarmored, carried weapons in the 4- and 5-inch-diameter range, and achieved high speeds exceeding 30 knots. The flexibility of these ships soon made them adaptable for other uses, most notably in World Wars I and II, when they served as escort ships to protect merchant shipping from **submarine** attack. Equipped with hydrophones that allowed the operator to detect the direction of emitted underwater sound, and later with active **sonar** and **radar** [VI], the destroyer could serve as a defense against both submarine and aerial attack. In the United States, larger destroyer vessels were classed as frigates, while smaller destroyer types were designated as destroyer escorts.

The high speed and flexibility of the destroyer extended its use to minesweeping, radar picket, pilot recovery, **helicopter** [VI] operation, and gunfire support of amphibious warfare. Many destroyers were made, with some 240 of the four-stacker model produced between 1918 and 1922. In 1940, the United States traded 50 of the four-stackers for long-term leases on British airbases in the Caribbean,

destroyer The guided missile destroyer USS *McFaul* of the Arleigh Burke class, on training in 1999. Descended from torpedo-boat destroyers of the 1890s, the destroyer class of ships has proven adaptable to many uses. *U.S. Navy*

Bermuda, and Canada. The destroyer-escort vessels, whose main purpose was antisubmarine warfare and convoy escort duty, were rapidly built during World War II, with some 450 turned out in a few years.

More than 240 of the full-scale Fletcher and Allen M. Sumner destroyer classes were built during World War II, in the 2,000-to-2,200-ton range. In general the destroyer evolved from its single-purpose beginning to be a high-speed, lightly armed, and lightly armored ship with a crew of about 300, capable of serving a widely varied set of missions.

diesel engine

In the early 1890s, Rudolf Christian Karl Diesel (1858–1913) sought to design an engine that would approach the ideal engine suggested by Nicolas Sadi Carnot (1796–1832). In the 1820s, in developing some of the basic concepts of the **laws of thermodynamics [IV]**, Carnot had conceived of a notional engine that would operate at 100 percent efficiency, converting all consumed energy to work. By the standard of the Carnot engine, the typical steam engine operated at about 7 percent efficiency. Diesel sought to more closely approach the Carnot ideal standard.

Diesel worked as engineer at the **refrigeration** [IV] works of Paul Gottfried von Linde (1842–1934), moving from the Paris to the Berlin works of the company in 1890. There he obtained development patents on his proposed engine in 1892 and 1893 and published the concept in *Theory and Construction of a Rational Heat Motor.* With support from the Krupp works and others, he began to produce models of the engine, demonstrating in 1897 a 4-stroke, 25-horsepower engine. Under his original design, the engine would operate at a very slow rate, less than 200 revolutions per minute, and the intake air would be compressed above the ignition point of the fuel, which would be injected gradually into the combustion chamber during the power stroke of the piston. He experimented with various low-cost fuels, including powdered coal, eventually settling on light heating oil.

By compressing the air in a ratio of 25 to 1, the temperature could be raised to 1,000 degrees Fahrenheit. One advantage of the engine was that with spontaneous ignition of the fuel from the heat of compression, no complex electrical ignition system was needed to keep the engine running. Furthermore, with low-grade fuel oil, the engine was safer than a gasoline engine, especially in the confined volume of a ship or boat engine space, where the fuel fumes from more volatile gasoline fuel presented a severe safety hazard. Because of the necessity of high compression pressures, the engines had to be heavier and more solid than the four-cycle Otto gasoline engine. For equivalent power, the diesel also was more expensive than a gasoline engine.

Although the fuel efficiency of Rudolf Diesel's engine exceeded any other design at the time, it was not immediately adapted to automotive transportation. Diesel insisted that his original design of slow injection of fuel be used, limiting the speed of the engine. Despite the gradual adoption of the design, Diesel soon became very wealthy from royalties from applications in shipping and locomotives.

Diesel died in 1913, apparently accidentally falling from the deck of a steamer crossing the English Channel. During World War I his engine was adapted by German engineers for use aboard **submarines,** and in the interwar years, the power-to-weight ratio was sufficiently reduced and the speed of diesel engines improved to make it practical for a wide variety of applications, including **automobiles** [IV], trucks, railroad engines, and even **airplane** use. In the 1990s, design of improved compression-ignition, direct-injection (CIDI) engines advanced in the United States and Europe in an effort to further improve fuel economy.

differential analyzer

The differential analyzer or integraph was invented by Vannevar Bush (1890–1974). An electromechanical device later classed as an analog **computer** [VI], it weighed more than 100 tons and consisted of 150 motors and hundreds of miles of wire connecting **diodes** and electrical relays. Bush constructed a preliminary model of the integraph in 1930 at the Massachusetts Institute of Technology, and the first full-scale working model was completed in 1935.

Eventually some 19 differential analyzers were built for military use and for research. The device was used to solve differential equations and operated about 100 times faster than a human using a desktop **adding machine** [III]. By the standards of later, fully electronic computers, the device was primitive and slow, but it vastly improved the ability of engineers and physicists to achieve approximately correct answers to some mathematical questions arising in the course of research. Vannevar Bush later went into government service, establishing the Office of Scientific Research and Development, which initiated work on **nuclear weapons** [VI] before the project was taken over by the Army Corps of Engineers in the Manhattan Engineer District.

The term *analog computer* was only later developed to refer to the class of instruments that used mechanical devices to register numerical values of other phenomena, offering an indication of a number as an analog of the effect being registered or represented. By this broad definition, such diverse items as the indicator dials on bathroom scales and automobile speedometers are analog devices.

diode

John Ambrose Fleming (1849–1945) invented the diode when he patented what he called the *thermionic valve* in 1904. The device used the fact that a heated plate in a vacuum could emit and absorb electrons, and like a valve, it could be used to control the flow of electrons. It consisted of two metal poles, an anode and a cathode, encased in a glass tube from which the air had been evacuated. His valve became better known as a diode (because of the two metal poles) or, in the United States, as a *vacuum tube*.

When the cathode, usually made of cesium, was heated, it would release electrons, while the anode was made of a metal that did not release electrons when heated to the same temperature. When a negative current was applied, the cathode would emit electrons that would flow to the anode; but when a positive current was applied, the anode would not emit electrons. When subjected to alternating current of positive and negative charges, only a single, negative charge would flow. Thus the diode served as a "rectifier," converting alternating current to direct current. It was this principle that Fleming saw as valvelike, in that the diode would cut off flow in one direction but allow it in the other.

The diode could be used to replace the crystal in **radio** sets. Radio waves in alternating current could be converted to direct current for amplification by certain piezoelectric crystals that were known to transmit electricity better in one direction than in the other. However, crystals were inefficient at higher frequencies. With the diode valve, higher-frequency broadcasting and receiving became possible.

In 1906, the American Lee de Forest (1873–1961) improved on the diode with an amplifier grid, making it a **triode valve,** or, as he called it, an *audion,* but he was unable to secure clear rights to the invention, as the Marconi Company controlled the Fleming patent. The triode valve allowed both rectification and amplification of the signal in the same tube. Later, the invention of the **transistor** [VI] allowed the replacement of the vacuum tube diode with a solid-state rectifier, sometimes still referred to as a diode.

dirigible

The dirigible, or metal-framed lighter-than-air balloon, was a direct descendant of the **hot-air balloon** [III] first developed by Joseph and Étienne Montgolfier in 1783 and a contemporary hydrogen-filled balloon developed by the French scientist Césare Alexandre Charles.

After a series of unsuccessful attempts to develop powered balloons, several worked out. In 1852 Henri Giffard developed a cigar-shaped, steam-driven dirigible airship, probably the first successful dirigible. He concluded that a larger size would be more practical, but he went blind and committed suicide before developing that aircraft. An American inventor, C. E. Ritchell, built a steam-driven dirigible in 1878, and the Tissandier brothers in France tried an electrical propulsion system, powered by batteries, in 1883.

Several enthusiasts tried to apply gasoline-powered **internal-combustion engines** [IV] to lighter-than-air aircraft in the 1890s. One inventor, David Schwarz, from Zagreb, Croatia, developed a small airship powered by a 16-horsepower Daimler engine that was demonstrated in 1897, after his death. Among those witnessing the demonstration was the German count Ferdinand Zeppelin, who had already begun experiments with lighter-than-air machines. In July 1900

Zeppelin demonstrated a 420-foot long dirigible, which he called *zeppelin* at Lake Constance on the Swiss-German border. He continued to experiment, raising funds from the general public with lotteries and collections. Zeppelin and others often used **aluminum** [IV] and aluminum alloys for the structural members of the aircraft. Despite the accomplishments of predecessors, Zeppelin is usually regarded as the inventor of the dirigible, with the date of 1900.

The German military developed an interest in Zeppelin's work and, with the outbreak of World War I, requisitioned all existing zeppelin aircraft for dropping bombs over Britain. About 480 German airmen died, as many of their craft were shot down. Among the zeppelins shot down was one the British used as a model for design of their own craft. That ship, the *R-34*, built in Britain, was the first aircraft of any kind to cross the Atlantic Ocean, in July 1919. The Zeppelin works continued to produce airships, making some 116 between 1900 and 1926. In the 1920s and early 1930s, zeppelins operated on regular air-travel routes, carrying passengers between Europe and North and South America. In May 1937, the *Hindenburg* caught fire as it was landing at an airship port in Lakehurst, New Jersey, with the loss of 33 lives. At the time, most observers assumed that the fact that the ship was filled with hydrogen led to the fire, but later study of the catastrophe has revealed that the fire was due to the flammable metallic coating on the skin of the airship.

Several other accidents, including the crash of the British *R-101* in France on its way to India, and the U.S. Navy's *Shenandoah* and *Akron*, convinced both the public and governments that dirigible travel was not as safe as heavier-than-air **airplanes**. As a consequence, the dirigible alternative was not pursued as a means of passenger transport after the 1930s.

electric blanket

An American physician, Sidney I. Russell, developed the concept for an electric blanket in 1912. He made a heating pad that worked by running an electric current through insulated metal tape sewn inside a square patch of blanket fabric. It was the practice then for patients with tuberculosis to sleep in the open air, and Russell developed his pads to provide heat in bed. Some patients stitched the pads together, leading to a form of blanket for warming a whole bed. The product developed slowly, with a British Thermega underblanket manufactured in the mid-1920s. In the 1930s some British and American companies began to market electric blankets, with electric overblankets introduced in the United States in 1937.

In World War II, the need to provide electrically heated thermal suits for aircraft personnel at high altitudes led to some research into safety procedures and to the use of plastic to cover the electric elements. Continuing concern with safety led to the introduction of thermostatic systems that cut off the electricity when a blanket got too hot. In about 1970, the British Thermega company introduced a blanket heated by circulation of warm water.

electric chair

No sooner was a regular supply of electricity available than it was used for execution of criminals. The first known use of an electric chair was in the United States in 1890, in Auburn, New York, to execute a condemned murderer. In a system that spread widely to more than 20 states, electric chairs were simply wooden armchairs into which the condemned person was strapped. One electrode would be strapped to the head and another connected to one leg, and a current of 2,000 volts or more would be applied in bursts of half a minute or more. Usually the prisoner would be declared dead within a few minutes.

electric charge of the electron

In 1909 an American physicist at the University of Chicago, Robert A. Millikan (1868–1953), determined the electric charge of a single **electron** by experiment. In the oil drop experiment, drops of oil are slowly expelled into a metal chamber containing air. The weight of a specific droplet is determined by measuring its rate of fall against air resistance. If the droplet has an electric charge, it can be stopped by the application of an electric charge to the metal box, creating a magnetic field. The amount of electric force required to stop the droplet can be registered. Through repeated experiments, Millikan discovered that the charge of a droplet was always an integer times a basic unit of electrical charge. The absolute value of the electric charge for one electron was determined to be 1.602×10^{-19} coulomb. In 1923 Millikan won the Nobel Prize in physics for this work and for studies of the photoelectric effect.

electron

The discovery of the electron in 1897 by Joseph John ("J. J.") Thomson (1856–1940) was one of several that sprang from follow-on work stimulated by the discovery of **X-rays** by Wilhelm Roentgen in 1895 and by the development of the **Crookes tube** [IV] or cathode-ray tube. J. J. Thomson investigated the behavior of electricity in gases in cathode-ray-discharge tubes. He became convinced that the cathode rays were extremely small, negatively charged particles. He measured the deflection of these particles in a magnetic field as well as the heat they generated, and he concluded that the mass of the particles was about 1,000 times smaller than that of hydrogen ions. By 1899 he determined that the **electric charge of the electron** was a natural constant.

Earlier work in Thomas Edison's laboratory at Menlo Park, New Jersey, during the testing of **electric light** [IV] bulbs in 1882 had resulted in a blue glow around the positive pole in a bulb and a blackening of the wire and glass near the negative pole. Discovered by a young researcher, William J. Hammer, the effect through a vacuum was first called Hammer's phantom shadow. After 1883, when Edison patented the bulb, the blue glow and blackening became known as the Edison effect. After Thomson's work, it became clear that the Edison effect resulted from the emission of electrons from one electrode to the other. Thus, although the electron had been detected by Hammer and in a different device, used by Crookes, Thomson made its discovery as a separate atomic particle. Later the **diode** electronic tube or valve developed by John Fleming (1849–1945), used in receiving radio waves, relied on the Edison effect.

By deflecting the cathode rays (or electron beams) with magnetic and electric forces, Thomson demonstrated that the particles were negatively charged and that they all had the same ratio of charge to mass. In later years the charge of an electron and its mass were precisely measured, with the mass calculated at 9.10939×10^{-31} kilogram.

Thomson originally described a model of the atom that was later discarded, sometimes called the plum pudding model, in which negatively charged electrons were embedded in a sphere of positively charged electricity. Even so, his discovery of the electron greatly advanced the study of atomic physics. He won the 1906 Nobel Prize in physics for his discovery.

By studying positively charged ions of gases, he later concluded that neon represented two different gases, leading to the discovery of **isotopes**.

electron microscope

Regarded by some scientists as the most important invention of the 20th century, the modern electron microscope allows magnifications of up to 2 million times. While other inventions have had much greater social and cultural impacts, the electron microscope has become a crucial tool in numerous scientific fields.

With the development of the **Crookes tube [IV]** and the discovery of the **electron,** scientists began to speculate by the 1920s that it would be possible to employ electrons instead of light to improve over the limits of the optical **microscope [III]**. Images derived from light were limited by the wavelength of light, an observation made by Ernst Abbe, a cofounder of the Carl Zeiss optical works, as early as the 1870s. French physicist Louis de Broglie suggested in 1924 that electrons, with a wavelength as small as a millionth that of light, could lead to a valuable increase in resolution.

Ernst Ruska (1906–1988), a German electrical engineer, built the first electron lens in 1931, using an electromagnet to focus a beam of electrons on the target material. In 1933 he developed a system in which the electrons would pass through a thin slice of the target material and then be deflected to film or to a fluorescent screen, producing an image that could be magnified. Although his first images were within the range of those that could be obtained by optical microscopes, he had opened the way to electron imaging. His device, a transmission electron microscope, required that the beam penetrate through the target. By the mid-1930s, commercial production of electron microscopes began in Germany, the United States, and Britain. Later, scanning electron microscopes developed by Heinrich Rohrer and Gerd Binnig used reflected beams to cast the image. In 1986 Ruska shared the Nobel Prize in physics with Rohrer and Binnig for the invention. The scanning electron microscope has the advantage of allowing the examination of larger targets and does not require slicing the specimen. A microscopic coating of a conducting metal is sometimes required to allow reflection of the electronic beam.

Vladimir Zworykin (1889–1982), known for his contributions to the development of **television,** improved on the electron microscope, allowing its use for the study of viruses and protein molecules.

By the end of World War II, electron microscopes were widely used in laboratories in the United States and Britain, facilitating research in areas of biochemistry, microbiology, medicine, material sciences, plastic manufacturing, and related fields. As a measuring tool, the electron microscope played a part in late-20th-century biological sciences similar to that played by the microscope in 17th- and 18th-century medicine and by the **telescope [III]** in 17th-century astronomy.

encryption machine

With the development of **radio** and its use in World War I to send Morse code messages, systems of enciphering and deciphering messages became extremely important to the military services of the warring powers. In Germany, Britain, and the United States immediately after the war, cryptographic offices worked to develop a mechanical or electromechanical method to encipher and decipher messages. In 1919 Hugo Koch in Holland patented an enciphering machine. In 1923, in Germany, Arthur Scherbius set up the Cipher Machines Company, which made a machine to which he gave the brand name Enigma. Dr. Scherbius hoped to sell Enigma to private companies for use in business correspondence. However, he found few buyers and soon went bankrupt.

At about the same time, an American, Edward Hugh Hebern, set up a company to make a similar encryption machine. The U.S. Navy considered his Hebern Electric Code Machine for acquisition, but Hebern's company also went bankrupt. Hebern provided several machines to the navy, and some of his ideas became incorporated in other machines. Despite a lengthy patent claim, he died in 1952 without ever winning his case. Scherbius and Hebern had independently invented similar devices from 1923 to 1925, confronted with the problem of mechanically encrypting radio messages, and used similar systems to achieve the goal.

In 1926 the German navy introduced a Scherbius-type Enigma machine that they used until 1934, which they then replaced with an improved "Enigma-M" design. The German army adopted an Enigma machine in 1928, and the German air force, the Luftwaffe, established in 1933, began using a third Enigma design. The commercial Enigma machine was no longer sold, but

encryption machine The German Enigma machine was supposed to produce unbreakable codes, but its system of transposing letters was cracked, first by the Poles in the 1930s and then by the British, using early analog computers at Bletchley Park. *National Security Agency*

the German military continued to use the name through the evolving designs.

The Enigma machine looked like a sort of **typewriter [IV]** or cash register, with a keyboard on a box. When the operator typed a letter on the keyboard, a light would come on under another letter on a display. As each letter was typed, a different letter was displayed and typed out. The electrical pathway from the keyboard to the output through electrical wiring could be changed each time by a series of drums and rotors. Each time a key was pushed, the rotors would turn, yielding a different code for the next letter position in a message.

The Germans believed their encryption system provided unbreakable codes, but British experts developed an early electromechanical **computer [VI]**, known as a Turing machine, for cracking the codes. Using the **mathematical principle of computing [VI]** developed by Alan Turing, the so-called *bombes* they developed would replicate the process of the Enigma machine in reverse, providing material that the British then carefully distributed for military and diplomatic use without revealing the source. The term *bombe* derived from the name of a European dessert, which the machines resembled, rather than an explosive bomb.

Later encryption machines used by the U.S. military relied on insertion of punched cards to provide separate keys for codes to be used for a single day. Before the advent of more advanced computers, such encryption machines were used into the 1970s.

escalator

The first patent for a moving staircase was issued in 1859, but no successful models were installed. In the early 1890s, two inventors—Jesse W. Reno and George H. Wheeler—independently developed different workable designs. Charles D. Seeberger further developed the Wheeler design, patented in 1892. The Reno design was an inclined belt with grooved tread, while the Wheeler-Seeberger design had flat steps, entered from the side, with an angled shunt that turned passengers off as they departed.

The Otis Elevator Company, established by the sons of the inventor of the **elevator [IV]**, Elisha Graves Otis (1811–1861), built demonstration escalators in the Otis headquarters in Yonkers, New York, in 1899. A Reno design was installed in a New York City elevated train station in 1900. Otis applied the name "escalator" to the device operated at the International Exhibition in Paris in 1900, and retained rights to the name until 1949, when it was judged to have moved into common language. Otis popularized the escalator with installations in department stores in several American cities. The company moved the Paris International Exhibition escalator to Gimbel's in Philadelphia, where it operated from 1901 to 1939.

The Otis company acquired the rights to both the Reno and Seeberger designs by 1922 and developed a design that merged both the flat step and the grooved tread concepts, with a slightly combed ramp that safely lifted the passengers' feet off the tread at the end of the ride. As escalators evolved, they tended to run either at slow speeds—about 90 feet per minute—for store applications, allowing customers to overlook merchandise, or more rapid, 120-feet-per-minute designs for transportation settings such as subways, other train stations, and airports.

facsimile or fax machine

The concept of transmitting a picture over wire was first developed by a Scottish clockmaker, Alexander Bain, in 1842. His system used a combination of synchronized electrical pendulums. Several improvements to the system were developed by the 1870s, but the concept remained a novelty until 1904. In that year Arthur Korn, an American inventor, was able to transmit the first photograph by wire. Between 1904 and 1910 a number of newspapers in Paris, Berlin, and London began exchanging photographs, and in 1924 RCA developed a method of sending photographs by radio.

These early-20th-century facsimile systems were known as *wirephotos,* and by the 1930s had become the standard way for newspapers to exchange photographs. Wire service organizations and companies such as the Associated Press used them to send pictures to subscribers around the world. The machines used a **photoelectric cell** to detect picture elements or *pixels* as dark or light, then convert them to electric signals that could be sent by wire. The receiving machine would send an electrical charge to mark coated paper. In addition to newspaper use, these facsimile systems found applications in the transmission of weather condition photographs and photographs from **X-ray machines** and by police departments for the transmission of fingerprints and criminal photographs. In the late 1930s some news organizations expected to be able to publish newspapers that would be delivered directly to the home by fax machine, but such plans were made obsolete in the postwar years by the development of **television.**

In 1966, the leading manufacturer of **copy machines** [VI], Xerox Corporation, combined efforts with Magnavox to market a facsimile terminal for business use. Between 1966 and 1972 a number of other companies began to participate in the growing market, including Minnesota Mining and Manufacturing Company (now 3M Company). With about 30,000 machines in the United States, problems of compatibility among the technologies soon emerged. In addition, the machines were quite slow, taking up to 6 minutes to transmit a single page. Those systems using thermographic technology suffered from the problems of light, fading images, and brittle paper as well.

In 1962 Dr. Robert E. Wernikoff experimented with another method of transmitting signals, using a system that transformed the images into a digital record. He formed Electronic Images Systems Corporation in 1969, marketing a digital fax terminal.

A United Nations forum, the Consultative Committee for International Telephone and Telegraph (CCITT), was set up to establish standardization among the formats. Over the 1970s and 1980s, CCITT issued standards that established groups or generations of facsimile machines. By the 1980s third-generation digital machines achieved speeds of less than a minute per page. Later, fourth-generation digital machines had speeds limited not by the transmission speed but by the speed of the terminal printer. With the improved technology, fax machines spread widely among businesses, schools, governments, and even to private home use.

Freon

Freon is the trade name for a gaseous compound of chlorofluorocarbon, invented in 1930 by Thomas Midgley (1899–1944), a researcher for General Motors, who had earlier discovered that tetraethyl lead was an excellent compound for reducing the knock characteristic of gasoline-driven **internal-combustion engines** [IV].

The problem of finding a safe, nontoxic, and nonflammable gas that could be used in **refrigeration** [IV] systems was a high research priority through the 1920s. Methyl chloride, one usual coolant gas, was dangerous in that it would explode if it leaked. Ammonia was caustic and irritating. Sulfur dioxide, although useful, had a rotten-egg smell. Of course, the odor had the advantage of ease of detection in case of a leak, but manufacturers sought a safer and less offensive gas.

Using a cut-and-try method similar to that by which he had discovered the value of tetraethyl lead, Midgley and his team worked through a variety of carbon compounds containing fluorine and chlorine. Several proved useful, including dichlorodifluoromethane (called Freon 12) and chlorodifluoromethane (Freon 22). The du Pont company trademarked the term Freon.

Both of Midgley's major contributions—tetraethyl lead and Freon—were later centers of controversies and were withdrawn from the market. Leaded ethyl gasoline was phased out of use in the 1980s, and chlorofluorocarbons (CFCs) were withdrawn from the market as a propellant in spray cans in the 1970s and from refrigeration systems in the 1990s. Leaded gasoline tended to damage catalytic converters installed to control automobile emissions, while CFCs were discovered to damage the ozone layer of Earth's atmosphere. The reduction in ozone would lead to greater ultraviolet radiation and a possible increase in skin cancer.

Geiger counter

Hans Geiger (1882–1945), a German physicist, developed an instrument for detecting and measuring all kinds of ionizing radiation in 1928. In 1908, working with Ernest Rutherford (1871–1937), Geiger had developed an instrument to detect alpha particles. Then, with Walther Müller at the University of Kiel, Geiger continued to study radioactivity, developing the counter as a tool for measurement. The counter contained an evacuated tube, a tungsten anode, and a copper cathode, which accelerated the charged particles created by incoming radiation to create a measurable electric current.

The Geiger-Müller counter was later used as a measuring device to monitor the dangerous effects of ionizing radiation in work with **nuclear reactors [VI]** and **nuclear weapons [VI]** and as an exploration tool to find sources of naturally radioactive minerals.

gyrocompass

Jean Bernard Léon Foucault (1819–1868) invented the gyroscope itself in 1851. As a wheel turning within a set of concentric rings, it demonstrated the rotation of Earth, maintaining its axis in parallel to the axis of Earth. It occurred to many inventors that such a device held the possibility of improving on the traditional magnetic compass, which shows only magnetic north, which in many locations deviates several degrees different from true north. From 1906 to 1908 Herman Anschütz-Kempfe developed a workable gyrocompass that could be used aboard submarines, where the presence of a steel hull would interfere with a magnetic compass and where finding direction while submerged was a serious problem. However, his gyroscope required a constant supply of energy and had to be mounted so as to eliminate the motion of the sea.

In 1907 Elmer Sperry (1860–1930), a manufacturer and inventor in the United States who had worked with electric dynamos, arc lamps, generators, **streetcars** [IV], and other late-19th-century innovations, including an electric car, developed his own gyrocompass. An inveterate showman, Sperry was able to promote his device and have it widely accepted by the U.S. Navy. Sperry set up the Sperry Gyrocompass Company in Brooklyn, New York, and his first gyrocompass was installed aboard the **battleship** *Delaware* in 1911.

hearing aid

Eighteenth- and 19th-century hearing aids had consisted of various types of ear trumpets or horns that the partially deaf or hard of hearing individual could use to magnify sound. However, such devices were cumbersome, not very portable, and inefficient.

Alexander Graham Bell (1847–1922) sought to develop an electrical hearing aid, and his work led to the development of a **microphone** [IV] and to the **telephone** [IV]. The microphone was quite a workable system, since it captured sound and converted it into voltage that could then be amplified. A similar device could then convert the electrical current back into sound. However, without miniaturization and a reliable power supply, it was extremely difficult to design a fully portable system suitable for the individual.

In 1923 the Marconi company introduced the Otophone, using vacuum tubes to amplify the sound, with a battery electrical supply. However, the system was hardly convenient, although it was, strictly speaking, portable, weighing some 16 pounds. By the 1930s, reducing the size of electronic tubes and using smaller batteries, hearing aids that weighed about 4 pounds or more were marketed regularly. A. Edwin Stevens introduced one of the first wearable hearing aids, weighing about 2.5 pounds, in 1935, through the company Amplivox. Miniaturization of hearing aids continued in the 1950s with the introduction of the **transistor** [VI], eventually leading to devices that were entirely contained inside the ear cavity.

inert gas elements

The element argon was the first of the inert gases discovered, and the name derives from the Greek word *argos,* meaning inactive. In 1785, when Henry Cavendish (1731–1810) experimented with nitrogen, he found that it could not entirely combine with oxygen. The identification of the nonreactive part of the gas took place in 1894. John William

Inert Gas Discoveries

Inert Gas	Symbol	Year of Discovery	Discoverer
argon	Ar	1894	William Ramsay and Lord Rayleigh
helium	He	1895	Ramsay
krypton	Kr	1898	Ramsay and Morris Travers
neon	Ne	1898	Ramsay and Travers
xenon	Xe	1898	Ramsay and Travers
radon	Rn	1900	Friedrich E. Dorn

Strutt (1842–1919), known as Lord Rayleigh, described the fact to William Ramsay (1852–1916) that nitrogen prepared chemically is always lighter than nitrogen prepared from the air, and Ramsay hypothesized that there was a heavier, inert gas mixed with the nitrogen found in the air. Both he and Rayleigh isolated the gas, finding it nonreactive. They then asked William Crookes (1832–1919) to conduct a spectrographic examination of the gas. It produced a spectrum entirely different from the known elements, and they named it argon (atomic number 18, symbol A, changed by international agreement to Ar in 1957).

Helium had already been detected in the Sun by the use of the **spectroscope** [IV], but Ramsay identified it on Earth in 1895. He began a search with Morris W. Travers (1872–1961) for other inert gases. Taking liquid air and evaporating it, they removed both oxygen and nitrogen, finding three new elements in 1898: krypton, neon, and xenon. In 1900 Friedrich E. Dorn identified the last inert gas, radon, and Ramsay calculated its atomic weight in 1904. All of these discoveries of new elements derived from the development by William Crookes of the **Crookes tube** [IV] or cathode-ray tube and the earlier development of the spectroscope. Lord Rayleigh won the 1904 Nobel Prize in physics for the discovery of argon. Argon later proved useful in a variety of electronic tubes, including **Geiger counters** for detecting **radioactivity** and for **fluorescent lighting** [VI] tubes.

See the table "Inert Gas Discoveries" summarizing this group of discoveries.

instant coffee

According to legend, coffee beans were discovered in the 9th century A.D. in Ethiopia by a goatherd who noticed that his goats seemed energized when they ate the beans. The drink made from the beans spread slowly through the Muslim world, reaching Europe in the 17th century, where its introduction helped stimulate **sugar [II] refining.** Since coffee, like tobacco, was unknown in biblical times, religious leaders were at a loss to find scriptural reasons for seeking a ban on the stimulant.

A liquid extract of coffee as a form of instant coffee was used in the 19th century. However, powdered instant coffee was invented in 1901 by a Japanese chemist, Sartori Kato, who introduced it in Buffalo, New York, at the Pan-American Exposition. In 1906 G. Washington improved on the powder, with a product that had a better taste. Both methods involved brewing coffee and then drying the result to a powder through a spray-drying process. In 1960 the Nestlé company introduced freeze-dried coffee. That method involved freezing coffee extract and drying it in a vacuum, leaving a mass to be ground into water-soluble crystals.

insulin

Between 1889 and 1920, physiologists had learned that some chemical produced by the islet tissue of the human pancreas apparently regulated the metabolism of sugar in the body. Although it was known that some sort of chemical messenger was secreted into the blood, knowledge of the exact nature was first conceived in the early 1920s. Frederick G. Banting, a medical student in Toronto, Canada, reviewed the extensive literature on pancreatic secretions, and with C. H. Best, a graduate in physiology, began preparing extracts from the pancreatic glands of dogs and cattle. Together Best and Banting discovered a technique for isolating the hormone insulin, preventing its degeneration by ferments in the pancreatic gland through the use of low temperature and moderate concentrations of alcohol.

The first application of animal pancreatic extract to humans took place in Toronto in 1922, with some success. Banting shared the 1923 Nobel Prize for physiology or medicine with his mentor at the university, John J. R. Macleod (1876–1935). Later, the group at Toronto cooperated with the Eli Lilly

For more than a century chemists and biochemists have labored to try to learn [the composition of proteins] and solve their labyrinthine structure.... In the history of protein chemistry the year 1954 will go down as a landmark, for last year a group of investigators finally succeeded in achieving the first complete description of the structure of a protein molecule. The protein is insulin, the pancreatic hormone which governs sugar metabolism in the body.... The insulin achievement was due largely to the efforts of the English biochemist Frederick Sanger and a small group of workers at Cambridge University.... Sanger's task was not only to discover the over-all chain configuration of the insulin molecule but also to learn the sequence of all the amino acids in the chains.

—E. O. P. Thompson, "The Insulin Molecule," *Scientific American,* 192, no. 5 (May 1955): 36

company, developing methods of extracting and purifying insulin from pancreatic organs obtained from the butchering of animals. Insulin extract became a useful and widespread treatment in regulating the effects of diabetes. In 1954 a research group under Frederick Sanger at Cambridge University in Britain achieved a complete analysis of the molecular structure of insulin, the first protein molecule ever completely described.

ionosophere

The ionosphere was theorized by Oliver Heaviside (1850–1925), a British physicist, and by Arthur Kennelly (1861–1949), an American electrical engineer, in 1902. Both observed that the reception of **radio** signals sent by Guglielmo Marconi (1874–1937) across the Atlantic Ocean in December 1901 might have resulted from their reflection from a layer in the outer atmosphere. Proof of the existence of this "Heaviside layer" by observation came in 1926 when Gregory Breit and Merle Tuve reflected a radio signal off the layer, about 180 miles above the surface of Earth.

The ionosphere is now divided into three regions: the D, E, and F regions. The D region, which disappears at night, is between 35 and 50 or 55 miles above Earth. The E region, once called the Heaviside layer or the Kennelly-Heaviside layer, ranges from about 55 to 85 to 100 miles high. The F region was once known as the Appleton layer, named after Edward Appleton, who found evidence of its existence in 1925. The F region, which consists of two layers, extends out to about 150 miles above the surface. The layers are caused by the ionizing effect of solar energy on gas particles.

The reflective quality of the different layers differs with the time of day, and different radio wavelengths bounce off the layers at different points. Regular AM broadcasts tend to be reflected in the D level, while shortwave signals are reflected at the E and F levels. Very-high-frequency signals, including television, FM broadcasting, and citizens band radio, are not reflected at all, continuing into outer space. It is for this reason that both FM broadcasting and television broadcasting are either local in broadcast and reception or are relayed by **artificial satellite [VI]** or by cable.

isotopes

The discovery of isotopes in 1919 by Francis Aston represented the destruction of the last aspect of John Dalton's classic theory of atoms, originally published in 1808. Dalton (1766–1844) asserted that the atoms of each element were identical to each other, and the existence of isotopes, or alternate forms of the atoms, discovered by Aston demonstrated the error of that claim.

Aston (1877–1945) was born in Harborne, England, and educated at the University of Birmingham and also at Cambridge. Aston worked with J. J. Thomson in developing the mass spectrograph that discharged ions or charged atoms through a magnetic field, which would deflect the rays dependent on the mass of the elements. By projecting the resulting divergent rays on a photographic plate, a spectrum reflecting the different masses of the atoms could be visualized. In 1920, after serving as an aircraft engineer during World War I, Aston moved to Cambridge and worked under Lord Rutherford in the Cavendish Laboratory there. Using a larger mass spectrograph he himself developed, Aston determined that most elements occur with a mixture of similar atoms of different weights. He specifically identified the isotopes of neon and chlorine, clarifying why some elements have atomic weights that are not simple whole numbers.

The issue had arisen earlier, in 1907, when H. N. McCoy and W. H. Ross discovered that some radioactive decay products had the chemical characteristics of thorium but differed in atomic weight from thorium. Frederick Soddy coined the term *isotopes* to refer to substances that occupied the same place in the **periodic law** [IV] of the elements but had different atomic weights. Thomson concluded that neon was actually a mixture of two gases. One had the atomic weight of about 20 and the other the atomic weight of about 22. Aston confirmed the existence of the two separate isotopes with his mass spectrograph. Thus, although Aston is credited with the discovery of isotopes, his work actually confirmed the contributions of several prior researchers in the field. From 1920 to 1922, Aston subjected more than 50 elements to analysis with the mass spectrograph, and he published his findings in *Isotopes* in 1922.

Soddy won the Nobel Prize in chemistry in 1921 for investigating the origin and nature of isotopes, and Aston won the Nobel Prize in chemistry in 1922 for the discovery of isotopes.

jukebox

In 1928 Justus P. Seeburg, a manufacturer of player pianos, introduced a new electrostatic **loudspeaker** into a record player that was coin-operated and offered a choice of 8 records in an Audiophone. Other manufacturers soon began bringing out similar systems, with Wurlitzer Company of Chicago pioneering in the introduction of the machines into taverns, legalized with the repeal of Prohibition in 1933.

The term *jukebox* derived from a West African word, *dzugu,* meaning a wild time or disorder, and had been associated with "jook houses" or roadhouses and remote gathering places that served alcohol and provided music during Prohibition. With legalized drinking, the association between loud music and drinking established in the speakeasy jook houses was preserved with the commercially produced jukeboxes. The slightly daring or risqué associations with the speakeasy, however, were soon forgotten as jukeboxes were installed in coffee shops and lunch counters as well as bars. By 1938–1939 there were more than 200,000 jukeboxes installed in the United States.

Other companies entered the business during World War II, including Rock-Ola and Rowe. In 1941 Ed Andrews developed a vertical record-changing machine, purchased by Seeburg and introduced in a 1948 model. Jukeboxes reached a peak in popularity in the 1950s, with elaborate chrome and lighted decorations, making them highly collectible nostalgia items by the 1980s and 1990s.

lie detector

Some ancient civilizations reputedly used methods of lie detection that tested the ability of the subject to swallow or salivate, as both functions are minutely constricted by most people when lying. However, the first modern lie-detection machines used sensing devices to record physiologic changes in a person when answering questions or giving testimony. Cesare Lombroso (1836–1909) was an Italian criminologist noted for his identification of criminal personalities with certain facial and body characteristics in a work published in 1889, *The Delinquent Man.* In 1895 Lombroso developed a method of measuring blood pressure and pulse rate to determine whether a suspect was telling the truth. His *hydrosphygmograph* was a water-filled tank into which the suspect's hand was immersed. Changes in blood pressure and pulse changed the water level, recorded by a pen on a revolving drum of paper.

Vittorio Benussi conducted research in 1914, noting changes in breathing rates associated with truth telling. In 1917 W. M. Marston researched the reaction of systolic blood pressure when lying and developed a sphygmomanometer and used a **galvanometer** [IV] to measure the changes in resistance to electric current on the subject's skin. In 1921 an American policeman, John Larson, developed a method of recording several separate measurements on a drum of paper in a *polygraph,* the forerunner of later devices. Larson also developed the method of questioning that starts with nonthreatening and easily answered questions with known truthful answers to establish a baseline. Among the other measures of body changes, the polygraph registered changes in the galvanic conductivity of the skin. Leonard Keeler, associated with the Chicago-based Scientific Crime Detection Laboratory, used a similar polygraph to evaluate testimony, and his evidence was the first to be accepted in a court case in 1935. Later estimates of the accuracy of lie detectors range from about 70 to 90 percent. Since some questions or the manner of the questioner can arouse anxiety even when the subject is telling the truth, false readings are fairly common.

liquid-fueled rockets

Prior to the 20th century, nearly all rockets had been fueled with solid propellants, and the liquid-fueled rocket, which gave rocketry a rebirth, was developed independently in the United States, Germany, the Soviet Union, and Britain in the 1920s. Rockets with solid propellants were originally developed in China, with military uses identified as early as A.D. 1180, including multiple rocket launchers mounted in wheelbarrows. The rocket worked on a very simple principle, following Newton's second **law of motion [III]**. By burning propellant at one end of a tube, the tube would be forced in the opposite direction. Rather than aiming the flaming end of a tube of burning materials at the enemy, the tube would be turned around, ignited, and released. With proper design, the weight of the casing could be reduced and great range achieved. By 1400 the Chinese had perfected rockets with wings and some with two stages and a range greater than a mile.

Europeans were slow to develop rockets for military purposes, and many Western sources erroneously depicted the Chinese rocket use as strictly for religious or celebratory purposes. Until the early 19th century, the reverse was the case, with the Chinese employing rockets in warfare and Europeans mostly using them only as fireworks. However, some 14th- and 15th-century European battles saw the use of rockets, as in 1429, when French troops defended Orléans with them. In general, Europeans found rockets less accurate than cannons, and rockets found little use.

In 1804 a British army officer, William Congreve (1772–1828), made a field artillery rocket that could be fired in groups with a range up to

9,000 feet. Congreve rockets were metal-cased and launched from a stick pointed in the direction of the target. Their first use in battle was during a British naval attack on Boulogne in 1805 and 1806, against Danzig in 1806, and against Copenhagen in 1807. The phrase "the rockets' red glare" in the first verse of "The Star-Spangled Banner" refers to a Congreve rocket used by the British in the War of 1812. European forces used them in warfare thenceforth. Congreve's rockets became so well known that his name became the popular nickname for a type of **match [IV]** used in Britain early in the 19th century. Other 19th-century improvements to the rocket included the addition of wings for spin stabilization and more powerful propellants. Adaptations of the Congreve rocket were used to send rescue lines to stricken ships. A rescue car would be mounted on the line and passengers and crew pulled to shore or another ship.

In Russia and the United States, individual scientists became enthusiasts for rocket propulsion. Konstantin Tsiolkovsky (1857–1935) published a prophetic article in 1903 predicting space flight with rockets, and in 1919 the American Robert Goddard, (1882–1945), in a report, "A Method of Reaching Extreme Altitudes," calculated the force and velocity needed to escape from Earth's gravitation. Goddard worked in some isolation, receiving research grants and studying methods of guiding and regulating the flight of liquid-fueled rockets. Goddard's contribution consisted of methods of controlling the rocket's flight through deflecting the direction of the thrust and the use of liquid fuels. He developed a method of stabilizing the rocket flight with a gyroscope. He achieved altitudes of 1.5 miles and speeds approaching 700 miles per hour, working on a ranch near Roswell, New Mexico.

In Germany and Russia, rocket enthusiasts formed amateur clubs, which later provided personnel and leadership for World War II efforts to build weapons that could carry high explosives in a high ballistic arc to

liquid-fueled rocket   Robert Goddard, an American scientist who calculated the force required to escape Earth's gravity, prepares to fire his first liquid-fueled rocket in 1926. *National Aeronautics and Space Administration*

an enemy position. By 1944 Germany had produced more than 4,000 of the V-2 weapons, with a range exceeding 150 miles. More than 1,000 were targeted on London, with others targeted against Allied positions on the European continent. After the war, many pieces of salvaged equipment and most of the teams were sent to the United States, including the technical director of the program, Wernher von Braun (1912–1977).

In 1957, with the successful launch of an **artificial satellite [VI]** by the Soviet Union, the Space Age use of rockets was initiated.

Lorentz transformation

Hendrik Antoon Lorentz (1853–1928), a Dutch physicist, developed a theory of electromagnetic radiation that was confirmed in 1896. He suggested that the oscillation of electric particles in matter produced not only electromagnetic radiation, as suggested by James Clerk Maxwell, but also produced light. Lorentz proposed an experiment that would demonstrate this effect, and in 1896 his student Pieter Zeeman (1865–1943) showed that magnetism could affect the wavelength of light. The effect was known as the Zeeman effect. In 1902, Lorentz and Zeeman shared the Nobel Prize for physics for this discovery.

In 1904 Lorentz paved the way for Albert Einstein's theory of relativity by proposing mathematical formulas that describe the transformation of matter, including the increase in mass, the shortening of length, and the dilation of time as two objects or systems increase in speed relative to each other. The **speed of light [III]** in a vacuum is a constant independent of the position of an observer. The discovery of the Lorentz transformation represented a confirmation of Einstein's theory.

loudspeaker

The loudspeaker as a device for transforming electric waves into audible sound was the result of several developmental improvements. Patented in 1874 by Werner von Siemens, the first significant improvement came with the work of Lee de Forest and Edwin Armstrong, who, in 1912, developed methods of amplifying the signal, producing a workable loudspeaker that soon found applications in public-address systems. The two engaged in an extensive patent fight.

In 1924, the use of an electric coil surrounded by a magnet attached to a cone or diaphragm in a frame was developed by Chester W. Rice and Edward W. Kelley of General Electric Company. Known as a *hornless* loudspeaker, their device relied on minute movements induced in the diaphragm. In 1929 Harold Black invented a loudspeaker that used different-size cones with different frequency ranges to remove the problem of distortion.

Widely adopted for radios and phonographs in the late 1920s, the hornless loudspeaker allowed the proliferation for the home of compact audio devices that could be listened to without earphones. The social impact was to allow radio audiences to gather in small groups. Trumpet or horn loudspeakers mounted on trucks or in public plazas played recordings or radio broadcasts, a characteristic means of utilizing sound as a mass medium in the 1930s and 1940s. From their first introduction in 1912 and 1913, they were used at open-air rallies and political gatherings, and over the next two decades they had vast political consequences, as they magnified the political influence of orators such as Benito Mussolini in Italy and Adolf Hitler in Germany. Public-address systems continued to rely on loudspeakers to amplify speech and musical performances.

motion pictures

Like some other complex modern inventions, such as the **automobile** [IV], the development of cinematography or the making of motion pictures derives from the contributions of numerous individuals and is surrounded with some controversy over issues of priority. Thomas Edison (1847–1931) conceived the idea of a motion picture in 1889 and patented a motion picture device in 1891. He began to produce a simple viewing machine by 1893–1894. But behind those statements lies a very complex history.

In 1893, Edison assigned an associate, William Kennedy Laurie Dickson, the job of perfecting a camera to take a series of films using **celluloid** [IV] film produced by George Eastman. Dickson developed a system using perforated filmstrips that could be viewed in a box, a *kinetoscope*. Thus Dickson developed the kinetoscope and the perforated 35-millimeter film. Generally limited to a minute, the Dickson films could be viewed in a coin-operated box. Soon nickelodeon parlors, promoted by the Kinetoscope company, popularized the procedure.

An American inventor named Woodville Latham (1838–1911) developed a projection apparatus called an eidoloscope. Latham successfully exhibited a film projected on a wall on April 21, 1895, to an audience of reporters. Edison later patented his own projection equipment, using an intermittent mechanism developed by Thomas Armat (1866–1948) that emulated a design of the brothers Auguste Marie Louis Nicolas

Lumière (1862–1954) and Louis Jean Lumière (1864–1948) of France. The resulting projection device was known as the Edison Vitascope.

Edison and his assistants were by no means the first inventors to struggle with the question of capturing motion on film. The Lumière brothers had started manufacturing celluloid film for motion pictures as early as 1887. In 1885 Professor Étienne J. Marey in Paris had constructed an apparatus for producing serial films, although his methods were not bases for later development. Marey, as a physiologist, sought to capture complex motions, such as the flight of a bird, and he developed a "photographic gun" that caught successive images. However, Marey was not interested in commercial or market value of his idea.

In Bath, England, in 1889, William Friese-Green (1855–1921) used celluloid film for a cinema camera, building on earlier work by J. A. Rudge. Friese-Green filed a British patent on June 1, 1889. After a lengthy patent suit between the Edison interests and Friese-Greene, the U.S. Supreme Court declared Friese-Green's patent as the "master patent" of the world in cinematography, since his methods became contributions and since his patent preceded that of Edison by more than a year and half.

The French inventor L. A. A. le Prince (1842–1890) had perfected and patented an apparatus for producing animated pictures in 1888, using perforated film. After demonstrating his devices in Britain and Austria, he vanished under mysterious circumstances while on a train trip from Dijon to Paris in 1890. Although his friends and family had seen demonstrations of his equipment, no one was able to make practical use of it after his death.

In France, in the provincial town of Lyon, the Lumière brothers constructed a simple camera/projector, calling the device a *cinématographe* in 1895, and held the first public demonstration on the Salon Indien of the Grand Café in Paris on December 28, 1895.

Even earlier, Eadweard Muybridge, an English-born photographer working in California in the 1870s, had worked on attempting to capture the motion of a galloping horse on multiple photographs to determine whether all four feet simultaneously left the ground. Early publication of such photos in 1878 in a pamphlet and in a supplement to *Scientific American* in 1879 stimulated similar work by E. J. Marey in France, and Marey introduced Muybridge to a number of European scientists working on similar issues in 1881. Muybridge's *zoopraxiscope*, first demonstrated in 1879 and 1880, amounted to a projection stroboscope and gave the first synthetic reconstruction of motions photographed from life. Other early pioneers included Emil Reynaud, who secured an 1888 French patent for the invention of a perforated flexible picture strip and invented a device called a *praxinoscope,* displayed in Paris in 1889 and 1892.

With all of these claimants to one or another aspect of priority of invention, it is perhaps most accurate to attribute the invention of the first practical motion picture to the Lumière brothers in 1895.

neutron

The neutron was discovered in 1932 by James Chadwick (1891–1971), for which he earned the Nobel Prize in physics in 1935.

Chadwick bombarded beryllium with alpha particles, producing a form of penetrating radiation that he identified as a stream of particles with no charge and with mass number 1. His work did not fit the existing theory of the composition of the atomic nucleus, thought to consist of positively charged protons and negatively charged electrons. As a result of his discovery, Werner Heisenberg (1901–1976) concluded that the nucleus consists of a number of neutrons and protons that together added up to the atomic weight of the element.

The variation in number of neutrons in an element accounted for the difference in **isotopes.** Thus, when J. J. Thomson (1856–1940) had discovered in 1913 that neon is a mixture of two gases, one with atomic weight of about 20 and the other of about 22, the new atomic model explained the difference in that the two isotopes are chemically identical except that neon 22 has 10 protons and 12 neutrons, while neon 20 has 10 protons and 10 neutrons. Chadwick's discovery opened the field of nuclear research to a series of extremely rapid developments in the following decade.

The discovery of the neutron and the development of understanding of the behavior of free neutrons led scientists to experiment with the bombardment of various elements with neutrons. Such studies generated the discovery of nuclear fission and the construction of **nuclear reactors** [VI] and **nuclear weapons** [VI]. These technological developments probably did more than any others to convince policymakers such as Vannevar Bush (1890–1974) in the post–World War II era that important new technology derived from science and that scientific research had to be funded to ensure the progress of technology.

Norden bombsight

With the development of the concept of dropping bombs from aircraft during World War I, it soon became obvious that hitting a target on the ground by releasing a bomb from a high-flying aircraft was an extremely complicated matter. The forward motion of the airplane would be imparted to the released bomb, requiring that it be dropped in advance of reaching a point over the target. Winds encountered as the bomb fell, the speed of the plane over the ground, the shape of the bomb, and the altitude of the aircraft all represented complicating factors. If the target to be hit was a moving and maneuvering ship, that motion made high-altitude bombing extremely difficult.

To develop a device that would calculate the point at which a bomb should be released, the U.S. Navy hired Carl Norden, an employee of the Sperry Gyroscope Company, immediately after World War I. Through the 1920s, Norden worked on developing a gyroscopically controlled electromechanical device that would combine the factors and provide the bombadier with a simple means of aligning the target and determining the point at which the bombs should be released. After working at the Dahlgren Naval Proving Ground in Virginia, Norden formed his own company, based in New York City, and by 1931 had produced the Mark XV bombsight.

The army became interested in the bombsight, and during World War II, both the army and the navy purchased Norden sights, manufactured both by the Norden company and by other companies recruited to assist, such as Burroughs and Remington. However, the navy recognized that the bombsight was not suitable for bombing against moving ships and relied more on dive-bombing, in which the pilot would aim the aircraft at the target in a steep dive and release the weapon at the last possible moment to achieve a hit on target. Thus the navy-financed invention found more use on army long-distance bombing aircraft against stationary strategic targets such as airfields, train depots, refineries, and factories than on navy planes that targeted enemy ships.

The Norden bombsight was regarded as a highly secret device and, despite at least one espionage attempt to provide the design to Germany, was not successfully emulated by the enemy during the war. According to bomb damage estimates, however, the bombsight was quite inaccurate. Its reputation as a secret weapon was apparently a product of Carl Norden's own promotion efforts.

oxyacetylene welding

The French engineers Edmond Fouché (1861–1931) and Charles Picard (1872–1957) developed the first oxygen-acetylene welding machine in 1903. The torch burned ethyne (acetylene) in a pure oxygen environment, with a temperature about 5,400 degrees Fahrenheit or about 3,250 degrees Celsius. The method of producing pure oxygen on an industrial scale, developed by Karl Paul Gottfried von Linde (1842–1934) in 1893, made possible the torch, which was simply an instrument for bringing the two gases together. By regulating the mix of the gases, the operator could vary the temperature. With a neutral or 50-50 mix, the torch produced a blue flame with the maximum temperature at its tip. Oxyacetylene welding was ideal for joining sheet **steel [IV]** plates.

parking meter

Carl C. Magee, a newspaperman of Oklahoma City, was the first to file for a patent on the parking meter, in 1932. His simple device, shaped like a loaf of bread on a pole, was coin-operated and had a window for viewing the time expired. He filed a second patent in 1935 and received rights to the invention from the U.S. Patent Office in 1936. His 1936 design remained relatively unchanged over the next 60 years. One of several devices and systems that came in with **automobiles [IV]**, along with **traffic lights,** gas stations, state gasoline tax–financed highways, and a 1934 British-invented reflective device known as "cats' eyes," with shiny aluminum disks mounted in hard rubber for marking the edge of highways, the parking meter rapidly became a regular fixture of 20th-century technology.

penicillin

Penicillin was discovered by accident in 1928 by Alexander Fleming (1881–1955), a bacteriologist at St. Mary's Hospital in London. He noted that some bacteria he had been cultivating died around a bit of mold that had been introduced to the sample. He wrote up the phenomenon and published it in May 1929, but no one followed up on the work for a decade. From 1939 to 1943, Australian-born Howard Florey (1898–1968) and German-born Ernst B. Chain (1906–1979) found a way to produce purified penicillin, the active ingredient in the mold. First a team at Oxford under Florey worked from 1939 to 1941 on isolating penicillin and confirming the evidence for its therapeutic effects. Chain identified the chemical structure of crystalline penicillin and identified four different types. From 1941 to 1943 Florey was in the United States, where he worked with staff of the research laboratory of the U.S. Department of Agriculture (USDA) in Peoria, Illinois, to develop methods for production. In 1943 he took samples of penicillin to Tunisia and

Sicily, where he used it successfully on war casualties. In 1945 Fleming, Florey, and Chain shared the Nobel Prize for physiology or medicine for the discovery and isolation of the antibiotic. The federal government financed research on new production methods, recognizing the value the new antibiotic as a matter of military urgency. Pharmaceutical firms were regularly producing penicillin by the end of World War II.

photoelectric cell

In the 19th century, many experimenters noted the phenomenon of the photoelectric effect, in which certain metals appeared to increase their conductivity of electricity when exposed to light. Selenium, an element discovered in 1817 by the Swedish chemist Jöns Jakob Berzelius (1779–1848), proved most pronounced in this regard. In 1883 an American inventor, Charles Fritts, created a photoelectric cell that would use light-sensitive selenium to measure light intensity. In 1884 Paul Nipkow (1860–1940) patented a system utilizing a selenium photoelectric cell to record light from the "Nipkow disk," a forerunner of one type of **television** system. A bright light would penetrate square holes set in a spiral on the disk, projecting an image into twenty lines. A photoelectric cell would record the light and dark areas of each line and convert the measured intensity into a signal that could be transmitted. To view the signal, a reverse process illuminated a light, brightly or dimly according to the signal, projected through a similar perforated disk. In effect Nipkow had developed a method for television using the photoelectric cell, a method that was not taken up and developed further until 1929.

In 1902 Arthur Korn (1870–1945), a German researcher, developed a practical photoelectric cell that produced a small current when exposed to light and that could be adapted for the electric transmission of "telephoto" pictures. The current could be amplified to produce a signal for a variety of purposes. Such photoelectric cells found many uses; among the most common was the so-called electric eye, which sensed the passage of a body between a light source and the cell, allowing automatic operation of doors, streetlights, and burglar alarms and used as counting devices in factories.

photon

The photon, a quantum unit of light, derived from an idea suggested in 1900 by Max Planck (1858–1947), that energy came in discrete packets or quanta. Einstein proposed in 1905 that the photoelectric effect was due to quanta of light. In 1923 Arthur Holly Compton (1892–1963) showed the quantum nature of **X-rays** and showed that they moved at the **speed of light** [III]. The so-called Compton effect is the change in wavelengths of X-rays when they collide with electrons, caused by the transfer of energy from the X-ray photon to the electron. The X-ray would continue with diminished energy and a longer wavelength, while the electron would recoil from the collision. The photon as a unit of electromagnetic radiation acts as a wave and as a particle.

Although Compton described the effect, Compton did not coin the term *photon* until 1926. The photon has no electric charge and no mass at rest, but it has both momentum and energy. It cannot be subdivided. The photon carries electromagnetic force, one of the fundamental forces of nature.

For his discovery of the effect and the nature of the photon Compton won the 1927 Nobel Prize in physics, which he shared with Charles Thomson Rees Wilson (1869–1959), the inventor of the Wilson cloud chamber.

Planck's constant

Max Planck (1858–1947), a German physicist, developed the quantum theory of energy, for which he received the Nobel Prize in physics in 1918. In 1900 Planck introduced the concept of a constant as a measure of the energy emitted, transmitted, or absorbed in discrete quanta or particles. His constant, represented as h, times the radiation frequency, represents the energy of each quantum, including each **photon** of light energy.

With E representing the total energy and f representing the frequency, the following formula represents the energy of a quantum or photon: $E = hf$. The actual value of Planck's constant has been calculated as $6.6260755 \times 10^{-34}$ joule-second. A joule itself is the work performed in 1 second by 1 watt. Planck's constant provides a measurement and representation of the smallest elementary quantum of action or energy in the universe.

Pluto

The planet Pluto was discovered on February 18, 1930, by Clyde W. Tombaugh, working under a plan initiated by Percival Lowell (1865–1916) at the Lowell Observatory in Flagstaff, Arizona. Lowell had predicted the location of a ninth planet, based on perturbations in the orbits of **Uranus [III]** and **Neptune [IV]**, and initiated a systematic search of the sky in the predicted area. The planet was tracked down 14 years after Lowell's death. However, later calculations have demonstrated that Pluto does not have sufficient mass to affect the orbits of the larger inner planets, suggesting that its discovery was more a matter of diligent empirical searching than deduced prediction. With a diameter of about 1,428 miles, Pluto is smaller than Earth's Moon. In effect, Pluto's discovery was the result of rigorous observation, by contrast to the calculations that led to the discovery of Neptune.

Tombaugh's method, which he began in 1929, was very carefully planned. Assuming that a new planet would be extremely faint, he took photographs of particular sectors of the sky, comparing two photographs taken on different nights, alternately projecting one then the other to a screen. Any faint point of light that appeared to alternately jump forward and back would be a planet. Tombaugh's discovery of Pluto derived from plates he had taken three weeks previously. After the discovery of Pluto by this method, Tombaugh continued systematically searching for another such indication of a planet but found none.

James W. Christy discovered Pluto's satellite Charon accidentally on June 22, 1978. Charon is about half the size of Pluto, with a diameter of about 728 miles. The orbit of Pluto is quite unlike that of other planets, inclined by over 17 degrees to the plane of Earth's orbit and forming a quite pronounced ellipse, 2.76 billion miles by 4.58 billion miles in shape that sometimes brings it closer to the Sun than the orbit of Neptune. Furthermore, its axis of rotation is almost in its plane of orbit. The average surface temperature is more than 300 degrees below zero Fahrenheit.

The small size of Pluto, the inclination of its axis, and its inclined pronounced elliptical orbit that passes inside the orbit of Neptune at its closest approach to the Sun all suggest that Pluto falls in a special category of celestial body, neither planet nor asteroid. It is extremely faint and can be detected only by the largest telescopes, making its original discovery an even more remarkable achievement.

polyethylene

Polyethylene plastic was first developed by Imperial Chemical Industries (ICI) Limited, of Great Britain, in 1933. Using high pressures, ICI researchers R. O. Gibson and M. W. Perrin began working to create new compounds with materials that were not normally reactive at low pressure. A mixture of ethylene and carbon monoxide was subjected to intense pressure, and a solid polymer of acrolein was discovered. In March 1933, blending ethylene and benzaldehyde under pressure yielded a white and waxy solid that was regarded as a polymer of ethylene, but the process was dangerous, as the ethylene under pressure could explode.

In 1935, with new equipment, the experiment was tried again, and a polymerized ethylene was again produced, but it was discovered that it had to be made at low temperature and high pressure. The material showed great promise, as it could be drawn out cold like nylon and could be molded and formed into film. Further development work at ICI made slow progress. In 1950 Dr. Karl Zeigler of the Max Planck Institute in Mülheim, West Germany, discovered that ethylene could be polymerized at normal pressure by the use of an aluminum catalyst. He published his work in 1954.

Meanwhile, American scientists at the Phillips Petroleum Company had discovered a method of polymerizing ethylene using chromium as a catalyst. Both the German and the American processes produced a denser and more rigid plastic than that produced by the British high-pressure method. German and American production of polyethylene began in 1956, and British production by the new process began in 1960. Follow-up research in Italy, Germany, and the United States led to similar catalytic polymerization to yield polypropylene, an even harder and denser plastic.

positron

The positron is an elementary atomic particle with the same mass as an electron but with a positive charge. Since electrons have negative electric charges, the positron is the *antiparticle* of an electron. The existence of the positron was predicted by the British physicist Paul A. M. Dirac (1902–1984) in 1928 and was discovered and demonstrated by the American physicist Carl D. Anderson (1905–1991) in 1932. Anderson was studying cosmic rays, and he was able to produce positrons by gamma irradiation in a Wilson cloud chamber.

Dirac formulated a relativistic theory of the electron, noting statistical rules of behavior of the electron that suggested the existence of a positive antiparticle. Dirac won a 1933 Nobel Prize in physics for his mathematical work, and Anderson won a 1936 Nobel Prize in physics for the discovery and production of the positron.

radio

Although the invention of radio transmission of messages is often attributed to the Italian physicist Guglielmo Marconi (1874–1937), its development grew out of the work of a number of scientists and experimentalists from several different countries from 1865 to 1901. James Clerk Maxwell (1831–1879) had predicted as early as 1864 that an oscillating current would produce radiation with a long wavelength. In 1890 Elihu Thomson (1853–1937) worked with alternating current and designed generators and transformers. Using a spark induction coil, he created long-wave electromagnetic radiation. Thomson's company later merged with Edison's to form General Electric Company.

In Germany, Heinrich Hertz (1857–1894) from 1886 to 1888, developed an oscillator and spark gap that produced radio waves, and he developed an antenna that could detect the shape and intensity of the waves. In 1895 the Russian Aleksandr Stepanovich Popov (1859–1906) improved on the antenna. In 1893 Croatian-born American citizen Nikola Tesla (1856–1943) established the first *wireless telegraph,* sending messages over distances up to 25 miles. William Crookes (1832–1919), who had followed the experiments closely, predicted as early as 1891 that radio transmission of signals or wireless telegraphy would become possible and widely utilized within a decade. His writings inspired a group of researchers including Marconi, who conducted a series of experiments through the mid-1890s. Several patents on radio, particularly centering on the issue of *frequency selectivity* or *tuning,* were filed between 1897 and 1900, leading to patent disputes that lasted for decades. In 1901 Marconi transmitted a Morse code message over 2,000 miles. It was later discovered that such long-range transmissions were possible because they bounced off the Heaviside layer or **ionosphere.** Radio transmission and receiving involved many separate components and inventions, including the vacuum tube or **diode,** improved alternators, and improved transmitters.

By World War I, wireless telegraphy was widely installed on naval ships, stimulating the development of complex cryptographic methods and the ability to broadcast messages understandable only by those knowing the decoding methods. The development led to rapid innovation of several types of

Nobody knew early in 1921 where radio was really headed. Everything about broadcasting was uncertain. For my own part I expected that since it was a form of telephony, and since we were in the business of furnishing wires for telephony, we were sure to be involved in broadcasting somehow. Our first vague idea, as broadcasting appeared, was that perhaps people would expect to be able to pick up a telephone and call some radio station, so that they could give radio talks to other people equipped to listen.
—President Walter Gifford of AT&T, 1944

encryption machines, including the famous German Enigma device used in World War II.

In 1920, **radio broadcasting** of voice and music began with the first radio station in Pennsylvania. Radio became a popular home entertainment medium over the next two decades, improved with the development of frequency modulation (FM) in 1935 by Edwin H. Armstrong. FM broadcasting began in 1939.

radio broadcasting

The broadcasting of voice and music by radio, as distinct from wireless telegraphy, developed experimentally from 1906 to 1919, with established radio stations getting under way in the 1920s. Reginald Fessenden used a high-frequency alternator, a generator producing alternating current, to send amplitude-modulated signals. Coupled with a microphone, he was able to transmit, on December 24, 1906, the first voice message and recorded music by radio over a distance of 200 miles.

Lee De Forest (1873–1961) installed a **microphone [IV],** using his **triode valve** or tube, at the New York Metropolitan Opera and transmitted a broadcast to his laboratory in 1910. During World War I, all **radio** transmission for military purposes remained wireless telegraphy rather than voice transmission, although Germany experimented with a voice transmission system between two stations at the Western Front in 1917.

In 1919 Dr. Hans Bredow, a director with the Telefunken Company, gave a demonstration lecture by radiotelephony, and in the same year, Guglielmo Marconi (1874–1937) broadcast a voice message from a yacht to Lisbon, a distance of about 300 miles. In November 1920, the first public broadcasting station in the world broadcast from Pittsburgh, Pennsylvania, with a report on the U.S. presidential election. The station, KDKA, broadcast an hour a day and was owned by Westinghouse.

In 1920, Radio Corporation of America (RCA) was formed as Western Electric (of AT&T), Westinghouse, and General Electric pooled their patents and bought out Marconi's American rights. The first RCA broadcast was on July 2, 1921, of a heavyweight fight from Hoboken, New Jersey. RCA's first regular station, WDY, opened in Roselle Park, New Jersey, on December 15, 1921. Discovering that two of its first stations interfered with each other, RCA shut down WDY and opened a third, in Massachusetts, WBZ.

During 1922 the U.S. Department of Commerce issued 600 broadcasting licenses, with call letters beginning with K for stations west of the Mississippi River and W for those to the east. The first paid radio commercial was aired by AT&T-owned station WEAF in New York City on August 23, 1922. GE and Westinghouse continued to manufacture radio sets of slightly different designs under the RCA patent-sharing agreement and continued selling them through RCA. The first radio receiving sets ranged in price from $25 to $125. Gradually the firms began to establish standards for tubes and other equipment through the mid-1920s.

RCA made cross-licensing agreements with other companies, including United Fruit Company, which had developed ship-to-shore equipment for use in its banana transport business from Central America to the United States.

radioactivity

The French physicist Antoine Henri Becquerel (1852–1908) discovered radioactivity in 1896. Following up on Wilhelm Roentgen's discovery of **X-rays**, emitted from the fluorescing spot in a **Crookes tube [IV]**, Becquerel investigated whether various materials that fluoresced or phosphoresced also emitted X-rays. He placed a crystal of phosphorescent material on a photographic plate that he wrapped in black paper, placing both in sunlight. When light hit the crystal, it would glow, and he wanted to check to see if the glow penetrated the paper as X-rays. Indeed, he found an image on the plate.

He prepared further experiments and, waiting for a sunny day, stored a wrapped plate and crystal in a drawer. He later discovered that this photographic plate had an intense image on it, and he recognized that a new form of radiation, not caused by the phosphoresence of the material, had cast an image on the plate. He soon tracked down the source of the new radiation to uranium in the crystal sample.

Ernest Rutherford (1871–1937), born in New Zealand, studied in Britain under J. J. Thompson, and in 1898 Rutherford took up a professorship at McGill University in Montreal. Using the well-equipped laboratory there, he immediately worked on the new field of radiation uncovered by Roentgen. By

It is probable that all heavy matter possesses—latent and bound up with the structure of the atom—a similar quantity of energy to that possessed by radium. If it could be tapped and controlled, what an agent it would be in shaping the world's destiny! The man who puts his hand on the lever by which a parsimonious nature regulates so jealously the output of this store of energy would possess a weapon by which he could destroy the Earth if he chose.
—Frederick Soddy, on atomic energy after the discovery of radioactivity, in a lecture to the Corps of Royal Engineers, Britain, 1904

radioactivity One of the many physicists who took up work in the new field of radioactivity, Ernest Rutherford settled in Montreal at McGill University. There he developed the concept of radioactive decay and named the alpha, beta, and gamma radiations. *National Archives and Records Administration*

1900 Rutherford determined that the penetrating radiation that Becquerel had discovered is made up of three different kinds of radiation, which Rutherford designated as alpha, beta, and gamma rays. It was at McGill that Rutherford, working with Frederick Soddy (1877–1956), soon identified the process of radioactive decay by which thorium gradually transforms into uranium. Soddy demonstrated in 1903 that one of the decay products of **radium** is helium. Soddy and his contemporaries recognized that they were beginning to tap into the energy of the atom and could see that the unleashing of atomic energy held great potential for good and evil.

Becquerel received the 1903 Nobel Prize in physics, jointly with Pierre and Marie Curie for their work in radioactivity. The Curies, inspired by Becquerel's work, searched for other radioactive materials, discovering the elements radium and polonium in 1898. Rutherford received the 1908 Nobel Prize in chemistry for his work on atomic disintegration, and Marie Curie received the 1911 Nobel Prize in chemistry for the discoveries of radium and polonium.

radium

Following the discovery of **X-rays** by Wilhelm Roentgen (1845–1923) in 1895, a series of studies on related topics was stimulated, leading to the discovery of **radioactivity** by Antoine Becquerel (1852–1908) in 1896, the discovery of the **electron** by J. J. Thomson (1856–1940) in 1897, and the determination of three types of radiation by Ernest Rutherford (1871–1937) from 1898 to 1900. Associated with these discoveries was the work of Marie and Pierre Curie in discovering two new radioactive elements, radium and polonium, in 1898.

After the discovery of natural radio-activity by Becquerel, Pierre and Marie Curie began a systematic study of uranium-containing ores. Marie Curie, who was born Marie Sklodowska in Warsaw, Poland, in

1867, had married Pierre in 1895 after they had studied together at the Sorbonne.

Marie discovered that pitchblende, which is an impure form of the uranium mineral uranite, was much more radioactive than uranium. She proposed that it contained a new element. She and Pierre separated pitchblende into its components and isolated both radium and polonium, the latter named after her country of birth. She was credited with the discovery of radium, and she and her husband were regarded as jointly discovering polonium.

Radium was difficult to extract. The Curies processed some 10,000 distillations to produce 90 milligrams of radium. Various isotopes of radium were identified in the following years. Ernest Rutherford identified radium 224 in 1902, Friedrich Giesel isolated radium 223 in 1904, and Otto Hahn identified radium 228 in 1907. Marie Curie isolated metallic radium or radium 226 in 1910. An associate of the Curies, André Debierne, found the element actinium in pitchblende in 1899.

Radium, which is phosphorescent, was used by the end of World War I to make glow-in-the-dark watch faces. It was discovered in the 1920s in the United States that factory workers who painted the watch faces with radium paint developed often fatal cancers, induced by licking the paintbrushes to obtain a sharp point. This case and others led to the establishment of industrial health standards for the handling of radioactive materials.

Pierre Curie was killed in a traffic accident in Paris in 1905, two years after the Curies shared the Nobel Prize in physics with Becquerel for their work in radioactivity. Marie Curie continued her researches in

radium Co-discoverer of radium, Marie Curie became a role model for many aspiring women scientists in the first decades of the 20th century. *Library of Congress*

We have been forced to admit for the first time in history not only the possibility but the fact of the growth and decay of the elements of matter. With radium and with uranium we do not see anything but the decay. And yet somewhere, somehow, it is almost certain that these elements must be continually forming. They are probably being put together now in the laboratory of the stars. . . . Can we ever learn to control the process. Why not? Only research can tell.
—Robert A. Millikan, "The Significance of Radium," an address delivered at the National Museum, Washington, D.C., May 25, 1921, in connection with the presentation of a gram of radium to Madame Curie, *Science* v. 54, no. 1383 (1921)

the subject and became the first woman lecturer at the Sorbonne. During World War I she helped establish programs for mobile X-ray laboratories for the diagnosis of battle injuries. Marie Curie died in 1934 of leukemia, presumably induced from her work with radioactive materials. It is said that her notebooks remain so radioactive that they cannot be placed on public display.

As the most famous woman scientist of her day and as the first woman to win a Nobel Prize, Marie Curie became honored and revered as a role model for aspiring women scientists from the 1920s through the 1940s.

Raman spectroscopy

In 1930, Chandrasekhara Venkata Raman (1888–1970), an Indian physicist, received the Nobel Prize in physics for his 1928 discovery that when a transparent substance such as a gas is illuminated by light of one frequency, the beam emerging at right angles from the substance contains other frequencies that characterize the material. This effect, known as the Raman effect, means that the exact nature of a sample of chemical compounds in a gas can be determined. Since 1930, Raman spectra have been widely used in qualitative and quantitative analyses of molecular species.

Since gases have low molecular concentrations at ordinary pressures, only faint Raman effects can be determined. However, Raman spectroscopic equipment developed in later years has been used to identify the exact composition of gases that are products of combustion in **internal-combustion engines [IV]**. By substituting a solid crystal window for a section of an engine cylinder head and with the use of **laser [VI]** light beams, Raman spectroscopy in the late 20th century allowed for engine and fuels research that aimed at more complete combustion and the reduction of emissions from both gasoline and diesel engines.

Raman, a boyhood genius who received his university degree at age 16, also made contributions in the fields of the physics of music, color perception, and crystalline structure. He served as professor of physics at the University of Calcutta for 16 years, and from 1933 to 1948 was director of the Indian Institute of Science. He headed his own institute from 1948 until his death in 1970. Raman was a major figure in Indian intellectual and scientific life through much of the 20th century. He trained hundreds of students and founded the *Indian Journal of Physics*.

rayon

The development of rayon as a synthetic fiber that could be used in textiles grew out of work on the **electric light** [IV] by Joseph Swan (1828–1914) in Britain. Swan developed a carbon filament lamp and a process to make nitrocellulose fibers for lamp filaments. Swan worked on a filament lightbulb from 1845 to 1878, successfully producing a lamp with a filament in the later year. Swan kept seeking a better filament and introduced his squirting process for nitrocellulose filaments in 1883. In his process, nitrocellulose was forced through small holes after an acetic acid bath.

In 1889, Comte Hilare de Chardonnet in France sought a substitute for silk and began to manufacture an artificial cellulose silk using the Swan method under the name *rayon,* displaying it at the Universal Exposition in Paris that year.

In 1892, in Britain, C. F. Cross invented a process for making pure cellulose or viscose from wood pulp and showed it to a friend, C. H. Stearn, who operated a lightbulb factory. Stearn, like Swan, sought a process to make synthetic filaments. Realizing that the fiber might be suitable for textiles, Stearn cooperated with another lightbulb manufacturer, F. Topham, who developed a spinning box for the production of viscose. In 1904 Samuel Courtauld began the manufacture of viscose rayon on a large scale, and the Courtauld method is often taken as the beginning point for rayon production.

In 1910, the first rayon stockings were produced in Germany. Although used in its first years as a silk substitute, in the United States high-tensile-strength rayon was introduced in 1924 to reinforce automobile tires. After 1938, **nylon [VI]** began to replace rayon as a silk substitute, although rayon continued to have many uses in fabric and tire manufacture.

reinforced concrete

The Romans developed concrete, using a cement mixture of pozzolana, a volcanic earth found near Naples and Rome, with lime. The cement would be mixed with gravel and sand to achieve a hardened concrete. Concrete structures included the base of the Pantheon **dome [II]** and the theater at Pompeii. The art of mixing Roman cement (using pozzolana) with sand and gravel apparently was lost until revived in 1568 by a French architect, Philibert de l'Orme, who recommended its use for underwater construction.

John Smeaton (1724–1792), a British civil engineer, used a form of concrete with aggregate gravel in it that would harden underwater to construct the Eddystone Lighthouse in 1759.

A British bricklayer, Joseph Aspdin, patented Portland cement in 1824. It consists of a mixture of chalk and clay heated to a high temperature. It was so named because a common building material in Britain early in the 19th century was a very hard Portland stone, and the name suggested durability and hardness. When mixed with aggregate rock and sand, Portland cement makes concrete very similar to that made with the pozzolana-lime cement used by the Romans.

The concept of reinforcing concrete with metal wire or bars developed in the 19th century and was not implemented with reinforcing rods until the 1890s. W. B. Wilkinson, a British plasterer, patented a process for using metal wire to reinforce the concrete in 1854, but his scheme was not widely followed.

The invention of reinforced concrete is often attributed to a Parisian gardener, Joseph Monier, who used iron mesh to reinforce concrete garden pots and tubs. His method was patented in 1867. Monier licensed his patents, and gradually the technique spread to the construction of bridges, particularly by a German engineer, G. A. Wayss.

In France, François Hennebique patented reinforcing rods in 1892 and 1893 and popularized the technique over the next few years. Reinforced concrete was used from 1894 to 1899 to construct the church of St. Jean at Montmartre in Paris and was used in the construction of a wharf at Southampton, England, in 1897. Word of its virtues spread rapidly, and in 1898 the first model basin for testing ship hulls constructed in the United States, at the Navy Yard in Washington, D.C., was made with reinforced concrete. By the turn of the century, reinforced concrete was spreading in use throughout Europe, North America, and Australia. The advantages of reinforcement are several. Using a metal structure inside the concrete allows large expanses to be poured into wooden molds on the construction site rather than cast into blocks to be transported. The reinforcing rods greatly increase the strength and bearing load of the concrete as well.

safety razor

The safety razor, or shaving razor with a replaceable blade, was invented by King Camp Gillette (1855–1932), a traveling salesman, who conceived of the concept while shaving in 1895. Another inventor had suggested to Gillette that he invent something cheap that filled a need and that when used had to be thrown away. Realizing the danger and inconvenience of the current razor, which had to be sharpened for every use, he seized on the idea. With no knowledge of metals or of engineering, he began experimenting and formed a company to develop and manufacture the new blade. By 1901 he filed for his patent, and by 1902 he had worked out the details: a heavy handle that would allow the user to adjust the angle of application, a T-shaped head that would allow the blade to have two usable sides, the right grade of steel, and equipment for making and sharpening the blades themselves.

In 1903 the firm sold 168 blades; the next year, it sold more than 12 million. Gillette retired from active management of his firm in 1913 to support his advocacy of technocracy, a plan that would abolish competition, result in the equal distribution of goods and services, and put engineers in charge of society. Even before the invention of the safety razor, Gillette had published a work visualizing a utopian urban society, run by engineers, that would concentrate 60 million people (the then total population of the United States) in one large city based around Niagara Falls, which would provide electricity for the concentrations of apartments and factories there. The rest of the country would be devoted to agriculture and tourism. Gillette continued to fund technocracy as a political movement until his death in 1932.

sonar

The term *sonar* originally was an acronym introduced in the United States in the 1930s that stood for "sound navigation and ranging." Paul Langevin (1872–1946), a French physicist, first developed the concept, working with a Russian émigré scientist, Constantin Chilowski. Langevin had earlier worked with Marie Curie (1867–1934), who had discovered **radium,** and with J. J. Thomson (1856–1940), who had discovered the **electron.**

During World War I, the British sought a system for detecting submarines, and Chilowski, who had developed a sound-projecting system in 1912, sought to improve on it to make it practical for such detection. In 1915 Langevin worked with Chilowski in France, using a system of quartz compression to generate high-pitched ultrasonic waves, producing a system that could send a sound and receive an echo underwater from about 2 miles. The British/French/Russian collaboration on the system led to the code name *asdic,* reputedly based on the initials of an Allied Submarine Detection Investigating Committee, and for the next decade the British sonar system was known by that term.

The sonar system developed by Langevin and Chilowski was not workable until 1918, the last year of World War I, and relied on a piezo-electric or quartz-based sending and receiving system that came to be known as *active sonar*. A simpler system, *passive sonar,* used underwater **microphones [IV]**, known as hydrophones, to detect the direction of underwater sounds such as the engine, propeller, or water-flow sounds of a ship. Such hydrophone systems, using a simple stereophonic direction-finding principle, were in place as early as 1916. With the hydrophones widely spaced from each other, it was possible to achieve a good direction bearing on a noisy ship or submarine by gradually turning the array until both hydrophones emitted an equal volume.

During World War II, quartz crystals useful in the design of active sonar systems became highly valuable, and British scientists quickly hunted down and retrieved collections of quartz from France before the fall of the French government to Germany. Sonar underwent considerable improvements over the next decades, improving the range and sensitivity of detection and continuing to utilize piezoelectric systems.

stainless steel

Stainless steel is defined as an alloy of steel that contains 12 to 30 percent chromium and less than 1 percent carbon. Austenitic stainless steel contains 7 to 35 percent nickel.

Various hardened steel alloys were developed early in the 20th century. Léon Guillet, a French researcher, tried various low-carbon and chromium alloys in 1904. Although he overlooked the resistance to corrosion that characterized his alloys, his descriptions showed that he had developed the principal grades of two types of stainless steel: martensitic, which can be hardened by heat; and ferritic, which cannot be hardened. Other researchers, both French and German, studied the alloys as well, from 1904 to 1915.

However, stainless steel usually is attributed to Harry Brearley in Britain and Elwood Haynes in the United States. Brearley developed martensitic steel for naval guns beginning in 1912. After disputes with his employer, Thomas Firth and Sons, Brearley obtained an American patent in 1915 and worked with another company that produced the alloy after 1920, Brown Bayley. Elwood Haynes also developed martensitic stainless steel as well as other hard metal alloys. In 1912 he applied for a patent, which was finally granted in 1919, with the claim that the metal was incorrodible. Steel companies then bought up both the Haynes and Brearley patents. In 1912, Krupp in Germany patented austenitic stainless steel, with nickel as a major component.

submarine

Submarines entered the world's navies in the first decade of the 20th century, after successful efforts to combine electric motor underwater propulsion with **internal-combustion engines** [IV] for surface propulsion. Several developers simultaneously sold submarines, including Simon Lake in 1897 and the Irish-born American inventor John Holland (1840–1914) in 1898. However, the "first" submarines had been invented and employed long before the final work of Lake and Holland, and during the 19th century there had been considerable experimentation with *subaqueous* boats.

The original invention of the submarine as a workable ship useful in undersea exploration and as a weapons platform in warfare can be traced to two 18th-century American inventors, David Bushnell (1742–1824) and Robert Fulton (1765–1815). Fulton is better known for his success with the **steamboat** [IV] than with his effort to interest the French in his submarine he called *Nautilus*.

Bushnell had entered Yale at age 29, selling his family farm on the death of his father. While in college he discovered that gunpowder would explode underwater, and he conceived of the idea of an underwater explosive to be planted on enemy warships. In 1775 he began building a submarine to deliver the charge. The Bushnell submarine, named the *Turtle,* was a one-man machine made of wood and propelled by hand-cranked screw propellers. The *Turtle* was tried out in New York Harbor in 1776 and piloted by a Revolutionary War soldier, Ezra Lee. Lee attempted to attach one of Bushnell's mines, consisting of a keg of powder, to the British ship *Eagle*. The ship was sheathed in iron and copper, and the screw mechanism could not penetrate the material because the *Turtle* simply pushed back as Lee attempted to screw into the surface. The mine floated loose and exploded harmlessly downriver.

Several other attacks also failed, although Bushnell was able to sink a British schooner with a tethered mine of his own design. A full-scale replica of the *Turtle* was constructed in 1977 and could be operated quite successfully.

Fulton's *Nautilus* appeared to have been inspired by *Turtle,* but like its predecessor it could not maneuver well underwater, and the procedure for attaching a mine to the hull of an enemy ship was not perfected. The

French funded the construction of Fulton's submarine, but it failed to sink any British ships. Later Fulton proposed the idea to the British, but they, too, found it unsatisfactory, abandoning the project after 1805.

The first successful use of a submarine in warfare was the *Hunley,* employed by Confederate forces against a Union ship blockading Charleston, South Carolina, in 1864. Invented by Horace Hunley, the ship was privately built in 1863 by Park and Lyons of Mobile, Alabama, made from a cylindrical iron steam boiler, deepened, and lengthened with tapered ends. Like the *Turtle,* the *Hunley* was hand-powered, requiring a nine-man crew, with eight to turn the propeller and one to direct the vessel. Each end had ballast tanks that could be flooded by opening valves or pumped out by hand. Iron weights bolted to the underside of the hull provided extra ballast, which could be released by unscrewing bolts inside the submarine.

The *Hunley* sank three times in trials, killing its inventor. Nevertheless, it was successfully used on February 16, 1864, against the Union ship *Housatonic,* a 1,800-ton, 23-gun sloop of war. The *Hunley* rammed the Union ship with a spar **torpedo [IV]** attached to a long pole on the bow of the submarine. As the submarine backed away, riflemen aboard the *Housatonic* fired on it, possibly sinking it. The resulting detonation of the torpedo sank the *Housatonic,* the first successful use of a submarine in warfare. In August 2000 the *Hunley* was raised and carefully examined by archaeologists and other specialists, who were surprised at the many advanced features incorporated in the design that had escaped the written record for more than 100 years.

The Confederates built other submarines, about which little is known, and research continues. Interest in submarines persisted after the Civil War. Simon Lake built the *Argonaut* in 1897, and John Holland built the *Holland* in 1898. Holland used a gasoline internal-combustion engine that would charge batteries on the surface to provide electric power for propulsion while submerged. Later **diesel** engines would replace gasoline engines, powering 20th-century submarines in World Wars I and II. In 1954 the United States applied a **nuclear reactor [VI]** to submarine propulsion, naming the first such vessel the *Nautilus* in recognition of Fulton's first attempt and the later use of the name in Jules Verne's classic novel *20,000 Leagues under the Sea.*

subway trains

With the congestion of city life in the 19th century, both the **streetcar [IV]** trolley line (or tram) and the underground railway or subway were introduced to speed inner-city travel. Underground rail lines were proposed as early as 1843 in London, but approval and construction delayed the opening for about 20 years. The first subway was inaugurated in London in 1863, consisting of a steam railway operating on coke, running in a roofed-over trench. In 1870, London opened the Tower Subway, also using coke-powered steam engines. The early lines were immediate successes. However, the 20th-century subway train is usually associated with electric motor–propelled systems, far more successful than **steam engines [III]** because of the problem of ventilation.

In 1890 London began installing electric-powered tube railways, often buried more than 100 feet deep in the clay under the city, in a burst of construction that continued until 1907. London added no new lines until 1968. The London system was known simply as the *underground* or as the *tube*. Electric lines proved far superior to steam-driven systems and were utilized in the tube systems, not only because of the ease of ventilation but also because no refueling is required and because an electric-powered railcar only uses power when it is moving. Most of the systems utilized power from a high-voltage power rail.

Budapest, Hungary, was the first city on the continent of Europe to open a subway, in 1896. The Paris Métro opened in 1900, and New York City opened its first subway in 1904. Soon Berlin, Glasgow, Madrid, Tokyo, Buenos Aires, Prague, Moscow, and Mexico City, among others, developed efficient underground rail lines. Munich and Frankfurt opened new lines in the early 1970s. Of all the systems, the Moscow subway, decorated with marble, chandeliers, mosaics, and murals, is generally regarded as the most handsome in the world.

superconductivity

In 1911, the phenomenon of superconductivity was discovered by Heike Kamerlingh-Onnes (1853–1926) of the Netherlands, who won the 1913 Nobel Prize for physics for his work in the field of low-temperature physics. Kamerlingh-Onnes had achieved liquefied hydrogen in 1906 (at the temperature of 20 degrees Kelvin) and liquefied helium in 1908 (at 4.2 degrees Kelvin).

`Studying the changed electrical conductivity of various materials at low temperatures, in 1911 Kamerlingh-Onnes found that the conductivity of mercury greatly increased at very low temperatures. One of his students discovered that electrical resistance virtually vanished in mercury at a temperature just above absolute zero, or 4.15 degrees Kelvin (about minus 289 degrees Celsius). The point at which resistance declined Kamerlingh-Onnes called the critical temperature, and he dubbed the phenomenon *superconductivity*.

In the 1980s, the critical temperatures of a wide variety of metals were discovered, most in a range of 1 degree Kelvin to 15 degrees Kelvin. In 1986 some ceramics were discovered with much higher critical temperatures, up to 125 degrees Kelvin, holding out the possibility of practical applications of the phenomenon. That transition temperature was discovered for a compound of thallium, barium, calcium, copper, and oxygen. For the discovery of this phenomenon, two scientists, J. Georg Bednorz and K. Alex Müller, working at the IBM research laboratory in Zurich, Switzerland, received the Nobel Prize in physics in 1987.

tank

The concept of an armored vehicle to protect soldiers from enemy fire as they advanced was an old one, and Leonardo da Vinci (1452–1519) included a sketch of a hand-cranked and hand-propelled armored vehicle in his notebooks in 1482. An American, E. J. Pennington, designed on paper a steam-powered armored car, with metal covers to protect the wheels, in the 19th century. An English engineer, F. R. Simms, built a similar armored car with steel-tired wheels in 1902. The novelist H. G. Wells had depicted in detail a "land-destroyer" in a story published in *Strand Magazine* in 1903. However, the idea was not developed into reality until World War I.

When a combination of trench warfare, **machine guns [IV]**, and **barbed wire [IV]** brought the armies of the Allies and the Central Powers to a standstill in 1915, the need for a track-laying vehicle became pressing. To cross the no-man's-land between the opposing forces' trenches, armor was required, and wheeled vehicles were too easily destroyed or immobilized by the trenches themselves. The idea was credited to Lieutenant Colonel Ernest Dunlop Swinton, in Britain. Swinton had earlier considered the idea while in South Africa, corresponding with a mining engineer who suggested that the **caterpillar track** of the American Holt tractor would make an excellent war machine because of its "surprising powers of crossing rough country" and traversing trenches.

Swinton's idea of using the Holt caterpillar tractor was abandoned. Then Lieutenant W. G. Wilson of the British navy and an engineer, W. A. Tritton, employed by William Foster and Sons Limited, drew up a workable design—a rhomboid or lozenge, with a long tread that cir-

tank Introduced during World War I, the tank came into its own as a combat system in World War II. Here a Sherman tank of the 1st Infantry Division, 745th Tank Battallion, rolls through Gladbach, Germany, on March 1, 1945. *U.S. Army*

culated around the total length of the vehicle, capable of crossing an 11-foot, 6-inch trench.

The term *tank* was suggested by Swinton as a code word for the new device to prevent premature disclosure of the concept, and the word later entered common usage for armored, tracked vehicles. The first British tanks weighed 30 to 31 tons, proceeded at a walking speed, and were armed with machine guns and **cannons [II].** They used a 6-cylinder Daimler water-cooled **internal-combustion engine [IV].** Their range on a full 53-gallon tank of gasoline was 12 miles. The Mark 1 tank required four men to drive it and carried a crew of eight, including the drivers.

The first tank engagement was on September 15, 1916, and its first major success came with a 400-tank assault on November 20, 1917. The French designed a tank, led by Colonel Jean Estiene, and Renault began to produce light tank models. The U.S. Army developed a Tank Corps, which fought in 28 engagements in World War I. The first German tanks weighed some 148 tons and each contained 28 men involved in operating the engine, steering the vehicle, and firing weapons.

The tank helped break the deadlock on the war front, and between World Wars I and II some military commanders, notably the German general Heinz Guderian (1888–1954), recognized its value. The Germans successfully incorporated much-improved tanks in their mobile *blitzkrieg* (lightning war) attacks in 1939 and 1940. Massed tank battles in North Africa and later at Kursk in April 1943 in the Soviet Union proved the value of tanks, and they have remained features of land warfare into the 21st century.

tape recording

The principle of recording sounds on magnetic tape was developed by a Danish engineer, Valdemar Poulsen (1869–1942), who invented what he called a *telegraphone* in 1898. He understood perfectly that a steel wire, steel tape, or a paper tape coated with iron oxide would receive an electromagnetic impression from sound impulses converted into electric modulations and that by reversing the process, the sound could be played back. He anticipated that his device might be used in what we now call telephone answering machines, as well as for more general recording of voice and music. He also pointed out that his system had an advantage over wax-impression recordings of the **phonograph [IV]** in that the magnetic recordings could be erased and the wires or tapes reused.

Poulsen patented the device in Europe and the United States, but it was very slowly adopted. Several minor technical barriers stood in the way, such as getting reliable winding and rewinding machinery. Although some dictation equipment was built using his method, it was not revived until the 1920s. Kurt Stille, a German engineer, produced a wire recording system used in broadcasting, and an Austrian researcher, Dr. F. Pfleumer, began working on a magnetic tape with iron powder. A German company manufactured a *magnetophon,* using plastic tape with ferrous oxide and based on Pfleumer's concept, in 1937 and 1938. I. G. Farben manufactured the tape, and such tape recording was common during World War II in Germany. Allied teams examining German war technology in 1945 and 1946 were impressed with the system.

In the United States during that war, advances in magnetic recording were made by Marvin Camras, who invented an improved wire recorder while a student at the Illinois Institute of Technology. Camras became a researcher at a nonprofit research organization, Armour Research Foundation, where he developed a variety of improvements in coating materials for magnetic tape.

After World War II, several wire-recording and tape-recording machines were introduced by small companies. Binding the ferrous oxide to polyvinylchloride plastic tape, introduced first in Germany during the war, became the standard for many formats.

The magnetic tape recorder found thousands of uses in such diverse roles as speech therapy, music appreciation, foreign-language instruction, broadcasting and professional music recording, dictation recording, and in the "black box" recorders aboard airplanes to record both conversation and electronic information recoverable after accidents. Large tape reels were used to record data for **computers** [VI]. Not least of its uses has been the telephone answering device anticipated in 1898 by its original inventor, Valdemar Poulsen. Similar technology was later developed for the **video tape recorder** [VI].

television

Attempts to transmit pictures by wire had been proposed in the 1870s, but the invention of the **photoelectric cell** in 1913 paved the way for electronic transmission of pictures. With the invention of **radio** and of **motion pictures** within a few years of each other, it was inevitable that inventors and engineers would seek to combine the two with a system of broadcasting images. In Britain, John Logie Baird (1888–1946) filed a patent in 1923 on a mechanical system, and in the United States, Vladimir K. Zworykin (1889–1982) patented an electronic method in the same year.

The Baird system was based on a patent in Germany from 1884, by Paul Nipkow, and involved a perforated spinning disk in the transmitter and one in the receiver, sometimes called a Nipkow disk. Photocells would capture emitted light to transmit a low-quality picture, with 30 lines from top to bottom, transmitted ten times a second. The system required that the scene to be televised be extremely highly lit. Baird demonstrated a transmission in 1924, and on October 2, 1925, transmitted a moving picture of a person. The Nipkow disk system developed by Baird was used in the United States, Britain, the Netherlands, and Germany until the mid-1930s.

In 1907 Boris Rosing, a Russian inventor, proposed that an electronic beam be used with a cathode-ray tube. Zworykin, a student of Rosing, immigrated to the United States in 1919. At Westinghouse, in Pittsburgh, Zworykin demonstrated in 1923 an all-electronic system, a kinescope, using a cathode-ray tube for displaying the received image. The receiving system of a cathode-ray tube was improved over the next decades, and commercial television receivers continued to use the same type of tube through the rest of the century. Although it later became a standing joke that the Russians claimed to have invented numerous technologies later made popular in the United States, in the case of television, there was considerable merit to the observation.

At the sending end of the system, a number of improvements in cameras or image-capturing devices took place through the 1930s and 1940s.

RADIO NEWS FOR APRIL, 1930   905

TELEVISION *Through a* CRYSTAL GLOBE

New Cone-shaped Tube Reproduces 4 x 5-Inch Picture, Is Quiet in Operation and Does Away With Need of Mechanical Parts in Home Receiver

By V. Zworykin

Reprinted by courtesy of the Institute of Radio Engineers

THE problem of television has interested humanity since early times. One of the first pioneers in this field, P. Nipkow, disclosed a patent application in 1884 describing a scanning of the object and picture, for which purpose the familiar perforated disk was employed and at present the rotating disk is giving excellent results within the mechanical possibilities of our time. The cathode-ray tube, however, presents a number of distinct advantages over all other receiving devices. There is, for example, an absence of moving mechanical parts with consequent noiseless operation, a simplification of synchronization permitting operation even over a single carrier channel, an ample amount of light for plain visibility of the image, and indeed quite a number of other advantages of lesser importance. One very valuable feature of the cathode-ray tube in its application to television is the persistence of fluorescence of the screen, which acts together with persistence of vision of the eye and permits reduction of the number of pictures per second without noticeable flickering. This optical phenomenon allows a greater number of lines and consequently better details of the picture without increasing the width of the frequency band.

This paper will be limited to a description of an apparatus developed in Westinghouse Research Laboratories for transmission by radio of moving pictures using the cathode-ray tube for reception.

In the author's opinion, if a receiver is to be developed for practical use in private homes, it should be designed without any mechanically moving parts. The operation of such a receiver should not require great mechanical skill. This does not apply to the transmitter, since there is no commercial difficulty in providing a highly trained operator for handling the transmitter, which consists of a modified standard moving (*Continued on page* 949)

Fig. 1 (above)—A cathode-ray tube—the heart of the Zworykin receiver. Fig. 2 (left)—One type of cabinet receiver housing the Zworykin apparatus

Fig. 3 (above)—Cross-sectional view of cathode-ray tube, including an enlarged drawing of the electron gun. Fig. 4 (left)—Diagram of the band-pass filter which divides the local receiver output into the picture and synchronizing frequencies

television Several different methods of sending and receiving pictures by radio waves were created during the early stages of the development of the television. Ultimately, Zworykin's adaptation of the cathode-ray tube emerged as the basis for modern televisions. *Radio News. Library of Congress*

Zworykin joined RCA in 1930 and began working on improvements to the kinescope, developing by 1931 a practical camera tube, the *iconoscope,* which RCA demonstrated publicly in 1932. Meanwhile, in San Francisco, Philo T. Farnsworth worked on a different type of camera tube, an image dissector, patented in 1927. By 1939 Zworykin had perfected the image iconoscope by a factor of 10 over the earlier iconoscope.

In Britain in 1936, the British Broadcasting Corporation set up two competing systems, one based on Baird's design and the other on license from the Zworykin system, in Alexandria Palace in London. The systems operated on alternate weeks, broadcasting to a few receiving sets. After three months the BBC decided to close down the Baird system, relying on the Zworykin-designed electronic system.

Following World War II, the *image orthicon* emerged from the RCA labs. This tube featured an electron-imaging stage, a slow scan of the image, with internal amplification of the signal. Other improvements allowed color transmission and improvement of the quality of the captured image. Developments late in the 20th century included high-definition TV with more than 1,000 lines of image resolution rather than the standard 525-line picture in the United States and the 625-line European system. The later development of digital high-definition television systems, which send and receive the images in digital data codes, provided even sharper pictures and sound with few imperfections.

thermos bottle

The fact that a vacuum does not transmit heat or cold was understood since the 17th-century work by Evangelista Torricelli (1608–1647), the inventor of the **mercury barometer [III]**. In 1892 James Dewar (1842–1923), a Scottish scientist working on extremely-low-temperature liquids, developed for laboratory use a double-walled bottle containing a space from which the air had been evacuated and the glass sealed. He coated the inner wall of the glass with mercury to further reduce heat transfer. An early Dewar flask remains at the Royal Institution in London. Dewar patented the concept of pumping out the air, but the flask itself was not patented.

One of two German glassblowers employed by Dewar, Reinhold Burger, realized that the flask could have applications for the general consumer, and he applied for a German patent in 1903. In 1904 his Munich-based firm of Burger and Aschenbrenner (his partner was another Dewar glassblower) offered a prize for the best name, and Thermos won. Burger took out trademark rights on the brand name, licensing it from 1906 to

1910 to firms in the United States, Canada, Britain, and Japan. Publicity attended the use of the flask on several expeditions, including Theodore Roosevelt's African trip and Ernest Shackleton's Arctic explorations from 1907 to 1909.

In a court case in the United States in 1962 brought by Aladdin Industries, a judge ruled that the term *thermos,* with a lowercase *t,* had entered the language and could no longer be a protected trademark. The fact that the Thermos Company had advertised that "Thermos is a household word" injured their own claim. The German patent was not upheld in Britain because the courts there ruled that the idea derived from prior work by Dewar.

traffic lights

In the 19th century, **steam railroads [IV]** had installed traffic signals using semaphore arms similar to those developed in 19th-century France, predecessors to the **telegraph [IV],** and similar systems were used to control horse-drawn city traffic as early as the 1860s, and perhaps earlier. However, the first electric traffic light system using red and green lights was installed in Cleveland, Ohio, in 1914. An amber light in a three-color system, operated manually by a policeman, was set up in New York City in 1918, followed by similar midstreet systems in other cities, including Detroit.

The first traffic lights in Britain were installed in 1925, with automatic switching with a time interval set up by 1926. Los Angeles and other cities in California installed a system that combined lights with a semaphore signal as well as an audible bell. It was thought that such systems could assist those who were color-blind.

triode valve

Radio as developed by Guglielmo Marconi (1874–1937) was at first truly wireless telegraphy, capable of making bursts of sound that could transmit either on or off, ideal for the dot-and-dash codes developed by Samuel Morse for the **telegraph [IV].** But to make possible the transmission of voice or music—that is, wireless telephony—several new pieces of equipment had to be developed. The first of these was a **diode** valve or tube, developed by Ambrose Fleming (1849–1945). Fleming, who had worked with Marconi, discovered in 1904 that a vacuum tube with two electrodes could detect alternating radio waves and convert or rectify them to direct current as it heated.

The Austrian electrical engineer Robert von Lieben (1878–1914) worked on the issue of how to amplify telephone transmission, and he used the Fleming thermionic or heating valve. In 1906 von Lieben modified the valve by adding a third, grid-shaped electrode, which could be used to modulate or modify the flow of electrons. In the same year, a similar device was invented by the American physicist Lee de Forest (1873–1961), which he at first called an *audion*. De Forest worked with a team that soon recognized that a three-electrode vacuum tube or triode could be used in **radio broadcasting** or sending equipment as well as in the receiving equipment, allowing a carrier wave that could be filtered out on the receiving end by similar equipment. The de Forest patent rights were acquired by AT&T in the United States.

> Tiny ferryboats they were, each laden with its little electric charge, unloading their etheric cargo at the opposite electrode and retracing their journeyings, or caught by a cohesive force, building up little bridges, or trees with quaint and beautiful patterns.
> —Lee de Forest on observing the flow of electrons in a vacuum, 1907, in his autobiography, *Father of Radio,* 1950

Radio broadcasting of voice and music, which relied on combining many separate patents for both sending and receiving equipment, did not get under way until 1920.

uncertainty principle

The uncertainty principle was first enunciated by Werner Heisenberg (1901–1976) in 1927. Also known as the Heisenberg principle, simply stated, it holds that it is impossible to determine precisely both the position and the momentum of particles on the atomic scale. Although it is possible to measure one or the other, the act of measuring one introduces a degree of uncertainty into the measure of the other.

According to the classical **laws of motion** as developed by Newton, if both the position and the momentum of a system (such as the solar system) are known at a particular time, then the future action of that system can be predicted. Such a view of celestial mechanics led to the 19th-century discovery of **Neptune [IV]** and contributed to the 20th-century search that led to the discovery of **Pluto.** Such a system of logic is said to be deterministic in that future behavior is determined by, and is mathematically predictable from, its present status. However, the Heisenberg uncertainty principle, along with other discoveries in the world of atomic physics, undermined the clarity of Newtonian determinism. With the introduction of the uncertainty factor, the future of a system could only be predicted with statistical probability. Thus it is said that much modern science and some engineering have proceeded on a probabilistic basis. Part of the transformation in science and philosophy in the 20th century has been attributed to an increasing awareness of the uncertainty principle in many fields and the decline of determinism in favor or probabilism and statistical thinking.

The reason for the difference between the deterministic logic that applies quite well in classical celestial mechanics and the probabilistic logic applied at the level of quantum mechanics or atomic and sub-atomic particle behavior is the role of **Planck's constant.** This minute constant number, discovered by Max Planck, is so small that it does not affect objects in the everyday world or the behavior of huge objects such as planets and stars. However, at the subatomic level, the Planck constant number plays such a large role that an accurate measurement of the position of a particle decreases the accuracy of the measurement of its momentum. And conversely, an accurate measurement of the momentum of a particle decreases the accuracy of measurement of its position.

Niels Bohr (1885–1962) developed a related concept in 1927, which he defined as the principle of complementarity. Bohr applied the principle much more widely, not only to the study of subatomic particles but also all disciplines in which the units of measurement are discrete, whole units, such as in human affairs, where "units" are individual people. Bohr held that in such systems, to observe them, one must interfere to some extent in the status of the system. The act of measurement itself disturbs what is being measured. In systems where two complementary features are needed to explain phenomena, analogous to momentum and position in the atomic world, neither can be measured precisely without decreasing the accuracy of measurement of the other factor. Such issues surface frequently in public opinion polls, where hearing a question about a particular policy or problem probably changes to some unknowable extent the opinions of those responding.

Bohr and Heisenberg were contemporaries, both deeply engaged in nuclear physics research. Heisenberg headed the effort of Germany to develop a **nuclear weapon [VI]** during World War II, and Bohr, after a period of neutrality in Denmark, fled to Britain and the United States and consulted on the Manhattan Project.

vitamins

The role of vitamins in sustaining life and in preventing diseases that result from the lack of them in the diet was discovered in a series of studies between 1893 and 1913, continuing into the 1920s and later.

In prior periods it had been assumed that when such diseases went into remission from a change in diet, some factor in the food had poisoned the disease germ. James Lind, a British surgeon aboard HMS *Salisbury,* was stunned at the high rate of scurvy aboard a British naval vessel on a round-the-world voyage from 1740 to 1744. In 1747 he conducted a controlled experiment, discovering that the juice of oranges and lemons were specific against the disease. Captain James Cook (1728–1779) experimented with various items in diet, including sauerkraut, citrus juices, and scurvy grass, an herb later determined to be high in vitamin C. In a 7-month voyage from Plymouth, England, to Tahiti from 1768 to 1769, Cook was able to report absolutely no cases of scurvy. In 1795 the Royal Navy ordered that sailors imbibe lemon juice, and later required lime juice, since limes grew in their own West Indies colonies. The British had eliminated scurvy among their sailors by 1800 and made the issuance of limes compulsory by 1804, the origin of the term *limey* to refer to British seamen. The chemical mechanism remained unknown, however. It was logical although incorrect to assume that a chemical in lemons, limes, and scurvy grass poisoned the scurvy disease germ.

Christiaan Eijkman (1858–1930), a Dutch bacteriologist, conducted studies in the Netherlands East Indies (now Indonesia) from 1893 to 1897, suspecting that the disease beriberi resulted from a lack of some substance in the diet. Eijkman found that he could induce a condition resembling beriberi by feeding fowls a diet of polished rice and that including the rice polishings in the diet would cure the birds' condition. The results of this work stimulated the investigation by other researchers into the relationship between specific diseases and possible elements in the food. Axel Holst and Theodor Frölich in Norway produced scurvy-like symptoms in guinea pigs by diet in 1907, curing it with a supply of cabbage.

Between 1906 and 1912, British doctor Frederick Gowland Hopkins (1861–1947) conducted a series of dietary studies, feeding animals purified carbohydrate, fat, proteins, minerals, and water. Rats did not grow well on such a diet but responded when supplied with minute amounts of natural cow's milk. He assumed that the existence of cofactors or associated food factors separate from energy-containing foods, proteins, and minerals were required by animals for healthy life. He failed to isolate the specific factor in milk, but his work led to further investigations.

In 1906 Hopkins delivered a lecture suggesting that both rickets and scurvy were caused by the lack of trace cofactors or associated food factors and that detailed study of diet would be needed to track down the various specific substances needed by the body. This lecture did much to stimulate dietary research over the next few years.

In 1912 Casimir Funk (1884–1967), a Polish-born American biochemist, collected all published literature on the issue of deficiency diseases. Funk was the first to isolate niacin, latter called vitamin B_3. He coined the term *vital amine* to describe the class of chemicals that he and other researchers were studying, and the word was simplified to *vitamin* by 1920.

In 1913, independent studies by teams at the University of Wisconsin and at Yale University led to the discovery of the fat-soluble growth factor that eluded Hopkins, later described as vitamin A. Thiamine was the first B vitamin, later designated B_1, to be identified chemically, in 1926 by Robert Williams (1886–1965). It was a deficiency of thiamine resulting from a diet of polished rice that had produced beriberi in Asian diets. Funk analyzed the molecular structure of vitamin B_1 and developed a method for synthesizing it in 1936. The relationships between deficiency of vitamin B_3 or niacin and the disease pellagra and vitamin D deficiency with rickets also were established in this period. Vitamin B_{12} or cyanocobalamin, first isolated in 1948, was found to be involved in the body's manufacture of proteins and essential in preventing one type of anemia.

In 1925 Hopkins was knighted for his work, and in 1929 he and Eijkman shared the Nobel Prize for medicine or physiology for the discovery of vitamins. Hopkins' work had led to the identification of vitamin A, while Eijkman's studies had led to the identification of vitamin B_1.

X-ray machine

After the discovery of **X-rays** by Wilhelm Roentgen (1845–1923) in 1895, the potential of the X-ray for medical work was immediately recognized. Within weeks of the announced discovery, an X-ray machine was used in the United States to examine bone fractures. In 1896 Thomas Edison (1847–1931) invented an X-ray fluoroscope, used by Walter Cannon (1871–1945) to trace the movement of barium sulfate through animal digestive tracks. Later this technique was used to help visualize the human digestive system.

The American inventor William Coolidge (1873–1975) invented in 1913 an improved X-ray tube similar to that later used by doctors and dentists. Coolidge's machine allowed precise control over X-ray wavelength and the voltage used.

Meanwhile, the British father-and-son team of William Henry Bragg (1862–1942) and William Lawrence Bragg (1890–1971) made a number of contributions to the knowledge of X-rays. While teaching in Australia, William Henry Bragg grew interested in the newly developing field of radioactivity. During this period he constructed a home-built X-ray machine, which he once used to diagnose his son's injured elbow. Father and son worked together in Britain from 1909 to 1913, following up on the discovery of X-ray diffraction by Max von Laue (1879–1960). The son, William Lawrence Bragg, developed a set of equations known as Bragg's law, describing the diffraction of X-rays through crystals. A system of X-ray spectroscopy developed by the Braggs allowed for the analysis of crystals, establishing the field of X-ray crystallography. The technique became an important tool in mineralogy, metallurgy, and biochemistry. X-ray crystallography was used to isolate the molecular structure of **penicillin** and later, **DNA** [VI]. Both father and son shared the 1915 Nobel Prize in physics, with William Lawrence Bragg, at age 25, the youngest person to receive the prize to that date.

X-rays

On November 8, 1895, German physicist Wilhelm Conrad Roentgen (1845–1923) discovered the existence of X-rays. He noted that when he turned on a modified **Crookes tube [IV]** or cathode-ray tube, completely enclosed in a cardboard box, a group of barium platinocynide crystals across the room fluoresced. Even when he blocked the path between the tube and the crystals with thin metal sheets, the crystals lit up. He soon discovered that some objects cast shadows, especially heavy metals such as lead. The bones of his hand cast faint shadows, although the rays passed through the flesh. He rightly concluded that he had discovered a new, short wavelength ray, which he dubbed *X-ray*.

The news of his discovery spread rapidly, and he and others immediately began to develop applications, especially for medical diagnosis. The original cathode tube, as

> Within about two years of Roentgen's accidental discovery, the whole world of physics had split open. For any one toiler in the vinyard, awareness of the change must have been much more sudden and traumatic.... Matter was no longer stable outside the normal reactions of chemistry, and the almost holy law of conservation of energy was being flagrantly violated. Atoms were not the final and ultimate smallest building blocks of the universe, but still tinier, tender particles existed, linking the previously distinct realms of matter and of electricity.
>
> —Derek de Solla Price, *Science since Babylon* (New Haven, Conn.: Yale University Press, 1961)

developed by William Crookes (1832–1919) and Phillipp Eduard Anton Lenard (1862–1947), could be modified to stabilize the heat produced, and thus the variations in gas pressure within the tube, and to focus the emitted X-rays through one point. Higher applied voltages produced more penetrating X-rays, soon called hard X-rays, discovered to be those of the shortest wavelength. X-rays are now described as falling in the range of 10^{-2} to 300 Å (angstroms), overlapping at the upper range the ultraviolet light spectrum and at the lower end the gamma-ray portion of the electromagnetic spectrum.

Roentgen received the first Nobel Prize in physics, in 1901, for his work, and the *roentgen* or *R* as a unit of radiation was named in his honor.

zipper

The invention of the zipper is a classic story of a persistent inventor with an idea. Witcomb L. Judson, a mechanical engineer, conceived of an automatic hook-and-eye device and filed his first patents in 1891 and 1894, intending them for shoes. He continued to work on the concept of a slide fastener, filing several more patents by 1905. His 1905 fastener included hooks fastened to tape. However, his fasteners were not very workable, in that the hooks would not disengage easily. He formed a company, the Automatic Hook and Eye Company, in Hoboken, New Jersey.

Judson hired a Swedish electrical engineer, Gideon Sundback, and together, Judson and Sundback struggled with the practical details. In 1913 Sundback developed a fastener that had all the essentials: locking members that were interchangeable and identical and an efficient machine for stamping out the parts and attaching them to the tape. The trademarked Talon model worked well. Clothing manufacturers disdained the device, however.

A navy contractor purchased a large order for flying suits, and in 1923 the B. F. Goodrich company bought the fasteners for galoshes. By the 1930s they began to be included in clothing items. According to legend, the name *zipper* came up rather suddenly as the manufacturers racked their brains for an appropriate term. Conversationally, one said, "See how it zips," and the phrase caught on.

PART VI

The Atomic and Electronic Age, 1935 into the 21st Century

After the mid-1930s, the advances, of science and technology went rapidly hand in hand, with popular awareness of the changes heightened by spectacular developments in weaponry and in "weapons platforms." The predictions of Lewis Mumford, made in 1934, that the "neotechnic" age would see multiple cases of science producing technologies and technologies leading to further scientific advance, were rapidly borne out in the remaining decades of the 20th century.

One of the major developments of the period was the creation of nuclear weapons out of the work of nuclear physicists. In this case, the discoveries of science clearly preceded the technological application. In fact, in 1938, there was no such discipline as nuclear "engineer," and the first weapons designers and reactor builders were nuclear physicists themselves. As Vannevar Bush, the inventor of an early type of analog computer, was put in charge of the U.S. Office of Scientific Research and Development, he advocated and fostered the idea that scientific advance had to be well funded and that it had to precede technological development, greatly influencing government policy on science and technology in the postwar decades. His ideas of funding basic research as a stimulus to technological advance were incorporated in the U.S. Navy's Office of Naval Research, and later in the formation of the National Science Foundation. Under his leadership and reflecting ideas current at the time, the phrase *research and development* or *R & D* came to characterize the view that scientific research would precede engineering and would lead to development.

The war and immediate postwar years saw a burst of creativity in many fields, with technical advances flowing rapidly out of the wartime work. Liquid-fueled rockets, computers, microwave transmission of voice and data, jet engines, and radar were only a few technologies boosted by wartime funding and then followed up by efforts to turn the devices to peacetime purposes. Nuclear reactors propelled submarines for the U.S. Navy by the mid-1950s; they were soon employed to generate electric power, with about 500 in place for that purpose worldwide by the end of the century.

Nuclear weapons brought a period of uneasy peace between the major powers, although surrogate wars and civil wars around the world continued to rage. After 1950, with both the United States and the Soviet Union in possession of nuclear weapons, mutual deterrence prevented an outright clash of arms. The threat was too terrible, and those nations and others continually improved weapons-delivery systems in an unrelenting arms race. By the 1960s, nuclear and thermonuclear weapons far more powerful than the first ones used against Japan in 1945 could be taken to a distant target by jet-propelled aircraft, by intercontinental ballistic missiles, or on missiles launched from nuclear-powered submarines.

As inventors labored to turn the machines developed for warfare to peaceful purposes, hundreds of new products came into everyday use. The turbojet engines suitable for warplanes provided a means for civilian air travel; computers designed to decipher coded messages and to project artillery and missile trajectories became more and more common, eventually reduced in size, increased in speed, and reduced in cost to the point that they were found in millions of private homes. By the 1990s, linked together by the Internet, computers and their programs became synonymous with *high technology*. In discussions of financial investments, *technology stocks* came to represent holdings in computer-related firms rather than in companies producing the array of other technologies in machinery, engines, vehicles, medicine, and ships. Reductions in component size with the printed circuit, the transistor, the computer chip, the microprocessor, and the light-emitting diode made it possible to install minuscule computer components in hundreds of applications, from automobiles and auto repair shops to household appliances and factory equipment. Assembly lines, employing numeric-controlled machines, used computer parts, sensors, and articulated tools to approach robotlike production.

With the advances of technology came new concerns about their social consequences. New weapons demanded counterweaponry and defenses, and the vicious cycle of an arms race seemed self-generating.

Arms control agreements by the 1980s brought the worst threats under control, especially those between the United States and its allies on one side and the Soviet Union and its allies on the other. The ease of producing weapons of mass destruction using nuclear or chemical agents raised concerns that terrorists or rogue nations would employ such tools of war.

In addition, continuing problems arising from technology caused further dilemmas that had nothing to do with warfare or defense. Insecticides such as DDT and chlordane poisoned the planet; the refrigerant Freon could destroy the ozone layer in the upper atmosphere; sulfur dioxide and nitrogen oxide emissions from petroleum-fueled engines and from coal-fired electrical plants caused acid rain and greenhouse effects. Nuclear power plants generated radioactive hazards in their waste and, when they suffered accidents, threatened catastrophic disasters. By the 1970s, environmental regulations and controls came into place in the United States and other countries, addressing some of these and other concerns, yet technological hazards persisted. Some of the hazards were results of 19th-century technologies, created when many health and environmental consequences of developments were ignored or simply not known. For example, the gasoline internal-combustion-engine-propelled automobile and the cigarette regularly killed, every year, far more people than were killed in wars. Bringing those hazards and their related health costs under control became major policy concerns in the last decades of the 20th century and continued into the 21st century.

As the world's publics became more aware of the hazards of technology, many of the issues surrounding technical progress, innovation, and experimentation became highly controversial and politicized. Not only the regulation of environmental impacts but also continued experimentation in the areas of biology and the health sciences appeared likely to put unwanted power in the hands of those who controlled the tools and direction of science and technology. Various science fiction scenarios, in which governments sought to control human behavior and structure social relationships by systems of technological observation and biological manipulation, haunted public consciousness. On the positive side, medicine benefited greatly from the advance of science and technology, with machines such as cardiac pacemakers and kidney dialyzers, both of which had been conceived in the 19th century, being perfected and coming into regular use in the last decades of the 20th century. Other tools, such as fiber-optic opthalmoscopes and magnetic

Chronology: The Atomic and Electronic Age, 1935 into the 21st Century

Year	Invention/Discovery	Inventor/Discoverer
1930–1935	cyclotron	Ernest O. Lawrence
1931	deuterium	Harold Urey
1935	analog computers	multiple inventors
1935	nylon	Wallace Carothers
1935–1939	radar	multiple inventors
1937	digital computers	multiple inventors
1938	ballpoint pen	Ladislao and Georg Biro
1938	fiberglass	Owens-Corning company
1938	fluorescent lights	General Electric company
1938	mathematical theory of computing	Alan Turing
1938	xerography	Chester F. Carlson
1939	aerosol spray can	Julian S. Kahn
1939	DDT (used as insecticide)	Paul Hermann Müller
1939	helicopter (with tail rotor)	Igor Sikorsky
1939	jet engine aircraft	Hans von Ohain
1939	snorkel	multiple inventors
1939	streptomycin	Selman A. Waksman
1940	plutonium	Glenn Seaborg and Edwin McMillan
1941	Dacron	J. R. Whinfield and team
1942	nuclear reactor	Enrico Fermi; Leo Sziland, patent 1934
1942	proximity fuse	Applied Physics Laboratory
1942	uranium isotope separation	Manhattan Engineer District
1943	aqualung	Jacques-Yves Cousteau
1944	ejection seat	James Martin
1944	thermofax	Carl Miller
1944	chlordane	Julius Hyman
1945	nuclear weapons	Manhattan Engineer District
1946	microwave transmission voice and data	multiple inventors
1947	holography	Dennis Gabor
1947	radiocarbon dating	Willard Libby
1947	transistor	John Bardeen, Walter Brattain, and William Shockley
1948	Polaroid camera	Edwin Land
1948	printed circuit	multiple inventors
1950	fiber optics	Narinder Kapany
1951–1953	DNA model	Francis Crick and James Watson
1952	magnetic resonance imaging (MRI)	Felix Bloch and Edward Purcell
1952	cardiac pacemaker	Paul Zoll

Chronology: The Atomic and Electronic Age, 1935 into the 21st Century *(continued)*		
Year	**Invention/Discovery**	**Inventor/Discoverer**
1952	wide-screen motion pictures (Cinerama)	Fred Waller
1954	kidney dialysis (successful)	Willem Kolff, Karl Wulter, George Thorn
1954	thermonuclear weapons	multiple inventors
1955	microwave oven	Raytheon company
1955	music synthesizer	Harry Olson and Herbert Belar
1956	Wankel engine	Felix Wankel
1957	artificial satellite	development team in the Soviet Union
1957	intercontinental ballistic missiles (ICBMs)	development teams in the Soviet Union and the United States
1958	hovercraft	Charles Cockerell
1958	laser	Theodore H. Maiman
1958	radio thermal generator (RTG)	Martin company researchers
1958	Van Allen belts	John Van Allen
1959	charge-coupled device	George Smith and Willard Boyle
1959	fuel cell application	multiple developers
1959	computer chip	Jack Kirby
1960	carbon fiber	multiple developers
1960	oral contraceptive	Gregory Pincus
1961	laser disc	David Paul Gregg
1962	light-emitting diode (LED)	Nick Holonyak
1964	quark	Murray Gell-Mann and George Zweig
1965	Kevlar as plastic	DuPont corporation
1966	vertical takeoff and landing (VTOL) aircraft	British Aerospace company
1969	ARPANET (Internet processor)	Leonard Kleinrock and Douglas Englebart
1970	microprocessor	Marcian Hoff and Frederico Faggin
1973	genetic engineering	Paul Berg
1976	Kevlar fiber	Stephanie Kwolek
1977	human in vitro fertilization	Patrick Christopher Steptoe
1992	planets outside the solar system	Alexander Wolszczan
1995	Internet	multiple developers
2001	human genome draft sequence	multiple researchers

resonance imaging, allowed better diagnosis of problems inside the human body without incurring the risk of tissue damage from open surgery or cancer associated with earlier generations' X-ray machinery. New medicines, particularly penicillin and streptomycin, provided cures for many diseases. Oral contraceptives, introduced after 1960, allowed more effective methods of birth control and family planning.

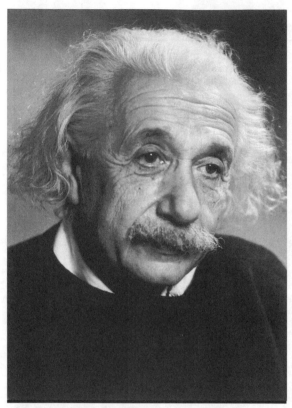

Albert Einstein neither invented devices nor made any major scientific discoveries, yet his theories regarding the relativity of matter and energy, as well as space and time, greatly influenced and shaped the development of late 20th-century science and technology. *Library of Congress*

The electron microscope aided research, and great breakthroughs came with the understanding of the roots of genetic imprinting on the DNA molecule.

Technologies being studied at the dawn of the 21st century held out the promise of solutions to long-standing human problems. With genetically modified food crops and study of the human genome, biology was clearly on the verge of new departures. The fuel cell went through a series of improvements that promised that it might replace the internal-combustion engine as a means for propelling automobiles and a wide variety of other machines, with no emission of noxious products and reduced risk to health. Nuclear fusion, without the radiation risks of nuclear fission, might provide a vast source of electrical power later in the century.

Yet suspicion of technology and fear of its consequences, spawned by the terrible effects and threats of warfare and the impersonal nature of its challenge to the environment, left humanity concerned: would invention and discovery provide tools for future benefit or pathways toward future planetary disaster and restrictions on liberty?

aerosol spray

The spray can demonstrating the principle of an aerosol was invented in 1926 by a Norwegian inventor, Eric Rotheim, who discovered that a liquid could be contained under gas pressure in an aluminum can. In 1939 American inventor Julian S. Kahn patented a disposable spray can, later perfected by Lyle David Goodhue, who in 1941 patented a refillable aerosol can. Goodhue is usually credited with the invention. During World War II, heavy-duty aerosols containing insecticides were constructed for use in delousing operations in refugee camps and as a form of extermination device. In the postwar era, surplus insecticide "bombs" introduced the concept to the general civilian consumer.

A New York City machine shop owner, Robert Abplanalp, developed in 1949 a plastic valve that led to the widespread use of the aerosol can in other applications. The propellant gas was originally **Freon [V]**, a chlorofluorocarbon (CFC), invented by Thomas Midgley, who also invented tetraethyl lead, the additive used from the 1920s through the 1980s to raise the octane of gasoline. However, controversy over the use of the CFCs developed in the 1970s as scientists came to conclude that these chemicals, when reacting with chlorine in the atmosphere, could break down the protective ozone layer in the upper atmosphere, which serves to protect animal and human life from the harmful effects of solar ultraviolet radiation. The use of CFCs in aerosols was banned in 1978 in the United States, and later in other countries. Manufacturers substituted hydrochlorofluorocarbons, carbon dioxide, and other gases as propellants.

There are three types of aerosol sprays. A space spray dispenses the ingredients in a fine spray of particles smaller than 50 micromillimeters in diameter. Larger, more liquid sprays, known as surface-coating sprays, produce a propelled mist appropriate for the application of paints, lacquers, and other films of material. Foam sprays are not truly aerosols, and shaving creams and other foams, such as whipped cream, are actually gases suspended in a liquid base. True aerosols are the reverse, liquids suspended in gaseous bases.

In construction of aerosol cans, the active ingredient is dissolved in alcohol. In the case of a common deodorant, aluminum chlorhydroxide, the active ingredient is not dissolved but suspended in particles in the alcohol. The concentrate of active ingredient and alcohol is measured into the can, and a valve and cap are fitted over the opening. The propellant is then injected under pressure through the valve in a standard amount. Then the spray nozzle is fitted onto the valve stem.

The term *aerosol* also has been applied to naturally occurring phenomena, such as cooled air that leads to the condensing of supersaturated water vapor in the form of fog or clouds. Smoke is a form of aerosol, with particles in the range of billions per cubic inch.

aqualung

Jacques-Yves Cousteau (1910–1997) developed the aqualung in occupied France in 1943 during World War II. An officer in French naval intelligence, Cousteau began working on free-diving equipment in 1936, experimenting with existing systems that used compressed air. The problem with compressed-air tanks for diving use was that they could not dispense the air at the changing pressures of various depths in water. Pressure increases about the equivalent of 1 atmosphere every 33 feet.

After several near-fatal accidents, Cousteau identified the pressure problem and took it to Émile Gagnan, an industrial gas equipment designer. Working with specifications developed by Cousteau, Gagnan developed a valve that linked the diver's exhaust breath with his intake breath. The outgoing breath opened an outlet valve and at the same time shut off the input air. When the diver breathed in, the outside pressure of the water would close the outlet valve and switch on the incoming air supply from the compressed air tank. A diaphragm kept the pressure equal to the outside water pressure.

The whole unit, including the control valve, the face mask, and the compressed-air tank, was known as the aqualung. Cousteau successfully dived for the first time with the system in June 1943, reaching a depth of about 60 feet. Aqualung dives tended to be limited to about 300 feet until the introduction of a helium-oxygen mixture, using the same valve system. With the removal of the nitrogen in regular air, the aqualung became capable of supporting divers safely to about 400 feet. In the postwar years, Cousteau used the revenue from the aqualung to support research and film voyages of the exploration ship *Calypso* and became known to television viewers all over the world with his popularization of undersea exploration. The aqualung had rapid impact on the scientific fields of underwater archaeology and marine biology.

By the 1950s and 1960s, the use of the aqualung for diving had become a popular sport. At the same time, many navies adopted aqualung equipment for special missions such as planting mines, intelligence-gathering, and for more routine tasks such as maintenance of underwater equipment.

artificial satellite

The concept of an Earth-orbiting artificial satellite had existed as scientific speculation and science fiction for centuries, stimulated by the publication by the Russian scientist Konstantin Tsiolkovsky (1857–1935) of the paper "Exploration of Cosmic Space by Reactive Devices" in 1903 and by the calculation of the energy required for liftoff from Earth gravity by the American Robert Goddard (1882–1945) in 1926. Wernher von Braun (1912–1977), the scientific director of the German V-2 program, anticipated that further development of rockets would soon lead to the placement of satellites. Von Braun, his team, and many captured rockets and related equipment were shipped to the United States at the end of World War II.

During the International Geophysical Year of July 1957 to December 1958, the United States anticipated launching a satellite but sought to use a nonmilitary rocket, the Vanguard, rather than a larger Jupiter rocket, developed at the Redstone Arsenal by the von Braun team.

The Soviet Union proceeded with its rocket development and successfully launched the first Earth-orbiting satellite, *Sputnik I,* on October 4, 1957. The Soviet team, headed by Sergei Korolev (1906–1966), operated from a *cosmodrome* near Baikonur in the Soviet republic of Kazakhstan.

In Russian *sputnik* means fellow traveler, suggesting that the satellite accompanied Earth in its travels through space and providing a rather coy allusion to the political term meaning a nonparty supporter of the

artificial satellite
The International Space Station was the largest artificial satellite attempted in the 20th century. It is shown here as it neared completion. *National Aeronautics and Space Administration*

Communist ideology. Worldwide teams of volunteers had been established to track satellites, and Soviets installed aboard the satellite a radio beacon that transmitted on ham radio frequencies, facilitating radio tracking. On November 3, 1957, the Soviets launched *Sputnik II,* with a dog, Laika, aboard. Laika survived 10 days aboard the satellite.

Meanwhile, the American Vanguard program was accelerated but continued to result in test failures. American politicians, journalists, and public opinion went into shock. Although the Soviets had announced in advance that they intended to launch a satellite, few in America believed the claims were more than propaganda. However, the Soviets had worked on a large military rocket for carrying **intercontinental ballistic missiles** with relatively heavy payloads of nuclear weapons, and Korolev used the R-7 rocket, powered by a cluster of 20 liquid-oxygen and kerosene-fueled engines, for the heavy lift.

The Soviet successes with *Sputnik I* and *II* in effect launched the *space race* between the two major powers and resulted in a boost of government spending for American science and technology. Investigations revealed that the military Jupiter rocket could easily launch a satellite, and the first successful American satellite, *Explorer 1,* was launched on a Jupiter booster on January 31, 1958. That satellite and *Explorer 3* detected the **Van Allen belts,** regarded as the most significant scientific findings of the International Geophysical Year.

The Soviets continued the Sputnik series with a total of eight satellites, and followed later with manned space flights and Sun-orbiting satellites. The first successful Vanguard launch was on March 17, 1958.

The first communications satellite, *Telstar,* was launched in 1962. It began operation by receiving and retransmitting television programs from the United States to Europe and back. *Telstar* circled at a height of 200 to 3,500 miles at a speed between 11,000 and 18,000 miles per hour.

autopilot

Experiments with drone aircraft and with systems of automatic piloting were conducted in the United States, Britain, Germany, and elsewhere from the 1920s through World War II. The first successful automatic flying and landing systems, useful in fog or other extreme weather, were introduced in several variations after World War II.

The first autopilots simply maintained an aircraft in a level and straight flight, controlling the movement of the aircraft to adjust back on course. In this simple form, autopilots allowed a period of rest for the human pilot. However, more complete systems have been designed to follow pre-programmed flight paths and even complete flights, from takeoff through landing.

A complete automatic pilot system is composed of four linked subsystems. First there must be a source of steering commands, either a computerized guidance program or a system linked to a radio receiver. Second, a variety of position and motion sensors have to be linked, such as airspeed indicator, an altimeter, and accelerometers. Third, there must be a computer system to compare the programmed flight to the sensed position and movement. Fourth, the computer must link to a system of servomotors that regulate the engines and the control surfaces to make the necessary adjustments.

The first aircraft to fly across the Atlantic on a complete automatic system was a British four-engined Skymaster, on a 2,400-mile trip in 1947. Using an analog **computer** system, the plane flew entirely without human intervention, from takeoff through climb, level flight, and landing, including braking after landing on the runway. By the 1960s, automatic pilot systems for level flight were installed on most long-distance military and civil aircraft, but automatic landing systems were used only rarely. In 1966, several fully automatic landings were made, in both London and New York, with airliners that carried passengers. Several systems for the "blind landings" were developed, including a British system that required cables laid on either side of the runway, sending signals to a receiver aboard the aircraft that connects to controls so that the plane is centered on the runway. More common was an instrument landing system, which fed the pilot needed information as to height, wind speed, descent speed, and orientation to the runway. Generally such systems would cut off during the last 250 feet of descent, requiring manual control for the final descent and touchdown.

ballpoint pen

Two Hungarian brothers, Ladislao J. Biro (1899–1985) and Georg Biro (1897–19??), invented the ballpoint pen in 1938. Ladislao was a sculptor, painter, and writer, while his brother was a chemist. The two brothers applied for patents in 1938, but during World War II they moved to Argentina. There they formed a company to produce the pens, and they patented several improvements that allowed the ink to be fed by capillary action rather than by gravity, reducing leakage. A British financier, Henry Martin, manufactured Biro pens for use by Royal Air Force pilots, as the pen did not leak with the changes in pressure of high altitude. Biro pens were in use throughout Europe by the end of the war.

In France, Baron Marcel Bich produced extremely cheap ballpoint pens in Clichy, outside Paris, achieving a production rate of 7 million pens a day in the late 1940s. Bich bought out the Waterman Company in the United States and imported the Bic.

The Biro pen was licensed to Eversharp and to the Eberhard Faber Company in the United States. However, in 1945, an American businessman, Milton Reynolds, visited Buenos Aires and brought back several Biros to the United States. Consulting a patent attorney, he found that an earlier ballpoint system for a marker had been patented in 1888 by John J. Loud. Working from that expired patent and developing an improved flow system, Reynolds developed a gravity-feed pen. The ink remained a problem, with early inks skipping, clogging, or fading under sunlight. In 1949 Fran Seech, an Austrian chemist, developed improved ink that dried quickly, and Frawley Corporation introduced a system employing his ink as the Papermate pen.

Other variations were introduced, with Pentel bringing out the felt-tipped ballpoint in 1973 and a ceramic nib in 1981. Gillette introduced an erasable ballpoint pen in 1979.

cardiac pacemaker

The concept of using electric pulses to stimulate or restart a heart had been suggested as early as 1862. By the 1930s, hospitals used outside electrodes to restart the hearts of patients on the operating table or in emergency situations. However, the first experimentation with a practical electrical device to regulate the beating of the heart began in the 1950s, when American doctor Paul Zoll attempted to pass an electrode down a patient's esophagus. In 1952 Zoll successfully introduced an experimental pacemaker that passed an electric current from electrodes mounted on the outside of a patient's chest. The patient had congestive heart failure, and after two days on the pacemaker, the patient's heart took over with a regular beat. Even though workable, the shock through the skin was painful, and the system required access to an electrical outlet.

Meanwhile, an American inventor, Wilson Greatbatch, had been experimenting with the concept of an implantable pacemaker, but with vacuum tubes and batteries, the system would be too large to implant. However, with the development of **transistors**, the possibility of a smaller system immediately occurred to Greatbatch. He worked with Dr. William Chardack at the Veterans Administration Hospital in Buffalo, New York. Greatbatch built a model in his own shop, and the model was successfully implanted in 1960.

Upgrades to the pacemaker continued, with improved and smaller batteries. In 1981 Ken Anderson and Dennis Brumwell of Medtronic, Incorporated, of Minneapolis developed a pacemaker that used piezo-electric crystals that reacted to different degrees of exertion of the patient, allowing the machine to vary its rate depending on the subject's activity level. Later models included a **computer chip** and a small radio receiver that allowed the pacemaker to be reprogrammed without removing it from the body. Most pacemakers were set to turn on and off, responding when the patient's heartbeat fell below 68 beats a minute.

charge-coupled device

At Bell Labs in New Jersey, two researchers, George Smith and Willard Boyle, were the coinventors of the charge-coupled device (CCD) in 1959. The heart of the CCD was a metal-oxide semiconductor, usually with silicon as a base material with a coating of silicon dioxide. Numerous electronic cameralike devices were made from CCD components over the next few years, including video cameras, telefax machines, scanners, and still-image recording cameras.

The core device was a set of discrete packets of electrically charged columns of semiconductor capacitors. The charged capacitors transformed light energy into electrical impulses. The electrical impulses were sent as signals to a **computer chip,** where they were transformed into digitally coded information. The microprocessor selected areas of light and dark from the image, breaking them down into points or pixels. Each pixel was assigned a numerical brightness value, ranging from white to black. Each pixel was also given an x-y coordinate, defining the pixel's location in the total image. In effect, the CCD allowed the transformation of an image into a numeri-

ccd camera This charge-coupled device camera was installed at the Lick Observatory on Mount Hamilton, California, to assist in the search for planets outside the solar system. *National Aeronautics and Space Administration*

cal sequence that could be stored or transmitted electronically. Among the first uses of this method were the transmission of images of the surface of Earth taken and sent from **artificial satellites.**

Bell Labs produced the first solid-state video camera with a CCD recording system in 1970, and by 1975 they had produced a television camera using the solid-state system that had image quality sufficient for broadcasting.

Because the CCD system both is highly sensitive and reduces images to data capable of being digitally transmitted, it found some of its first uses in surveillance satellites for environmental monitoring and defense purposes. The *KH-11* satellite, developed for the Central Intelligence Agency in 1972, represented a crucial early use of the Bell Labs technology, since it allowed high-quality images gathered in space to be sent to Earth without the necessity of returning film. The first telescopes to be outfitted with CCD cameras were developed in 1983. Extremely faint and distant stars and galaxies could be imaged more readily with CCD cameras than with traditional photographic plate or film cameras. One advantage of CCD photography is that it is quite easy to store the information in digital form in computers. By the mid-1990s, **digital cameras** for amateur use utilized CCD technology.

chlordane

Chlordane was a powerful insecticide discovered in the mid-1940s by an independent researcher, Julius Hyman. Hyman had studied at the University of Chicago and at Leipzig and worked in the United States. Together with a cousin, who provided financing, he set up a company, Velsicol, to develop drying oils for paints and varnishes as well as for other chemicals. With the discovery that **DDT** had insecticide properties, Hyman began to examine a wide variety of polymers manufactured by his company to see if they worked as insecticides.

After experimenting with cyclopentadiene, a by-product of the manufacture of synthetic rubber, Velsicol sent a sample to the Entomology Department of the University of Illinois in 1944. It had mild insecticidal effect, and Hyman modified the chemical and produced a new compound. After a dispute with Velsicol, Hyman set up his own company, Julius Hyman and Company, to manufacture the resultant compound, chlordane. Two other chemicals that could also be used as insecticides, aldrin and dieldrin, were developed by researchers who left Velsicol. A court in Colorado ruled that the patents to these chemicals belonged to Velsicol. As a result of the contributions by different scientists to the development of the three chemicals, extensive patent fights ensued. After Velsicol won several cases from 1950 to 1952, Shell Development Company purchased Julius Hyman and Company and also secured the rights to aldrin and dieldrin from Velsicol. Shell produced chlordane, aldrin, and dieldrin as insecticides.

Like DDT, chlordane's use in agriculture was prohibited in 1972 because of its persistence in the environment.

composite materials

Fiberglass and carbon fiber, new composite materials developed in the late 20th century, began to revolutionize the manufacture of a wide variety of products.

Glass-reinforced plastic (GRP) or fiberglass was developed in 1938 by Owens-Corning, and in the post–World War II years the material began to find multiple applications in boat construction, automobile bodies, luggage, and other settings where a strong and lightweight material that could be made in sheets was needed. The principle of layering glass fibers and bonding them with a resin led to a wide variety of composite materials. In the United States, the introduction by Chevrolet of the Corvette automobile, in 1954, with an all-fiberglass body remained the only such use for a full automobile body for many years.

In military applications, fiberglass was used to construct radar domes for aircraft and submarines, and the U.S. Navy experimented with some applications in deckhouses, piping, exhaust ventilation, and minesweeper hulls. Several methods of heating and impregnating the fibers with resin developed, including an ultraviolet-cured *prepreg* and a low-temperature prepreg method. In the 1990s, a vacuum-assisted resin-transfer molding process was used for large structures and applications, in which the layers of fiber cloth and resin would be encased in a plastic bag from which the air would be evacuated, ensuring the absence of air bubbles in the finished product.

Carbon fiber had been developed by Thomas Edison in 1879, for filaments in **electric light** [IV] bulbs, but by the 1920s, carbon filaments had been replaced by metal filaments, particularly tungsten. The use of carbon fiber derived from **rayon** [V] in fireproofed textiles for the aerospace program in the United States began in about 1958.

In 1960, Standard Oil of Ohio developed a method for producing carbon fiber products, bonding the carbon with a plastic known as polyacrylonitrile (PAN). Simultaneously, experimenters in Japan, the United States, Britain, the Soviet Union, and France began working on carbon fiber composites, known as carbon-fiber-reinforced plastics or CFRPs. Experiments in the 1960s with the use of carbon fiber instead of glass fiber produced a much stronger but more expensive product that found application in a wide variety of products in which cost of materials was a minor consideration and where a lightweight material as strong as metal would be desirable. The high cost of the product tended to restrict its application, but as the cost was reduced and its virtues became better known, it was sought for new uses.

By the 1990s, carbon fiber found applications in such varied products as guitars and cellos, boats, rugged mountain bicycles, automobile parts, ski poles, tennis rackets, golf-club shafts, and snowboards. Depending on the heat of preparation, carbon fiber composites approach steel in strength, but with a fraction of the weight.

computer

There are numerous contributors to the development of the first computers from 1935 to 1945 in the United States, Great Britain, and Germany. The concept of a computing machine had predecessors, with the theoretical ideas of Charles Babbage with his **analytic engine [IV]** in the 19th century, the developments by Herman Hollerith of the **card punch machine [IV]** in the 1880s and its improvement over the next decades,

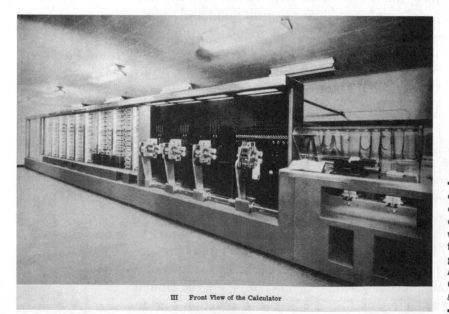

III   Front View of the Calculator

computer Among the many claimants to inventing the computer was Howard Aiken, who worked with colleagues from IBM to develop this first, general purpose computer, the Aiken Automatic Sequence Coltrolled Calculator (ASCC). *Library of Congress*

and the design by Vannevar Bush of a **differential analyzer [V]**. In 1938, working on breaking the codes of the German **encryption machine [V]**, Alan Turing proposed the **mathematical principle of computing.**

In the late 1930s and through World War II, a wide variety of electromechanical machines were developed that used relays, vacuum tubes, and electrical wiring to accomplish mathematical computations. Since the machines made mechanical moves to register numerical quantities, some were later referred to as *analog computers,* to distinguish them from machines that registered the numbers in purely electronic form as on-or-off states or 0s and 1s, known as *digital computers.*

Some of the simultaneous or near-simultaneous inventions are listed in the table "Early Analog Computers."

By the standards of later computers that utilized **transistors, computer chips,** and **printed circuits,** these early electromechanical and vacuum tube computers were slow, large, and had very little memory or speed. For example, the ENIAC weighed 30 tons and contained 17,480 vacuum tubes, 70,000 resistors, and 6,000 switches. The ENIAC operated for about ten years, first at the Moore School in Philadelphia, and then at the U.S. Army proving ground in Aberdeen, Maryland.

Although the potential of all of these machines for use on scientific problems was apparent, their first uses were all for the military. Early uses included decryption, as with the Colossus machines at Bletchley Park in Britain; calculation of ballistic trajectories, with the MARK I and the ENIAC; and later calculations of missile trajectories, fuel

Early Analog Computers

Year	Machine	Inventor	Detail
1927–1928	differential analyzer	Vannevar Bush	Massachusetts Institute of Technology; analog calculator
1937	ABC	John Atanosoff	Atanasoff-Berry computer, Iowa State, first-generation digital computer, using vacuum tubes
1938	Z1	Konrad Zuse	Berlin, independent inventor; first programmable binary analog computer
1943	Colossus machines	M. Newman	Bletchley Park, Britain; implemented Alan Turing concept for cracking Enigma codes
1943	ASCC/MARK I	Howard Aikin	Automatic sequence controlled calculator, Harvard University and IBM Corporation; used by navy for ballistic trajectories
1943–1945	ENIAC	John Mauchley	Moore School of Electrical Engineering of Pennsylvania; for army ballistic trajectories
1949	EDSAC	Maurice Wilkes	Electronic delay storage automatic calculator; Cambridge University, Britain

requirements for **nuclear reactors,** and targeting commands for long-range missiles.

computer chip

The microprocessor is a general-purpose logic computer chip, or printed circuit, that can perform the generic functions of a computer. When combined with separate chips for memory, data, and a program, the microprocessor could perform all the functions of a larger computer. Sometimes the terms *computer chip* and *microprocessor* are used interchangeably, although, strictly speaking, the term *microprocessor* refers to a chip that can perform the central processing functions of a computer. When combined with a memory chip, a data chip, and a program chip, the microprocessor performs as a complete central working part of a computer.

A British radar engineer, G. Dummer, had first suggested the integrated circuit in 1952. His concept would have implanted miniature electronic equipment in a layered block of insulating material, with connectivity achieved by cutting out the insulation. Although the British did not take the idea to fruition, two American companies proceeded with the concept in 1958 and 1959. Texas Instruments developed the integrated circuit, while a printed circuit along similar lines was developed at Fairchild Semiconductors.

In 1958, Jack Kilby of Texas Instruments in Dallas, Texas, invented the integrated circuit or printed circuit that allowed several **transistors** and connections to be mass manufactured. Kilby's 1959 patent combined transistors, resistors, and capacitors on a single chip, with the connections made directly on the insulating layer of the chip without connecting wires.

Robert Noyce invented a different type of integrated circuit in 1959. Robert Noyce and Gordon Moore, who had first worked with William Shockley, one of the inventors of the transistor at Bell Laboratories, formed Shockley Industries, and then Fairchild Semiconductors, with funding from Sherman Fairchild, who owned a camera company in Syosset, New York.

Noyce and Moore established their own firm, Intel, in 1968, with the concept of developing a standardized chip that would replace the bulky magnetic core memory in computers. The microprocessor that served to replace the central processing unit of a computer was developed from 1969 to 1971 by Marcian Hoff and Frederico Faggin at

Intel Corporation. The first microprocessor, the 4004, sold in 1971, was equivalent in computing power to the 1945 computer ENIAC, which had occupied a room. The assemblage of microprocessor with related program, storage, and data chips would be about ¼ inch square. At first utilized in handheld calculating machines, as the devices were improved they found many applications, including microprocessor-controlled ignition systems in automobiles; small notebook computers capable of high-grade performance; and a wide variety of games, electronic equipment, and programmable equipment such as thermostats, ovens, and video recording devices. Commercial applications included automatic teller machines and point-of-purchase terminals in retail stores.

The name Intel was a contraction of *integrated electronics,* reflecting the concept that the new **printed circuit** represented the integration of several transistors into a unit. Intel's work at large plants in Mountain View, California, and in a new factory built in Santa Clara County, on orchard land, began the move of much of the U.S. electronics industry to that region, soon to be called Silicon Valley.

Working with Andrew Groves, who had immigrated to the United States from Hungary after the 1956 uprising there, Intel worked on a secret process to develop a new chip. The firm relied on secrecy rather than patents to preserve its niche in the expanding computer business.

The Pentium II chip, introduced by Intel in 1994, was seriously flawed. Although many computer users had the new chip in electronic games or word-processing machines and never encountered the problem, others who used the computer for mathematical computations found that the Pentium chip frequently caused errors in long division. Intel recognized the mistake and offered to replace the original chips with new ones that rectified the problem. The error cost Intel some $475 million, about half of the company's earnings during the last quarter of 1994. Later iterations of the Pentium chip did not suffer the same problem.

copy machine

The Scottish inventor James Watt (1736–1819), who is better remembered for his work on the **steam engine [III],** invented an early version of a device to reproduce letters. In 1780 Watt patented a letter-copying machine that pressed a very thin piece of paper over a handwritten text written with special ink, producing a copy in reverse on one side. The paper was so thin that the copy could be read properly from the other side. Watt's letterpress was widely used over the 19th century, and letterpress copies are encountered in archival research.

Other devices for copying included one invented by Thomas Edison (1847–1931) in 1877. Edison's device was a stencil machine that squeezed ink onto waxed paper, which in turn could produce a positive copy. In 1884 A. B. Dick invented the mimeograph, using a gelatin material to pick up the ink from the back of a specially coated paper. The gelatin reverse copy could then be used to produce multiple paper copies from a hand-cranked or motorized paper feeder. When the original was produced from a **typewriter [IV]**, up to 100 or more legible copies could be produced. Mimeograph machines with their gelatin stencils were in common use, particularly in educational settings, into the 1970s. Meanwhile, other methods were developed for various purposes.

Photocopies involved one or another device employing photography. In 1906 the development of *photostats*—simply photographs made on photosensitized paper, rather than on film—allowed for a quick method of reproducing a readable copy. Blueprint involved a form of contact photography that required the original to be written or drawn on translucent paper. Copies of white on blue would be produced from a paper containing iron compounds that react to light, developed with ammonia vapor. The process was limited to the production of a few copies, excellent for precise and permanent copies limited in number, ideal to meet the requirements for architectural plans or copies of legal contracts.

Thermography was a different process, involving the action of heat on heat-sensitive materials. In 1944 Carl Miller developed a method based on the fact that a dark surface absorbs heat more readily than a light surface, leading the Minnesota Mining and Manufacturing Company (now the 3M Company) to produce the *thermofax*. That process required infrared rays to burn a copy directly onto a thermosensitive paper. The thermofax copy was useful in situations in which one or more temporary copies of a document were needed. The images faded, however, and the paper grew brittle, leading to continuing the search for a better procedure.

Xerography, invented in 1938 by the American inventor and physicist Chester F. Carlson (1906–1968), was a method employing electrostatic charges, a dry powdered ink, and heat that would fuse the print onto paper. Carlson's first successful transfer of image occurred on October 22, 1938. Carlson experimented at home, and by 1948 he was able to interest a small photographic company, Haloid, in his product. After improvements, Haloid changed its corporate name to Xerox and began to market the device as a Xerox machine in 1951, with a plain-paper model introduced in 1959. The method employed an aluminum drum coated with selenium, which is a poor electrical conductor but which will hold an electric charge in the dark. A mirror reflects a negative image to the drum, and the dark portions with an electrostatic charge pick up ink pigment. Then rolling an electrostatically charged sheet of paper along the drum releases the image to the paper. The superior quality of xerography over mimeography supplanted the earlier method by the last decade of the 20th century.

cyclotron

The cyclotron for the acceleration of subatomic particles was invented by Ernest O. Lawrence (1901–1958), an American nuclear physicist working at the University of California. The Hungarian émigré Leo Szilard

cyclotron This cyclotron at the University of California was built under the supervision of E. O. Lawrence, right. *National Archives and Records Administration*

[The first cyclotron was] an ordinary wooden kitchen chair on top of a physics laboratory table at the University of California at Berkeley. On either side of this chair stood a clothes tree, with wire hanging on the hooks which normally would hold hats and coats. Between the two poles the wire was suspended in loose hoops. The loops went all around the chair, on the seat of which was an object about the size and shape of a freshly baked pie. It was made of window glass, sealing wax, and brass.

—L. A. Schuler, "Maestro of the Atom," *Scientific American* 163, no. 8 (August 1940): 68

(1898–1964) applied for a patent on the concept on January 5, 1929. Ernest Lawrence thought up the same idea on about April 1, 1929, and built his first working model a year later. He used it to discover a number of transuranic (heavier than uranium) elements and won the Nobel Prize in physics for his work in 1939.

The first cyclotron was only 4.5 inches in diameter, and a 27-inch model was used to produce artificial **radioactivity [V]**. **Electrons [V]** or protons received repeated synchronized accelerations by electrical fields as the particles spiraled outward from their source. The particles were kept in the spiral by a powerful magnetic field and deflected toward a target to measure effects.

Dacron

Dacron is the U.S. trade name for a synthetic fiber first invented or discovered in Britain under the name Terylene in 1941. Dacron is tough, resistant to abrasion and sunlight, and highly resilient. In many regards superior to **nylon,** Dacron was a follow-up to the discovery of nylon and a result of the interest in polymers stimulated by that discovery. J. R. Whinfield, J. T. Dickson, W. K. Birtwhistle, and G. G. Ritchie, research workers at Calico Printers Association in Britain, worked to find a polyester with a higher melting point than nylon.

Whinfield had graduated from Cambridge in chemistry and had worked on fibers before joining Calico Printers Association. After pressuring the firm to allow research in the field from 1935, he assembled a small staff by 1940. By 1941 he had produced polyethlyene terephthalate, the crude raw material for the fiber, on a small scale. Whinfield and Dickson submitted the patent application in 1941, and it was finally granted in 1946. Meanwhile, du Pont in the United States began work on a similar fiber, and in 1946 purchased for American production the British patent that Calico Printers Association had sold to Imperial Chemical Industries (ICI).

The Dacron fiber is made by taking an amorphous polymer and extruding it through spinnerets, then stretching the filament to approximately four times its original length, causing the molecules to take up a parallel structure in a crystalline form of the polymer. The material is opaque and insoluble in most organic solvents, making the resultant fabric highly useful for rugged use.

ICI went into production of Terylene on a large scale in 1955, while du Pont began experimental production in 1950 and full-scale production in 1953, first as Fiber Five and later as Dacron. Dacron has found many uses, not only in clothing but also in lines and sails for small boats; electronic tape; photographic film base; webbing; filter cloth; and in upholstery and carpeting, where toughness and resistance to temperature change and wear are important considerations.

deuterium

An American chemist, Harold Urey (1893–1981), discovered deuterium, an **isotope [V]** of hydrogen, in 1931, and he was awarded the Nobel Prize in chemistry for the discovery in 1934. Deuterium is a naturally occurring isotope, usually about 0.015 percent of the total natural hydrogen.

Urey identified the isotope in the spectrum of residue from evaporating liquid hydrogen. Beginning his work in 1929, Urey noted discrepancies between the value for the atomic weight of hydrogen determined chemically and that determined with a mass **spectroscope [IV]**. He assumed that a hydrogen isotope with a mass of 2 would account for the discrepancy. Working with F. G. Brickwedde and G. M. Murphy, Urey put liquid hydrogen through successive distillations. The next year, Gilbert Lewis was the first to separate deuterium from ordinary water. By subjecting ordinary water to electrolysis to liberate hydrogen gas, less deuterium is released as gas, leaving slightly more in the remaining water. Nearly pure *heavy water,* composed of deuterium atoms and oxygen, can be obtained by repeatedly subjecting water to electrolysis.

Further work allowed accurate measurement. A hydrogen atom, with only a proton at its nucleus, has an atomic mass of 1.008142, whereas deuterium, with a proton and a **neutron [V]**, has an atomic mass of 2.014735. Because the atomic mass is about twice that of regular hydrogen, the chemical behavior of deuterium is quite different from that of hydrogen. Deuterium in the form of heavy water has been used as a moderator in **nuclear reactors.**

During World War II, the expansion of a factory by the Germans to conduct electrolysis of water to produce heavy water at Rjukan, Norway, convinced the Allies that Germany had embarked on a program to build a nuclear reactor moderated with heavy water. Such a reactor, if successful, would have been capable of producing plutonium for the construction of nuclear weapons. Consequently, the deuterium facility was repeatedly attacked by commando and bombing raids until the project was abandoned.

DDT

DDT (dichloro diphenyl trichloroethane) as a chemical was first synthesized in 1874 by Othmar Zeidler, a German chemist, while he was a student. However, its effectiveness as an insecticide remained unknown for decades. In 1939 the Swiss chemist Paul Hermann Müller (1899–1965) discovered that it could be used as an effective insecticide, bringing paralysis to insects after being absorbed through the body surface. Müller worked as a research chemist for the J. R. Geigy Company in Basel, Switzerland, until the year of his death. Geigy was known for its production of dye materials and first got into the field of insecticides in the search for a mothproofing agent for fabrics. Müller won the Nobel Prize in physiology or medicine in 1948 for his discovery of the insecticidal properties of DDT.

DDT, with the chemical formula $(ClC_6H_4)_2CHCCl_3$, turned out, however, to be extremely resistant to decay from light, air, or water, and remains effective for as long as 12 years. Thus it would pass from crops dusted with it, through seeds, to the digestive tracts of birds, which are quite susceptible to it. It nearly led to the extinction of the American falcon before its use was outlawed. In humans it is stored in body fat and can even be passed from mother to nursing baby. Another difficulty was that surviving insects with resistance to the insecticide tended to breed and replace the more susceptible ones, passing on the resistance to later generations.

During World War II, Geigy informed the British legation in Switzerland of the effectiveness of DDT. During and after World War II, DDT was widely used to eliminate human lice (which carry typhus) and mosquitoes. In some areas, both malaria and yellow fever, carried by insects, were practically eliminated. By the 1970s and 1980s many countries outlawed the use of DDT as an insecticide due to its long-lasting persistence in the environment. In the United States, the Environmental Protection Agency banned DDT and other chlorinated hydrocarbons, such as **chlordane**, in agricultural use in 1972.

digital camera

The digital camera was an outgrowth of the **video tape recorder,** first developed in 1951, and the **charge-coupled device** (CCD), invented in 1959 at Bell Labs. Texas Instruments patented the first filmless digital camera for still images in 1972. In 1981 Sony Corporation produced the Sony Mavica electronic still camera, the first commercially produced electronic camera. The images were recorded to a small disk and then put into a video reader, to be linked either to a television monitor or to a printer. Basically, the Mavica was a video camera that took still pictures as freeze-frames.

A digital camera takes in light energy through a lens and focuses it on a CCD, which then transforms the light energy into electrical impulses. The impulse signals are sent to a **computer chip,** where they are sampled and then transformed into digitally coded information. The computer chip or *digitizer* samples areas of light and dark from across the image, breaking them down into points or *pixels*. The pixels are quantized or assigned brightness values. In black-and-white quantizing, each pixel is placed on a numerical scale ranging from pure white to pure black, while in color imaging the process includes registry of chromatic intensity and color resolution. Each pixel is also assigned an x and a y coordinate that corresponds to its location in the image. The more pixels, the greater the range of tone in the image or the higher quality. Both the spatial density, measured in number of pixels, and the optical resolution of the lens contribute to the spatial resolution of the image. The digital data are stored on a small disk and later transferred to a computer, where the image can be manipulated and viewed.

In 1986 Kodak invented the first megapixel sensor, which allowed the recording of 1.4 million pixels, allowing a 5-by-7-inch picture with quality equivalent to or better than a film picture. In 1987 Kodak released several products for recording, storing, manipulating, transmitting, and printing electronic still images, and by the early 1990s Kodak produced professional-grade digital cameras.

Many early digital cameras aimed at the home market, producing images that could be displayed and manipulated on a home computer. Consumer-market cameras were produced by Apple, Kodak, Casio, and Sony from 1994 to 1996. By the turn of the century they had revolutionized home photography, with several computer programs on the market for cropping and retouching photos, and with facilities for sending them by electronic mail through the **Internet.** With the fact that the picture could be reviewed before being stored or saved, digital cameras eliminated one of the advantages of the **Polaroid camera**—the potential for immediate viewing—without the expense; in the digital camera discarded pictures could simply be erased, freeing the memory disk for other images.

DNA

The model for the molecular structure of deoxyribonucleic acid or DNA was discovered from 1951 to 1953 by the team of Francis H. C. Crick and James D. Watson. They, along with DNA researcher Maurice H. F. Wilkins, shared the Nobel Prize in physiology or medicine in 1962 for the discovery.

The Watson-Crick model described the complementary double helical structure of the molecule that itself had been known as *nuclein* since its original discovery by Johann Miescher in 1869 along with other nucleic acids. For decades, biochemists had speculated that the genetic code through which traits were passed in living organisms from generation to generation resided in the nucleic acid. However, the acid is relatively simple, with a molecule made up of a sugar deoxyribose "backbone" linked to four bases: adenine, cytosine, guanine, and thymine. How the code for all the complex inherited characteristics could be embedded in the four bases remained a mystery.

In effect, Watson and Crick proposed a model for a structure that contained a code that translated into various sequences of amino acids in proteins, showing the physical mechanism by which characteristics of living creatures were transmitted from generation to generation. Although Charles Darwin's theory of evolution had been supplemented by studies conducted by Gregor Mendel showing genetic transmission patterns of dominant and recessive genes, the proof of physical transmission of characteristics as based in molecular chemistry was only deciphered with the development of the DNA model by Watson and Crick. The understanding of DNA provided a physical explanation for both standard inheritance of characteristics and the mutation of species into others.

The Watson-Crick model consisted of two corkscrewlike spirals or "helixes" wrapped around each other, in a double helix. Each helix had a back chain of alternating sugar and phosphate groups. Attached to each sugar was one of the four bases, known by their initials: adenine (A), cytosine (C), guanine (G), and thymine (T). Working from X-ray crystallography, in which X-rays are refracted and reflected from the atoms that make up a crystal, Maurice Wilkins and Rosalind Elsie

Franklin at King's College in London had produced a photographlike image of the structure of the DNA molecule. When Watson saw the photograph, he believed he had the answer to how the molecule could carry the genetic code. He worked with Crick, literally building Tinkertoy models to try to replicate the X-ray image.

> When James D. Watson and Francis H. C. Crick developed their two-strand model for the structure of DNA, they saw that it contained within it the seeds of a system for self-duplication.
> —John Cairns, "The Bacterial Chromosome," *Scientific American* 214, no. 1 (January 1966): 37.

By their model, a sequence of the bases, such as A-T-T-C-G and so on, could reflect one inherited sequence. Watson and Crick proposed that particular sequences would act as a template for the structure of a second kind of nucleic acid—ribonucleic acid, or RNA. The RNA would leave the nucleus of the cell, then act in turn as a template for later building of protein structures. That concept was at the heart of the genetic mechanism proposed by Watson and Crick.

The findings of the structure and later discoveries of how to strip one strand of a partial double helix and replace it with another led to processes of **genetic engineering,** finding applications in the creation of new plants, new breeds of animals with improved resistance to disease, and new drugs. Genetically engineered plants, especially such common food plants as corn and wheat, became matters of international controversy by the turn of the 21st century.

ejection seat

In 1944, with the increased speed of fighter aircraft and the introduction of jet aircraft, the British Ministry of Aircraft Production invited a team headed by James Martin to design a means of assisting the escape of a fighter pilot from the aircraft in an emergency. The ejection seat, propelled from the aircraft by an explosive charge, was Martin's design. Since that time, all improved ejection seats have relied on similar cartridge-actuated or propellant-actuated devices, known as CADs or PADs. Not only is the ejection seat literally fired from the aircraft, but also the canopy is blown from the plane by small charges set under its frame.

The ejection seat itself was designed to fly clear of the aircraft by at least 300 feet, allowing sufficient height for a parachute to deploy and open, even if the seat was ejected when the aircraft was on the ground. In the original Martin design, the pilot would pull a cord between his legs to start a sequence, then pull another set of cords over his head to

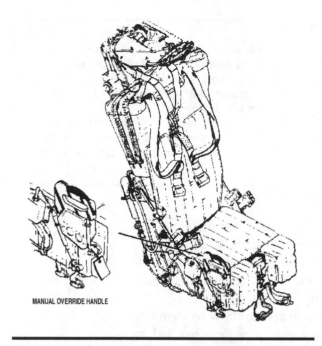

MANUAL OVERRIDE HANDLE

ejection seat The Martin-Baker company perfected many models of the ejection seat that use cartridge-actuated devices to propel a pilot from a damaged aircraft so he or she can parachute to safety. *U.S. Army*

bring down a shield over his face to protect him during the blastoff. Pulling the face shield released the first charges, ejecting the canopy. Then the seat would begin to eject, and at the same time a charge to fire small drogue parachutes would be primed. Half a second after the seat was ejected, the drogues would be shot out and begin to stabilize the seat and reduce its forward speed. A time release would transfer the load of the drogues to pull out the main parachute, and at the same time the pilot would be freed from the seat. The pilot would then descend by parachute and land at a safe distance from both the seat and the aircraft.

If the ejection took place above 10,000 feet, a time-release unit would be delayed by a barostat, a barometer-based altitude meter. The barostat delays the opening of the main parachute until 10,000 feet, allowing descent in a quick fall stabilized by the drogue chute through the cold upper atmosphere. The seat also has a small oxygen supply that is automatically turned on during the ejection sequence, to supply the pilot during the descent. Although the whole system operates automatically, the pilot can override the system and open the parachute manually by pulling a rip cord in the event of a malfunction in the mechanism.

Ejection seats of the Martin and later designs have saved more than 8,000 pilots' lives.

fiber optics

The use of glass fiber to serve as a pipe for light spread widely in both medicine and in communication after the 1950s. Glass fiber is made by coating the inside of a silica glass tube with many layers of thin glass that serve as cladding, reflecting any light entering the tube back into the tube rather than letting it escape out the sides. The tube is then heated and stretched into a minutely thin strand, a quarter the thickness of human hair.

The first medical use of fiber optics was apparently developed by Dr. Narinder Kapany, who in about 1950 built an endoscope that could be inserted into a patient's body and used to examine internal conditions or injuries. The endoscope had two bundles of glass fiber. One bundle would transmit light; another, a coherent bundle, would be used by the physician to view the color image. Earlier endoscopes, using inflexible tubes, had been developed as early as 1854 to allow examination of the throat.

Progressive work by Bausch & Lomb further reduced the diameter of fiber optics, allowing more applications. Endoscopes with specialized functions, such as examining the interior of the human body, were developed in the 1960s, including the arthroscope for joints, the bronchoscope for examining the inside of the lungs, and the laparoscope for examining inside the abdomen. Tiny forceps are sometimes attached to the tip of the endoscope to permit removal of samples for laboratory analysis or for conducting minute surgical operations such as the removal of calcifications or gallstones.

By 1966, experimenters began working with the use of optical fibers for communication. Internal flaws in the glass could result in loss of light, and a system of boosters would amplify the light signal at a wavelength of 1.54 micrometers, which allowed the light to travel the farthest in glass fiber. Optical fiber turned out to be ideal for the transmission of digital data, as the lines were not subject to electromagnetic interference, leading to much higher accuracy in transmission. Furthermore, glass fiber is much lighter than copper wire, making fiber optics far preferable in internal wiring in aircraft and spacecraft.

fuel cell

British jurist and physicist William R. Grove (1811–1896) made the first fuel cell in 1839. In his experiment he used water and sulfuric acid between two platinum electrodes to electrolytically break down water into hydrogen and oxygen. He placed a tube over one electrode to collect the hydrogen and another tube over the other to collect the oxygen. When he lowered the tubes containing the gases back down over the electrodes, he observed that the system generated, rather than consumed, electricity. Thus he discovered that electrolysis could be a two-way process—with electricity into water, he could make gases, but by feeding gases into the system to make water, he could produce electricity. Grove was later knighted for his scientific work and served as a justice of Britain's high court.

Sometimes fuel cells are regarded as a type of "refuelable battery," as they can serve to generate electricity when supplied with chemicals. Fuel cells tended to remain a laboratory curiosity until 1959, when English chemists Francis T. Bacon and J. C. Frost of Cambridge University developed a fuel cell using nickel electrodes and pressurized gas. The so-called Bacon cell could produce 6 kilowatts of power, using a potassium hydroxide electrolyte. Bacon cells were used in spacecraft, with *Apollo* 6 using 31 cells in series, producing 29 volts and up to 1,420 watts for 400 hours. In addition, the Apollo fuel cells produced drinkable water. By contrast to **internal-combustion engines [IV]**, which typically convert less than 20 percent of the energy in gasoline into power, fuel cells can convert 40 to 60 percent or more of a fuel's energy, making them appear to be a possible alternative for automobile propulsion. For this reason, fuel cells have come under serious consideration early in the 21st century as a means to propel the automobile in the future, without

fuel cell Deriving electricity from reverse electrolysis, the fuel cell has proven to be a valuable source of power, with potential applications in space travel and for automotive transportation. This model is used to power an experimental "robotic assistant" for future manned missions to the **Moon or Mars.** *National Aeronautics and Space Administration*

any emissions (other than pure water) to the environment. Intensive research into improved and less expensive electrodes, better structure of catalysts to reduce the amount needed, and improved membranes progressively brought down the cost, reduced weight, and improved the efficiency of fuel cells from the 1990s into the first decade of the 21st century.

Companies such as Ballard in Vancouver, British Columbia, International Fuel Cells (IFC) in Connecticut, and Plug Power near Albany, New York, were among the leading firms in development during this period. Large fuel cells, made by IFC and capable of producing 200 kilowatts and using natural gas from pipelines as a fuel source, were sold as reliable backup power systems for use in businesses and hospitals where an assured power supply was crucial. Meanwhile, smaller units suitable for installation in buses and automobiles were produced by Ballard and installed by automobile manufacturers in demonstration vehicles.

fluorescent lighting

Fluorescent lightbulbs were first marketed in the United States in 1938 by General Electric (GE) and represented the culmination of sporadic advances since 1900 by a number of European and American inventors. Scientists were aware that some materials would fluoresce when excited by ultraviolet rays. In about 1900, an American individual inventor, Peter Cooper-Hewitt, developed a low-pressure mercury discharge lamp, and he experimented with various dyes to attempt to obtain white light rather than blue light. The red neon lamp, introduced by the French industrial chemist Georges Claude (1870–1960), caught on in the 1920s. In the early 1930s a German company, Osram, and a French firm, Claude, pioneered in introducing fluorescent lamps for floodlighting and advertising, with a coating of powders applied on the inside of glass tubes. Albert Hull of GE patented improvements to the cathodes, and GE experimented with improved fluorescent powders. The powders would give off colored light when an arc passed through the gas in the tube.

George Inman of GE received a series of patents on combined elements for a practical lamp, including a mercury discharge lamp that produced a maximum of ultraviolet radiation that would cause fluorescence in especially responsive materials such as silicates, borates, and tungstates to generate varieties of white light. GE acquired prior patents in the area by the German company Rectron, and the fluorescent tube was introduced to the market at the 1939–1940 World's Fair in Flushing, New York. In Britain, fluorescent lights were first introduced during World War II. The strip or tube light tends to be 50 times more efficient than the incandescent lamp, making the former cost-effective in industrial and commercial settings. However, the diffuse and bright white character of the light was such that it took much longer to be integrated into home lighting decor, finding its first applications in kitchens and bathrooms.

Related gas vapor lamps, such as the mercury-vapor light with a greenish tinge and the sodium vapor lamp with an orange-yellow light, are similar in that they emit a fluorescing light, and they were introduced in the 1950s. However, they differ from fluorescent light tubes in that the gas conducts the electrical charge and fluoresces, rather than a coating on the inside of the tube. The vapor lamps are used in outdoor illumination for settings such as parking lots, highways, and city streets.

genetic engineering

After the discovery of **DNA** as the basic building block of genetics, a number of biochemists began working on the concept of joining two DNA molecules from two organisms to develop special qualities. The first to succeed was American biochemist Paul Berg, in 1973. He joined DNA from two separate viruses. Later that same year, Stanley Cohn at Stanford and Hubert Boyer at the University of California discovered an enzyme that improved on the Berg process.

In 1990, a research team headed by W. French Anderson at the National Institutes of Health attempted human gene therapy on a patient with severe combined immune deficiency, a condition in which children are born with no immune system due to the lack of a single gene.

Although human gene therapy has proceeded with caution, genetic engineering has had a much greater impact in agriculture. One of the first successes with transgenic plants was an experiment in Cologne, Germany, with a tobacco plant with high resistance to an antibiotic. The first genetically engineered food plant sold for consumption was a tomato produced by the American company Calgene in 1994. By the early 21st century there were hundreds of genetically engineered crops, including types of corn, rice, soybeans, wheat, and other crops resistant to herbicides and to insects.

The use of genetically engineered crops has led to several controversies. Some countries have banned the import of genetically engineered grains out of fear that they will be used as seeds and that the plants will naturally pollinate local crops, causing mutation of the local crops with unknown consequences.

helicopter

Like the **automobile [IV]** and the **airplane [V]**, the helicopter represents the achievement of a concept that had existed in the human imagination for centuries. After many prior experiments, Igor Sikorsky (1889–1972) succeeded in flying a test model, the VS-300, on September 14, 1939, that became the basis for follow-up production.

Ancient Chinese toys employed the same principle, in which a string-pulled spindle with four small blades on the top would be spun and rise into the air. Leonardo da Vinci (1452–1519) created several sketches of an aerial-screw machine in 1483, and his notes with the Greek words for helix (*heliko*) and wing (*ptèron*) appear to be the origin of the name of the device. Prior to Sikorsky's 1939 design, many attempts by others failed or had very limited success.

In the 19th century, several efforts to achieve vertical flight failed largely because available engines were simply too heavy to provide sufficient lift for the engine and aircraft, let alone pilot, passengers, or cargo. In 1907 the Frenchman Paul Cornu lifted off about 5 feet in a helicopter of his own design, and in 1909 Sikorsky, then living in Kiev, in the Ukraine, built an early model that he tested in 1910, but he could not achieve liftoff. Sikorsky then pursued regular aircraft production in Russia and supplied airplanes to the Russian army in World War I. Following the war, he immigrated to the United States and set up a company on Long Island, New York. In 1929 he reorganized his firm as United Aircraft Corporation and established it in Bridgeport, Connecticut. There he continued building aircraft, including the Clipper, a large, passenger-carrying seaplane employed by Pan American World Airways in transatlantic flights.

Meanwhile, the Spanish designer Juan de la Cierva developed the *autogyro* in 1923. Designed like a small conventional airplane with engine and propeller at the front, the craft replaced the regular wings with a four-bladed, nonpowered hinged rotor for lift. As the autogyro moved forward, the rotor blades would rotate like a **windmill [II]**, providing lift and flexibly adjusting to speed and angle. De la Cierva was able to descend vertically and fly at speeds as low as 30 miles per hour, but the autogyro could not hover or take off vertically.

Experimental helicopters designed in the interwar years had some

The helicopter is a bundle of paradoxes. Its charm as a means of travel lies in the fact that it makes haste slowly, although so far it has been of greatest interest to the military, it is not essentially an engine of destruction; even for military purposes it is simply a means of transport and rescue. On the technological side, the helicopter seems the simplest of flying machines, but it is actually the most complex. Although it was one of man's earliest conceptions for mechanical flight, it was about the latest to get into the air in a practical form.

—Lawrence P. Lessing, "Helicopters," *Scientific American* 192, no. 1, (January 1955): 37

limited success. The most serious design problem was that of torque. When the rotors spun to lift the craft, the body of the machine would turn in the opposite direction. A solution was to mount two blades spinning in opposite directions, either one above the other or side-by-side, separated by a tandem boom. A French design of 1936 had two rotors mounted one over the other and reached a speed of 65 miles per hour. In 1938 a German design built by the Focke aircraft works had two counterrotating rotors in tandem and achieved a speed of 76 miles per hour. However, neither design was taken to production in the period. Another Focke design, built in 1940, remained secret during World War II and achieved an altitude of 23,400 feet.

Sikorsky's breakthrough was to mount a small propeller on an outrigger tail, facing against the direction of torque to keep the craft stable. That design in the VS-300 of 1939 proved simple and successful. Using the same principle in an experimental model for the Army, the XR-4, he developed the design into a regular production model, the R-4B, by 1942. By the end of World War II, he had produced more than 400 helicopters for military use, at first only for transport and rescue, and his firm remained a leader in the field in the postwar era. By the 1960s helicopter gunships regularly carried heavy machine guns. Helicopters are limited in their speed to not much more than 200 miles per hour by the fact that at higher forward speeds, the retreating blade creates a reverse effect of **Bernoulli's principle [III]**, eliminating lift and resulting in a stall.

holography

Holography, or the depiction of three-dimensional objects in three dimensions so that the viewer's perception of the image changes with the viewer's position, was first invented in 1947 by Dennis Gabor (1900–1979), a Hungarian-born physicist who had become a British subject. Gabor had fled Germany during the Nazi regime, but as an alien from an enemy country, he could not work on any of the wartime physics programs such as **radar** or **nuclear weapons.** He turned his attention to improving the electron microscope, and in 1947 he developed the theoretical basis and named the procedure of generating a *holo* (Greek for whole) *gram* (meaning recording).

Gabor's method essentially required that an intense beam (electronic, light, or X-ray) be bounced off the object being imaged and that the beam itself be photographed. To display the image, the photograph would be illuminated, reconstructing the pattern of light so accurately that the image appeared to be hovering in space at the same distance from the photographic plate as the original spacing of the object. To achieve the process, an intense, coherent beam must be used, and ordinary light did not meet the requirement. In effect, Gabor's concept remained a notional one, as various attempts were made to implement it with X-rays, electronic beams, and bright light. However, no satisfactory hologram was produced until after the invention of the **laser** beam in the 1950s. Emmet Lieth and Juris Upatnieks at the University of Michigan applied the laser to the problem of holography, producing a successful image by 1963. Gabor was awarded the Nobel Prize in physics in 1971 for his contribution.

Holograms are produced using both continuous-wave lasers and pulsed lasers. The latter are particularly useful in capturing still holographic images of objects in motion. Gabor's original intention, to improve electron microscopes, has been achieved, and holography has found many scientific applications in studying small objects of complex shape where differences in depth preclude a clear image using conventional means.

hovercraft

The first air cushion vehicle was successfully conceived and built by a British inventor, Charles Cockerell, and flown for the first time in May 1959. Dubbed the SR N1, Mark 1, the test vehicle was powered by a 435-horsepower engine that compressed air and sucked it in by an axial fan and expelled it through jet orifices on the two sides of the craft. The driver or pilot could control vanes, so the jets could drive the vehicle forward or backward. Earlier experiments with air cushion vehicles by an American inventor, Douglass Warner, in the 1930s and a Finnish inventor, T. J. Kaairo, from 1935 to the 1950s had resulted in liftoff, but neither had taken his device to production. Cockerell's system went into production with financial support from the British National Research and Development Corporation. That corporation set up Hovercraft Developments Limited (HDL), which held the patents.

Cockerell was an electronics engineer rather than an automotive or aeronautics specialist. However, his hobbies were building, designing, and sailing small boats. He had experimented with introducing a layer of air between the hull of his vessels and the water to reduce drag, thinking the air would produce what he called *lubration*. His first experiments had involved fixed sidewalls and hinged end doors to contain the air layer, but by experimentation, he decided on a system of water curtains. Some of his first experiments consisted of a vacuum-cleaner pump, an array of tin cans, and a kitchen scale to measure the air pressure. He discovered that by distributing the ejected air from the vacuum-cleaner pump through a concentric assembly of tin cans, he could increase the air pressure from 1 pound to 3 pounds. This was the principle of the annular jet, which became key to the lift principle employed in later air-cushion vehicles.

The SR N1 was an oval-shaped vehicle 30 feet long, 24 feet wide, and weighing 3.5 tons. Cockerell used an Alvis Leonides air-cooled radial engine driving a fan that drew in air through a bell-shaped intake 8 feet in diameter. The air provided both lift and directional movement. His vehicle could carry three passengers at very slow speeds but would not work over high waves in water. Tested at first over dry land, the SR N1 successfully crossed the English Channel on July 25, 1959, commemorating the first crossing of the Strait of Dover by Louis Blériot in a monoplane 50 years earlier.

Eventually Cockerell worked out five basic components to the air cushion vehicle: a containing skirt, a hull, an engine, a system for lift, and a system for propulsion. Commercial hovercrafts were built capable of carrying more than 400 passengers and 70 vehicles at speeds of up to 80 miles per hour, operating as ferries across the English Channel and elsewhere in the world. Separate engines were used to provide lift and propulsion, and the hover height was raised from a few inches to more than a foot. Although experiments were conducted in the United States and in other countries, only Britain engaged in regular production of the commercial air-cushion vehicle. Other terms have been used to describe the type of vehicle, including the *surface effect ship (SES)* or *ground effect machine (GEM)*.

Two types of containing skirts were developed for the early hovercraft. An HDL engineer, Denys Bliss, developed a "finger" skirt in 1962, consisting of strips of rubber held in place by the air-cushion pressure. A small British flying-boat manufacturer, Saunders-Roe, had developed an inflated sausage-shaped bag. Later models combined a Saunders-Roe bag made of tough nylon with a secondary finger skirt

hovercraft The Landing Craft-Air Cushion (LCAC) has proven to be a formidable vehicle for the U.S. Marines, because it can cross from sea inland over beach and marsh without stopping.
U.S. Marine Corps

that protected the bag skirt from friction damage. Saunders-Roe built several hovercrafts, and its interests were later taken over by Vickers to form the British Hovercraft Corporation in 1966. At that time, the HDL, in which Cockerell was a shareholder, was made a division of the National Physical Laboratory over his protests. Later models of hovercraft tended to follow the Saunders-Roe design for skirts.

The U.S. Marines adopted a vehicle based on the principle, known as a *landing craft—air cushion (LCAC)*, capable of carrying a main battle tank or several smaller vehicles and troops from standoff positions at sea through the surf zone and across beaches and lagoons for amphibious landings. By the beginning of the 21st century the LCAC had become a major Marine Corps asset and was employed in numerous small military operations.

human genome

Mapping the human genome was an international project initiated in 1990 with funding from the U.S. Department of Energy and the National Institutes of Health and scheduled to last 15 years. However, breakthroughs in the project yielded a draft of the human genome map in 2003. As the project proceeded, it was determined that the human being's nature is governed by an estimated 30,000 to 35,000 genes. The term *genome* refers to the complete set of **DNA** or deoxyribonucleic acid. In the human genome, some 24 chromosomes, or physically separate molecules, range in length from about 50 million to 250 million base pairs of DNA. The approximately 30,000 genes are specific sequences spaced along the genome, with many other base pairs with other functions, such as controlling the production of proteins in the body. The largest chromosome has nearly 3,000 genes, while the Y chromosome has the fewest, 231.

In June 2000 Craig Ventor of the company Celera Genomics announced the completion of a basic working draft of the DNA sequence of the human genome, and in February 2001 the project authorities and Celera published a first working draft DNA sequence. The draft has been published on the **Internet.** Around the world, 16 different genome sequencing centers detailed the sequences, with scheduled duplication of work to ensure that most parts were covered eight or ten times so that errors and gaps could be detected and eliminated.

It is likely that the completed genome map may lead to the identification and pinpointing of genetic effects such as blindness, deafness, and conditions that lead to cancer, arthritis, diabetes, and other common diseases. The scientific findings from the project are likely to produce a wide range of medical and pharmaceutical applications over the first decades of the 21st century.

intercontinental ballistic missiles

During the 1960s, the United States and the Soviet Union began to develop and deploy long-range ballistic missiles, both intercontinental ballistic missiles (ICBMs) and intermediate-range ballistic missiles (IRBMs). The former would have ranges of more than 5,000 miles, and the latter would be limited to about 1,500 miles. The German V-2 rocket built during World War II represented what would later be called a short-range or medium-range ballistic missile. The efficiency and long range of the 1960s missiles derived from the fact that they required fuel only to be launched up through the atmosphere and directed toward the target. They used virtually no fuel traveling through near outer space. They were ballistic rather than guided missiles because they fell at their target, after a ballistic arc, like an artillery shell. Such missiles were intended to carry **nuclear weapons** and **thermonuclear weapons.**

From 1963 to 1988 the U.S. Air Force deployed more than ten dif-

U.S. Air Force ICBMs	
Thor	1957–1975
Atlas	1950s–1975
Jupiter C	1950s–1960s
Titan I	1950s–1960s
Minuteman I	1962–1969
Titan II	1963–1987
Minuteman II	1965–
Minuteman III	1970–
Peacekeeper	1981–
Midgetman	canceled

ferent models or modifications of ICBMs. Sometimes the same warhead would be employed on different missiles, leading to some confusion of terminology among sources. See the table showing the U.S. Air Force ICBMs in order of introduction.

The Thor, Atlas, Jupiter C, and Titan missiles were liquid-fueled, presenting serious hazards for handlers and requiring considerable advance notice for fueling time. The Minuteman and Peacekeeper missiles, by contrast, were solid-fueled, safer, and could be kept on alert at all times. The Titan II, although liquid-fueled, could be maintained in a ready state. The explosion of a Titan II missile in its silo near Damascus, Arkansas, in 1980 demonstrated the hazards associated with that type of missile and contributed to the decision to retire it.

By the mid-1980s the United States and the Soviet Union each had more than 1,000 such missiles, some with multiple independently targetable reentry vehicles (MIRVs) on them. A 1985 estimate indicated that the United States had 2,130 warheads on ICBMs, while the Soviet Union had 6,420 warheads on similar missiles. The apparent lead of the Soviet Union in this regard was more than compensated for by the large fleet of long-range bombing aircraft maintained by the United States and by a large number of U.S. submarine-launched ballistic missiles.

The Soviet SS-20 was an intermediate-range nuclear missile introduced in 1978 that could threaten targets all across Western Europe. The United States responded by the deployment to Europe of Pershing II (an IRBM) and ground-launched cruise missiles (GLCMs) in 1983–1984 that could reach targets well inside the Soviet Union. These "Euromissiles," although they could hold at risk strategic targets, were officially regarded by the United States as long-range theater nuclear forces rather than as strategic weapons for purposes of arms control discussions. With ranges of 1,000 to 1,500 miles they were not intercontinental, but since they could target facilities within Soviet borders from bases in Britain and Western Europe, they altered the Soviets' perception of the nuclear balance of power. The Pershing II and the SS-20 were sometimes called medium-range ballistic missiles (MRBMs) to distinguish them from ICBMs. Under arms control agreements signed in the late 1980s and the 1990s, both the Soviet Union and the United States first removed the intermediate-range (or medium-range) missiles and later disarmed the intercontinental missiles.

The development of intermediate-range and tactical missiles by North Korea, Pakistan, India, and China and the modification of theater missiles by Iraq, Iran, and other nations, all capable of carrying weapons of mass destruction (chemical, biological, or nuclear warheads) continued to present threats to international stability in the 21st century.

Internet

The Internet, a system of connecting computers in remote locations by telephone and cable networks, was developed as a U.S. government–sponsored project of the Defense Advanced Research Projects Administration (DARPA) in the 1960s. To make such communication possible, researchers at RAND and at the British National Physical Laboratory recognized that a new system of switching known as "packet switching" needed to be developed. The first application of an interface message processor (IMP), between two computers over a telephone line, was achieved in October 1969, between Leonard Kleinrock at the University of California at Los Angeles and Douglas Englebart at the Stanford Research Institute. This first use of an IMP is sometimes considered the inauguration of the Internet. The first system was known as ARPANET, for the funding by ARPA. In 1983 the original Network Control Program was replaced with the Transmission Control Protocol/Internet Protocol (TCP/IP), developed by Robert Kahn. By 1985 the military applications were separated from the civilian research applications. Funding and encouragement from the National Science Foundation increased the numbers of universities connected, and by 1988 the proliferation of systems and connection improvements was well under way.

The ARPANET was decommissioned, and the TCP/IP system had largely supplanted the other wide-area computer network protocols by 1990. In 1995 the Federal Networking Council defined the term *Internet* as a global information system logically linked together by the TCP/IP or its follow-up systems. The system continued to grow and be modified, but it remained characterized by a free-access orientation. By the turn of the 21st century, hundreds of millions of personal computers and larger institutional computers were connected through the Internet, creating a communications and information revolution.

in vitro fertilization

In 1978 an English obstetrician, Patrick Christopher Steptoe (1913–1988), developed the technique of in vitro fertilization. In vitro (meaning in a glass container) fertilization had been seriously suggested in medical literature as early as 1937. Even earlier, in 1932, Aldous Huxley published a novel, *Brave New World,* depicting a future society in which human beings would be produced on an industrial scale through in vitro fertilization and would be hatched with certain characteristics that would allow them to fill different social roles. His vision of governmental control of the reproductive process became a haunting refrain in later science fiction works and reflected popular fears that technology could yield nightmarish social consequences.

Successful experiments with rabbits in the 1960s led to more detailed knowledge of the physiology of fertilization. Working with Robert Edwards, a biologist, Steptoe succeeded in implanting in a woman's womb an egg that had been fertilized outside the woman's body.

Steptoe and Edwards had worked since 1971 on techniques using fertility drugs to stimulate the development of eggs within the ovaries of volunteer patients. In 1977 they decided to remove the egg at the proper stage of maturity and attempt outside fertilization, replacing the egg within the womb two days later. The child, Louise Brown, was the first human conceived outside the human body and was regarded by the press as a "test tube baby."

By 2003, more than 3,500 babies conceived in this fashion had been born. Despite the dire predictions of Huxley and others, the first applications of human in vitro fertilization had nothing to do with social control but were simply methods by which couples desiring to have a child could achieve their family goals.

jet engine

The jet engine, or more properly, the turbojet engine, is a clear case of simultaneous invention, with at least two inventors working in complete isolation from each other. In Britain, Frank Whittle (1906–1996) filed a British patent in 1930 that lapsed in 1935. The first aircraft to be propelled with an engine of his design was the Gloster E28/39, on May 15, 1941. Meanwhile, in Germany, Hans von Ohain designed the Heinkel He-178, which flew in 1939.

Limitations on the speed of the **airplane [V]** imposed by the airframe had been reduced with streamlining, enclosure of engines and cabins, retractable landing gear, and the elimination of struts and wires characteristic of biplane designs. By 1928 it became apparent to aircraft engineers that the limit to airspeed was no longer the drag of the machine but the ability of the engine to propel the craft faster. Whittle wrote a paper in 1928 suggesting combining the rocket engine with the gas turbine for faster propulsion than possible with a reciprocating piston engine. He began work on the concept of a turbine for propulsion and filed his patent for a turbine device on January 16, 1930. Lacking funds and encouragement, he had more or less given up on the concept when he took it up again in 1935–1936, forming a company, Power Jets, to exploit the ideas.

He remained in the Royal Air Force, on special duty from 1937, working on the designs and solving problems with turbine blades that tended to shear off at high speeds. His first successful test of the jet engine on a stand was on April 12, 1937. By 1939 the British Air Ministry provided support.

In Germany, Hans von Ohain also encountered a lack of interest, although Ernst Heinkel was impressed with the concepts, and von Ohain began work on developing an engine in 1936. The He-178 flew successfully on August 27, 1939, the first successful flight of a jet-powered aircraft, with a top speed of 435 miles per hour. Germany, after initial hesitation, became the first nation to fly operational military jet aircraft. Both Heinkel and Messerschmitt produced jet aircraft based on the von Ohain designs during World War II. The Messerschmitt Me-262 achieved speeds of more than 550 miles per hour.

Later aircraft took the concept further, such as the Gloster Meteor, in 1944, and by the 1950s, the American F-86 Sabre and the Soviet Union's Mig-17. Both the Sabre and the Mig-17 achieved the speed of sound (Mach 1). The jet engine revolutionized civilian air travel, beginning with the 1949 British Comet. The American Boeing 707 entered regularly scheduled jet air travel in 1959.

Kevlar

Kevlar is a trademarked fiber first manufactured by Du Pont in 1965. The polymer plastic is extremely strong and has a high melting point. In

1976 a Du Pont researcher, Stephanie Kwolek, invented a process for spinning the polyamide into fiber, which led to an increase in its longitudinal strength. She received an award for creative invention from the American Chemical Society and a patent on the spinning process. The molecular structure of the polymer is such that long carbon chains are linked together. The resultant fiber or fabric is resistant to chemicals and heat but needs to be protected against ultraviolet light with a coating. Kevlar variants are used in the manufacture of tires, lines and ropes, fabrics, tapes, and bulletproof vests.

kidney dialysis machine

The process of cleaning the blood for a patient whose kidneys had failed was suggested by a Scottish chemist, Thomas Graham (1805–1869), in 1861, in a procedure he called *dialysis*. In the method he suggested, the blood could be cleansed by allowing the ions (or molecules of small dimensions) with the impurities to pass through a parchment or colloidal membrane, leaving the blood behind. An attempt to build a practical dialysis machine was made by John Jacob Abel, a professor of pharmacology at Johns Hopkins University, in 1912. Working with colleagues, he developed a machine that circulated blood through tubing immersed in a saline and dextrose solution. Oxygen passed into the blood and toxins passed out. After successful tests on animals, the team published their findings in 1914.

Various technical barriers stood in the way of successful application to humans. One was the tendency of the blood to clot in the tubing. With the development of a successful anticoagulant agent, heparin, in the 1930s, several researchers began to experiment with an improved dialysis machine during World War II (1939–1945).

Willem Kolff, a Dutch physician, first began experimenting with a crude dialyzing machine of his own design in 1937. In 1941, after the German occupation of the Netherlands, Kolff constructed a machine out of tubes made of **cellophane** [V] with beer cans to contain the cleansing solution. He used his device on patients in 1943 and 1944, but almost all of them soon died. At the end of the war, Kolff built improved machines and took them to Amsterdam and London. Meanwhile, another inventor, Nils Alwall, in Sweden, experimented with an alternate design. The application of the technique that had been understood for some 80 years to humans required the development of both the anticoagulant and an appropriate cellophane membrane that allowed the ion transfer.

In 1947 Kolff took his machines to the United States and presented the blueprints to doctors at Harvard Medical School. There, Karl Walter and George Thorn began to employ kidney dialysis machines of the Kolff design in kidney transplantation operations in 1954. In 1960 Dr. Belding Scribner at the University of Washington initiated long-term dialysis with the use of a connection made of **Teflon** to the patient's arteries and veins.

laser

The first working laser that was built in the United States was the 1958 creation of Theodore H. Maiman (1927–), an American physicist, who had earned his doctorate at Stanford University. While there he studied the *maser,* which was an acronym for microwave amplification by the stimulated emission of radiation. The first maser had been constructed in 1954. In 1958 Charles Townes, who had developed the maser, delivered a paper suggesting that light waves could be similarly stimulated. Several researchers worked on the concept simultaneously, including Gordon Gould. Two scientists in the Soviet Union, F. A. Butayeva and V. A. Fabrikant, amplified light with a similar technique.

The *laser,* or light amplification by the stimulated emission of radiation, worked by wrapping a flash tube around a ruby rod. The flash would start a reaction that ended up producing a beam of coherent red light from the end of the rod. Coherent light is light of a single frequency or color. Within 18 months of Maiman's invention, several hundred companies and universities began laser research. When focused through a lens, laser beams could be used to cut materials, even diamonds.

Lasers have been used to spot-weld retinas in eyes and in other bloodless surgery. In business, lasers have been used to read universal product codes in supermarkets. Data can be stored in digital form on discs that have been struck by laser beams to burn precise holes in a protective coating, leaving a digital code that can then be read back by

> From the invention of the laser in the 1950s it was recognized that intense laser beams might be a good way to deposit large quantities of energy in materials for manufacturing purposes. That potentiality has now become a mature technology. Over the past decade high power lasers have found a place in many manufacturing processes: the welding of automobile parts, electronic devices and medical instruments; the heat-treating of automobile and airplane parts to improve their surface properties; the cutting of sheet metal in the punch and die industry, and the drilling of small cooling holes (.007 to .05 inch) in airplane parts. In all these operations laser systems have made production lines more efficient and have reduced costs.
>
> —Aldo V. La Rocca, "Laser Applications in Manufacturing," *Scientific American* 246, no. 3 (March 1982): 94

another laser on a **laser disc.** Heavy-duty laser beams have been used in industry for welding ceramics and metals and for the cutting, drilling, and heat treatment of metals. Other uses have included surveying, range finding, and targeting of weapons and as a tool and measuring device in a wide variety of scientific fields.

With its numerous applications, the laser has been regarded as one of the most significant technological inventions of the late 20th century.

laser disc

David Paul Gregg envisioned and patented an optical disc for recording a video record, conceiving the idea in 1958 and receiving patents on the process in 1961 and 1969. He was employed at the time at Westrex Corporation in Hollywood, California. The system he developed was acquired by MCA, which produced the first consumer optical disc player in 1978. Later, Pioneer trademarked the term LaserDisc. The technology was superseded in 1997 by the introduction of the digital versatile disc or DVD, also trademarked by Pioneer. These and other optical and laser disc devices are manufactured under license from the Gregg patents.

The original optical disc method required that picture and audio information be recorded by embossing the data on a transparent disc by bringing together a recorded die and a transparent record blank. A light beam then read the embossed data. The original optical disc technology was improved by the use of **laser**-generated light beams.

Laser discs are usually coated with a thin layer of a sensitive material such as tellurium about 100 nanometers thick, protected with a layer of transparent plastic. A high-powered recording laser focuses on the tellurium, melting it and pulling it back from the transparent surface. Each burst of the laser makes a microscopic pit. The finished disc is played back by focusing a laser beam on the line of pits as the disc rotates on a turntable. The variation in the amount of laser light reflected is a faithful copy of the frequency, spacing, and length of the pits recorded. The reflected signal is converted into electronic form and sent for display on a screen.

light-emitting diode

The light-emitting diode or LED is a refinement of the vacuum tube or diode valve developed by John Fleming in 1906.

Nick Holonyak Jr., a General Electric researcher, made the first practical demonstration of an LED, in 1962. Similar to the **transistor,** the LED consists of a piece of semiconducting material such as silicon or germanium coated differently at each end. Thus one end is biased positively and the other is biased negatively. Depending on direction of electric flow, the transistor can be either on or off, with two positions, thus known as a diode. A common type of LED is made of gallium arsenide and coated with a phosphorescent material and is known as a GaAs P type LED. When in the on position, the diode gives off radiation in the visible light spectrum. The development of the LED followed quickly on the heels of the transistor, and some of the first LED work was done at Bell Labs in the 1950s. LEDs have been widely adapted for instrument displays, and by the turn of the century were widely employed in high-resolution flat screens for both television receivers and computer uses.

long-playing record

After the development of the **phonograph [IV]**, the recording industry settled by 1913 on a standard disc as the recording medium, rotating at 78 revolutions per minute (rpm) and made of a solidified shellac or resin. One drawback for serious lovers of music was the time limit of such recordings: about 5 minutes. Although suitable for single popular or folk songs or short pieces, a recording of a movement of a symphony or a full scene from an opera would require interruptions and record changing through the course of the work. A successful long-playing (LP) record would require that several separate inventions be brought together: a slower rotational speed, a larger-diameter disc, finer grooves, a vinyl record, a steady and silent turntable, and a lightweight pickup device to be able to sense the variations in the finer grooves.

The innovations were brought together by Peter Goldmark of Columbia Broadcasting System (CBS) laboratories, the same organization that had been responsible for some improvements in **television [V]**. Beginning experiments right after World War II, Goldmark established the basic concept of a narrow-groove vinyl record. He found that vinyl would accept finer grooves, from 224 to 300 per inch.

Working with assistants at the CBS laboratories, Goldmark developed a lightweight pickup arm as well as slow-speed, silent, and relatively inexpensive turntables. Working within a budget of about $250,000 for three years, CBS was able to announce the new type of record in 1948. Soon, RCA tried to counter the 33⅓ rpm record with a smaller disc that ran at 45 rpm. The 45 rpm was widely accepted for popular music, while the 33⅓ cornered the market for classical music. Later innovations in small cassettes that utilized magnetic **tape recording** [V] and the still later introduction of audio recordings on **laser disc** rendered the LP nearly obsolete by the early 21st century, except for collectors.

magnetic resonance imaging

The phenomenon upon which magnetic resonance imaging (MRI) is based was first observed in 1946. The nuclei of atoms in which at least one proton or one neutron is not paired act like magnets, and a strong magnetic field applies a force that causes the nucleons to precess, or tilt, at a certain frequency. When the frequency of the precessing nucleons corresponds to an external radio wave striking the material, energy is absorbed from the radio wave. This selective absorption or resonance varies considerably from material to material, depending on its molecular structure. A similar phenomenon can be detected in which the absorbing particles are electrons, known as electron paramagnetic resonance (EPR).

In 1952 two scientists, Felix Bloch and Edward M. Purcell, shared the Nobel Prize in physics for their development of a device for studying chemical composition using nuclear magnetic resonance. Work by other scientists, including Richard R. Ernst, a Swiss physical chemist, led to the development of a practical nuclear magnetic resonance machine for use in diagnosing and detecting areas of disease inside the body. Ernst received the 1991 Nobel Prize in chemistry for his work. By the 1980s nuclear magnetic resonance was being regularly used in medical diagnosis.

Since MRI relies on radio signals rather than X-rays, it is not harmful to the subject. The modern MRI scanner consists of a large tube into which the patient slides, completely surrounded by the magnetic field. Radio signals and a computer allow the analyst to obtain pictures of thin slices of the patient's body, providing excellent resolution of joints such as the shoulder or knee, or organs such as the brain, heart, liver, pancreas, or breast.

mathematical principle of computing

The logical basis for the modern **computer** was developed in 1936 by Alan Turing (1912–1954), a young British logician. He published a paper, "On Computable Numbers, with an application to the Entscheidungsproblem" (the decidability problem). Turing developed the concept of the Universal Turing Machine. By this he meant a machine designed so it could read and replicate all instructions given to other machines. This idea was very similar to what a later generation came to call a computer program or algorithm. In effect, the ideal Universal Turing Machine was a computer that would be able to turn to different tasks. It was this concept that lay behind the development of so-called *bombes,* or electromechanical computers that would replicate the instructions used in German and Japanese **encryption machines** [V], and it is this concept that is at the heart of later computers that operate by reading instructions from programs to execute actions.

Although computers in the modern sense did not exist when Turing did his work, he foreshadowed the concept that instructions for machines could be represented by numbers, and that instructions for a specific task for an individual machine could ideally be deciphered by a universal machine as long as it had the language of the instructions. Although it cannot be said that Turing "invented" the computer any more than Charles Babbage did with his **analytic engine** [IV], Turing's contributions to the basic principles became widely adopted in the next decade as computer builders began to adapt special-purpose machines to a variety of functions by allowing for varied instructions and as they developed common languages so that computing machines could be programmed in similar ways.

While studying for his doctorate at Princeton University, Turing became interested in ciphers. In 1938 he returned to Britain and was recruited to work at the government's Code and Cipher School. There he assisted in developing machines that were known to his colleagues as Turing Machines for breaking the Enigma code of German encryption machines. However, the concept of a Turing Machine was more basic and was at heart the model of a digital computer.

Widely recognized as an eccentric genius in his own time, Turing later committed suicide after he revealed he was a homosexual.

microwave oven

The microwave oven is a civilian application of a military technology perfected in World War II. According to legend, a technician at Raytheon was experimenting in 1945 with a magnetron, an electronic tube that emitted extremely short (or micro) radio waves, the type of tube used to generate **radar** waves. While running a test, Percy Spencer noted that a candy bar in his shirt pocket melted, even though he had not felt heat. He speculated that the effect might be useful in cooking.

In 1955 Raytheon produced its first *radar range*. The microwave oven used a magnetron to generate short-wavelength radio waves in the 1-millimeter to 3-centimeter range. With a magnetic field, the waves oscillate some 2.45 million times every second (2,450 megahertz). The heating effect is caused by the fact that when the radio waves strike a molecule of water, the molecule is aligned in the direction of the wave. With the oscillation of the wave, the molecule is reversed in alignment. With several million realignments per second, the vibration heats the water molecule, and the hot water then cooks the surrounding food. Since only water is affected in this way, the containing plastic, glass, or paper container is not heated directly. Compared to electric heating, microwave heating is far more efficient, as only the food or liquid is directly heated rather than heating the container first in order to heat the food.

Typically microwave ovens have a window through which the cooking process can be observed or checked. A metal screen deflects the waves back to the food while allowing the much smaller short-wavelength visual light to be seen.

microwave transmission of voice and data

During World War II, the development of **radar,** with wavelengths in the 0.3-to-30-centimeter range, using radio frequencies between 300 and 300,000 megahertz, spurred a new technology. The microwaves, in the electromagnetic range that falls between very-high-frequency radio waves and infrared light, represented a virtually unused part of the electromagnetic spectrum when they were first used in military and commercial applications.

In the United States, the Federal Communications Commission (FCC) ruled in 1946 that experimentation in the use of microwaves for transmission of data could go forward, but there was no provision for commercial use of that part of the spectrum. Development of new devices to make use of the technology was slow over the next decade. However, in 1959, the FCC ruled in the so-called Above 890 Decision that the frequency range over 890 megacycles could be used for privately built and privately operated microwave networks, although the FCC still did not allow its use for common carriers. Between 1959 and 1970 a number of companies established private networks, using microwave transmitters and receivers mounted on towers, to transmit data or voice. At the same time, American Telephone & Telegraph (AT&T) tried to forestall such development by offering discounted rates on its long-distance landlines to large subscribers.

An investment group of radio salesmen formed in 1963 and developed the idea of establishing a microwave tower route that would follow highway Route 66 from Chicago to St. Louis, with the idea that the system would be open for subscribers, primarily truck drivers who could use it for communication. The group incorporated in 1968 and later changed its name to Microwave Communications Incorporated (MCI). MCI secured FCC permission in 1969 to construct the proposed system between Chicago and St. Louis. In 1972 MCI opened that microwave route to private-line service and began expanding microwave linkages elsewhere.

In the microwave system, towers had to be constructed in line of sight, usually on raised ground to allow the farthest possible interval between towers to reduce land-lease costs. At the base of each tower, repeater equipment, air conditioners, and other equipment would be housed in a small metal shack. By the 1980s such systems were widely installed not only in the United States but also worldwide, as the transmission system was far less expensive to put in place than construction of landlines.

In the United States, the rapid expansion of microwave communication, and its use for both voice and computer data transmission, soon forced the end to the monopoly held by AT&T over telephone communication. The Justice Department filed an antitrust suit against the massive phone company, leading to its breakup and divestiture of holdings in 1982. MCI and other telephone companies were granted full access to the existing system. By 1988, MCI and other microwave transmission companies participated with the National Science Foundation, IBM, and others to form NSFNET, one of the steps to the creation of the **Internet**.

music synthesizer

Musicians had experimented with the concept of music produced by electromechanical means in the 1920s and 1930s, but the first development of a true music synthesizer occurred in 1955, when Harry Olson and Herbert Belar developed a machine at RCA laboratories in Princeton, New Jersey. They fed preprogrammed music on a punched tape, allowing experimentation with a wide variety of sounds.

During the 1960s, practical musical synthesizers for use by musicians were developed, first by Robert Moog, who used a keyboard like that of a piano to signal the sounds to the **loudspeaker** [V] output. He continued to improve on the machine. His Moog III had two 5-octave keyboards controlling voltage and regulating pitch, attack, tone decay, timbre, and tone. Like the early computers of the 1930s through the 1960s, the Moog was an analog device.

Donald Buchla developed a synthesizer that had a keyboard, with keys that were touch-sensitive without moving. Special music was composed for it, including Morton Subotnick's "Silver Apples of the Moon" in 1967. Another 1960s synthesizer was built by Al Perlemon.

Both the Buchla and Moog machines used a system of subtracting components from an electronically generated sound. At the University of Illinois, James Beauchamp introduced a synthesizer that reversed the process by adding or building tones from sound components.

Later digital synthesizers used a system based on sampling—that is, by taking snapshots of digitally recorded sound, the machine can store the signals and then release them on command, allowing reproduction of the sound of drums or other instruments from a keyboard. This system is used by synthesizers produced by Kurzweil and by the New England Digital Synclavier.

nuclear reactor

Scientists in the United States, Canada, Germany, and the Soviet Union first constructed nuclear reactors during World War II. Their goals were to establish the nature of nuclear chain reactions in uranium and to make reactors that could produce plutonium, a new element, which could serve to fuel nuclear weapons. The first successful reactor was called an *atomic pile* because of its structure of graphite bricks. The operation of Chicago Pile-1 by Enrico Fermi (1901–1954) on December 2, 1942,

nuclear reactor Enrico Fermi can be credited with building the first successful nuclear reactor, Chicago Pile-1, at the University of Chicago in 1942.
National Archives and Records Administration

represented the first demonstrated sustained and controlled nuclear chain reaction. In this sense, Fermi was the "inventor" of the nuclear reactor, although Leo Szilard (1898–1964) had patented the concept of the nuclear chain reaction in Great Britain in the mid-1930s. The first operating reactor was built on a squash court underneath the Stagg Field stands at the University of Chicago. Fermi's design used graphite bricks to slow down or moderate neutrons emitted from fissioning natural uranium, to increase the probability of collision with other uranium atom nuclei to initiate and sustain a chain reaction.

The original reactor contained separate inventions that were to become part of the design of future reactors in different forms. To control the reaction, Fermi had wooden rails inserted from above into open spaces left through the stack of graphite bricks. Thirteen-foot-long strips of cadmium metal were nailed to the rails; cadmium absorbed neutrons and would prevent the reaction from progressing as long as the rails were in the pile. By removing the cadmium-covered rails slowly, the reaction would get under way. The principle of control through removing and inserting such control rods became standard on later nuclear reactors. For a decade, even as reactor designs took different forms, nuclear reactors were still called atomic piles. As a joke, the red button switch to electrically insert an emergency control rod was labeled *scram,* and that word became permanent reactor jargon for an emergency shutdown. Cooling was provided by the fact that the pile operated in an unheated room in Chicago; in effect, it had an air-cooled system. A large square rubber envelope was constructed to provide some minimal protection to workers as a form of shielding. Thus Fermi's reactor provided the ancestral systems of fueling, control rods, emergency shutdown, cooling, and shielding that characterized all later reactors.

In the United States, the first reactors to produce plutonium were built beginning in 1943 at Hanford, Washington, to take advantage of the constant flow of cold water from the Columbia River for cooling the reactors. Beginning in November 1944, small amounts of plutonium were collected from the first production reactor; they were later used in the first nuclear explosion at Alamogordo, New Mexico, and also used to make the atomic bomb dropped on Nagasaki, Japan.

In the United States, power reactor development began in the late 1950s. The first reactor to produce power commercially went operational at Shippingport, Pennsylvania, in 1957. That reactor followed the design of the pressurized water reactor (PWR) used earlier for submarine and ship propulsion. After some experimentation with other types of reactors, nearly all reactors for power in the United States used water for both coolant and moderator.

Reactors were identified by the types of coolant and moderator they used. Most of the commercial reactors in the United States have followed one of two design patterns. The PWR, designed to fit into the small confines of a nuclear submarine, was a proven power plant. Beginning in the 1950s, Argonne National Laboratory developed the boiling water reactor (BWR). In the PWR, pressurized water is circulated through the reactor, heated, and then piped through a steam generator, where a separate water system is heated to produce steam to drive the turbine generators. In the BWR, a single loop takes water through the reactor, raising it to steam temperature. The steam then proceeds through pipes and valves to the turbine, where it is condensed back to water to be pumped back into the reactor. In both PWRs and BWRs the steam is cooled by an outside water supply, usually circulated through a characteristic cooling tower.

Following 1974, construction of new reactors in the United States declined, partly due to a slowing in the rate of growth of electric demand. Following a reactor accident at Three Mile Island in Pennsylvania in 1979, many power companies canceled orders for new reactors and applications for reactor licenses further declined. By the beginning of the 21st century there were about 100 commercial power reactors in the United States.

In the United States, the proportion of electricity generated from nuclear reactors climbed rather rapidly in the early 1970s and then more gradually in the 1980s, approaching 20 to 25 percent of total U.S. power production by the end of the century. Nuclear power made up a greater proportion of the power resources of some other countries, with an estimated 75 percent of France's electricity and about 30 percent of electric power in Japan.

The total number of nuclear power reactors in operation worldwide increased gradually, to nearly 500 by the late 1990s. At the end of the 20th century, the United States had more than 20 percent of the power reactors in the world and more than a third of

> The decisions we make about nuclear energy will determine in large measure the kind of world our grandchildren will inherit. The issues this technology has created constitute a remarkable microcosm of the present predicament of our planet. The nuclear predicament raises a host of social, political and even ethical problems, many of them with long-term implications beyond any foreseeable horizon. Clearly such issues demand the fullest public consideration and the widest possible participation in the crucial decisions to come.
> —Walter C. Patterson, *Nuclear Power* (New York: Penguin, 1976)

the total nuclear power–generating capacity. With the closure of some older U.S. reactors, the proportion of the world's nuclear power generated in the United States is expected to decline.

Public controversies arose in nearly all countries that built or planned nuclear power plants. Advocates of nuclear power pointed to the relatively low cost compared to oil- and coal-driven steam turbines, to the fact that nuclear power did not pollute the atmosphere, and to the fact that nuclear power could provide an important resource to meet expanding energy needs. Opponents emphasized the hidden costs of nuclear fuel fabrication and waste management, as well as the risk to public health and safety from nuclear power plant accidents.

The tendency of government officials to conceal information about nuclear accidents also fueled opposition. A release of radioactivity at the Windscale reactor in Britain in 1957 was not immediately disclosed, and details of the 1986 Chernobyl accident in the Soviet Union were temporarily suppressed.

nuclear weapons

Weapons based on nuclear fission were developed under the auspices of the Manhattan Engineer District in the United States from 1942 to 1945. Although the invention cannot be attributed to any single individual, General Leslie Groves (1896–1970) headed the project, and the technical director who brought together the scientific team was J. Robert Oppenheimer (1904–1967). Crucial contributions were made by many other scientists, including Enrico Fermi (1901–1954), who first demonstrated a controlled nuclear reaction in an atomic pile or **nuclear reactor** on December 2, 1942.

The Soviet Union successfully tested its first nuclear weapon, based on espionage information from the American program, in 1949. The scientific head of the Soviet program was Igor V. Kurchatov (1903–1960), and the administrative head was the director of the secret police, Lavrenti Beria (1899–1953). Britain, France, China, India, Israel, Pakistan, and North Korea each developed their own nuclear weapons over the decades 1950 to 2000. Iraq initiated a nuclear weapons project that appeared to be successfully destroyed during and after the brief Persian Gulf War of 1991. South Africa developed the weapons and then abandoned the project.

The first atomic or nuclear weapons relied on two separate designs. The weapon that used **plutonium** was designated Fat Man and was designed at Los Alamos and detonated at the Trinity test in Alamogordo, New Mexico, on July 16, 1945. That weapon relied on implosion of wedge-shaped high-explosive lenses to bring the critical mass of plutonium together. Another design, using uranium 235, employed explosive charges to fire a plug of uranium down a barrel into another plug, known as the gun-type or Little Boy design. Two days after the aircraft *Enola Gay* dropped the uranium-fueled weapon, Little Boy, on Hiroshima on August 6, 1945, another aircraft, *Bock's Car*, dropped a Fat Man–design weapon over Nagasaki. These were the only two operational uses of a nuclear weapon during the 20th century.

In August 1949 the Soviet Union tested a duplicate of Fat Man in a test called Joe 1 by the American press. The first Soviet nuclear weapon used information gathered through espionage on the American program provided by the German-born British physicist Klaus Fuchs and other agents, although later designs were largely based

nuclear weapon One of the first two nuclear weapons ever designed, this model, Little Boy, awaits loading on an aircraft in a specially built pit. *National Archives and Records Administration*

on research by Soviet scientists. The British decided to build nuclear weapons in January 1947, and the first British test of a nuclear device was held in Australia in 1952. France tested its first weapon in 1960, China in 1964, and India and Pakistan in 1998. Neither North Korea nor Israel openly admitted to testing nuclear weapons, but both were known to possess them.

numerical controlled machines

The term *numerical controlled machine* was applied by Victor Scheinmann of the Massachusetts Institute of Technology (MIT) Artificial Intelligence Laboratory to a class of industrial robots in the 1960s. The concept of a robot can be traced back in the history of technology to early automatons often connected with clocks that performed a variety of entertainment functions such as dancing, playing musical instruments, or imitating other human motions.

The concept of developing a machine to do human work was a staple of literature through the ages and found fruition in the 1922 classic of science fiction by the Czech author Karel Capek (1898–1938), *R.U.R.* In that play, Capek introduced the term *robot,* derived from the Czech word *robota* (forced work). Throughout the Austro-Hungarian Empire from 1514 to 1789, a system of *robota* had been installed to extract labor from peasants, similar to obligatory work systems elsewhere in Europe.

The dream of a mechanical robot came closer to reality with the development in 1946 by the American inventor George Devol of a magnetic machine capable of storing and replaying program instructions. In 1954 he developed a manipulator arm which, when combined with the recording machine, represented a first programmable industrial robot.

In 1961, as MIT was experimenting with numerical control, Jerome Lemelson began producing a machine he had patented in 1954. His Unimate 2000 was a hydraulically powered industrial robot that found its first uses in Japan. By 1980 there were some 5,000 industrial robots in the United States, and by 1993 there were nearly 500,000 worldwide.

Several companies began to develop programming languages and robots with a variety of "skills." Most common was a stationary machine to which the work would be brought on an assembly line; the machine would then follow programmed steps to conduct tasks such as spot or line welding, assembling with screws or bolts, or lifting heavy parts to hold in place while another machine connected the parts, sometimes using sensory feedback in the form of television images or another form of sensory perception.

A major goal of early industrial robot designers was to imitate some of the many complex functions of the human hand and arm. From the elbow through the fingertips, the human hand can make 36 articulations or movements in independent directions; accomplished robots achieve up to 10 or so articulations. Instead of a hand, an industrial robot may have only a 2-position gripper, either open or closed, set to the size of the part being moved.

nylon

Nylon was the first synthetic fiber, developed at the E. I. Du Pont laboratories in Wilmington, Delaware, by a research team under the direction of Wallace Carothers (1896–1937) in 1935.

After teaching briefly at the University of Illinois and at Harvard University, Carothers took a research position at Du Pont in 1928, looking into polymer chemistry. Carothers theorized that polymers were large molecules with a repeating structure, and his research confirmed this concept by 1930. One of the first products of his laboratory was neoprene, a rubber substitute relatively impervious to oil, gasoline, and temperature changes, making it ideal for hoses, gaskets, and belts in automotive engines.

Carothers obtained what he called superpolymers by reacting acids with alcohols. Assistant Julian Hill discovered that long-chain molecules could be drawn out, strengthening them. One assistant, Donald Coffman, drew out a chain molecule of a type of polymer based on an alcohol/acid formula in 1934, representing a type of nylon. The Carothers team developed an extrusion machine that removed water from the product and allowed for thin, continuous threads. On February 28, 1935, the Carothers group developed a resin made from an acid and an alcohol both containing six carbon atoms. Dubbing the material *nylon 6, 6,* they found that it could be stretched to a very thin thread, making it an ideal substitute for silk.

Carothers, who suffered from acute depression, committed suicide in 1937. However, Du Pont went on to manufacture nylon. The first nylon stockings were sold in 1938. During World War II, nylon found uses in parachutes, sutures, mosquito netting, lines for boats, and other applications. Because of these military demands for the product, nylon stockings virtually disappeared from the market until after the war.

In the postwar years, nylon became widely used in the production not only of stockings but also of a wide variety of fabrics and fabric blends, affecting the world of fashion. At the same time, the more durable **Dacron,** developed first in Britain, was widely introduced in applications requiring exposure to chemicals or weathering.

oral contraceptive

The search for a simple and safe oral contraceptive began with the work of Margaret Sanger (1883–1966), an advocate of family planning. After serving time in prison for operating a family planning advice clinic in New York, she founded the International Planned Parenthood Federation in 1948. She influenced Dr. Gregory Pincus (1903–1967) of the Worcester Foundation in Massachusetts to extend his work on fertility in animals to a search for an oral contraceptive. In 1955 he reported success with a hormone called norethisterone. To prevent an irregular menstrual cycle caused by the hormone, he combined it with estrogen. The result was an effective oral contraceptive, soon dubbed *the pill* and introduced in 1960. By combining synthetic estrogen and synthetic progesterone, the hormone dosage simulates the biochemical effect by which ovulation is held up during pregnancy.

Most birth-control pills have to be taken on a regular schedule for 20 or more days during each menstrual cycle, depending on the composition of the pill. Depo-Provera, a progesterone type of contraceptive, can be administered by injection, while *Norplant* was a small tube that would be implanted to release the synthetic hormone progestin. All of the hormonal contraceptives had a number of negative side effects, including disorders of the blood, and sometimes led to weight gain and nausea.

The decline in unwanted pregnancies was so pronounced that some social analysts regarded the introduction of the pill in the United States as the end of the baby-boom generation. Some have defined that generation as born after the bomb and before the pill—that is, between 1945 and 1960.

The use of the oral contraceptive as an artificial means of birth control was opposed by the Catholic Church and by some other religious organizations as an unnatural means of interfering with the creation of human life. Despite such objections, the pill revolutionized social and family relationships throughout the world.

planets outside the solar system

Astronomers assumed that planets surround many stars, but because of their distance and small size, direct observation of planetary systems surrounding distant stars was never possible. However, in the 1960s astronomers began attempts to detect the presence of extra–solar system planets by their effects on the host star. Peter van de Kamp claimed to have detected two planets around Barnard's Star in 1960, but later observations could not confirm the discovery.

Alexander Wolszczan discovered the first confirmed extra–solar system planet in 1992 at Pennsylvania State University. He had found two large planets and possibly one small planet in orbit around a pulsar, a very small and dense star that emits pulses of radio signals. Variations in the pulsed signal gave proof of the planets' existence, as they "eclipsed" the beam at regular intervals. In 1994 Wolszczan presented a paper describing his methods and a computer model calculating the orbits of the two planets. Wolszczan received the 1996 Tinsley Prize of the American Astronomical Society for his discovery.

In 1995 Michael Mayor and Didier Queloz in Geneva, Switzerland, found a planet whose mass was 0.5 to 2 times that of Jupiter orbiting around the star 51 Pegasi. Later that same year, Geoffry W. Marcy and Paul Butler, of San Francisco State University and the University of California at Berkeley, respectively, found two more planets. Over the next few years, more than 50 planets were detected. Improvements in spectrometers, electronic sensors, and computer software all contributed to the new finds. Almost all of the newly discovered planets were extremely large, about the size of Jupiter.

The discovery of numerous extra–solar system planets fueled speculation that life exists elsewhere in the universe. Although life as known on Earth would not be likely on a planet with the gravity of Jupiter, several possibilities were suggested, such as the possibility of smaller planets or satellites of the Jovian-scale planets that would be capable of hosting Earth-like life.

plutonium

Plutonium was the second "artificial" element (created after neptunium), manufactured in a chain reaction in an atomic pile or **nuclear reactor,** although there is evidence that it occurred naturally at a uranium mine in Oklo, Gabon, in West Africa in prehistoric times. The nature of one plutonium **isotope [V]**, plutonium 239, meant that it was readily fissionable, and it was therefore made for use in nuclear weapons. The first nuclear weapon, tested in New Mexico in 1945, was fueled by plutonium, and the second weapon dropped on Japan, at Nagasaki on August 8, 1945, was of a plutonium Fat Man design.

Two American scientists, Glenn Seaborg (1912–1999) and Edwin McMillan (1907–1991), shared the Nobel Prize in chemistry in 1951 for their 1940 discovery of plutonium. Seaborg bombarded uranium with deuterons in a **cyclotron** at the University of California and succeeded in synthesizing plutonium in 1941. Since fissionable plutonium could be manufactured from uranium 238, the more plentiful isotope of that element, it was used as an alternate fuel for nuclear weapons. Uranium 235, which was a fissionable isotope of uranium, also could be used for a weapon, but it was extremely rare and difficult to separate from the more plentiful uranium 238, leading to the construction of elaborate **uranium isotope separation** facilities at Oak Ridge, Tennessee, and elsewhere. During and after World War II, plutonium 239 was made in reactors built during the Manhattan Project at Hanford, Washington. Later, British, French, Soviet, Chinese, and other scientists built similar reactors for its production and use in weapons. After arms control agreements between the Soviet Union and the United States in the 1980s and 1990s reduced nuclear weapons stockpiles, plutonium became "surplus," and Japan and other countries have considered developing plutonium-fueled reactors for the generation of electrical power.

Seaborg, McMillan, and another team member, Albert Ghiorso, discovered several other transuranic elements—that is, elements heavier than uranium—including americium and curium (1944), berkelium (1949), californium (1950), einsteinium (1952), fermium (1953), mendelium (1955), and nobelium (1958).

Because of the unstable nature of the plutonium 239 atom, when enough of it is assembled to achieve critical mass, the chain reaction proceeds so rapidly that the production of X-rays and neutrons will blow apart the pit of a weapon extremely quickly, so that not all of the plutonium present is fissioned. To increase the amount of critical mass involved in fission, designers at Los Alamos hit on the idea of imploding the critical mass, or compressing it suddenly, by surrounding wedges of the material with wedges of high explosive. When the high explosive was simultaneously detonated, the focused explosions would drive the plutonium together, compressing and compacting it and increasing the proportion fissioned. The complex spherical design required a weapon 6 feet in diameter, and hence the entire plutonium weapon was dubbed Fat Man. Furthermore, the complexity of the design suggested that it needed to be tested before use, requiring the August 16, 1945, test at Alamogordo, New Mexico.

Details of the design of Fat Man were smuggled out by several atomic spies who provided the information to the Soviet Union. Researchers there followed the American plans in precise detail to detonate their first atomic bomb four years later, in August 1949.

Polaroid camera

Edwin Land (1910–1991) invented the Polaroid camera, capable of developing a picture within the camera. Land had voluntarily dropped out of Harvard in 1932, intent on making a fortune, hoping to build on the idea of embedding a sheet of plastic with lined-up crystals that would polarize, or line up, light rays. He promoted a number of ideas using polarized light, with only one success, sunglasses. Eventually he sold the same principle for polarized glass to reduce glare through windshields. He developed a concept for **three-dimensional motion pictures** that relied on polarized eyeglasses to resolve a double image into a three-dimensional image. Although briefly a fad during the 1950s, such 3-D movies did not have lasting success.

In 1947 Edwin Land staked his future on the concept of an instant-developing camera. According to a story he later told, the idea came to him when, in 1943, he took a picture of his three-year-old daughter Jennifer. After he snapped some pictures his daughter asked him, "Why can't I see them now?"

He put the engineers at his company to work on the problem, and the design, using a developer inserted between two layers of film in a "pod," was ready for demonstration early in 1948. The key to the system was that when the film was squeezed between rollers, the processing fluids in the pod spread over the picture between the negative and positive sheets. Much to the surprise of his employees, but fulfilling Land's expectations, the camera was an "instant" success, with orders soon outstripping supplies. Land kept making improvements. The first camera had produced a brownish, sepia-colored picture, but soon the Land camera could produce black-and-white prints. Color Polaroid was introduced in 1963, and in 1973 the company brought out a new SX-70 color film. With this film, the picture developed inside the camera, rather than being pulled out and the negative peeled off by hand. Furthermore, the processing time for some Polaroid film was reduced from 60 to 15 seconds.

An additional feature of the Polaroid camera was the ability to instantly determine whether a shot was satisfactory. With the advent of the **digital camera** that contained a small display showing the picture in the 1990s, the Polaroid lost one of its competitive edges, and sales declined.

printed circuit

First invented in 1936 by the Austrian scientist Paul Eisler, the printed circuit did not come into widespread use until about 1948. The printed circuit represented an improved method for wiring equipment such as **radio [V]** and **television [V]** sets and later, **computers** and control systems for guided missiles. Printed circuits became particularly useful in missiles and aircraft, as they were far smaller and lighter than equivalent hand-soldered circuits and were generally much more reliable. Even before the introduction of **transistors,** printed circuits allowed for reduction in size and for the cheaper, mass production of electronic components.

Strictly speaking, the printed circuit itself is simply a board with a wiring pattern, usually in thin, copper paths sandwiched in thin, flexible plastic film that serves to insulate the paths. A printed circuit board is complete with electronic components such as resistors, capacitors, or diodes. The simplest printed circuits are single-sided, with components and wiring on one side of the sheet, and are commonly used in consumer electronic devices. Two-sided boards, with copper paths on both sides, can handle more complex equipment and more components and interconnections. Sandwiched or multilayer boards consist of a stack of several two-sided boards, used in large computers and in navigation systems. Although several processes have been developed, the basic principle has remained the same. It consists of reducing and transferring a photographed image of a circuit to a plastic sheet and then impregnating the lines with conductive material, to which components are directly soldered rather than plugged in.

When combined with transistors, printed circuit boards allowed for the reduction in size and reliability of many electronic devices and were key to the revolutions in communication that characterized the last decades of the 20th century.

proximity fuse

The idea of a fuse that would detonate an artillery shell when it was close to its target had occurred to weapons designers in Britain, Germany, and the United States in the 1930s. The rapid flight of airplanes and their high altitude made it extremely difficult to score direct hits, and a detonation of the shell with a near miss within a few feet of the aircraft could greatly increase the effectiveness of antiaircraft fire. The first successful proximity fuse, based on a radio-detection or **radar** mechanism, was invented in the United States in 1941 and perfected for production in 1942. The project was supported by the National Defense Research Committee, later supervised by the Office of Scientific Research and Development, and initiated at the Carnegie Institution in Washington. Much of the work was done at a new laboratory established by Johns Hopkins University in Silver Spring, Maryland. The lab was later known as the Applied Physics Laboratory of Johns Hopkins University.

The U.S. Navy concentrated on work on the 5-inch antiaircraft shell. The major development problem was developing vacuum tubes small enough and rugged enough to fit inside a shell and survive being fired from a gun. Tests on the proximity fuse were conducted at the Naval Proving Ground at Dahlgren, Virginia. Full-scale production of the tubes was undertaken by the Crosley Corporation, and the Mark 32 fuse was delivered to the fleet in 1942. The first enemy plane destroyed by a proximity-fused shell was fired from the cruiser *Helena* in January 1943.

The rapid production of the proximity fuse during the war, for use in the war as a "secret weapon," was widely hailed in the postwar years as an example of research and development on order. Together with advances in radar, **sonar [V],** and **nuclear weapons,** the proximity fuse was regarded by Vannevar Bush, a leading advocate of military research and development, as one of the major achievements demonstrating the need to turn science and technology to issues of defense.

quark

The quark is an elementary particle whose existence was independently theorized in 1963–1964 by two physicists, Murray Gell-Mann (1929–) and George Zweig (1938–). Although quarks may be "unobservable," physicists accept indirect evidence of their existence. Both Gell-Mann and Zweig proposed that a class of elementary particles made up the other subatomic particles and that their qualities explained the behavior of the others. The name itself was derived from a line in a novel by James Joyce, *Finnegans Wake,* in which the dreaming hero hears seagulls crying, "Three quarks for Muster Mark."

The original theory of quarks described them as falling into three types, known as up, down, and strange (*u, d,* and *s*). Later, other quark varieties, also named in the same whimsical fashion, were discovered, known as the beauty (or *bottom*), the truth (or *top*), and the charmed (or *c*) quarks. Each of these six types was said to represent a "flavor" of a quark, and in addition they possess another quality, known as "color." The six flavors of quarks come in three colors, called green, red, and blue.

Whether or not quarks can be said to have been "discovered" or only postulated is a difficult philosophical question. However, since the qualities ascribed to quarks describe effects later produced in particle accelerators, most nuclear physicists now accept their existence as providing an explanation for the behavior of a wide variety of subatomic particles. The 1990 Nobel Prize in physics was awarded to three scientists whose work with linear accelerators demonstrated the existence of quarks: Jerome I. Friedman, Henry W. Kendall, and Richard E. Taylor.

radar

The origin of radar (originally an abbreviation for radio detection and ranging) can be traced to the German researcher Christian Hulsmeyer, who in 1904 demonstrated a radio wave warning system that detected the interruption of a radio beam for ships. At the Naval Research Laboratory in the United States in 1922, A. Hoyt Taylor and Leo C. Young proposed a system that would require two ships, one sending and one receiving, using radio waves to detect any ship that passed between the two. In 1924–1925, Gregory Breit and Merle A. Tuve succeeded in using radio wave echoes to determine the height of Earth's **ionosphere [V]**. In 1930, L. A. Hyland demonstrated a radar capable of detecting aircraft. Taylor and Young patented their system.

Since the velocity of radio waves, equivalent to the **speed of light [III]** (approximately 186,000 miles per second), is known with some accuracy, it is easy to calculate the distance to and from an object. When a radio signal is bounced off a target, the distance to the target can be calculated by measuring the time elapsed between sending and receiving the signal. Furthermore, the precise angle of direction can be determined. The information can be combined and displayed on a cathode-ray screen (itself derived from the 19th-century **Crookes tube [IV]**). The complete system of pulse generator, sending and receiving aerial (or transponder), linkages, and display system, while readily conceived, required considerable development over a period of years.

With mounting international tensions in the 1930s, radar improvements proceeded in relative military secrecy in Britain, Germany, France, the United States, and Japan. By 1939 both Germany and Britain had developed ground-based radars capable of detecting air-

radar Large radar antennas like this one at Edwards Air Force Base can track distant aircraft, while much smaller, microwave radar mounted on aircraft and on missiles are used to detect targets on the ground. *U.S. Air Force*

craft. The British system, known as the Home Chain, was designed by Robert Watson-Watt and was characterized by coordinating information from widely separated radar installations rather than relying on a single-point radar. A committee headed by Henry Tizard took the lead in sponsoring such research. In 1935 Watson-Watt first described his system in two memoranda in which he suggested a system of short-pulse transmission and sets of receivers. The Tizard committee approved, and the Air Ministry constructed a set of five interconnected stations. The Home Chain was effective in warning of German air raids in 1940 and 1941. Radar could detect motion by noting the Doppler shift in wavelength as aircraft moved closer to or away from the stationary radar installation. By subtracting images that remained stable, reflected "ground clutter" could be eliminated, leaving only images of moving objects.

The device suggested for the high-power microwave transmitter was the klystron, first described by W. W. Hansen, of Stanford University, in 1938. However, researchers settled on using the magnetron, invented by A. W. Hull of General Electric in 1921, but modifying it as the cavity magnetron, which could produce high-power signals on the centimeter-wavelength level. Britain developed a high-power microwave cavity magnetron, a device that allowed for a very small radar sending and receiving unit that could be mounted aboard aircraft. In Britain the microwave effort was headed by M. L. Oliphant of the University of Birmingham, and in the United States, by R. H. and S. F. Varian of Stanford University.

Highly secret during the war, microwave radar installed on British and American aircraft was instrumental in locating German **submarines** **[V]** when they surfaced to run their **diesel engines [V]**, necessary to charge their battery systems.

Since the microwave radar had destroyed the main advantages of the submarine—its stealth and ability to remain concealed—German and American researchers immediately sought a solution. One was the **snorkel,** which allowed the diesel to run while the submarine remained submerged. Another, introduced in the 1950s, was a submarine propulsion plant that did not require surfacing to obtain oxygen, using a **nuclear reactor.**

radio telescope

In 1932 Karl Jansky, an American researcher at Bell Labs, received radio waves apparently coming from empty interstellar space. Over the next 15 years these radio waves remained a mystery, until more sensitive radio equipment, developed during World War II, was applied to the phenomenon. The first radio telescopes, designed to sort out and register such signals, were built in 1948, and researchers immediately found that the most powerful sources were two specific spots, in the constellations of Cygnus and Cassiopeia. Two "radio stars," without visible light, were discovered.

Over the next five decades, several thousand more radio sources have been discovered. The nearest galaxy to ours, the Andromeda nebula, contains many suns that emit on the 1.80-meter radio wave frequency but not on the visible light frequencies. Other sources emit on wavelengths varying over a wide range of frequencies, from a few centimeters up to 20 meters or more. A source in Cygnus was identified as resulting from the collision of two nebulae about 200 million light-years away. In 1951, radiation at a constant wavelength of 21 meters was discovered coming from clouds of hydrogen in interstellar space.

Bernard Lovell of Manchester University in Britain did some of the pioneering work in radio telescopy. After working in the 1930s on **radar,** in the postwar years he took over surplus equipment and set up

radio telescope
The Arecibo radio telescope uses a natural depression in the hills of Puerto Rico to support a dish of reflector panels.
National Aeronautics and Space Administration

a team at Jodrell Bank, in Cheshire, where Manchester University already maintained a botany facility. Lovell's original plan was to use radar beams to bounce off the Sun and planets. His first radio telescope, designed to bounce radar beams off the Sun and Venus, was built from 1947 to 1957. In 1957 it was used to track the first **artificial satellite,** *Sputnik I,* launched by the Soviet Union. The Jodrell Bank receiver was 80 feet in diameter, suspended from a circular track 350 feet in diameter that could be angled to point to various positions in the sky.

Soon, however, the Jodrell Bank receiver was used not to receive radar echoes but to assist in mapping the new radio wave–emitting stars, nebulae, and galaxies. Another British radio telescope receiver system, constructed at Cambridge University in Britain, was known as the Mullard Radio Astronomy Observatory. Rather than using a bowl-shaped receiver, the Mullard system consisted of two cylindrical structures 2,400 feet apart. One was fixed to the ground, while the other moved along a 1,000-foot track.

The largest dish-type radio receiver or radio telescope in the world was completed in 1963 and operated by Cornell University at Arecibo in Puerto Rico. Funded by the National Science Foundation, Arecibo is built in a natural hollow on a mountain peak. One of the first discoveries at Arecibo was a corrected calculation of the rotation of the planet Mercury. In 1996 Arecibo received an upgrade at the cost of $25 million, increasing its sensitivity by a factor of 10.

radiocarbon dating

The process by which the radioactive decay of the **isotope** [V] carbon 14 or ^{14}C can be used to date the point in time when a particular piece of vegetable or animal matter died was invented by Willard Libby (1908–1980), an American physicist, in about 1947. He had worked on methods of separating uranium isotopes for the construction of **nuclear weapons,** and in the postwar period he turned his attention to carbon isotope decay. Carbon 14 is continually formed by the interaction of neutrons with nitrogen 14 in the atmosphere, with the neutrons produced by the interaction of cosmic rays with the atmosphere. Plants absorb the ^{14}C with carbon dioxide from the atmosphere, and it is passed from plants to animals as they eat vegetable matter.

Once a plant dies, it ceases to absorb carbon from the atmosphere, and the ^{14}C in the tissue begins to decay to nonradioactive carbon. The isotope has a half-life of about 5,730 years. Thus if a piece of wood is found in a structure with half the normal distribution of ^{14}C, the tree from which the wood was cut died about 5,730 years ago. With careful measurement it is possible to determine quite accurately the age of ancient remains. Libby and his team applied the technique to Egyptian relics. They were able to use the technique to date the age of artifacts and remains up to 50,000 to 60,000 years old. The technique has been widely used as a tool in a number of sciences, including paleontology, archaeology, Pleistocene geology, and by historians.

However, when findings through radiocarbon dating contradict fondly held assumptions or theories regarding the dating of a particular artifact, defenders of the alternate chronology have challenged the reliability of the method, leading to heated controversies. For example, in the case of the Shroud of Turin, a large piece of fabric that carbon 14 dated to A.D. 1300 to 1400, some researchers had assumed the cloth to be the shroud in which the historical Jesus was wrapped after the Crucifixion. The cloth bears a strange photographlike image that is difficult to explain. Defenders of the view that the shroud could be dated to the time of Christ suggested that contamination of the sample by later impregnation of atmospheric carbon, perhaps as the result of exposure to smoke, would account for the later dating by the carbon 14 method.

radio-thermal generator

A radio-thermal generator or RTG (also known as radioisotope thermoelectric generator) is a device that uses the heat derived from the decay of a radioactive **isotope** [V] to generate electricity. The first demonstrated RTG was produced in 1958 and displayed by President Dwight D. Eisenhower in January 1959. The Martin Company of Baltimore accomplished the research and design work, and Minnesota Mining and Manufacturing Company (now the 3M Company) developed the conversion system for transforming the heat into electricity. The RTG used the concept developed by Thomas Seebeck (1770–1831) in the **thermocouple** [IV] in which an electrical current is produced by keeping two metals at different temperatures in contact. In most RTGs the heat from the decay of ^{238}Pu, an isotope of plutonium, was used to heat a thermocouple.

Larger and larger RTGs with greater power production were utilized in the space program of the National Aeronautics and Space Administration (NASA) from 1961 to about 1986. They were employed in the Pioneer, Viking, and Voyager space exploration vehicles. However, with the larger power demands of the space vehicles that transported humans, the RTG was insufficient, and NASA adopted the use of **fuel cells** as a source of electric power. As a measure of the efficiency of RTGs, the ratio of watts produced per kilogram (or power density) reflected incremental improvement to the RTGs. Some of the first RTGs employed in Transit satellites in 1961 had a watt/kilogram ratio about 1.29, while those on the Voyager space vehicle in 1980 produced 158 watts with a 4.2 watt/kilogram ratio.

In the mid- and late 1980s NASA phased out the use of RTGs, replacing them with improved fuel cells. In addition to the higher power requirements, RTGs caused concern that a satellite or space vehicle that accidentally crashed to Earth and that had an RTG for the production of electricity could create a radiological hazard on the ground. It was recognized that the adverse publicity from such an event could harm the space program more generally, so NASA discontinued their use.

radio-thermal generator   Launched aboard a Thor-Able rocket, this Transit satellite contained equipment powered by a radio-thermal generator that worked on a principle discovered by Thomas Seebeck in the 19th century. *National Archives and Records Administration*

snorkel

The snorkel was invented in 1938 in Holland, as a device to allow **submarines [V]** to take in air that would allow diesel engines to operate while submerged. When Germany invaded Holland in 1940, they acquired the device and gave it the name *schnorkel* or *schnorchel*. In English the term has usually been spelled *snorkel*.

At first the Germans had little use for the system, but in 1943 they began applying it on the VIIC and IXC classes of submarines. The first was installed on *U-58*, which experimented with the equipment in the Baltic Sea in the summer of 1943. Operational installations were made in early 1944, and by June 1944 about half of the German submarines based in France had snorkel installations.

The advantage was that the system allowed the submarine to remain submerged, whereas previously, the time submerged had been limited because the submarine operated on electric motors and batteries while underwater, usually surfacing at night to operate the diesels to recharge the batteries. The snorkel air-breathing connection required a large pipe to the surface with a closing mechanism at the top that would seal when a wave hit the opening. With the pipe raised, the speed of the submarine was reduced to about 6 knots underwater. Other disadvantages included the fact that when the snorkel closed, the diesel engines would suck in air from the atmosphere in the submarine, causing a great drop in air pressure and damage to the sailors' eardrums. Furthermore, although the submarine could remain submerged for days, garbage would accumulate, making living conditions in the cramped quarters even worse. The German type XXI and XXIII submarines, constructed in the last months of the war, were designed to take full advantage of the snorkel. In the first applications on the type VII and type IX, the mast containing the snorkel folded down along the side of the submarine. In the type XXI and XXIII submarines, the snorkel was mounted on a telescoping air mast that rose vertically through the conning tower, next to the periscope.

With the development of improved **radar,** submarines lost the ability to evade detection by surfacing at night for battery recharging; hence the German use of the snorkel. In the postwar period the U.S. Navy experimented with fitting snorkels to existing submarines, and at the same time the navy sought other means to achieve long submergence periods to evade radar. Among the most promising techniques was the use of **nuclear reactors** for underwater propulsion.

Meanwhile, with the development of the **aqualung** and the increased popularity of diving and undersea exploration, simple air-closure mechanisms called snorkels were developed to attach to face masks for sports diving. By the 1960s, snorkeling had become a popular sport.

streptomycin

Streptomycin was discovered by Dr. Selman A. Waksman (1888–1973), a Ukrainian-born biochemist who immigrated to the United States. Waksman worked as a professor of soil microbiology at Rutgers University in New Jersey from 1918 to 1958. He had studied microorganisms in soil, and after one of his students, R. Dubos, isolated the powerful antibacterial tyrothricin from an organism in soil in 1939, Waksman began a research program to isolate other organisms with antimicrobial action, studying more than 10,000 soil cultures. Waksman developed the generic term *antibiotic* to cover these organisms.

Contemporary work with **penicillin** [V] stimulated the search, and in 1943 Waksman identified a strain of organism, *Streptomyces griseus,* from which he extracted streptomycin. This chemical was found to be active against various bacteria, some of which were insensitive to the effect of penicillin. In particular, streptomycin was found effective against the bacillus of tuberculosis. The Merck & Company laboratories found it was effective against additional diseases and began pilot production of the drug in 1944.

Waksman received the 1952 Nobel Prize in physiology or medicine for the discovery of streptomycin.

Together with penicillin, streptomycin was widely regarded as a "wonder drug" in the 1950s as numerous infectious diseases, such as tuberculosis, pneumonia, and sexually transmitted diseases, that resulted from bacterial infection succumbed to treatment by one or the other of the new drugs.

Technicolor

With the development of **motion pictures** [V] and their popularity during the 1910s and 1920s, many experimenters sought to improve the medium with the development of color photography. From 1926 to 1933, three researchers at the Massachusetts Institute of Technology (MIT)—D. F. Comstock, H. T. Kalmus, and W. B. Westcott—worked on perfecting a system that was finally marketed as Technicolor. The first Technicolor motion picture was a Walt Disney cartoon, *Flowers and Trees,* released in 1933. Over the next decade, Technicolor became more and more widely adopted, and during World War II some Technicolor newsreels of naval engagements in the Pacific were released.

The Technicolor system uses a special camera in which the light beam is split into three beams, each filmed separately. The red, green, and blue films are each developed separately, along with a fourth, black-and-white film. The four films are then printed into a single film for projection.

Other processes that were less complex than the Technicolor process were introduced in the 1930s. One, developed by Leo Godowsky and Leopold Mannes, two inventive music students in New York, was first developed in 1923 and marketed by Kodak as Kodachrome in 1935. The German Agfa company developed Agfacolor in 1936. Both of these systems were monopack systems in which a single film contained four color-sensitive layers. The negative records the colors in the complementary color—that is, blue is recorded as orange, green is recorded as purple, and red is recorded as green. The positive prints are made through the use of color filters. In the United States, the MIT-developed Technicolor, despite its complexity, remained the dominant form of color motion pictures.

Teflon

Teflon, a plastic material technically known by its abbreviation PTFE, for polytetrafluoroethylene, was discovered by a Du Pont chemist, Roy Plunkett, in 1938. The material is an excellent insulator and is extremely slippery, resisting adherence. Applications proliferated after World War II, with the concept of developing nonstick cookware occurring in several locations in about 1952. In France, Marc Gregoire founded a company, Tefal, in 1955 to market such cookware, and in Britain, Philip Harben produced Harbenware, a popular brand of nonstick cooking pan. Du Pont entered large-scale production of Teflon cookware in 1962. Along with several other chemicals and plastics, the marvelous qualities of PTFE appear to have been discovered accidentally.

Teflon is soft, waxy, opaque, and inert to nearly all solvents. It stands up well under temperature variations, although it breaks down at temperatures above about 1,100 degrees Fahrenheit. It is very resistant to weathering and is nonflammable. Its durable and unique qualities have led to a constantly proliferating set of applications, not only in cookware but also in medical and industrial uses. Teflon is used in industry for tubing, valve packing and lining, gaskets, stopcocks, belt coatings, insulation, sockets, plugs, and other parts. Since it resists breaking down under exposure to chemicals, is lightweight, and does not adhere, it has found uses in **kidney dialysis machines** and in the construction of artificial organs.

thermonuclear weapons

Thermonuclear weapons or hydrogen bombs were first built in both the United States and the Soviet Union from 1953 to 1955, and a dispute as to which country developed such a weapon first soon ensued, a symptom of the Cold War competition between the two powers. Both the United States and the Soviet Union considered several designs of thermonuclear bombs that relied on nuclear fusion rather than on fission for the majority of their energy release.

One reason for confusion as to priority is the fact that in some **nuclear weapon** designs that used **plutonium,** a small amount of the tritium **isotope** [V] of hydrogen was used to create a partial fusion effect, to more thoroughly fission a plutonium core. The Americans tested this method, which they called a *boosted* weapon, as early as 1951. Since the process of boosting was kept classified for some years, outside commentators remained unaware of the distinction between a boosted fission weapon and a true hydrogen bomb. In the latter, the majority of the energy release was due to fusion.

In the late 1940s, American weapons designers thought through and abandoned a preliminary design for a true hydrogen fusion weapon. In that design, a fission weapon would ignite a deuterium fusion fuel. The device could be known as a fission-fusion-fission weapon, with layers of fissionable material, including some ^{238}U that would fission in the intense neutron environment. The yield of such a weapon would be the equivalent of several hundred thousand tons of TNT, or several hundred kilotons.

Instead of using that abandoned design, the United States built a cooled device using liquid deuterium, detonated in the Ivy-Mike test on October 26, 1952 on Enewetak, an island in the Pacific Ocean. Later, another design worked out by Edward Teller (1908–2003) and Stanislaw Ulam (1909–1984) was far smaller and could be transported by aircraft.

The Soviets pursued a weapon similar to the abandoned early U.S. design and called it the *sloyka* (layer cake) design, first conceived by Yakov Zel'dovich (1914–1987) as the "first idea" and later modified by Andrei Sakharov (1921–1989). Sakharov's "second idea" surrounded deuterium with a ^{238}U shell in the layers or *sloyka,* to increase the neutron flow during the detonation. The Soviet scientists tested the *sloyka* on August 12, 1953, with a yield of 400 kilotons. About 15 to 20 percent of the weapon's force derived from fusion. The United States already had larger boosted weapons, and American analysis of the Soviet 1953 test was that it was simply a boosted fission weapon. The Soviet authorities, however, continued to maintain that the August 1953 test was the world's first hydrogen bomb. American analysts continued to call it a boosted fission weapon.

The first test by the United States of a thermonuclear device that could be dropped as a weapon from an airplane was the Teller-Ulam design tested in 1954, with a yield estimated at 15 megatons.

The Soviet "third idea," developed by Igor Tamm (1895–1971) and Sakharov, could produce, like the Teller-Ulam device, much higher yields than using the *sloyka* idea. The Soviet Union tested the "third idea" weapon on November 22, 1955, producing an estimated yield equivalent to 1.6 million tons of TNT. American scientists claimed that the test of Ivy-Mike in 1952 was the first true thermonuclear detonation and that the Teller-Ulam weapon tested in 1954 was the world's first deliverable thermonuclear weapon. The Soviets continued to regard their 1953 test as the first thermonuclear test.

three-dimensional motion pictures

In the 1950s, the increasing popularity of television caused movie producers to search for new technologies that might lure TV viewers back into theaters. Two of the efforts were three-dimensional motion pictures (3-D) and **wide-screen motion pictures.**

Three-dimensional effects could be achieved in a number of ways. One had been developed as early as 1915, in which twin images, one red and one green, were superimposed and viewed with special red and green eyeglasses. In 1937, after the development of polarized lenses by Edwin Land, the inventor of the **Polaroid camera,** the first feature film using polarized lenses to achieve three-dimensional effects was released in Germany. Produced by Arch Oboler, the 1952 film *Bwana Devil* was the first of about 100 feature films using the polarized lens method to achieve the 3-D effect. Oboler made an improvement in 1966, using a special polarizator lens, which allowed for a single projector.

When viewed without the lenses, the double image on the screen simply looked blurred at the edges. With the glasses, each eye received a single image, each photographed from a separate lens. One twin camera, developed by Nigel and Raymond Spottiswoode, two English brothers, worked with two lenses polarized in different planes. Audiences found the effect an interesting novelty, but the inconvenience of wearing the eyeglasses, which were extremely uncomfortable for those wearing corrective eyeglasses, and the tendency of the films to be low-budget thrillers soon led to the fad wearing out.

Although Dennis Gabor achieved a more realistic effect in experiments with **holography** in 1947, holographic motion pictures were never developed in the 20th century for theater projection.

transistor

The transistor was invented in 1947 by a team that included William Shockley (1910–1989), Walter H. Brattain (1902–1987), and John Bardeen (1908–1991). In 1956 the three men received the Nobel Prize in physics for the development.

Later close study of the invention has tended to minimize the role of Shockley, suggesting that he insisted on remaining in charge of the project to share credit, even though his contributions were minimal and superseded by the contributions of the other two researchers. Shockley later published and spoke about his views on the question of genetically acquired intelligence, leading many critics to charge that he was a racist. Shockley's unpopular views on this sensitive issue contributed to the reexamination by historians of his role in the transistor invention.

Before World War II, the director of research at Bell Labs established a team to study solid-state physics, putting Shockley and S. O. Morgan in charge of the research. That team speculated that it would be possible to control the flow of electrons inside a semiconductor by imposing an outside electric field. Semiconductors included materials such as germanium and silicon that could strengthen a current, conduct a current, or resist a current. If turning on and off an outside field could prevent the current flow, small semiconductor pieces controlled by surrounding electric fields could be used as switches and amplifiers in electric circuits. During World War II, silicon was used in **radar** as a detector, and some knowledge about the material was accumulated.

After the war, Shockley assembled a team to work on the project. The original idea to use silicon as an amplifier did not work, as the surrounding electric field did not penetrate into the interior of the semiconductor. John Bardeen, who was trained as a theoretical physicist, proposed that the surface of the material accounted for the lack of penetration and developed a set of predictions about the behavior of semiconductors. Walter Brattain observed that a field would penetrate the semiconductor if it were applied through an electrolyte on the surface of the semiconductor. Later experiments and a series of unexpected results led to coating the semiconductor with a rectifying metal directly wired to the switching system and to the conclusion that a point-contact system worked properly, quite different from the electric-field system originally proposed. The "point contact resistor," dubbed the transistor, was the result, and Bell Labs announced and patented the invention in 1948.

The transistor was able to achieve the same function as the vacuum tube with less space and without generating heat. Later refinements and miniaturization led to even smaller integrated circuits and to the **computer chip.** By the early 1970s, the transistor had largely superseded the vacuum tube and had become a staple product of the electronics industry, with literally billions produced annually. Transistors not only replaced the vacuum tube in radio and television equipment but also were employed in computers, logic and switching circuits, in medical electronic devices, in miniaturized calculators and computers, and in **numerical controlled machines** employed in manufacturing industries.

uranium isotope separation

When the researchers in the Manhattan Project sought to build an atomic or nuclear bomb from 1942 to 1945, they soon recognized that there were two elements that could represent the fissionable core of the weapon. One fissionable material was **plutonium,** which could be made in **nuclear reactors.** The second was an **isotope [V]** of uranium with an atomic weight of 235. However, uranium 235 or ^{235}U was extremely rare, found with the more common ^{238}U in refined uranium. Since the two isotopes were chemically identical and only different at the atomic level, separating the fissionable ^{235}U from the ^{238}U became a major technical and scientific challenge of the project.

Like many other aspects of the work on the Manhattan Engineer District in the design of the weapons and the manufacture of the materials needed, it is often not realistic to attribute the invention of particular processes and gadgets to specific individuals. Some individuals were given major credit for various aspects of the work, but it is more accurate to regard all of the research as the product of team effort.

Three methods of uranium separation were developed during the war. When the proportion of fissionable ^{235}U was increased above its naturally occurring rate of 0.7 percent, the product was designated as enriched uranium. An enrichment that approached 90 percent ^{235}U was regarded as sufficient for a fissionable core. Thus, even though the methods were thought of as separation, they were actually designed to separate the isotopes only partially, reducing the proportion of ^{238}U and increasing the proportion of the fissionable ^{235}U.

The three methods were electromagnetic separation, gaseous diffusion, and thermal diffusion. In the electromagnetic separation method, large devices called *calutrons* were built at Oak Ridge, Tennessee, so named because the concept had been developed at the University of California. In a calutron, a sample of uranium with both isotopes was heated and vaporized and then projected in a stream of ionized gas through a strong magnetic field. The heavier ^{238}U isotope stream would hit the wall of the calutron chamber lower than the lighter ^{235}U isotope. On the wall of the chamber, separate slots would collect the two isotopes into two collection containers. Use of the calutron was an extremely expensive and slow process, yielding only small amounts of the fissionable material.

In the second method, gaseous diffusion, uranium was compounded with fluorine into uranium hexafluoride, and as a gas it was passed through a series of pipes, or cascades, each of which had a filter made of a classified material. On each pass through the filter, the amount of ^{238}U would be slightly diminished. As the material circulated through another cascade, the proportion of ^{235}U would increase slightly. The gaseous diffusion plant at Oak Ridge was a vast building, not completed during the war. However, the product of the first cascades, put in operation during the summer of 1945, was somewhat enriched and could be used. It was calculated that if the ideal separating effect between the feed and the output of a single stage were approached, to achieve 99 percent pure ^{235}U hexaflouride, roughly 4,000 stages would be required. Of the gas that passed through the barrier at any stage, only half would pass to the next higher stage, and the other half returned to an earlier stage in the cascade. Most of the material that emerged had been recycled over and over.

The third method, thermal diffusion, was initiated outside the Manhattan Project at the Philadelphia Navy Yard. There researchers participated in a completely independent research project that had as its focus the development of enriched uranium for small reactors that could be used to power ships and submarines. In thermal diffusion, the concept

uranium isotope separation The huge K-25 complex at Oak Ridge, Tennessee, used a gaseous diffusion method to separate uranium 235 from the more plentiful uranium 238. Part of the facility, long abandoned because of radioactive contamination, is shown here during clean-up. *U.S. Department of Energy*

was very much like that in a fractionating tower used in **catalytic cracking of petroleum** [V]. A column would be filled with gaseous uranium, and the lighter compounds would be removed at the top of the column. Partially enriched uranium produced by this method was used as feedstock for the other two methods.

All three methods were very slow and cumbersome. By July 1945 the project had enough enriched uranium for one weapon and enough plutonium for two. Since the plutonium design was technically more complicated and had a greater chance of failure, one of the two plutonium weapons was tested on July 16, 1945, and the second held for use against Japan. The uranium weapon was dropped on Hiroshima on August 6, and the second plutonium weapon was dropped on Nagasaki on August 8. It was estimated that the United States would be able to produce one new weapon about every five or six weeks through 1946.

Van Allen belts

In 1958, the American scientist John Van Allen (1914–) discovered two belts of radiation encircling Earth, now known as Van Allen belts. The belts are shaped like concentric doughnuts, with the doughnut hole over the North and South Poles of the planet. Van Allen, who had earned a doctorate in physics in 1939 from the University of Iowa, served as a naval officer during World War II and assisted in the development of the **proximity fuse** for naval artillery ordnance. He worked with captured V-2 rockets in upper-atmosphere tests and in developing research rockets for further exploration.

The inner belt, discovered first on the basis of data received from **Geiger counters** [V] carried aboard the early spacecraft *Explorer 1* and *Explorer 3,* is within about 1 radius (4,000 miles) of the surface of the Earth. In December 1958 Van Allen's group of researchers detected the second belt 9,000 to 12,000 miles beyond Earth. The radiation belts required that electronic equipment aboard spacecraft be shielded from their effects.

The charged particles in the belts derive from a variety of sources, including cosmic ray particles that collide with atmospheric atoms, and protons from magnetic storms. The inner belt largely consists of protons, and the outer belt, of protons with lower charges and electrons. At first it was assumed that the Van Allen belts accounted for the effect of the aurora borealis, but later research has suggested that the belts do not produce effects in the visual range and that the aurora derives from

Sun-emitted plasma or charged particles of the solar wind that have avoided capture by the belts and have been trapped closer to Earth, in its magnetic field. Later studies of the belts with satellites showed that their shape is less regular than originally assumed, being deformed by the effect of the solar wind.

vertical takeoff and landing aircraft

The vertical takeoff and landing (VTOL) aircraft has been an elusive goal, but by the end of the 20th century, a British and an American model were in military service. The **helicopter** can achieve vertical takeoff and landing but is limited in its forward speed because at more than 200 miles per hour it will stall. The reason is that the helicopter blade is shaped like an airfoil, and as the speed of the aircraft increases, the blade on the retreating half of its rotation is carried forward sufficiently fast to cause the blade to create a downward thrust rather than an upward thrust. Because of this feature, the helicopter is limited in its missions and range, not being capable of engaging jet fighters in air combat and being highly vulnerable to ground fire from shoulder-held missiles when flying in a combat-support role.

The search for a VTOL included the requirement that the aircraft achieve high speed and long range and at the same time be capable of vertical takeoff and landing, like a helicopter. The British solution is the Harrier (AV-8), first known as a jump jet. Developed by British Aerospace, the Harrier first successfully flew on August 31, 1966, and entered into service in 1969. It was powered by a Rolls-Royce **jet engine** and saw successful combat service in the Falklands War in 1982 and in the Persian Gulf War in 1991. The engines themselves remain stationary, but their thrust is directed downward or rearward by pivoting vectored-thrust nozzles.

The American design was the tilt-rotor V-22 Osprey, developed by Bell and Boeing. Delivery of the V-22 began in 1999, but the aircraft was troubled by accidents and poor performance. The aircraft is lifted by two 38-foot propellers known as proprotors, which serve much as helicopter blades for direct lift and descent. The proprotors, interconnecting shafting, and their engines are mounted on a wing, which is tilted vertically during liftoff and descent. The wing and its rotors are tilted horizontally for forward flight. The limit to forward speed is reported at just over half that of the Harrier, under 400 miles per hour, considerably faster than a helicopter but far lower than a jet-propelled

aircraft. The Osprey remained extremely controversial because of a series of accidents and because of excessive cost overruns in its manufacture. Originally planned to cost about $24 million each, the cost escalated to more than $80 million apiece. An Osprey accident in 1992 killed 7 troops; another, in 2000, killed 19 marines.

Alternative VTOLs have been considered, with experiments on a tail-sitting jet and an X-wing aircraft. In the latter, the aircraft rises like a conventional helicopter and proceeds to near-stall speed in forward flight. Then the trailing blades are converted to airfoils by the ejection of airjets from within, and freezing the rotation to convert the total X-shaped blade assembly into a set of wings that create lift, and the engine power is shifted from rotating the rotors to forward jet propulsion. The mechanical difficulties and funding decisions with that design have prevented it from reaching a production stage.

videotape recorder

The first videotape recorders were developed by Bing Crosby Laboratories, under a research team headed by engineer John Mullin, in 1951. The camera converted the visual information into electrical impulses and saved the digital information onto magnetic tape. Over the following years the video recorder was perfected. By 1956, the VR 1000, invented by Charles Ginsburg and manufactured by the Ampex Corporation, came into common use in the television industry.

Video recording utilized the same principle as audio **tape recording** [V], storing a varying magnetized line in a track on a plastic tape covered with iron oxide or other magnetic material. However, video recording requires the storage of much more information than audio recording, and the solution was achieved by arranging the video track in diagonal strips across the width of the tape. The recording drum turns at an angle to the wrapping tape, allowing the diagonal striping to record the image data. The sound track is recorded as a separate straight track along one edge of the tape.

Two Japanese firms competed for the world market by introducing videocassette systems. The Sony Betamax standard (or Beta) was first introduced in 1965 and had superior visual quality. However, Matsushita introduced the VHS standard early in the 1970s, and with a larger marketing effort, succeeded in dominating the market. A "Super VHS" system introduced in 1989 improved on the quality of the early VHS system. Although neither system was ideal for permanent recording due to the gradual decay of image quality, the videocassette and home-recording machines changed television viewing habits, as audiences taped sports events and even serial soap operas for replay at more convenient hours. Video rental stores became big business enterprises, at first offering both Beta and VHS formats, but by the 1990s, almost entirely abandoning the Beta format for the more popular VHS.

Following 1959, the development of a solid-state **charge-coupled device** allowed for miniaturized and much improved video recording, by recording the image in digital form rather than by magnetic alignment of iron oxide.

Wankel engine

A major improvement over the piston-driven **internal-combustion engine** [IV], the Wankel engine was developed over the early 1950s and perfected in 1956 by Felix Wankel (1902–1988). Wankel was a German engineer who sought for many years to find a way to harness internal combustion without the necessity of transforming linear motion into rotary motion, a major source of the inefficiency of piston-driven internal-combustion engines. Wankel began working on the concept of a rotary engine in 1924. He formed his own engineering company in the 1930s and worked for the German Air Ministry during World War II. Following the war he worked on the engine design at his own risk. He began in 1945 with a pump design, improving it by trial and error. When he sought a patent, he discovered that some elements of his final design had been tried out previously in Sweden, France, and Switzerland.

His revolutionary solution to the internal-combustion chamber was a lighter and more efficient engine. Inside a cylinder, a triangular-shaped rotor rotates around a central axis. The engine can fire three times on each rotation, allowing the engine to do three times as much work as a conventional piston engine with the same internal engine volume. The rotor itself serves to clear the intake and exhaust openings, removing the necessity for valves and the complex camshaft drive for the valve mechanism. In some of Wankel's designs the outside casing would rotate in one direction while the rotor rotated in the opposite direction. However, in the automotive application, Wankel decided on a stationary casing.

The Wankel engine operates at high efficiency and at very high revolutions per minute, up to 17,000. A number of manufacturers bought the rights to the Wankel design. Mazda was the first to introduce a production car powered with the new engine, in 1968. Its smooth operation at high speed won it praise from automotive journalists and observers. However, difficulties with worn seals and the lack of mechanics trained to repair the engines worked against their commercial success.

wide-screen motion pictures

As part of the effort to make motion picture theaters more competitive with television in the 1950s, several systems of wide-screen cinematography came into use in that decade, along with the brief fad of **three-dimensional motion pictures**. In some systems, the same 35-millimeter film was employed; in others, a 70-millimeter-wide film was used. A French scientist, Henri Chrétien, developed a special *anamorphic* lens, which allowed the camera to squeeze the wide image onto the film and then expand it when the film was run through the projector. Cinerama was a system developed by an American inventor, Fred Waller, between 1937 and 1952. His system worked with three synchronized cameras and projectors and used a semicircular screen that partially surrounded the audience. In the Soviet Union and Germany, more complete systems were developed in 1959; these completely surrounded the audience with a circular screen, with an optical system in the middle of the theater, using multiple projectors to display the picture on the surrounding screen. Neither the Russian system, known as Circorama, nor the German Cinetarium ever became widely adopted.

INDEX

Page numbers in **boldface** refer to entries.

Abbe, Ernst, 347
A. B. Dick, 416
Abel, John Jacob, 439
Abplanalp, Robert, 402–403
absolute zero, 272
Academia del Cimento, 201
Academia Secretorum Naturae, 201
Accademia del Lincei, 201
Acheulian stone tool culture, 67
achil dye, 41
acid rain, 399
acupuncture, 236
Adams, John Couch, 279, 280
adaptation of an existing design to
     an alternate technology, 177
adding machine, **155–156**, 191, 192
Ader, Clément, 327
aerosol spray, **402–403**
Aeschylus, 70
Afghanistan
     copper in, 33
     windmills in, 146
African Americans, 324
African slaves, 87, 140
African weaving, 69
Age of Scientific Revolution. See
     Scientific Revolution, Age of
Agfa company, 467
Agincourt, battle at, 120
agriculture, 10, **15–17**, 38, 82–83
     cities, as precondition for, 31
     crop diversification, 15
     horse collar and, 40, 78–79, 83,
        116–117, 118, 142
     horseshoe and, 118, 142
     McCormick reaper, 277–278
     the plow, 82–83, 117, 118,
        130–131
     productivity during Industrial
        Revolution, 225–226
     three-field system, 82, 117,
        141–142
Aiken, Howard, 412
air brake, **231**
aircraft carrier, **325–326**, 332
airplane, 157, 178, 227, 321,
     326–329, 438

aluminum in, 232
autopilot, **406–407**
Norden bombsight, **364–365**
air pump, 195
AK-47, 198
Altamira, Spain, cave paintings, 56
Alamogordo, New Mexico, 448,
     450–451, 456
Al-Biruni, 123
alchemy, 85–86, 105, 122
alcoholic beverages. See beer;
     distilled spirits; wine
Alexandria, Egypt, 31
     library at, 47, 49, 203
     mosaic glass in, 45
algebra, 13, **90**
algorithm, 90
Alhazen, 119
al-Khwarizmi, Mohammed ibn-
     Musa, 90, 104, 148
Allen, Horatio, 301
alphabet, **17–18**
aluminum, **231–232**, 344
Alwall, Nils, 439
Amdahl Corporation, 42
American Civil War, 96, 102, 126,
     292, 382
American Philosophical Society, 201
American Standard for steam
     locomotices, 300, 301
American System of Manufactures.
     See interchangeable parts
Amontons, Guillaume, 214
Ampère, André-Marie, 232, 259,
     260, 264–265, 311
Ampère's law, 223, **232–233**, 276
Ampex Corporation, 477
amphorae, 51
Anacin, 329
analytic engine, **233–234**, 283, 412
anamorphic lens, 478
*Anatomical exercises concerning the
     motion of the Heart and of the
     Blood in Animals* (Harvey),
     169
anatomy, **90–91**, 109
Anaximander, 98

Ancient World through Classical
     Antiquity, 8000 B.C. to A.D.
     300, 2, 4, **9–79**
     eras of the prehistorical and
        classical world, 10
     overview, **9–14**
Anderson, Carl D., 370
Anderson, Ken, 409
Anderson, W. French, 428
Andrews, Ed, 358
Andromeda nebula, 462
anemometer, 166
anesthetics, 227, **234–236**, 329
Anikythera computer, 12, 99
animals, domestication of. See
     domestication of animals
Anschütz-Kempe, Herman, 352
answering machines, 386
antibiotics, 366–367, 401, 466–467
anvil, 77
Apollo, 426
Appert, Nicolas, 247
Appian Way, 62
Apple Computer, 421
Appleton, Edward, 356
Applied Physics Laboratory, Johns
     Hopkins University, 458
aqualung, **403–404**
Aqua Marcia, 18
Aqua Tepula, 19
aquavit, 106
aqueducts, **18–19**, 58, 94
Arabian horses, 40
Arabic alphabet, 18
Arabic numerical system, 86
Arab world, 86
     astrolabes, 92
     decimal numbers in, 103
     painkillers in, 234
     zero, concept of, 148
Arago, François Dominique Jean,
     255, 258, 279, 280
arbalest, 35
arch, **19–20**, 62
Archer, Frederick Scott, 255
Archimedes, 20–22
Archimedes (ship), 290

Archimedes' principle, **20–21**
Archimedes' screw, **21–22**, 102, 289
Arecibo radio telescope, 462, 463
argon, 353, 354
Argonaut (submarine), 382
Argonne National Laboratory, 449
Aristarchus of Samos, 114
Aristophanes, 70
Aristotle, 114, 115, 180
Arizona, USS, 332
Arkwright, Richard, 209
Arkwright water frame, 209
Armat, Thomas, 362
Armenia, 74
armor, 83
   chain mail, 30, 83, **97–98**
   plate, 30, 83, 97, 98
arms race, 398–399
Armstrong, Edwin, 361, 372
Army Corps of Engineers, 342
ARPANET, 436
arquebus, 120
arthroscope, 425
artificial satellite, 356, 361,
   404–405, 410
Artists of Dionysius, 71
Ascent of Man, The (Brownowski),
   92
asdic, 379
asha, 25
Aspdin, Joseph, 378
asphalt paving, 62, **236–237**
aspirin, **329–330**
Assayer, The (Galileo), 193
assembly line method, 239
asses, 38
Associated Press, 350
Assurbanipal, King, 48–49
Assyrian Empire, 76
   cuneiform system, 17
   magnifying lenses in, 187
Aston, Francis, 356–357
astratrium, 99
astrolabe, 85, 89, **91–92**, 132, 202
astronomy, 87, 201, 202, 208–209,
   214
Aswan High Dam, 36
Athens, theater in, 70, 71
atlas, **92–93**, 123
*"Atlas: or cosmological meditations
   upon the creation of the
   universe, and the universe it
   created"* (Mercator), 93

Atomic and Electronic Age, 1935
   into the 21st Century, 3, 4,
   320, 325, 397–479
   chronology of inventions and
     discoveries, chart of, 400–401
   overview, 397–402
atomic energy, 374
atoms, 320, 325
   Dalton's classic theory of,
     356–357
AT&T, 372, 373, 445, 446
auditorium, 71
auger drills, 22
aulos, 131
Aurignacian phenomenon, 67–68
Australia, 451
autogyro, 429
Automatic Hook and Eye Company,
   396
automobile, 62, 159, 217, **237–239**,
   267, 333, 366
   battery, 243
   Bourdon oil pressure gauge, 289
   carburetor, 238, 268, **334–335**
   drive chain, 30, 255
   electric self-starter, 239
   fuel cell and, 402, 426–27
   interchangeable parts and, 239,
     267
   internal combustion engine. *See*
     internal combustion engine
   pneumatic tire. *See* pneumatic
     tire
   social impact of, 319
   steel in body of, 305
autopilot, **406–407**
Avery, William, 302
Avogadro, Amedeo, 240
Avogadro's law, **240**
ax, **22**, 27, 77
axle, 73
Aztec metalworking, 53

Babbage, Charles, 227, 233–234,
   283, 412
Babylon, 31
Babylonia
   algebra, 90
   astrological system, 36–37
   beer in, 24
   cities of, 32
   dams in, 35
   days of the week, 36–37

   plumbing in, 58
   road surfacing in, 236
   soap developed in, 65
Bach, Johann Christian, 195
Bacon, Francis, 151, 173, 200
Bacon, Francis T., 426
Bacon, Roger, 109, 113–114
"Bacterial Chromosome, The," 423
Baekeland, Leo Hendrik, 330
bagpipe, 131
Bain, Alexander, 350
Baird, John Logie, 387
Bakelite, **330**
balance spring, 99–100, 144,
   156–157, 166, 180, 194
Ballard, 427
ball bearings, **240–241**
ballista, 137
ballpoint pen, 306, **407–408**
Banting, Frederick G., 355
barbed wire, 148, **241–242**, 321
Bardeen, John, 470, 471
Barker, Robert, 254
barometer
   mercury, 166, 176, 178,
     186–187, 213
   water, 186
Barron, Robert, 183
baskets, 16, **22–23**, 68, 72
baths, public, 29, 58
Baths at Caracalla, 58
Batter, F. W., 337
battery, **242–243**
Battle Creek Sanitarium, 333, 334
battleship, 321, 325, **330–332**
Bausch & Lomb, 425
Bayer and Company, Friedrich, 329

Beauchamp, James, 447
Becquerel, Antoine Henri, 373, 374,
   375
Bedorz, J. Georg, 384
beer, **24**
"Beginnings of Wheeled Transport,
   The," 73
Beighton, Henry, 210
Belar, Herbert, 446
Belisarius, 144
Bell, Alexander Graham, 278, 287,
   312–314, 353
   hearing aid and, 227, 353
Bell, Chichester A., 287
Bell-Beaker, 75

Bell Labs, 409, 410, 414, 421, 442, 461, 471, 475
Benga, Bernadino, 91
Benjamin, H., 338
Benny, Jack, 325
Benussi, Vittorio, 358
Benz, Karl, 238, 334, 335
Berg, Paul, 428
Beria, Lavrenti, 450
beriberi, 393, 394
Berliner, Emile, 278, 287, 313
Bernoulli, Daniel, 157, 161, 201, 289
Bernoulli, Jakob, 157
Bernoulli, Johann, 157, 161
Bernoulli, Nicholas, 201
Bernoulli's principle, **157**, 327, 334, 430
Berthollet, Claude Louis, 167
Bertrand, Joseph, 263
Berzelius, Jöns Jakob, 367
Bessemer, Henry, 296, 304–305
Bessemer process, 261, 296, 304–305
Best, C. H., 355
Best Friend of Charleston, 301
Bevan, Edward John, 338
Bible, 30, 116, 122
Bich, Baron Marcel, 408
bicycle, 62, 236, 239, **243–244**, 267, 288
   drive chain, 30, 244, 255
   pneumatic tire. *See* pneumatic tire
Bier, August K. G., 235
Bing Crosby Laboratories, 477
Binney, Edward, 251
Binnig, Gerd, 347
binomial theorem, 160
Birdseye, Clarence, 339
Biro, Ladislas and Georg, 306, 407–408
birth control, 401, 453–454
Birtwhistle, W. K., 418
Bissell, George H., 282
Black, Harold, 362
Black, Joseph, 161, 205
"black box" recorders, 387
blacksmiths, 30
Blackwell, Richard J., 179
Blériot, Louis, 328, 432
Bliss, Denys, 432
Bloch, Felix, 443

Bloch, Marc, 139
block printing, 124, 128, 129
blood
   circulation of, 143, **169–170**, 189
   red corpuscles of, 189, **196**
Blucher steam engine, 299
board games, **24–25**
boat-keel design, 157
Bock's Car, 451
Bode, Johann Elert, 158, 218
Bode's law, 152, **157–158**
Boeing, 475
Bohr, Niels, 392
boiling water reactor, 449
Boltzmann, Ludwig, 244–245
Boltzmann constant, 223, **244–245**
bombes, 444
Bonsack, James Albert, 250
books
   engravings and illustration of, 85, 108
   in the vernacular, 125
   See also movable type; printing ink; printing press
boosted weapon, 469
"boozah," 24
boring mill, **158–159**, 266
Bottger, R. C., 276
Bouguer, Pierre, 205
Boulton, Matthew, 177
Boulton and Watt, 211, 225
bourbon, 106
Bourdon, Eugène, 288–289
Bourgeoys, Marin le, 111
bow drill, 77
Boyer, Hubert, 428
Boyle, Charles, 189
Boyle, Robert, 150, 159, 166, 176, 195, 275
   barometer and, 186, 187
   vacuum pump and, 219
Boyle, Willard, 409
Boyle's law, **159**, 166, 170
Bradley, James, 207–208
Bragg, William Henry and William Lawrence, 395
Brahe, Tycho, 100, 179
braille, **245–246**
Braille, Louis, 245–246
Bramah, Joseph, 135, 153, 158, 174, 183, 184, 191, 265, 266
Branca, Giovanni, 302

Brandenberger, Jacques E., 338
brandies, 105
brass, **25–26**
brassiere, 102, 320, **332–333**
Bratsk Dam, 36
Brattain, Walter H., 470, 471
Braun, Karl Ferdinand, 253–254
Brave New World, 437
Brayton, George, 237, 268
Brazilwood dye, 41
bread wheat, 15, 65
breakfast cereal, **333–334**
Brearley, Harry, 380, 381
Bredow, Dr. Hans, 372
Breit, Gregory, 460
Brickwedde, F. G., 419
bridges, 62
   steel cable suspension, 147, **305–306**
Briggs, Henry, 184
Bright, Timothy, 136
Britain
   in Age of Scientific Revolution, 151–152
   beer in, 24
   brass in, 25
   canal systems, 94
   inks, colored, 47
   Norman Conquest, 139
   nuclear weapons, 451
   radar system, 461
   Stonehenge, 12
   toll roads, 62
British Aerospace, 475
British Broadcasting Corporation, 389
British National Physical Laboratory, 436
British National Research and Development Corporation, 431
British Science Museum, 234
bronchoscope, 425
bronze, **26–27**, 34, 51, 52, 65
   cannon, 26, 95
   chains, 30
   weapons, 27, 64
Bronze Age, 11
Brooklyn Bridge, 306
Brown, Louise, 437
Brown, Samuel, 30
Brownian movement, 245
Brownie camera, 247

Brownowski, J., 92
Brumwell, Dennis, 409
Brunel, Isambard Kingdom, 290
Brunelleschi, Filippo, 106–107
Buchla, Donald, 447
Budding, Edwin, 269
buffalo, domestication of, 39
Bunsen, Robert Wilhelm, 189, 246, 299
Bunsen burner, 189, **246**
buoyancy, Archimedes' principle of, 21
Burdin, Claude, 317
Burger, Reinhold, 389–390
burial customs, 28
Burroughs, William, 156
Burroughs Company, 365
Burt, William A., 315–316
Bush, Vannevar, 342, 397, 413, 459
Bushnell, David, 381
Butayeva, F. A., 440
Butler, Edward, 335
Butler, Paul, 454
Byron, Augusta Ada, 234

cabinetry, 84
cable car, 308
Cadillac Company, 239
Cairns, John, 423
calculus, 149, 155, 157, **159–161**
Calgene company, 428
Calico Printers Association, 418
calutrons, 472
Calypso (exploration ship), 404
Camden and Amboy Rail-Road and
    Transportation Company, 300
camel, 39
camera, **246–247**, 249, 255
    digital, 410, 421, 457
    Polaroid, 422, **456–457**
Cameron, Michael, 42
Campbell, Henry, 301
Camras, Marvin, 386
canal lock, 28, **93–94**
canals, 12, **27–28**
canned foods, 225, **247–248**
Cannizzaro, Stanislao, 240
cannon, 83, **95–96**, 113, 114
    bronze, 26, 95
    cast iron balls, 165
    rifling of large ship cannons, 198
Cannon, Walter, 394
can opener, 248

Capek, Karel, 452
capitalism, 153, 225
Capitol building, 107
carbon dioxide, **161**
carbon fiber, 411–412
carbon-fiber-reinforced plastics
    (CFRPs), 411–412
carburetor, 238, 268, **334–335**
Carbutt, 246
Cardano, Girolamo, 90
cardiac pacemaker, 399, **408–409**
card punch machine, 233, **248–249**, 412
Caribbean islands, 86–87, 139–140
Carlson, Chester F., 416–417
Carnegie Institution, 458
Carnot, Nicolas Leonard Sadi,
    270–271, 272, 340–341
Carothers, Wallace, 452–453
carpenter, 77–78
Carré, Ferdinand, 291
Cartesian mechanistic science,
    161–163, 194, 201
cartridge brass, 26
Caruso, Enrico, 287, 325
Casio, 421
Cassegrain telescope, **163–164**, 212
Castile soap, 66
cast iron, **164–165**
Çatal Hüyük, 30–31
catalytic cracking of petroleum,
    106, **335–337**
caterpillar track, **337–338**, 384
Caterpillar Tractor Company, 337
cathedral, 84, **96–98**
    colonnades, 20
    Gothic, 20, 96
cathode-ray tubes, 253–254, 320,
    346, 347, 354, 387, 389,
    395–396, 460
cats, 38
Catskill aqueduct system, 19
cattle, domestication of, 38–39
Cavendish, Henry, 150, 177, 182,
    220–221, 353
Cavendish Laboratory, 357
cave paintings, 9, 56
Cayley, George, 326
Celera Genomics, 434
celestial body names, language
    equivalents of, 37
Cellini, Benvenuto, 240
cellophane, **338**

cellular structure of plants,
    165–166, 188, 200, 330
celluloid, 246, **249**
celluloid film, 362
Celsius, Anders, 215
cemeteries, **28–29**, 64
central heating, **29**
Central Intelligence Agency (CIA),
    410
cereal, breakfast, **333–334**
Ceredi, Giuseppe, 103
Cesalpino, Andrea, 169
cesspools, 58
C. F. Martin Company, 113
Chadwick, James, 364
chain, **29–30**
Chain, Ernest B., 366–367
chain mail, 30, 83, **97–98**, 139
Chalcolithic Age or Copper and
    Stone Age, 11, 34
Chaldeans, 104
Chambers, B., 96
Chaplin, Charlie, 325
Chappe, Claude, 294–295, 310
Charch, William Hale, 338
Chardack, Dr. William, 408–409
charge-coupled devices, 213,
    409–410, 421
chariots, 73
Charlemagne, 82, 141–142
Charles, Césare Alexandre, 343
Charles, Jacques Alexandre César,
    166–167, 177, 178
Charles' Law, 177
Charles's law, **166–167**
Chartres cathedral, 97
Chaucer, Geoffrey, 92
chemistry, modern, 159, 189, 320
Cheops, pyramid of, 61
Chernobyl accident, 450
Cherokee Indians, 18
Chevrolet Corvair, 411
Chilowski, Constantin, 379, 380
chimney, **98**
China, 40, 86
    acupuncture in, 236
    armor, 97, 98
    cast iron in, 165
    central heating in, 29
    clocks, 99
    coal mining, 55
    compass in, 101
    crossbow in, 34

China (*continued*)
  dams in, 35
  eyeglasses in, 109
  gunpowder in, 113
  helicopter concept and, 429
  ink in, 46, 47
  man-carrying kites in, 178
  movable type in, 124
  paddlewheel boat in, 127
  paper in, 128
  pictographic alphabet, 18
  pottery, 59
  rockets in, 359
  silkmaking, 65, 69
  steam-powered cart, 237
  stirrup in, 138
  suspension bridge in, 305
  umbrella in, 72
  windmills in, 146
chinaware, 59
chlordane, 399, **410–411**, 420
chlorine bleach, 41, **167**
chlorofluorocarbons (CFCs), 352,
  403
choloroform, 235
Chrétien, Henri, 478
Christy, James W., 369
*Chronicles of the Discovery and
  Conquest of Guinea* (de
  Zurara), 132
chronometer, 100, 136, 152, 157,
  168–169, 212
Chrysler Building, 297
Chubb, Jeremiah, 183
Church of Santa Maria dei Fieori,
  Florence, 106–107
cigarette, **249–250**, 333, 399
Cinerama, 478
Cipher Machines Company, 348
circulation of blood, 143, **169–170**,
  189
circulation of water, **170**
cities, 16, **30–32**, 142, 319
City of Dionysius plays, 70, 71
Civetta (ship), 290
Civil War, American, 96, 102, 126,
  292, 382
clarinet, 131
Clark, Alvan, 213
Clark, William, 235
Claude, Georges, 427
Clausius, Rudolf, 270, 271, 272
Clay, Henry, **191**

Clayton, John, 171
Clement IV, Pope, 109
Clerk, Dougald, 268
Clermont, 303
clippers (seaplanes), 329, 429
clipper ships, 134
clock, 84, **98–99**, 143, 156, 180,
  193, 266–267
  pendulum, 99–100, 156,
    182–183, 192, **193–194**, 195,
    199
clock watch. *See* watch
coal gas, **170–172**, 238, 251, 267,
  268, 334, 335
coal oil, **251**, 282
cocaine, 235
cochineal dye, 41
Cockerell, Charles, 431, 432
Code of Hammurabi, 74
coffee, instant, **354–355**
Coffey, Aeneas, 105
Coffman, Donald, 453
Cognac, 105
Cohn, Stanley, 428
Collins, John, 160
Collins, Samuel and David, 22
Colombo, Realdo, 169
colonnade, 19–20
color blindness, 283
Colossus of Rhodes, 50
Colt, Samuel, 224, 266, 292
Colt Patent Arms factory, 292
Columbia Broadcasting System,
  287, 442
Columbus, Christopher, 139, 203
column, 29, **32–33**
comedy, 70
comet paths as orbits, **100**, 202, 214
Commentariolus (Copernicus), 115
*Commentary on the Effect of
  Electricity on Muscular
  Motion* (Galvani), 264
Commercium epistolicum, 161
compact disc, 288
compass, 85, 86, 89, **101**, 132, 133,
  136
  gyrocompass, **352–353**
composite materials, 191, **411–412**
Comptoir de Textiles Artificels, 338
Compton, Arthur Holly, 368
Compton effect, 368
computer, 274, 387, 398, **412–414**,
  444

analog, 342, 407, 413
digital, 413
origins of, 233, 412–413
computer chip, 398, 409, **414–415**,
  471
Comstock, D. F., 467
concrete, reinforced, **377–378**
Condamine, Charles Marie de La,
  205
Condit, Carl W., 296
condom, **172**
Congreve, William, 359–360
Congreve rockets, 359–360
conservation of energy, 85, 273,
  283
Constitution, U.S.S., 133
*Construction of the Marvelous Rule
  of Logarithms, The* (Napier),
  184
Consultative Committee for
  International Telephone and
  Telegraph, 351
Conte, N. J., 119
*Continuation of New Experiments,
  Physico-Mechanical* (Boyle),
  219
*Conversations on the Plurality of
  the Worlds* (de Fontenelle),
  163
Cook, Captain James, 168, 393
Cooke, William, 311, 312
Coolidge, William, 258, 394
Cooper-Hewitt, Peter, 427
Copernicus, 87, 114–116, 150,
  183
copper, 11, 27, **33–34**, 52, 53, 57
  mining, 54, 55
copy machine, **415–417**
Corinthian columns, 33
Cornell, Ezra, 312
Cornu, Paul, 429
corset, **101–102**, 332
Cortés, Hernán, 140
corvettes, 134
Coryate, Thomas, 112
cotton, 16, 68, 69
cotton gin, **251–253**, 292
Coughlin, Charles, 324
Coulomb's law, **172–173**, 276
Coulomb, Charles Augustin, 172
counting, 104
Courtauld, Samuel, 377
Cousteau, Jacques-Yves, 403–404

crank, 81, 85, **102–103**, 127, 211
  pedal, 244
Crapper, Thomas, 174
Crécy, Battle of, 120
Crete
  maritime trade, 50, 51
  plumbing in, 58
Crick, Francis H. C., 422–423
Cristofori, Bartolomeo, 194–195
Crombie, Alistair C., 96, 110
Crompton, Samuel, 174, 209
Crookes, William, 253, 346, 354,
  396
Crookes tube, **253–254**, 320, 346,
  347, 354, 373, 395–396, 460
Crosley Corporation, 458
Cross, C. F., 377
Cross, Charles, 287
Cross, Frederick, 338
crossbow, **34–35**, 83, 102
cross-plowing, 130
cross-staff, 85, **103**, 132, 168, 202
crumhorn, 131
Crusades, 132, 146
Cugnot, Nicolas, 237
Cullen, William, 291
Cummings, Alexander, 174
cuneiform system, 17
Curie, Marie, 324, 374–376, 379
Curie, Pierre, 374, 375
Curtis, Charles, 303
cyanocobalamin (Vitamin B_{12}), 394
cyclotron, **417–418**, 455
Cyrillic alphabets, 18

d'Abbans, Claude de Jouffroy, 303
Dacron, 418–419
Daguerre, Louis Jacques Mandé,
  246, 254–255
daguerreotype, 246, **254–255**
Dahlgren, John, 95
Daimler, Gottlieb, 238, 268, 335
Dalton, John, 356–357
D'Amato, Salvino, 109
dams and dikes, **35–36**, 142
Daniel, John Frederic, 243
Danner, Hans and Leonard, 134
Darby, Abraham, 165
Darius I, Emperor, 28, 62
d'Arrest, Heinrich, 279
d'Arsonval, J. A., 265
Darwin, Charles, 422
Davenport, Thomas, 259

Davis, Alice Catherine, 285
Davy, Humphry, 232, 235, 245,
  262
Day, Joseph, 268
days of the week, **36–37**
DDT, 399, 410, **420**
Debierne, André, 375
de Boisbaudran, Lecoq, 286
de Broglie, Louis, 347
de Chardonnet, Comte Hilare, 377
decimal numbers, **103–104**, 148
de Dondi, Giovanni, 99
de Dondi, Jacopo, 99
deductive scientific discovery, 6
deep-freezing food, **338–339**
Defense Advanced Research
    Projects Administration
    (DARPA), 436
de Fermat, Pierre, 160
deficiency diseases, 393–394
de Fontenelle, Bernard, 163, 201
de Forest, Lee, 343, 361, 372, 391
De humani corporis fabrica
    (Vesalius), 91
de la Cierva, Juan, 429–430
de la Condamine, Charles Marie,
    293
De Laval, Gustave, 302
Delaware (battleship), 353
Delinquent Man, The (Lombroso),
    358
de l'Orme, Philibert, 377
de Milameete, Walter, 113
de Moleyns, Frederick, 257
de Morgan, Augustus, 161
Denis, Robert and Claude, 112
Denner, J. C., 131
Densmore, James, 316
Depo-Provera, 454
De re militari (Valturio), 128
De Revolutionibus orbium
    celestium (Copernicus), 91,
    115
de Rochas, Alphonse Beau, 268
Descartes, René, 44, 90, 160,
    161–163, 191, 197
de Sivrac, 243
de Solla Price, Derek, 12, 395–396
destroyer, 321, 331, **339–340**
deuterium, **419–420**
De Venarum Ostiolis (Fabrizio),
    143
Devol, George, 452

Dewar, James, 389, 390
de Zurara, Gomes Eanes, 132
dice, ancient, 25
Dickson, J. T., 418
Dickson, William Kennedy Laurie,
    362
Diesel, Rudolf Christian, 340–341
diesel engines, 159, 272, **340–341**,
    382, 461
differential analyzer, **342**, 413
diffusion of science and technology,
    13
Digges, Leonard, 141
Digges, Thomas, 141
digital cameras, 410, **420–421**, 421,
    457
digital computers, 413
diode, **342–343**, 346, 390, 442, 471
Dionysius, 69–71
Diophantus, 90
dioptra, 92
Dirac, Paul A. M., 370
dirigibles, 178, 232, 321, **343–344**
Discourse on Method (Descartes),
    162, 197
discovery and invention,
    relationship between, 152
Discovery in the Physical Sciences
    (Blackwell), 179
distilled spirits, **104–106**, 140
diving bell, **173**
DNA, 395, 402, **422–423**, 428, 434
dogs, 28
  domestication of, 16, 38, 39
dome, **106–107**
Domesday Book, 144
domestication of animals, 11, 15,
    16, 31, **38–39**, 56
domestication of the horse, 38,
    39–41
donkey, 39
Doppler, 461
Doric columns, 33
Dorn, Friedrich E., 354
Draisine, 243
Drais von Sauerbronn, Baron Karl
    Friedrich Christian Ludwig,
    243
Drake, Colonel E. L., 282
Dreadnaught (battleship), 331
dreadnaughts, 332
Drebbel, Cornelius, 188
Dresser, Hermann, 329

drive chains, 30, 244, **255**
Dubos, R., 466
DuCros, Harvey, 288
Duke, James B., 250
Dummer, G., 414
Dunlop, John Boyd, 239, 244, 288
Dunlop Rubber Company, 288
Du Pont, 292, 338, 351, 418,
    438–439, 452, 468
Durand, Peter, 248
Duryea, Charles James, 238–239
dye, **41–42**, 57
dynamite, 36, **255–257**

Earhart, Amelia, 324–325
Earth
    gyrocompass and rotation of,
        352
    size and shape of the, 42, 47,
        202–205
Easter Island, 12
East Indiamen, 134
Eastman, George, 246–247, 249,
    362
Eberhard Faber Company, 408
Ecuador, baskets of, 23
Eddy, Oliver T., 316
Edgworth, Richard, 337
Edison, Thomas, 172, 224,
    257–258, 278, 286–287, 313,
    316, 324, 346, 362–363, 371,
    394, 411, 416
Edison Electric Light Company,
    257–258
Edison Vitascope, 363
Edward III of Britain, King, 120
Edwards, Robert, 437
Egypt, and ancient Egypt, 12, 72
    agriculture in, 16
    algebra, 90
    the arch, 19–20
    armor, 98
    baskets in, 22
    board games, 24
    canals, 12
    chains in, 30
    clocks, 98–99
    columns, 32
    dams in, 35
    domestication of animals in, 16
    elevator in, 260
    gold mining, 55
    hieroglyphics, 18

ink in, 46
lighthouse in, 50
metalworking, 53
musical notation in, 125
paint in, 57
papyrus, 128
pottery, 58–59
pyramids, 12, 60–61
winemaking, 74
eidoloscope, 362
Eiffel Tower, 261
Einstein, Albert, 206, 324,
    367–368, 402
    theory of relativity, 320, 361
Eisenhower, Dwight D., 464
Eisler, Paul, 457
ejection seat, **423–424**
Ejikman, Christiaan, 393, 394
elasticity, law of, 165, **180**, 200
Electrical Age, 1891 to 1934, 3, 4,
    319–396
    chronology of inventions and
        discoveries, 322–323
    overview, 319–325
electrical transformer, 262
electric blanket, **344–345**
electric chair, **345**
electric charge of the electron, **345**,
    346
electric eye, 367
electric generator, 212, 258,
    258–260, 262, 268
electricity, 227, 242
Electricity and Magnetism
    (Maxwell), 276
electric light, 148, 172, **257–258**,
    287, 320, 321, 346
    power supply systems, 259, 260,
        321
electric motor, 212, **258–260**, 262,
    268
electrolysis, Faraday's laws of,
    262–263
electromagnetic radiation, 371
electromagnetism, 223, 232–233,
    258, **260**, 277
electron, 253, 320, **345–346**, 347,
    374
    electric charge of the, **345**, 346
Electronic Images Systems
    Corporation, 351
electron microscope, 188, **346–347**,
    402

electron paramagnetic resonance
    (EPR), 443
elements, **107–108**
elephant, domestication of the, 39
elevators, 191, **260–262**, 296, 305
Eli Lilly, 355–356
Ellis, Evelyn, 239
Ellwood, Isaac L., 241
Ely, Eugene, 325
embalming, 60
embroidery, 84
"Emergence of Elites: Earlier Bronze
    Age Europe, 2500–1300 B.C.,
    The," 75
emmer wheat, 15, 65
Empedocles, 85, 107
Empire State Building, 297
empirical scientific discovery, 4
encryption machines, **348–349**,
    372, 413, 444
endoscope, 425
Enfield rifle, 198
Engels, Friedrich, 226
engineering schools, 224
Englebart, Douglas, 436
engraving, 90–91, 93, **108–109**
    for book illustration, 85, 108
ENIAC, 413, 415
Enigma machine, 348–349, 373
enlightenment, 149
Enola Gay, 451
Enterprise, U.S.S., 325, 326
entropy, 272, 273
Environmental Protection Agency,
    420
environmental regulations, 399
eotechnology, 1, 13–14
Epic of Gilgamesh, 74
Equus caballus przewalskii, 40
Eratosthenes, 42, 47, 203
Eratosthenes' sieve, **42**
Ericsson, John, 256, 290
Erie Canal, 94
Ernst, Richard R., 443
escalator, **349–350**
escort carrier, 326
Essen, Gordon-Smith, 208
Estiene, Colonel Jean, 385
ether, 235
Etruscans, 17
Euclid, 44
Euler, Leonhard, 44, 161, 201
Euripides, 70

European explorers, 7, 86, 87, 119, 133–134
Eustachian tube, 90, **109**
Eustachio, Bartolommeo, 109
Evans, Oliver, 211, 291
Eve, Joseph, 251
Eveleth, Jonathan G., 282
Eversharp, 408
evolution, theory of, 422
Evolution of Useful Things, The (Petroski), 68
Experimental Researches in Electricity (Faraday), 262
Experimenta nova anatomica (Pecquet), 216
"Experiments on Air," 220
"Experiments to Determine the Density of the Earth," 221
"Exploration of Cosmic Space by Reactive Devices," 404
explorers, 7, 86, 87, 119, 133–134
external-combustion engine, 307
eyeglasses, 45, 81, 84, 85, **109–110**, 113, 119
    bifocal lens system, 110
    prescription, 227

Faber, John, 188
Faber company, 119
Fabrikant, V. A., 440
Fabrizio, Girolano, 142–143
facsimile or fax machine, 324, **350–351**, 409
factory working class, 153, 226
Faggin, Federico, 414–415
Fahrenheit, Daniel Gabriel, 214–215
Fairbank, John, 316
Fair Building, 297
Fairchild, Sherman, 414
Fairchild Semiconductor, 414
Fallopian tubes, 90, 91, **110–111**
Falloppio, Gabriele, 91, 110–111, 143, 172
Faraday, Michael, 45, 221, 258, 259, 260, 262, 276
Faraday's laws, 223, **262–263**, 276
Farnsworth, Philo T., 389
Fascist Party, Italian, 22
Father of Radio (de Forest), 391
Faure, Camille, 243
fax. *See* facsimile or fax machine

Febonacci, Leonardo (Leonardo of Pisa), 104
Federal Communications Commission (FCC), 445, 446
Federal Networking Council, 436
felt, 69
Ferdinand II, Duke, 214
Fermat, Pierre de, 173–174, 191
Fermat's Last Theorem, **173–174**
Fermi, Enrico, 447–448, 450
Fernel, Jean, 204
Ferris, A. C., 282
Ferro, Scipione del, 90
Fessenden, Reginald, 372
feudalism, 139
Feudal Society (Bloch), 139
fiberglass, 411
fiber optics, 399, **424–425**
Field, Cyrus, 290, 309
field theory, 263
fifth essence, 85, 107–108
Finley, James, 30
fire, **43–44**, 190
firebacks, 165
fire drills, 43
fire piston, 43
fire plows, 43
Fischer, Philipp Heinrich, 243
fishing, 23, 29
Fitch, John, 289, 303
Fizeau, Armand, 208
flax, 16, 68
Fleming, Alexander, 366–367
Fleming, John Ambrose, 342, 346, 390, 441
flintlock, 83, 95, **111**, 113, 120, 125
floodlighting, 324
Florey, Howard, 366–367
fluorescent lighting, 354, **427–428**
flush toilet, 58, 153, **174**
flute, 131
flux, 66
flying buttresses, 20, 84, 96, 97
flying shuttle, 69, 152, **174–175**
Focke aircraft works, 430
food-storage systems, 15
    canned foods, 225, **247–248**
    deep-freezing food, **338–339**
    refrigeration, 159, 227, **291–292**, 338–39, 351
Ford, Henry, 239, 324
fork, 85, **112**

Forsyth, Alexander, 111
fortune-telling, 129
Foucault, Jean Bernhard, 352
Foucault, Léon, 208
Fouché, Edmond, 365
Fourneyron, Benoît, 317
Franklin, Benjamin, 110, 177–178, 201
Franklin, Elise, 422–423
Fraunhofer, Joseph von, 298
Fraunhofer lines, 298–299
Frawley Corporation, 408
freezing food, deep-, **338–339**
French Bibliotèque Nationale, 49
French Revolution, 140, 150, 190, 224, 265, 294, 295, 310
Freon, 292, **351–352**, 399, 403
Frères, Daum, 45
Freud, Sigmund, 235
Friedman, Jerome I., 459
Friese-Green, William, 363
frigates, 134
Frigidaire, 291
Fritts, Charles, 367
Froben, Johann, 121
Frölich, Theodor, 393
Frost, J. C., 426
Froude, William, 263
Froude number, 223
Fuchs, Klaus, 451
fuel cell, 402, **425–427**
Fulton, Robert, 254, 303–304, 381–382
Fund, Casmir, 394
Furious, HMS, 325
fusee, 143–144

Gabor, Dennis, 430–431, 470
Gagnan, Émile, 404
Galen, 66, 169
Galileo, Vincenzio, 193
Galileo Galilei, 114, 116, 150, 152, 156, 181, 183, 185, 192–193, 198, 218
    compound microscope, 187, 188
    Inquisition and, 201
    Saturn observed by, 198
    on speed of light, 207
    telescope, 212, 213
    thermometer and, 214
Galle, Johann Gottfried, 279–281
galleon, 134

Galvani, Luigi, 232–233, 242, 264, 265
galvanometer, 232–233, **264–265**, 358
Garand, John C., 198
Garnerin, Jacques, 284
gases
    Avogadro's law, 240
    Charles's law, **166–167**
    coal gas. *See* coal gas
    inert gas elements, **353–354**
    Raman spectroscopy, **376**
    relationship between pressure and temperature in, 159, 166
Gas Light Company of Baltimore, 171
gas lighting, 170–171
gasoline, 238, 283, 334, 335
    aviation, 337
    high-octane, 337
    leaded ethyl, 352
    sales taxes, 239, 366
Gatling, Richard Jordan, 274
Gauss, Karl Friedrich, 265, 310
Gay-Lussac, Joseph, 167
Geiger, Hans, 352
Geiger counter, **352**, 354, 474
Geigy Company, J. R., 420
Gell-Mann, Murray, 459
General Electric, 258, 260, 303, 321, 371, 372, 373, 427, 461
Genesee Aqueduct, 94
genetically modified food crops, 402, **428–429**
genetic engineering, 402, 423, **428–429**
Genghis Khan, 40
Geométrie, La (Descartes), 90
geometry, 13, **44**
Gerhardt, Charles F., 329
Gesner, Abraham, 251
Ghiorso, Albert, 456
Ghirlandaio, Dominico, 110
Gibson, Orville, 113
Gibson, R. O., 369–370
Giesel, Friedrich, 375
Giffard, Henri, 343
Gifford, Walter, 371
Gilbert, William, 184–185
Gillette, 408
Gillette, King Camp, 378–379
Ginsburg, Charles, 477
Girard, Phillipe de, 63

girdle, 102
Giza, pyramids of, 61
glass, **44–45**, 84, 109
    laboratory glassware, 150
Glauber, Johan Rudolf, 176
Glauber's Salt, **176**
Glen Canyon Dam, 36
Glidden, Carlos, 316
Glidden, Joseph Farwell, 241
Gloire (ship), 330
goats, 38
goat's foot lever, 35
Goddard, Robert, 360, 404
Godfrey, Thomas, 202
Godowsky, Leo, 467
gold, 27, 34, 52, 53, 55
Goldmark, Emile, 287
Goldmark, Peter, 442
goldsmiths, 53, 108
Good, John, 63
Goodhue, Lyle David, 402
Goodrich Company, B. F., 396
Goodwin, Rev. Hannibal, 246
Goodyear, Charles, 294
Gorrie, John, 291
gouache, 57
Gould, Gordon, 440
governor for engines, **176–177**, 211
Graham, George, 189
Graham, Thomas, 439
Gramme, Zénoe-Theophile, 259
granola, 333
Grape-Nuts, 334
grappling hooks, 21
grave slabs, 165
gravity, law of. *See* laws of motion and gravity
Gray, Elisha, 312, 313
Greatbatch, Wilson, 408–409
Great Britain (ship), 290
Great Eastern (steamship), 265, 309
Greece, ancient
    alphabet in, 17, 18
    Anikythera computer, 12
    armor, 98
    beer in, 24
    central heating in, 29
    columns, 32–33
    maritime trade, 50, 52
    metalworking, 52
    mirrors, 55
    painkillers in, 234

pulleys, 59
siege weapons, 137
theater, 69–71
Greek siege ballista, 34
Green, Robert, 46
Greene, Catherine, 252
greenhouse gases, 399
Greenwich Observatory, 152
Gregg, David Paul, 441
Gregg, John, 136
Gregg system of shorthand, 136
Gregoire, Marc, 468
Gregory, James, 164
Greit, Gregory, 356
grenade, 113
Greyhound (ship), 264
Griffith, Harold, 236
grindstone, 102
ground effect machine (GEM), 432
ground-launched cruise missiles, 435–436
Grove, William R., 425
Groves, Andrew, 415
Groves, General Leslie, 450
Guardian company, 291
Guderian, General Heinz, 386
Guericke, Otto von, 218–219
Guido of Arezzo, 125
guilds, 83–84
Guillet, Léon, 380
Guillotin, Dr. Joseph Ignace, 265
guillotine, **265**
Guilmet, André, 244
guinea pig, 16
guitar, 84, 85, **112–113**
Gulf Refining Company, 336
gunpowder, 35, 83, 86, 95, **113–114**
Gunter, Edmund, 204
Gusmão, Bartholomeu Lorenço de, 177
Gutenberg, Johannes, 124
gyrocompass, 249, **352–353**

hachbuss, 120
Hadley, John, 202
Hadrian, Emperor, 19
Hagia Sophia, dome of, 106
Hahn, Otto, 375
Haish, Jacob, 241
Haiti, 140
Hales, Stephen, 216
Hall, Charles Martin, 232

Hall, Frank, 246
Halley, Edmond, 100, 173, 202
Halley's comet, 100
Hallidie, Andrew S., 308
Haloid, 417
Halsted, William Stewart, 235
Hammer, William J., 258, 346
Hancock, Thomas, 294
hand crank, 81, 85, **102–103**
hand weaving, 175
Hanford, Washington, 448, 455
Hansen, W. W., 461
Harben, Philip, 468
Hargreaves, James, 174, 209
Harpers Ferry arsenal, 266,
    292–293
harpoon, **45–46**, 54
harpsicord, 194, 195
harquebus, 35
Harrier (AV-8), 475
Harrison, John, 168–169
Harvey, William, 143, 169, 189
Haukweicz, Godfrey, 275
Haynes, Elwood, 380, 381
hearing aid, 227, **353**
heat, specific and latent, **205**
Heaviside, Oliver, 356
heavy plow. *See* plow
heavy water, 419, 420
Hebert, Edward Hugh, 348
Hebrew alphabet, 18
Heinkel, Ernst, 438
Heisenberg, Werner, 364, 391, 392
Heisenberg principle, 320, 391–392
Helena cruiser, 458
helicopter, 157, **429–430**, 475
heliocentric solar system, 36–37,
    114–116, 150, 179, 182, 183,
    212
    orrery used to represent, 189
helium, 354
Helmholtz, Hermann von, 272,
    273, 283
hemp, 63
Henfrey, Benjamin, 171
Henlein, Peter, 99, 143
Hennebique, François, 378
Henry, B. Tyler, 198
Henry, Joseph, 258, 259, 311, 312,
    313
Henry the Navigator, Prince, 87,
    132, 133
hens, 38

heparin, 439
*Hermetic and Alchemical Writings
    of Aureolus Phillipus
    Theophrastus...,* 122
Herodotus, 74
Héroult, Palu Louis Toussaint, 232
Herschel, Caroline, 217
Herschel, John, 217
Herschel, William, 158, 212–213,
    217, 218
Hertz, Heinrich, 277, 371
Hevelius, Johannes, 185
Heyerdahl, Thor, 12
hieroglyphics, 18
highways, 239, 366
Hill, Julian, 453
Hindenburg, 344
Hindu culture, 86
    decimal numbers in, 103
    zero, concept of, 148
Hiroshima, Japan, 474
Hisab al-jabr w 'al muqabala, 90
Hitler, Adolf, 324, 362
Hobbs, A. G., 183
hoe, 16
Hoff, Marcian, 414–415
Hoffmann, Felix, 329
Holabird, William, 297
Holland
    canal lock, 28, 93–94
    dikes in, 35–36, 142
    movable type in, 124
    windmills in, 36, 147
Holland (submarine), 382
Holland, John, 381, 382
Hollerith, Herman, 233, 248, 412
holography, **430–431,** 470
Holonyak, Nick, Jr., 441–442
Holst, Axel, 393
Holt, Benjamin, 337
Home Insurance Company
    Building, Chicago, 296–297
Hooftman, Aegidius, 93
Hooke, Robert, 149, 152, 156,
    165–166, 182, 195, 200, 216,
    310
    barometer and, 186–187
    cellular structure of plants,
        165–166, 200
    compound microscope, 187,
        188
    law of elasticity, 165, **180,** 200
    vacuum pump and, 219

Hoover Dam, 36
Hopkins, Dr. Frederic Gowland,
    393–394
Hornsby and Sons, Ruston, 337
Horologium Oscillatorium
    (Huygens), 194
horse armor, 98
horse collar, 40, 78–79, 83,
    116–117, 118, 142
horses
    breeding of, 40–41
    domestication of, 38, **39–41**
    -mounted cavalry, 39, 40
    stirrup. *See* stirrup
horseshoe, 40, 83, **117–118,** 142
Hosho (aircraft carrier), 325
hot-air balloon, 113, 167, **177–178,**
    284, 343
Houdry, Eugène, 336
Housatonic, USS, 315, 382
hovercraft, **431–432**
Hovercraft Developments Limited
    (HDL), 431, 433
Howe, Elias, 295–296
howitzer, 95
Hubbard, Gardiner, 313
Hubble Space Telescope, 213
Huddart, Joseph, 63
Hughes, Howard R., 283
Hull, Albert, 427, 461
Hulls, Jonathan, 303
Hulsmeyer, Christian, 459–460
human genome, 402, **433–434**
humors, four, 121, 163
Hunley (submarine), 315, 382
Hunley, Horace, 382
Huns, 39
Hunt, Walter, 295
hunting-gathering societies, 9
Huxley, Aldous, 437
Huygens, Christiaan, 149, 160,
    163, 180, 182–183, 193–194,
    195, 197
    balance spring, 156, 166, 194
    pendulum clock, 156, 182–183,
        194, 195, 199
    Saturn's rings detected by, 198,
        199
    on speed of light, 207
    wave theory of light, 206
Hyatt, John Wesley, 249
hydraulic elevators, 261
hydraulic jack, 191, **265,** 266

hydraulic press, 153, 158, 191, 265, 266
Hydrodynamica (Bernoulli), 157
hydrogen bomb. *See* thermonuclear weapons
hydrophones, 380
hygrometer, 166
Hyman, Julius, 410–411
hypocaust, 29

iatrochemistry, 121
Ibañez, Salvador, 113
IBM, 248, 316, 412, 446
Ibn-an-Nafis, 169
Ibn Yunus, 192
Ice Ages, 9, 46
  end of, 10
ice cream, **46**
ice-cream maker, hand-cranked, 46
ice cream soda, 46
iconoscope, 389
I. G. Farben, 386
image orthicon, 389
impasto, 57
Imperial Chemical Industries Limited, 369, 418, 419
Impressionism, 127
incandescent lightbulb, 257–258, 346, 411
Incas
  agriculture and, 16
  pull toys, 12, 73
  rope suspension bridge, 305
India
  dams in, 35
  umbrella in, 72
India ink, 46–47
Indian Journal of Physics, 376
India rubber, 293
induction, Faraday's law of, 262, 263, 276
inductive category of scientific discovery, 6
Industrial Revolution, 1790 to 1890, 3, 4, 209, 223–317
  chronology of inventions and discoveries, chart of, 228–230
  setting the stage for, 152–153
  social consequences of, 225–226
Indus Valley, ancient cities of the, 31
ink, 46–47
Inman, George, 427
insecticides, 399, 410–411, 420

instant coffee, **354–355**
insulin, **355–356**
"Insulin Molecule, The," 355
integrated circuit, 414
Intel, 414–415
interchangeable parts, 135, 158–159, 183, 224, 239, 253, **266–267,** 292
intercontinental ballistic missiles (ICBMs), 406, **434–436**
intermediate-range ballistic missiles (IRBMs), 434, 435
internal combustion engine, 212, 217, 227, 320, 327, 335, 381, 385, 399, 426
  dirigibles and, 343
  four-stroke, 238, 239, **267–268**
  two-stroke, 267, **268–269**
International Business Machines. *See* IBM
International Fuel Cells, 427
International Planned Parenthood Federation, 453
International Space Station, 405
Internet, 398, 421, **436–437,** 446
Interstate Commerce Commission, 302
invention
  government aid and sponsorship for, 224–225
  priority of. *See* priority of invention, claiming
  relationship between discovery and, 152
  simultaneous, 153, 202, 231–32, 239, 300, 310, 312
in vitro fertilization, **437**
Ionic columns, 33
ionosphere, **356**
Ipswich company, 269
Iran, copper in, 33
Iraq, baskets in, 22
Irish whisky, 106
iron, 30, 52, 65, 330
irrigation, 15
isotopes, 320, 346, **356–357,** 364
Israel, 451
Italy, eyeglasses in, 109

Jacob, Mary Phelps, 332–333
Jacquard, Joseph Marie, 233
Jacquard power loom, 233
Jansky, Karl, 461

Janssen, Zacharias, 187–188
Janzoon, Laurenz, 124
Japan
  man-carrying kites in, 178
  nuclear power and, 455–456
  pottery in, 59
Jazz Age, 334
Jenney, William Le Baron, 296–297
Jericho, 30
Jervis, John B., 301
jet engine, **437–438,** 475
jet engine aircraft, 326, 328, 398
  ejection seat, **423–424**
jewelry, 84
Johnson, Eldridge, 287
Johnson, Nancy, 46
Johnson, Torrence V., 199
joule, 271
Joule, James Prescott, 270, 271
Judson, Witcomb, 396
jukebox, **357–358**
Jupiter rocket, 406
Justinian, Emperor, 65
Juzan, G., 244

Kaairo, T. J., 431
Kahn, Julian S., 402
Kalmus, H. T., 467
Kamerlingh-Onnes, Heike, 383–384
Kapany, Dr. Narinder, 425
Kato, Sartori, 355
Kay, John, 174, 175
Kay, Robert, 175
KDKA, 372
Keeler, Leonard, 359
Keill, John, 160
Kelley, Edward W., 361
Kellogg, Dr. John Harvey, 333
Kellogg, Keith, 333
Kellogg Toasted Corn Flake Company, 333
Kelvin, Lord. *See* Thomson, William (Lord Kelvin)
Kelvinator company, 291
Kendall, Henry W., 459
Kennelly, Arthur, 356
Kepler, Johannes, 185
  compound microscope, 187, 188
Kepler's laws of planetary motion, 152, **178–180**
kerosene, 251, 268, 269, 334, 335
Kettering, Charles F., 239
Kevlar, **438–439**

KH-11 satellite, 410
Khufu (Cheops), 61
kidney dialysis machine, 399, **439–440**, 468
Kier, Samuel, 282
Kilby, Jack, 414
Killer, Carl, 235
kinescope, 387, 389
kinetoscope, 362
Kirchhoff, Gustav Robert, 246, 298–299
Kleinrock, Leonard, 436
Knapp, Ludwig, 48
knives, 27, 68
Koch, Hugo, 348
Kodak camera, 247, 249
Kodak Company, 330, 421
Kolff, Willem, 439–440
Korn, Arthur, 350, 367
Korolev, Sergei, 405
Kruesi, John, 286
Krupp, 381
krypton, 354
Kuhn, Thomas, 218
Kuiper, Gerard, 218, 280
Kurchatov, Igor V., 450
Kurzweil, 447
Kwolek, Stephanie, 439

labor unions, 226–227
Laënnec, René Théophile Hyacinthe, 307
Lake, Simon, 381, 382
Lalique, René, 45
Land, Edwin, 456–457, 470
landing craft-air cushion (LCAC), 433
Langen, Eugen, 267
Langevin, Paul, 379, 380
Langley (aircraft carrier), 325
Langley, Samuel Pierpont, 326, 328
Langmuir, Irving, 258
laparoscope, 425
Laplace, Pierre, 190, 220
Lapps, 39, 54
Lardner, Dionysius, 234
La Rocca, Aldo V., 440
Larson, John, 359
Lascaux, France, cave paintings, 56
laser, **440–441**
"Laser Applications in Manufacturing," 440
laser disc, 288, **441**, 443

Lassell, William, 218, 279, 280
lateen sails, 86, 89, **118–119**, 132, 133
Lateran Council of A.D. 1139, 35
Latham, Woodville, 362
lathe, 85
latitudinal zones, 42, **47**
laudanum, 121, 234
Lavoisier, Antoine, 150, 159, 161, 190, 220
lawn mower, **269–270**
law of elasticity (Hooke's law), 165, 180, 200
Lawrence, Ernest O., 417, 418
laws of motion and gravity, 114, 149, 159–160, 163, 166, **180–183**, 194, 205, 279, 359, 391
laws of thermodynamics, 205, 211, 223, 244, 267, **270–273**, 283
leaded crystal, 45
lead pencil, **119**
Leakey, L. S. B., 67
leather, **47–48**
  upholstery, 75
Lebon, Philippe, 171
Lee, Ezra, 381
Leeuwenhoek, Anton van, 152, 166, 187, 188–189
Leibniz, Gottfried Wilhelm von, 155, 156, 157
  dispute over development of calculus, 149, 159–161
Leiter Building, Chicago, 297
Leland, Henry, 239
Lemelson, Jerome, 452
Lenaea festival, 70
Lenard, Phillipp Eduard Anton, 396
Lenoir, Jean Joseph Étienne, 237, 267, 268
lenses, **119**
Leonardo da Vinci, 30, 128, 134, 135, 143, 244, 255, 283, 326, 384, 429
Leon cathedral, 97
Leopoldo de Medici, Cardinal, 214
le Prince, L. A. A., 363
Lessing, Lawrence P., 430
Leverrier, Urbain J., 279, 280
levers, Archimedes' work on, 21
Lewis, Gilbert, 419
Libby, Willard, 463–464

library, **48–49**
  at Alexandria, 47, 49, 203
lie detector, 358–359
Lieth, Emmet, 431
light, 361
  electric. *See* electric light
  fluorescent lighting, 354, **427–428**
  gas lighting, 170–171
  photons, 206, **367–368**
  refraction of, 162, **196–197**
  spectrum of, 119, **205–206**
  speed of, **207–209**, 214
lightbulb, incandescent, 257–258, 411
light-emitting diode (LED), 398, **441–442**
lighthouse, **49–50**
Lilienthal, Otto, 326
Lind, James, 393
Lindbergh, Charles, 324–325
linen, 68
Linneaus, Carolus, 215
linotype machine, 273–274, **273–274**, 324
Lippershey, Hans, 212
Lippizanner horses, 40
liquid-fueled rockets, **359–361**, 398
literacy, 227
Livingston, Robert, 300, 303
llama, 16, 39, 79
locks, **183–184**
logarithms, **184**
Lombroso, Cesare, 358
London, England
  aqueducts in, 19
  gas-lighting system, 171
  subway, 383
London Bridge, 144
Long, Crawford, 235
longbow, 35, 83, **120**, 139
longhouse, 64
long-playing records, 287, **442–443**
looms, 69
Lorentz, Hendrik Antoon, 361
Lorentz transformation, **361**
Los Alamos, 450–451, 456
"Lost Generation," 321
lost-wax method, 53
Loud, John J., 408
loudspeakers, 324, **361–362**
Lovell, Bernard, 462–463
Lowell, Percival, 368–369

Lubbock, John, 26–27
Lumière, Auguste Marie and Louis Jean, 362–363, 364
Lundy, John, 236
Luppis, Captain, 315
Lusitania, 303
Luther, Martin, 125
Lyman, William, 248
lymphatic system, 216

Macadam, John Loudon, 236
McAfee, A. M., 336
McCormick, Cyrus, 277–278
McCormick reaper, **277–78**
McCoy, H. N., 357
McFaul, USS, 340
Machine, Le (Branca), 302
machine gun, 241, **274–275**
machine tool industry, 135, 266
Machiony, Italo, 46
Macintosh, Charles, 293
Macleod, John J. R., 355
McMillan, Edwin, 455, 456
Macmillan, Kirkpatrick, 243
"Maestro of the Atom," 418
Magdalenian peoples, 45
Magee, Carl C., 366
Magnavox, 350
magnetic resonance imaging, 399–401, **443**
magnetism, **184–185**
magneto-optical effect, 263
magnetophon, 386
magnetron, 461
Maiman, Theodore H., 440
Manhattan Building, 297
Manhattan Project, 342, 392, 450, 455, 472
Mannes, Leopold, 467
manorial system, 130
mapmaking, 108–109
    atlas, **92–93**, 123
    Mercator projection, 92, **122–123**
    of the Moon, **185–186**
map of the Moon, **185–186**
Marconi, Guglielmo, 277, 356, 371, 372, 390
Marconi company, 353
Marco Polo, 46, 109, 178
Marcus, Siegfried, 237
Marcus Claudius Marcellus, 21
Marcy, Geoffry W., 454

Marey, Étienne J., 363
Margraf, Andreas, 140
Mariotte, Edmé, 160, 170
maritime commerce, 11, **50–52**, 131–132
Marston, W. M., 358
Martin, C. F., 113
Martin, George, 407
Martin, James, 423–424
Martin, Pierre and Emile, 305
Martin-Baker company, 424
Martin Company, 464
Martini, Frederick, 198
Marx, Karl, 226
maser, 440
Massachusetts Institute of Technology, 342, 451
Massachusetts lighthouses, 50
mastaba, 60
matches, 43, **275–276**, 360
matchlock, 83, 111, 113, **120**
mathematical principle of computing, 349, 413, **443–444**
mathematical zero. *See* zero
Matsushita, 477
Maudslay, Henry, 134–135, 158, 183
Mauritania, 303
Maxim, Hiram Stevens, 274, 275, 297–298, 326
Maxwell, James Clerk, 220, 223, 232, 260, 263, 270, 272, 276–277, 361, 371
Maxwell's equations, 260, 263, **276–277**
Mayans
    mathematical zero, 12
    pull toys, 12, 73
    pyramids, 61
    tobacco use, 249–250
    zero, concept of, 148
Maybach, Wilhelm, 268, 334, 335
Mayor, Michael, 454
Mazda, 478
MCA, 441
MCI, 446
mead, 75
Mead, Thomas, 176–177
medicine, **121–122**, 227, 399–401
    anesthetics, 227, **234–236**
    antibiotics, 366–367, 401, 466–467

aspirin, **329–330**
    fiber optics' uses in, 424–425
Medieval and Early Modern Science (Crombie), 96, 110
Medieval Technology and Social Change (White), 103, 147
medium-range ballistic missiles (MRBMs), 436
Medtronic, 409
Melville, David, 171
Menabrea, Luigi, 234
Mendel, Gregor, 422
Mendeleev, Dmitri, 108, 285–286, 298
Menshutken, Professor, 285
Mercator, Gerardus, 92, 93, 123
Mercator, Rumold, 93
Mercator projection, 92, **122–123**
Mercedes-Benz, 238
Merck & Company, 467
mercury barometer, 166, 178, 180, **186–187**, 213
mercury thermometer. *See* thermometer and temperature scale
mercury-vapor light, 428
Mergenthaler, Ottmar, 273
meridional armillary, 92
Mesolithic Age, 11
Mesopotamia, 72
    agriculture in, 15
    board games, 24
    chains in, 30
    mysteries about technologies of, 12
    pottery, 59
    roads in, 62
metal lathe, 158
metal planer, 158
metalworking, **52–53**, 84, 158
"Method of Reaching Extreme Altitudes, A," 360
Mexico, agriculture in, 15, 16
Meyer et Cie, 244
Michaux, Pierre, 243–244
Michelangelo Buonarroti, 107
Michelson, Albert, 208, 277
Micrographia (Small Drawings) (Hooke), 165, 166
microphone, **278–279**, 353
microprocessor, 398, 414–415
microscopes, 45, 84, 119, 152, **187–188**, 196, 213, 220
    electron, 188, **346–347**, 402

microscopic organisms, 166,
188–189
Microwave Communications
Incorporated (MCI), 446
microwave oven, 444–445
microwave transmission of voice
and data, 445–446
Middle Ages through 1599, 2, 4,
81–89
chronology of inventions,
88–89
overview, 81–89
Midgley, Thomas, 292, 351, 403
Miescher, Johann, 422
migratory herd following, 39, **54**
Mill, Henry, 315
Miller, Carl, 416
Miller, Phineas, 252
Millikan, Robert A., 375
mimeograph, 416
Minckelers, Jean Pierre, 171
Minié, Claude-Étienne, 197
mining, **54–55**, 151–152, 212
Minnesota Mining and
Manufacturing Company, 351,
416, 464
mirrors, **55–56**
Mitchell, John, 306
Model T, 239
Mohenjodaro civilization, 57–58
molasses, 140
Monarch (ship), 330
M-1 or Garand rifle, 198
Monet, Claude, 127
Mongols, 39
Monier, Joseph, 378
Monitor, 290
Montgolfier brothers, 167,
177–178, 343
Moog, Robert, 446–447
Moon, map of the, **185–186**
Moon, William, 246
"Moons of Saturn, The," 199
Moore, Gordon, 414
Morgan, J. P., 257–258
Morgan, S. O., 471
Morley, Edward W., 277
Morse, Samuel, 309, 310–312, 390
mortise lock, 183
Morton, William Thomas, 235
Moscow subway, 383
motion, laws of. *See* laws of motion
and gravity

motion pictures, 249, 325, 334,
362–364, 387
Techicolor, **467–468**
three-dimensional, 456, **470**, 478
wide-screen, 470, **478–479**
Moulin Rouge (Red Mill) cabaret,
147
movable type, 53, 85, 110, **124–125**
MRI (magnetic resonance imaging),
443
M-16 rifle, 198
mule, 41
Mullard Radio Astronomy
Observatory, 463
Müller, K. Alex, 384
Müller, Paul Hermann, 420
Müller, Walther, 352
Mullin, John, 477
multiple independently targetable
reentry vehicles (MIRVs), 435
Mumford, Lewis, 1–2, 13–14, 124,
267, 397
mummies, 60
Muntz metal, 26
Murdock, William, 171
Murphy, G. M., 419
Musashi (battleship), 332
music, 334
phonograph, 257, **286–288**, 321
phonograph record. *See*
phonograph record
musical notation, 85, **125**
music synthesizer, 195, **446–447**
musket, 83, 113, 120, **125–126**,
158, 197, 252–253, 266
Mussolini, Benito, 324, 362
mutual deterrence, 398
Muybridge, Eadweard, 363
Mycenaean maritime trade, 51

Nagasaki, Japan, 448, 451, 455,
474
Napier, John, **184**
National Academy of Engineering,
202
National Academy of Sciences, 202
National Advisory Committee for
Aeronautics, 328
National Aeronautics and Space
Administration (NASA),
464–465
National Cathedral, Washington,
D.C., 97

National Defense Research
Committee, 458
National Geographic Society, 314
National Institutes of Health, 428,
433
National Institutes of Medicine, 202
National Research Council, 202
National Science Foundation, 397,
436, 446, 463
Native Americans
baskets, 23
relay communications, 310
rubber used by, 293
tobacco use, 249–250
Natufian peoples, 65
Natural History (Pliny), 66
Nautilus, 381–382
navigational aides, 84–85
Anikythera computer, 12
astrolabe, 85, 89, **91–92**, 132,
202
chronometer, 100, 136, 152, 157,
168–169, 212
compass, 85, 86, 89, **101**, 132,
133, 136
cross-staff, 85, **103**, 132, 168,
202
Mercator projection, 92,
122–123
sextant, 103, 136, 168, **202**
ship's log, 85, **135–136**
Neckham, Alexander, 101
Neolithic Age, 9, 10, 16
neon, 354, 357
neoprene, 452–453
neotechnology, 1, 3, 397
Neptune, 183, 218, **279–281**, 369
Nestlé, 355
neutron, **364**
Newcomen, Thomas, 210, 212
New England Digital Synclavier,
447
New Experiments (Boyle), 219
New Jersey (ship), 332
Newlands, John Alexander Reina,
285
New Orleans (steamboat), 304
newspapers, 324
New Stone Age. *See* Neolithic Age
Newton, Isaac, 198
analytic geometry and, 44
calculus and, 149, 157, 159–161
comet paths of orbits, 100

laws of motion and gravity, 114, 116, 149, 150, 163, 166, 179, 180, 181–183, 194, 205, 279, 359, 391
motion of planets, 116, 150, 180, 182–183
optics and, 119, 205–206
on speed of light, 207
telescope, 152, 164, 212
New York City subway system, 383
New York Public Library, 49
niacin (Vitamin B$_3$), 394
Niemann, Albert, 235
Niépce, Joseph Nicéphore, 246, 254–255
Nile River valley, 16
Nineveh, 27, 31, 32, 48–49
Nipkow, Paul, 367, 387
Nipkow disk, 387
nitrocellulose, 297–298
nitrogen oxide, 399
nitroglycerine, 256, 297–298
nitrous oxide, 235
Nobel, Alfred Bernhard, 255–257, 297–298
nomadic herd tending, 10, 16–17, 56
Norden, Carl, 365
Norden bombsight, **364–365**
Norman Conquest of Britain, 139
Norplant, 454
Norse sailing vessels, 132
North Korea, 451
Norwood, Richard, 204
Notre Dame cathedral, 97
Novum Organum (Bacon), 200
Noyce, Robert, 414
NSFNET, 446
nuclear fission, 364
nuclear fusion, 402, 468
nuclear power, 449–450
nuclear reactors, 326, 352, 398, 419, 420, **447–450**, 466, 472
for submarine propulsion, 382, 461
nuclear weapons, 342, 352, 392, 397, 398, 434, **450–451**, 459
plutonium. *See* plutonium
See also thermonuclear weapons
numerical controlled machines, **451–452**, 472
Nuremberg eggs, 144

Nuremberg method of drawing wire, 147
nylon, 377, **452–453**

Oak Ridge, Tennessee, 455, 472, 473
oboe, 131
Oboler, Arch, 470
odometer, 156
Oersted, Hans Christian, 231, 232, 258, 264, 310
Office of Naval Research, 397
Office of Scientific and Research Development, 342, 397, 458
Ohm, Georg Simon, 281
Ohms law, 223, **281–282**
oil drilling, 251, **282–283**, 335
oil painting, 57, **126–127**
Oklahoma, USS, 332
Olds, Ransom, 239
Olduvai Gorge, Tanzania, 67
Oliphant, M. L., 461
Oliver, Rev. Thomas, 316
Olson, Harry, 446
Opera Omnia Chymica (Glauber), 176
ophthalometer, 283
opium, 234
Oppenheimer, J. Robert, 450
opthalmoscope, **283**
optical fiber, **424–425**
Optica Promota (Gregory), 164
Opticks (Newton), 206
optics, science of, 119
Opus Majus (Bacon), 109
oral contraceptives, 401, **453–454**
Orata, Caius Sergius, 29
orchestra, 70
Oresme, Nicholas, 114, 115
Orontes Dam, 35
orrery, **189**
Ortelius, Abraham, 92–93, 123
oscilloscope, 254
Osprey V-22, 475–476
Osram company, 427
Otis, Charles and Norton, 261
Otis, Elisha Graves, 261, 296, 349, 350
Otis Elevator Company, 261, 349, 350
Otto, Nikolaus A., 238, 267, 268
Otto-cycle (Otto silent) engine, 238, 239, 267, 335

Owens-Corning, 411
oxyacetylene welding, **365–366**
oxygen, 150, **189–190**
ozone layer, 352

pacemakers, 399, **408–409**
packet switching, 436
paddlewheel boat, 103, **127–128**, 196, 225
Pagés, Fidel, 236
paint, **56–57**, 126–127
"Palace Civilizations of Minoan Crete and Mycenean Greece, 2000–1200 B.C.," 51
Paleolithic period, 22
paleotechnology, 1, 2
Palestine, brass in, 25
Palladio, Andrew, 71
"Panama hats," 23
Pan American World Airways, 329, 429
Panhard, 238
Pantheon, 106
Pantometria (Digges), 141
paper, **128**, 129
Papermate pen, 408
papier-mâché, **191**, 249
Papin, Denis, 195, 196, 303
papyrus, 128
Paracelsus, 87, 121–122, 234
parachute, **283–284**
parallax effect, 207–208
parallel invention, 12
parchment, 128
Paris, France
aqueducts in, 19
Metro, 383
Paris Academy of Sciences, 151, 152, 163, 170, 195, 201, 205, 207, 255
Park and Lyons, 382
Parkes, Alexander, 249
parking meter, 320, **366**
Parsons, Charles, 302
Pascal, Blaise, 150, 155, 160
barometer and, 186, 187
Pascal's law, 152, **191–192**, 265, 266
Pasteur, Louis, 284, 285
pasteurization, **284–285**
patents, 149
Patterson, Walter C., 449
Peale, Rembrandt, 171

Pearl Harbor, 325
Pecquet, Jean, 216
peer review, 151
Peking man, 43
pellagra, 394
Pelouze, Théophile Jules, 256
pen
    ballpoint, 306, **407–408**
    steel-tipped, **306**
pencil, lead, **119**
pendulum, **192–193**
pendulum clock, 99–100, 156,
        182–183, 192, **193–194**, 195,
        199
penicillin, **366–367**, 395, 401,
        467
Pennington, E. J., 384
Pennsylvania (cruiser), 325
Pensées (Pascal), 192
Pentel, 408
Pentheon, 377
Pentium II chip, 415
Pepys, Samuel, 136
Percherons, 40
periodic law, 108, **285–286**
Perkin, William Henry, 41–42
Perkins, Jacob, 291
Perky, Henry D., 333
Perlemon, Al, 447
perpetual motion machine, 85
Perrin, M. W., 369–370
Perry, James, 306
Persian Royal Road, 62
Peru, 12
    cotton in, 16, 69
    domestication of animals in, 16
Petroski, Henry, 68
Pfleumer, Dr. F., 386
Philadelphia Navy Yard, 473
Philips Company, N. V., 307
Phillips Petroleum Company, 370
phlogiston theory, 190
Phoenicia, 55
    alphabet in, 17
    winemaking in, 74, 75
phonograph, 257, **286–288**, 321
phonograph record, 324, 325, 334
    long-playing records, 287,
        442–443
photocopies, 416
photoelectric cell, 350, **367**, 387
photography
    camera, **246–247**, 249, 255

daguerreotype method of, 246,
    254–255
photons, 206, **367–368**
photostats, 416
physics, 320
piano, **194–195**
Picard, Charles, 365
Picard, Jean, 204
Piggott, Stuart, 73
pill, the, **453–454**
Pincus, Dr. Gregory, 453
Pioneer LaserDisc, 441
Pitman, Isaac, 136
Pixii, Antoine-Hippolyte, 259
Planck, Max, 206, 367, 368
Planck's constant, 320, **368**, 392
planetarium, 189
planets. *See* solar system; *individual
    planets*
planets outside the solar system,
    454–455
Planté, Gaston, 243
Plantin, Christophe, 93
plants
    cellular structure of, **165–166**,
        188, 200
    domestication of, 15, 16
    transpiration of water in, **216**
plastics
    Bakelite, **330**
    celluloid. *See* celluloid
    glass-reinforced, 411–412
    Kevlar, **438–439**
    polyethylene, **369–370**
    Teflon, **468**
plate armor, 30
playing cards, 81, 85, 124, **128–129**
Pliny, 55, 66
plow, 82–83, 117, 118, **130–131**
Plucknett, Thomas, 269
Plug Power, 427
plumbing, 18, **57–58**
Plunkett, Roy, 468
Pluto, **368–369**, 391
    satellite Charon, 369
plutonium, 447, 448, 450,
    455–456, 472
pneumatic tire, 236, 239, 244, **288**,
    294
Poitiers, battle at, 120
Polaroid camera, 422, **456–457**
polders, 36, 142
polonium, 320, 374, 375

polyethylene, **369–370**
polygraph, 359
Polynesians, mysteries about
        technologies of ancient, 12
Pomo of California, 23
pom-pom gun, 275
Popov, Aleksandr Stepanovich, 371
population growth, 64, 87, 118
Portland cement, 378
Portuguese explorers, 119, 123
positron, **370**
Post, Charles W., 334
Post Toasties, 334
Potter, Humphrey, 210
pottery, 16, **58–59**, 72, 84
Poulsen, Valdemar, 386, 387
power loom, 175
Powers, James, 248–249
praxinoscope, 364
prayer wheel of Tibet, 146
Pre-Historic Times (Lubbock),
        26–27
pressure cooker, **195–196**
pressure gauge, **288–289**
pressurized water reactor, 448, 449
Priestley, Joseph, 150, 153, 159,
        161, 190, 293
prime numbers, 42
Prince of Wales, HMS, 332
Princeton (ship), 290
Principia (Newton), 181, 182
principle of complimentarity, 392
Principles of Chemistry
        (Mendeleev), 286
Prindle, Karl Edwin, 338
printed circuit, 415, **457–458**
printing ink, 47
printing press, 47, 53, 85, 124–125
priority of invention, claiming, 150
    in Age of Scientific Revolution,
        149–150, 156, 166, 180, 182,
        194
    during Industrial Revolution, 311
    rise of individualism in the
        Renaissance and attitude
        toward, 110
privies, 58
propeller-driven ship, **289–290**, 304
proscenium, 71
Protestant Reformation, 125
protozoa, discovery of, 189
Proudhomme, E. A., 336
proximity fuse, **458–459**, 474

Przewalski, Nikolai Mikhailovitch, 40
Ptolemy, 115, 122–123, 181
publication of scientific papers, 150
Puffing Devil, 237
pulley, 21, **59–60**, 71, 72, 78
pulsilogia, 193–194
Purcell, Edward M., 443
Pyramid of the Sun, 61
pyramids, 12, **60–61**
Pythagoras, 44
Pythagorean theorem, **61**

quadrant, 204
quantum mechanics, 392
quantum theory of energy, 368
quark, **459**
Queloz, Didier, 454
quercitron dye, 41
quintessence, 85, 107–108

R-34 (zeppelin), 344
rabies vaccine, 284
radar, 339, 398, 445, **459–461**
radiation
    electromagnetic, 361
    Geiger counter for measuring ionizing, **352**
radio, 164, 260, 276, 277, 319, 321, 325, 334, **371–372**, 387, 390
    AM broadcasting, 356
    broadcasting, 277, **372–373**
    diode in, 343
    FM broadcasting, 356, 372
radioactivity, 320, 373–374, **373–374**, 399
radiocarbon dating, **463–464**
Radio Corporation of America (RCA), 287–288, 350, 372, 389, 446
radio telescope, **461–463**
radio-thermal generators (RTGs), 314, **464–465**
radium, 320, **374–376**
radon, 354
railroads, 225, 300
    air brake, 231
    steam. *See* steam railroad
    steel rails, 305
    transcontinental, 225, 301
rainwear, 293–94

Raman, Chandrasekhara Venkata, 376
Raman spectroscopy, **376**
Ramirez, José, II, 113
Ramsay, William, 354
Ramsden, Jesse, 141, 164
RAND, 436
Ranger (aircraft carrier), 325
ranked society, 28, 64
Ransome, 269
Ravenscroft, George, 45
rayon, 338, **377**
Raytheon, 444–445
razor, safety, **378–379**
records, phonograph. *See* phonograph record
Rectron company, 427
red corpuscles of blood, 189, **196**
Reech, Ferdinand, 263
reed instruments, 84, 85, **131**
refraction of light, 162, **196–197**
refrigeration, 159, 227, **291–292**, 338–39, 351
refrigerator, 321
Reichenbach, H. M., 247
reindeer, 39, 54
reinforced concrete, **377–378**
Reis, Phillip, 312
religion, cities and, 31
Rembrandt, Harmens van Rijn, 127
Remington Arms Company, 316, 365
Remington typewriter company, 249
Renaissance. *See* Middle Ages through 1599
Renne, John, 177
Reno, Jesse W, 349, 350
Renold, Hans, 244, 255
Renwick, James, 97
repeating rifle, 198
replication of scientific findings, 151, 200
Repulse, HMS, 332
research and development (R&D), 397, 459
Ressel, Joseph, 290
revolutionary (word origin), 115–116, 150
revolver, 159, **292–293**
Reynaud, Emil, 363–364
Reynolds, Milton, 408
Reynolds, Richard J., 250

Rhind papyrus, 104
Rice, Chester W., 361
rickets, 393–394
rifle, repeating, **197–198**
rifled gun barrels, 113, 126, **197–198**
Ritchell, C. E., 343
Ritchie, G. G., 418
RNA, 423
roads, asphalt paving of, **236–237**
Robert, Richard, 158
Roberts, David, 337
robots, industrial, 451–452
rockets
    artificial satellite. *See* artificial satellite
    liquid-fueled, **359–361**
    with sold propellants, 359
Rock-Ola and Rowe, 358
Rodman, T. J., 95
Roebling, John A., 305–306
Roebling, Washington Augustus, 306
Roentgen, Wilhelm, 253, 345, 373, 374, 394, 395, 396
Rohrer, Heinrich, 347
Roman Catholic Church
    index of prohibited books, 116
    indulgences, 125
    oral contraception and, 454
Roman numerals, 104
Rome, ancient, 31
    alphabet, 17–18
    aqueducts, 18–19
    the arch, 19, 20
    armor, 98
    brass, 25
    central heating, 29
    chains, 30
    concrete, 377, 378
    days of the week, 36
    decline of, 81
    ice cream, 46
    lateen sails, 118
    lighthouse, 50
    magnifying lenses, 187
    metalworking, 52
    mining, 55
    mirrors, 55
    planting of legumes for soil improvement, 142
    plumbing, 58
    pulleys, 59

Rome, ancient (*continued*)
   roads, 62
   siege weapons, 137
   soap, 66
   theater, 71
   waterwheels, 144
Römer, Ole (Olaus), 207, 214–215
Roosevelt, Franklin D., 324
rope, **63**
ropewalks, 63
Rose, Henry M., 241
Rosing, Boris, 387
Ross, W. N., 357
Rotheim, Eric, 402
Royal Astronomical Society, 217
Royal Society, 152, 159, 160, 180,
   189, 195, 200, 201, 209,
   281
Rozier, Pilâtre de, 177
R-7 rocket, 406
rubber, **293–294**, 309
Rubens, Peter Paul, 127
Rudge, J. A., 363
Rudolf of Nuremberg, 147
rum, 106, 140, 234
Rumsey, James, 303
Ruska, Ernst, 347
Russell, Sidney I., 344
Ruth, Babe, 325
Rutherford, Lord Ernest, 357,
   373–374, 375

safety razor, **378–379**
saffron dye, 41
sailing vessels, 51, 84, **131–134**
   carvel-built ships, 133
   lateen sails, 86, 89, **118–119**,
     132, 133
   stern rudder, 119, 132
St. Patrick's Cathedral, 97
St. Paul's Cathedral, 97, 107
St. Peter's basilica, Rome, 107
Sakharov, Andrei, 469
sand casting, 53
San Francisco cable car system,
   308
Sanger, Frederick, 356
Sanger, Margaret, 453
Sargon, king of Akkad, 72
satellite, artificial. *See* artificial
   satellite
Saturnian System, The (Huygens),
   163

Saturn's rings and moons, 163,
   198–199, 214, 217
Saunders-Roe, 432–433
Savery, Thomas, 210, 212
Sax, Adolphe, 131
saxophone, 131
Scaliger, Julius, 108
scanners, 409
scene shop, 71
Sceptical Chemist, The (Boyle),
   159
Scheele, Carl Wilhelm, 189–190
Scheele, Karl, 167
Scheinmann, Victor, 451
Scherbius, Arthur, 348
Scheutz, Georg and Edvard, 234
Schilling, Baron Paul, 310–311
Schmidt, Gustav, 268
Schuler, L. A., 418
Schultz, Augustus, 48
Schwarz, David, 343
Schweigger, Johann S. C., 265
science and technology, relationship
   between, 119, 152, 223, 224,
   320
Science since Babylon (de Solla
   Price), 395
Scientific American, 313, 330
scientific discovery, explanation of
   periodization of ages of, 1–8
scientific journals, 200
scientific method, 149, 163
*Scientific Papers of James Clerk
   Maxwell, The,* 260
Scientific Revolution, Age of, 2–3,
   4, 87, 125
   characteristics of, 149–150
   chronology, chart of, 154–155
   overview, 149–155
   priority of invention, interest in.
     See priority of invention,
     claiming
   prominent scientists and their
     contributions, chart of, 151
scientific societies, 150, 151, 159,
   199–202
   See also individual societies
Scotland, distilled spirits in, 105
screw lathe, **134–135**, 183
screw pump, 21–22
Scribner, Dr. Belding, 440
scurvy, 393–394
Seaborg, Glenn, 455, 456

Secret of the Universe, The (Kepler),
   179
Seeback, Thomas, 314, 464
Seeback effect, 314
Seeberger, Charles D., 349, 350
Seeburg, Justus, 357, 358
Seech, Fran, 408
Segovia, Andres, 113
Selden, George Baldwin, 238
Selenographia, 185
semaphore, 294–295, 301, 310, 390
Semitic alphabet, 17
Senf, Christian, 94
Sennacherib, King of Assyria, 32
Sequoyah, 18
Serveto, Miguel, 169
sewing machine, 227, 266–267,
   295–296
sextant, 103, 136, 168, **202**
Sharps, Christian, 197
Shaw, Joshua, 111
shawm, 131
sheep, 38, 69
Shell Development Company, 410,
   411
shelter, **63–65**
Sherman tank, 385
Sherratt, Andrew, 75
Shih Huang Ti, Emperor, 34–35
ships, 51, 227
   aircraft carrier. *See* aircraft
     carrier
   battleship. *See* battleship
   flag signaling system, 295
   Froude number, **263–264**
   oceangoing steamships, 225
   planes launched from. *See*
     aircraft carrier
   propeller-driven, **289–290**
   screw-propelled, 21, 225
ship's log, 85, **135–136**
ships of the line, 134, 330, 331
Shockley, William, 414, 470–471
Shockley Industries, 414
Sholes, Christopher Latham, 315,
   316
shorthand, **136**
Shroud of Turin, 464
sickle, 15, 16, **65**
siege weapons, **136–137**
Siemens, 321
Siemens, Ernst Werner von, 259,
   278, 308, 313, 314, 361

Siemens, William and Friedrich, 305
"Significance of Radium, The," 375
Sikorsky, Igor, 429, 430
Silicon Valley, 415
silkmaking, **65**, 69
Silk Road, 62
Simms, F. R., 384
Singer, Merritt, 295–296
Sisson, Jonathan, 141
size and shape of the Earth, 42, 47, **202–205**
Sketch of the Analytic Engine (Babbage), 234
skyscrapers, 261, **296–297**, 305
Slater, James, 30, 255
slavery, 87, 132, 140, 253
sloops, 134
Smeaton, John, 50, 378
Smith, Bessie, 325
Smith, Francis Pettit, 290
Smith, George, 409
Smith, Horace, 293
Smith & Wesson, 293
Smithson, James, 201
Smithsonian Institution, 201–202, 312, 313
smokeless powder, 96, 274, 286, **297–298**
Snell, Willebrord van Roijen, 119, 162, 196–197, 204
snorkel, 461, **465–466**
Snow, Dr. John, 235
soap, **65–66**
Sobrero, Ascanio, 256
Socony (Mobil), 336
soda water, 153
Soddy, Frederick, 357, 373, 374
Soderblom, Laurence, A., 199
sodium pentothal, 236
sodium vapor light, 428
solar system
    Bode's law, 158
    heliocentric. *See* heliocentric solar system
    Kepler's laws of planetary motion, 152, **178–180**
    laws of motion and gravity and, 116, 150, 180, 182–183
solder, 53, **66–67**
Solomon, King, 55
Sömmering, Thomas von, 310
sonar, 321, 339, **379–380**, 459
Sony Betamax, 477

Sony Corporation, 421
Sophocles, 70
Soulé, Samuel W., 316
Soviet Union, nuclear weapons and, 435–436, 450, 451, 455, 456, 468–470
Spain
    guitar in, 112, 113
    winemaking in, 75
sparking devices, flint-steel, 43–44
specific and latent heat, **205**
spectroscope, 246, 253, **298–299**, 354, 419
spectrum of light, 119, **205–206**
speed of light, **207–209**, 214, 361
Spencer, Herbert, 273
Spencer, Percy, **444**
sperm cells, discovery of, 189
Sperry, Elmer, 353
Sperry Gyrocompass Company, 353, 365
Sperry Rand, 249
spices, 86
spinning jenny, 69, 152, 174, 175, **209–210**
spinning mule, 174
spinning wheel, 69, 85, **137–138**
Sprague, Frank J., 259, 308–309
Sprengel, Herman, 257
Springfield, Massachusetts arsenal, 266, 293
Springfield rifle, 198
Sputnik I, 405, 463
Sputnik II, 406
Sri Lanka, dams in, 35
Stahl, George, 190
stainless steel, 304, **380–381**
Standard Oil Company, 337
Standard Oil of Ohio, 411
Starley, John Kemp, 244
steamboat, 196, 225, **303–304**
steam engine, 55, 78, 152, 153, 158, 159, **210–212**, 225, 237, 288, 303, 305
    governor for, **176–177**, 211
steam locomotives, 237, **299–300**
steam railroad, 62, **300–302**, 305
    signaling system, 295, 390
steam turbine, 212, **302–303**
Stearn, C. H., 377
steel, 52, 84, 302, **304–305**
    basic oxygen method, 305
    battleships, 331

Bessemer process, 261, 296, 304–305
    open-hearth method, 305
    stainless, **380–381**
steel cable suspension bridges, 147, **305–306**
steel-tipped pen, **306**
Steinheil, 310
Stenographic Shorthand (Pitman), 136
Stephan, Heinrich, 313
Stephenson, George, 299
Stephenson, Robert, 299–300
Step Pyramid, 61
Steptoe, Patrick Christopher, 437
sterilization of surgical tools, 227
stern rudder, 119, 132
stethoscope, **307**
Steuben Glass, 45
Stevens, A. Edwin, 353
Stevens, John, 211, 300
Stevens, Robert, 301
Stille, Kurt, 386
Stirling, Robert, 307
Stirling engine, 307–308
stirrup, 39, 40, 83, **138–139**
Stonehenge, 12
stone tools, **67–68**
Stourbridge Lion, 301
Strabo, 144
streetcars, 239, 259, **308–309**
streptomycin, 401, **466–467**
Strowger, Almon, 314
Structure of Scientific Revolutions, The (Kuhn), 218
Strutt, John William (Lord Rayleigh), 353–354
stud-link chain, 30
Sturgeon, William, 259
subaqueous boats, 381
submarine, 227, 321, 339, 352, **381–382**, 398
    antisubmarine warfare, 331
    snorkel, 461, 465
    sonar detection, 379–380
submarine cable, 225, 265, 290, 309, 312
subway trains, 239, **382–383**
Suez Canal from 510 B.C., 27–28
sugar, **139–141**, 355
sulfur dioxide, 399
Sullivan, Louis, 297
Sumerians, 104

Sumerian wheels, 73
sundae, ice cream, 46
Sunday, Billy, 324
Sundback, Gideon, 396
sundial, 98
Sun Oil Company, 336
superconductivity, **383–384**
superpolymers, 453
surface effect ship (SES), 432
surveyor's theodolite, 85, **141**, 220
suspension bridge, 30
Swammerdam, Jan, 189, **196**
Swan, Joseph, 257, 258, **377**
Swinton, Lt. Col. Ernest Dunlop,
    384, 385
Sylvania Industrial Corporation,
    338
syphilis, 172
Syria
    agriculture in, 15
    dams in, 35
    glassblowing in, 45
System Saturnium (Huygens), 199
Szilard, Leo, 417–418, 448

Tabulae anatomicae (Eustachio),
    109
Tainter, C. S., 287
Tamm, Igor, 469
tank, 128, 321, 337, **384–386**
tanning of leather, 48
tape recording, **386–387**
tarmac, 236
tarot cards, 124, 128–129
Tartaglia, Niccolò (Fontana), 90
Taylor, A. Hoyt, 460
Taylor, Admiral David, 264
Taylor, Richard E., 459
Teatro Olimpico, 71
Techicolor, **467–468**
Technics and Civilization
    (Mumford), 1, 124, 267
technology and science, relationship
    between, 119, 152, 223, 224,
    320
technology stocks, 398
Teflon, **468**
telegraph, 148, 224–225, 259, 265,
    277, 301, **310–312**, 320, 390
    wireless, 371–372
telephone, 148, **312–314**, 321, 353
telephone answering machines,
    386

telescopes, 45, 84, 116, 119, 152,
    212–214, 347
    Cassegrain, **163–164**, 212
    reflecting or Newtonian, 206,
        212
television, 276, **387–389**
    broadcasting, 356
    cathode ray tube screens, 254,
        320, 388, 389
    high-definition, 389
    photoelectric cell, **367**, 387
television camera, 410
Teller, Edward, 469
Telstar communication satellite, 406
tempered steel, 304
Tennant, Charles, 167
Tennessee walking horses, 40–41
tequila, 106
Tesla, Nikola, 260, 371
tetraethyl lead, 352
Texas Instruments, 414, 421
textiles, 16, **68–69**, 84, 152,
    174–175, 209–10
    Dacron, **418–419**
    nylon, 377, **452–453**
    rayon, **377**
Thales, 44
thallium, 253
theater, **69–71**
theodolite, 85, **141**, 220
theoretical scientific discovery, 4
*Theory and Construction of
    Rational Heat Motor* (von
    Linde), 341
theory of relativity, 320, 361
Thermega company, 345
thermocouple, **314**, 464
thermodynamics. *See* laws of
    thermodynamics
thermography, 416
thermometer and temperature scale,
    214–215
thermonuclear weapons, 434,
    468–470
thermos bottle, **389–390**
thespian, 70
Thespis of Icaria, 70
thiamine (Vitamin B$_1$), 394
Thimonnier, Barthéley, 295
Thompson, E. O. P., 356
Thompson, William (Lord Kelvin),
    215
Thomson, Elihu, 371

Thomson, Joseph John ("J. J."),
    245–246, 3353–354, 357,
    364, 373, 374, 379
Thomson, William (Lord Kelvin),
    263, 270, 271, 272, 309
thoracic duct, **216**
Thorn, George, 440
Thoroughbred horses, 40
three-dimensional motion pictures,
    456, **470**, 478
three-field system, 82, 117,
    141–142
3M Company, 351, 416, 464
Three Mile Island accident, 449
Thurber, Charles, 316
Tibet, windmills in, 146
Tiffany, Louis Comfort, 45
Tiffin, William, 136
Tigris and Euphrates civilizations,
    16, 31
Tintern Abbey Works, 147
Tissandier brothers, 343
Titius, Johann Daniel, 158
Tizard, Henry, 461
Tokarev rifle, 198
Toltec pull toys, 12, 73
Tombaugh, Clyde W., 368, 369
tomb markets, 25
Tommasa da Modena, 110
Tompion, Thomas, 189
Topham, F., 377
torpedo, **315**
torpedo-boat destroyers, 331
Torres, Antonio, 113
Torricelli, Evangelista, 186, 187,
    218, 389
town clocks, 99, 100
Townes, Charles, 440
tractors, **337–338**
traffic lights, 366, **390**
Traité de l'auscultation médiaté,
    307
transatlantic communication, 309
transhumance way of life, 39, 54
transistor, 343, 353, 398, 408, 458,
    470–472
transit, 141
Transmission Control
    Protocol/Internet Protocol
    (TCP/IP), 436
transpiration of water in plants,
    216
Travers, Morris W., 354

Treatise on the Heavens (Aristotle), 114
trebuchet rock-throwing siege machine, 83, 137
Tretis of the Astrolabe (Chaucer), 92
Trevithick, Richard, 211, 237, 299
triangular trade, 89, 140
triode valve, 343, 372, **390–391**
triquetrum, 92
Tritton, W. A., 384–385
trolleys. *See* streetcars
Tropic of Cancer, 47
Tropic of Capricorn, 47
Tsiolkovsky, Konstantin, 360, 404
tungsten, 411
Turbina (ship), 302–303
Turing, Alan, 349, 413, 443–444
Turing machine, 349, 443–444
Turkish baths, 29
turnpike, 62
turpentine, 57, 127
Turtle (submarine), 381
Tuve, Merle, 356, 460
20,000 Leagues under the Sea (Verne), 382
typesetting machine, automatic, 273–274
typewriter, 227, 266–267, **315–316**
   braille, 246
Tyrian purple dye, 41

Ulam, Stanislaw, 469
umbrella, 72
uncertainty principle, **391–392**
Underwood, John T., 316
Underwood Typewriter Company, 316
unions, 226–227
U.S. Department of Energy, 433
U.S. Justice Department, 446
universal joint, **216–217**
universities, 84
University of California, 455, 472
University of Chicago, 447, 448
Upatnieks, Juris, 431
Upton, Francis, 258
Ur, 31, 49
uranium isotope separation, 455, **472–474**
Uranus, 158, 183, 213, **217–218**, 279, 369

Urey, Harold, 419
Uruk, Mesopotamia, 31

vacuum pump, 180, 210, **218–219**, 257
vacuum tube. *See* diode
Vail, Alfred, 312
Vai people of West Africa, alphabet of, 18
Valentino, Rudolph, 325
Valturio, Robert, 128
Van Allen, John, 474
Van Allen belts, 406, **474–475**
van de Kamp, Peter, 454
Vanderbilts, 258
van Eyck, Jan, 127
van Gesner, Konrad, 119
Vanguard rocket, 405, 406
van Leyden, Lucas, 108
Varian, R. H. and S. F., 461
Vasco da Gama, 123
Vaughan, Philip, 240
veinous valves, 90, **142–143**
Velazquez, Diego, 127
vellum, 128
Velsicol, 410
Venice, 66
   glassmaking, 45, 84, 119
Ventor, Craig, 434
Venturi, G. B., 157
Venturi effect, 157, 334
Verne, Jules, 382
vernier, **219–220**
Vernier, Pierre, 141, 219–220
vernier scale, 141
vertical takeoff and landing aircraft (VTOL), **475–476**
Vesalius, Andreas, 90–91, 110–111, 143
Vespucci, Amerigo, 123
Vetruvian wheels, 144
VHS standard for videotape recording, 477
Vickers, 433
video cameras, 409
videotape recorder, 387, 421, **477**
Vienna burner, 251
Viète, François, 90
village life, 15, 64
viola, 112
vitamins, **392–394**
Vitruvius, 33, 144
Voboam, Jean and René, 112

Volta, Alessandro, 242, 264, 276
von Baeyer, Adolf, 42
von Braun, Wernher, 361, 404–405
von Guericke, Otto, 186, 187
von Lieben, Robert, 391
von Linde, Karl Paul Gottfried, 291, 341, 366
von Ohain, Hans, 438
Voyager 2, 280
V-2 rocket, 361, 434, 474
vulcanized rubber, 294

Wagner, Franz X., 316
Waksman, Dr. Selma A., 466
Waldensian heresy, 124, 129
Waldo, Peter, 129
Waldseemüller, Martin, 123
Wales, longbow in, 120
Walker, John, 276
Waller, Fred, 478
Walter, Karl, 440
Wankel, Felix, 477–478
Wankel engine, **477–478**
Wardle, K. A., 51
Warner, Douglass, 431
Warner, Ezra, 248
Warner Brothers Corset Company, 333
Washburn and Moen Manufacturing Company, 241
Washington, G., 355
Washington, Ichabod, 147
watch, 84, **143–144**, 156–157, 180, 194
water
   circulation of, **170**
   as a compound, 190, **220–221**
   transpiration of, in plants, **216**
water barometer, 186
water clock or clepsydra, 98–99, 100
water closet, **174**
watercolors, 57
Waterman, Lewis Edson, 306
water pump, 152
   Archimedes' screw and, 21
water turbine, 260, **317**
waterwheels, 53, 78, 84, 103, 118, **144–145**
   aqueducts with, 19
   mining systems and, 55
Watson, James D., 422–423
Watson, Thomas, 313

Watson-Watt, Robert, 461
Watt, James, 153, 158, 159, 177, 205, 210–211, 225, 299, 303, 415–416
Watt and Boulton, 270
Wayss, G. A., 378
weapons. *See specific weapons*
weapons of mass destruction, 399
week, days of the, **36–37**
Wells, H. G., 384
Wells, Horace, 235
Wernikoff, Dr. Robert E., 351
Wesson, Daniel, 293
Westcott, W. B., 467
Western Electric, 372
Western Union Telegraph Company, 313
Westinghouse, 260, 321, 372, 373, 387
Westinghouse, George, 231, 260
Westminster Abbey, 97
Westrex Corporation, 441
whalebone corset, 101–102
Wheatstone, Charles, 311, 312, 313
wheel, 62, **72–73**
    domestication of cattle and development of the, 38
    pull toys, 12
Wheeler, George H., 349
Whinfield, J. R., 418
whiskey, 83, 105–106
Whiskey Rebellion, 106
White, Lynn, Jr., 103, 147
Whitehead, Robert, 315
Whitney, Eli, 224, 251–253, 266, 292
Whittle, Frank, 437–438
wide-screen motion pictures, 470, **478–479**

Wiles, Andrew, 174
Wilkins, Maurice, 422–423
Wilkinson, John, 158, 210–211
Wilkinson, W. B., 378
Williams, Robert, 394
Wilson, Charles Thomson Rees, 368
Wilson, Mitchell, 231
Wilson, Lt. W. G., 384–385
Winchester Company, 198
"Wind Bracing of Buildings, The," 296
windlass crank, 35
windmills, 36, 53, 78, 81, 84, 118, **145–147**, 176
    ball bearings in, 240
wine, 24, 69–70, **73–75**
Winzer, Friedrich Albrecht, 171
wire, **147–148**
wirephotos, 350
Wisconsin, USS, 331
Wistar, Caspar, 45
Wöhler, Friedrich, 232
Wolszczan, Alexander, 454
women, liberation of, 321–323, 333
wooden furniture, **75–76**
Woodstock company, 316
woodwind instruments, 131
woodworking tools, 22, 43, **76–78**
wool, 69
working class, 153, 226
World War I, 241, 320, 321, 325, 337, 338, 364, 372, 379, 380, 384, 385
World War II, 325–326, 332, 337, 365, 386, 406, 420, 453
Wren, Christopher, 97, 107

Wright brothers, 324, 327–328
wrought iron, 164
Wurlitzer Company, 358

xenon, 354
xerography, 416–417
Xerox Corporation, 350, 417
Xerox machine, 417
X-ray machines, 350, **394–395**
X-rays, 253, 320, 345, 373, 394, **395–396**
    quantum nature of, 368

Yale, Linus, 183
Yale, Linus, Jr., 183
Yale lock, 183, 184
Yamato (battleship), 332
Yerkes, Charles, 213
yoke, **78–79**
Young, James, 251
Young, Leo C., 460
Yuan Hao-wen, 113

zalabia, 46
Zech, Jacob, 143
Zeeman, Pieter, 361
Zeeman effect, 361
Zeidler, Othmar, 420
Zeigler, Dr. Karl, 370
Zeiss optical works, Carl, 347
Zel'dovich, Yakov, 469
Zeppelin, Ferdinand, 343–344
zero, 12, 86, 104, **148**
    absolute, 272
zipper, 320, **396**
Zoll, Paul, 408
Zweig, George, 459
Zworykin, Vladimir, 347, 387–389